STUDY GUIDE
FOR
CAMPBELL
BIOLOGY IN FOCUS

LISA A. URRY • MICHAEL L. CAIN
STEVEN A. WASSERMAN • PETER V. MINORSKY
ROBERT B. JACKSON • JANE B. REECE

MARTHA R. TAYLOR
Ithaca, New York

PEARSON

Boston   Columbus   Indianapolis   New York   San Franciso   Upper Saddle River
Amsterdam   Cap Town   Dubal   London   Madrid   Milan   Munich   Paris   Montréal   Toronto
Delhi   Mexico City   São Paulo   Sydney   Hong Kong   Seoul   Singagore   Taipei   Tokyo

Vice President/Editor-in-Chief: Beth Wilbur
Senior Acquisitions Editor: Josh Frost
Senior Editorial Manager: Ginnie Simione Jutson
Senior Supplements Project Editor: Susan Berge
Executive Marketing Manager: Lauren Harp
Managing Editor, Production: Michael Early
Production Project Manager: Jane Brundage
Image Manager: Donna Kalal, Travis Amos
Manufacturing Buyer: Michael Penne
Production Management and Composition: S4Carlisle Publishing Services
Cover Production: Seventeenth Street Studios
Text and Cover Printer: Edwards Brothers

Cover Photo Credit: Chris Hellier / Photo Researchers, Inc.

Photo Credits: **SYK 9.1, page 75** USDA/ARS/Agricultural Research Service; **FQ 24.2,
page 172** Susan M. Barnes, Ph.D.; **FQ 28.4, top, page 202** Ed Reschke; **FQ 28.4, bottom,
page 202** Chuck Brown/Science Source/Photo Researchers, Inc.

ISBN-10: 0-321-86499-9; ISBN-13: 978-0-321-86499-4

Contents

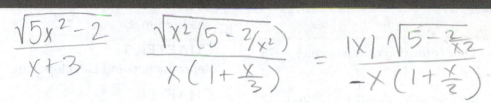

$$\frac{\sqrt{5x^2 - 2}}{x + 3} \qquad \frac{\sqrt{x^2\left(5 - 2/x^2\right)}}{x\left(1 + \frac{x}{3}\right)} = \frac{|x|\sqrt{5 - \frac{2}{x^2}}}{-x\left(1 + \frac{x}{2}\right)}$$

Another name for the *Study Guide for Campbell Biology in Focus* could be "A Student Structuring Guide." This guide is designed to help you structure and organize your developing knowledge of biology and create your own personal understanding of the topics covered in the text. Take a few minutes to learn about some strategies of successful students, as well as some strategies for using this study guide.

Strategies of Successful Students. What does it take to be a successful student? How can you learn more efficiently and earn better grades? And how can you enjoy your biology course more?

Every successful student tackles learning in his or her own unique way. But there are commonalities that can be shared and results from educational research that can inform you as you develop your own approach. Here is a short list of suggestions that may help.

Make it interesting. Biology is fascinating and so much easier to learn when you love it. Try to bring a sense of wonder to your classes and your studying. Appreciate how much has been discovered and how much we have yet to learn about the living world.

Focus in lecture. What does it mean to focus? Obviously, sleeping or texting are out, but so is taking down every word your professor says without thinking. If a lecture outline is provided, take notes on it. You can also try taking notes on one page of a notebook and then, on the facing page, rework, elaborate, or at least make legible what you wrote down in class. Doing this shortly after class helps you review the material while it is fresh and develop an organized set of notes to use in studying for exams.

Make reading active. Research has shown that simply rereading material does not improve test scores, but interacting with the reading assignment does. How do you interact with the printed word? Ask yourself questions and look for answers. Reword every concept heading as a question before you start reading. Make questions out of subheadings. If you are trying to

answer a question, you are looking for meaning. Your brain is not just decoding words, it is working toward understanding those words.

Use your resources. What does your course provide—review sessions, office hours, tutors, study groups? Science is a social endeavor; interact with people as well as the material. Use the resources provided by your course and by your textbook: websites, practice exams, tutorials. When possible, prepare for exams by practicing with questions from previous exams. Each professor has his or her own approach to writing questions. One way to improve your grade is to study how your professor asks questions.

Learn from your mistakes. Carefully review your exams—the questions you missed and even those you got right. If possible, go over the exam with an instructor. Review both *what* you missed and *why* you missed it, so you can learn *how* to choose the correct answer on the next exam. Work to improve your test-taking skills. (See the section on tips for taking tests.)

Try something new. If what you are doing isn't working, don't just keep trying to do it harder. Experiment with new strategies. Are you a visual learner? Try drawing diagrams. If you are an auditory learner, listen to lecture or study tapes as you commute or exercise. Do you need to interact physically with ideas? Build models and be active in your learning.

Strategies for Using This Study Guide. This study guide is not a replacement for your textbook or biology class. However, it should help support and even streamline your learning process. Some students read the text chapters before lecture and then read the study guide chapters after class to reinforce and review. Others reverse that order, skimming the study guide before lecture and then carefully dissecting the text after the professor's presentation. Use this study guide to prepare for tests. With the textbook open so you can refer to essential diagrams, read the relevant study guide chapters for a quick review.

Each study guide chapter has the following six sections:

- The **Chapter Focus** identifies the overall picture; it provides a conceptual framework into which the chapter's information fits.
- The **Chapter Review** is a condensation of the textbook chapter, including a brief summary for each concept heading. All the bold terms in the text are also shown in boldface in this section. Interspersed in this review section are **Focus Questions** that help you stop and actively review the material just covered.
- The **Word Roots** section presents the meanings of key biological prefixes, suffixes, and word roots. Examples of terms from the chapter are then defined. Breaking a complicated term into identifiable components will help you to recognize and learn many new biological terms.
- The **Structure Your Knowledge** section directs you to organize and relate the main concepts of the chapter. It also helps you to piece together the key ideas into a bigger picture.
- In the **Test Your Knowledge** section, you are provided with objective questions to test your understanding. The multiple choice questions ask you to choose the best answer. Some answers may be partly correct; almost all choices have been written to test your ability to discriminate among alternatives or to point out common misconceptions.
- Suggested answers to the Focus Questions, the Structure Your Knowledge section, and the Test Your Knowledge questions are provided in the **Answer Section** at the end of the book.

Using Concept Maps. What are the concept maps that appear throughout this study guide? A **concept map** is a diagram that organizes and relates ideas. The *structure* of a concept map is a hierarchically organized cluster of concepts, enclosed in boxes and connected with labeled lines that explicitly state the relationships among concepts. The *function* of a concept map is to help you structure your understanding of a topic and create meaning. The *value* of a concept map is in the thinking and organizing required to create it.

Developing a concept map requires you to evaluate the relative importance of a group of concepts (Which are most inclusive and important? Which are subordinate to other concepts?), arrange the concepts in meaningful clusters, and draw and label the connections between them.

This book uses concept maps in several ways. A map of a chapter may be presented in the Chapter Focus section to show the organization of the key concepts in that chapter. A Focus Question may provide a skeleton map, with some concepts provided and some empty boxes for you to label. This technique is intended to help you become more familiar with concept maps and to illustrate one possible approach to organizing the concepts of a particular section.

You will also be asked to develop your own concept maps on certain subsets of ideas. In these cases, the Answer Section will present a suggested map. A concept map is an individual picture of your understanding at the time you make the map. As your understanding of an area develops, your map will evolve—sometimes becoming more complex and interrelated, sometimes becoming simplified and more streamlined. Do not look to the Answer Section for the "right" concept map. After you have organized your own thoughts, look at the answer map to make sure you have included the key concepts (although you may have added more), to check that the connections you have made are reasonable, and perhaps to see another way to organize the information.

Tips for Taking Multiple Choice Tests. Interact with each question. Read the stem of the question carefully, underlining or using a highlighter to identify the key concept. Read each answer slowly. Cross out the ones you know are wrong. Circle the key idea that you think identifies the correct answer. Now read the question and the answer you chose together, making sure your choice really does answer what is asked. If you aren't sure about a question, try rereading the question and each choice individually. Don't do all of your thinking in your head. Write in the margins and blank spaces of your test. Draw yourself diagrams and pictures. Write down what you do know, and it may jog your memory. If you still are not sure, mark the question and come back to it later. As you work through related questions, you may find information that helps you figure out that question. And remember, there is no substitute for good preparation, proper rest and nutrition, and a positive attitude.

Biology is a fascinating, broad, and exciting subject. *Campbell Biology in Focus* is filled with information organized in a manner that will help you build a conceptual framework of the major themes of biology. This study guide is intended to help you learn and recall information and, most importantly, to encourage and guide you as you develop your own understanding of and appreciation for biology.

Martha R. Taylor
Ithaca, New York

Chapter 1

Introduction: Evolution and the Foundations of Biology

Chapter Focus

This chapter outlines the broad scope of biology, describes themes that unify the study of life, and examines the scientific construction of biological knowledge. A course in biology is neither a vocabulary course nor a classification exercise for the diverse forms of life. Biology is a collection of facts and concepts structured within theories and organizing principles. Recognizing the common themes within biology will help you structure your knowledge of the fascinating and challenging study of life.

Chapter Review

Biology is the scientific study of life, with **evolution,** the process of change that has shaped life from its origin on Earth to today's diversity, as its organizing principle.

1.1 Studying the diverse forms of life reveals common themes

Theme: New properties emerge at successive levels of biological organization The scale of biology extends from the biosphere to molecules.

FOCUS QUESTION 1.1

Write a brief description of each of the following levels of biological organization.

a. biosphere

b. ecosystem

c. community

d. population

e. organism

f. organs and organ systems

g. tissues

h. cells

i. organelles

j. molecules

Interactions among components at each level of biological organization lead to the emergence of novel properties at the next level. These **emergent properties** result from the structural arrangement and interaction of parts.

Biology today combines the powerful and pragmatic strategy of *reductionism*, which breaks down complex systems into simpler components, with **systems biology,** which studies the interactions of the parts of a biological system and models the system's dynamic behavior.

The form of a biological structure is usually well matched to its function. Form fits function at all of life's structural levels.

The cell is an organism's basic unit of structure and function—the lowest structural level capable of performing all the activities of life. The simpler and smaller **prokaryotic cell,** unique to bacteria and archaea, lacks both a nucleus to enclose its DNA and other membrane-enclosed organelles. The **eukaryotic cell**—with a nucleus containing DNA, and numerous organelles—is typical of all other living organisms.

Theme: Life's Processes Involve the Expression and Transmission of Genetic Information The genetic information of a cell is coded in **DNA** (deoxyribonucleic acid), the substance of genes. **Genes** are the units of inheritance that transmit information from parents to offspring. Genes are located on chromosomes, long DNA molecules that replicate before cell division and provide identical copies to daughter cells.

The biological instructions for the development and functioning of organisms are coded in the arrangement of the four kinds of nucleotides on the two strands of a DNA double helix. Most genes program the cell's production of proteins, and almost all cellular structures and actions involve one or more proteins.

Gene expression is the process by which a gene's information is transcribed to RNA and then translated into a protein. Genes also code for RNAs that serve other functions, such as regulating gene expression. All forms of life use essentially the same genetic code of nucleotides.

FOCUS QUESTION   1.2

Describe the pathway from DNA nucleotides to proteins.

All the genetic instructions an organism inherits make up its **genome.** One set of human chromosomes contains about 3 billion nucleotide pairs, and codes for the production of about 75,000 proteins and a large number of non-protein-coding RNA molecules.

The sequences of nucleotides in the human genome and the genomes of many other organisms have been determined. Using a systems approach called **genomics,** scientists study whole sets of genes in one or more species.

Three research developments contribute to genomics: "high-throughput" technology that can analyze biological materials rapidly; **bioinformatics,** which provides the computational tools to process and analyze the resulting data; and interdisciplinary research teams with specialists from many diverse scientific fields.

Theme: Life Requires Transfer and Transformation of Energy and Matter   Life requires energy. Producers transform light energy to the chemical energy in sugars, which powers the cellular activities of plants. Consumers eat plants and other organisms, using the chemical energy in their foods to power their movement, growth, and other activities. In each energy transformation, some energy is lost to the surroundings as heat.

FOCUS QUESTION   1.3

Compare the movement of chemical nutrients and energy in an ecosystem.

Theme: Organisms Interact with Other Organisms and the Physical Environment   Both organisms and the environment are affected by interactions between them. Interactions between organisms may be mutually beneficial or may harm one or both participants.

Evolution, the Core Theme of Biology   Evolution explains how diverse organisms of the past and the present are related through descent from common ancestors, and how organisms become adapted to their environment.

1.2   The Core Theme: Evolution accounts for the unity and diversity of life

Classifying the Diversity of Life: The Three Domains of Life   Of an estimated total of 10–100 million species, only about 1.8 million species have been identified and named. Biologists have grouped species into ever broader categories, from genera to family, order, class, phylum, and kingdom.

The number of kingdoms is an ongoing debate, but all of life is now grouped into three domains. The prokaryotes are divided into domains **Bacteria** and **Archaea.** All eukaryotes are placed in domain **Eukarya.** The mostly unicellular protists are being reorganized to better reflect evolutionary relationships.

Within this diversity of life, organisms share many similarities, including a universal genetic language of DNA.

FOCUS QUESTION   1.4

What is a commonly used criterion for placing plants, fungi, and animals into separate kingdoms?

Charles Darwin and the Theory of Natural Selection In *On the Origin of Species*, published in 1859, Charles Darwin presented his case for "descent with modification," the idea that present forms have diverged from a succession of ancestral forms. Darwin proposed **natural selection** as the mechanism of evolution by drawing an inference from three observations: Individuals vary in many heritable traits, the overproduction of offspring sets up a competition for survival, and species are generally matched to their environments. From this, Darwin inferred that individuals with traits best suited to the environment leave more offspring than do less-fit individuals. This natural selection, or unequal reproductive success within a population, results in the gradual accumulation of favorable adaptations to the environment.

The Tree of Life   The underlying unity seen in the structures of related species, both living and in the fossil record, reflects the inheritance of those structures from a common ancestor. The diversity of species results from natural selection acting over millions of generations as populations adapted to different environments. The tree-like diagrams of evolutionary relationships reflect the branching genealogy extending from ancestral species. Similar species share a common ancestor at a more recent branch point on the tree of life. Distantly related species share a more ancient common ancestor.

FOCUS QUESTION   1.5

Describe in your own words Darwin's theory of natural selection as the mechanism of evolutionary adaptation and the origin of new species.

1.3   Biological inquiry entails forming and testing hypotheses based on observations of nature

Science is an approach to understanding the natural world that involves **inquiry,** the search for information by asking questions and endeavoring to answer them.

Making Observations   Careful and verifiable observation and analysis of data are the basis of scientific inquiry. Observations involve our senses and tools that extend our senses; **data,** both *quantitative* and *qualitative*, are recorded observations. Using **inductive reasoning,** generalizations can often be drawn from collections of observations.

Forming and Testing Hypotheses   Observations and inductions lead to the search for natural causes and explanations. A **hypothesis** is a tentative answer to a question or an explanation of observations, and it leads to predictions that can be tested. **Deductive reasoning** uses *"if . . . then"* logic to proceed from the general to the specific—from a general hypothesis to specific predictions of results if the general premise is correct.

In science, the ideal is to frame two or more alternative hypotheses and design experiments to test each candidate explanation. A hypothesis cannot be *proven* true; the more attempts to falsify it that fail, however, the more a hypothesis gains credibility.

Science seeks natural causes for natural phenomena; it does not address questions of the supernatural.

A Case Study in Scientific Inquiry: Investigating Coat Coloration in Mouse Populations   Beach and mainland mice are found in two distinct habitats in Florida and differ in coloration, although they are members of the same species. H. Hoekstra and her students tested the hypothesis that coloration patterns evolved as adaptations that protect mice from predation. To test the prediction that mice with coloration that did not match their habitat would be preyed on more than mice that were camouflaged by their coloration, they set out models of beach and mainland mice in both habitats. After recording signs of predation, they calculated the proportion of attacked mice in each habitat. In both cases, the mice whose coloration did not match their habitat had a higher predation rate. Thus, their experiment supports the camouflage hypothesis.

This experimental design illustrates a **controlled experiment** in which subjects are divided into an *experimental* group and a *control* group. Both groups are alike except for the one variable that the experiment is trying to test.

FOCUS QUESTION   1.6

a. Identify the control and experimental groups in the mouse camouflage experiment.

b. Why were the results of this study presented as the proportion of attacks on camouflaged and non-camouflaged mice in each area rather than as the total number of attacks on non-camouflaged mice?

Theories in Science   A **theory** is broader in scope than a hypothesis, generates many specific hypotheses, and is supported by a large body of evidence. Still, a theory can be modified or even rejected when results and new evidence no longer support it.

Science as a Social Process: Community and Diversity Most scientists work in teams and share their results with a broader research community in seminars, publications, and websites. Scientists often attempt to confirm the observations and experimental results of other colleagues. Science is distinguished by adherence to the criteria of verifiable observations and hypotheses that are testable and falsifiable.

Science and technology are interdependent: The information generated by science is applied by **technology** for specific purposes, and technological advances are used to extend scientific knowledge.

Women and many racial and ethnic groups have been underrepresented in scientific professions. A diversity of backgrounds and viewpoints is important to the progress of science.

FOCUS QUESTION 1.7

a. Compare hypotheses and theories.

b. Compare science and technology.

Word Roots

bio- = life (*biology:* the scientific study of life; *bioinformatics:* the use of computers, software, and mathematical models to process and integrate biological information from large data sets)

eu- = true; **karyo-** = nucleus (*eukaryotic cell:* a type of cell with a membrane-enclosed nucleus and organelles)

pro- = before (*prokaryotic cell:* a type of cell lacking a membrane-enclosed nucleus and organelles)

Structure Your Knowledge

1. Briefly describe in your own words each of the five unifying themes of biology presented in this chapter:
   a. emergent properties and levels of biological organization
   b. expression and transmission of genetic information

c. the transfer and transformation of energy and matter
d. interaction with other organisms and the physical environment
e. evolution

Test Your Knowledge

MULTIPLE CHOICE: *Choose the one best answer.*

1. The core idea that makes sense of the unity and the diversity of life is
   a. the scientific method.
   b. inductive reasoning.
   c. deductive reasoning.
   d. evolution.
   e. systems biology.

2. Suppose that, in an experiment similar to the mice study described in this chapter, a researcher found that more total predator attacks occurred on model beach mice placed in a beach habitat than in a mainland habitat. From this the researcher concluded that
   a. the camouflage hypothesis is false.
   b. the predators in the beach habitat were hungrier than the predators in the mainland habitat.
   c. model beach mice do not resemble living beach mice enough to protect them from attack.
   d. the data that should be compared to draw a conclusion must include a control—a comparison with the number of attacks on model mainland mice in both habitats.
   e. more data must be collected before a conclusion can be drawn.

3. Why can a hypothesis never be "proven" to be true?
   a. One can never collect enough data to be 100% sure.
   b. There may always be alternative untested hypotheses that might account for the results.
   c. Science is limited by our senses.
   d. Experimental error is involved in every research project.
   e. Science "evolves"; hypotheses and even theories are always changing.

4. In a pond sample, you find a unicellular organism that has numerous chloroplasts and a whiplike flagellum. In which of the following groups do you think it should be classified?
   a. plant
   b. animal
   c. domain Archaea
   d. one of the proposed kingdoms of protists
   e. You cannot tell unless you see if it has a nucleus.

5. What is DNA?
   a. the substance of heredity
   b. a double helix made of four types of nucleotides
   c. a code for protein synthesis
   d. a component of chromosomes
   e. all of the above

6. Which of the following sequences correctly lists life's hierarchical levels from lowest to highest?
   a. organ, tissue, organ system, organism, population
   b. organism, community, population, ecosystem, biosphere
   c. molecule, organelle, cell, tissue, organ, organism
   d. tissue, cell, organ, organism, community
   e. Both **b** and **c** are correct sequences.

7. Which of the following themes of biology is most related to the goals and practices of systems biology?
   a. Evolution accounts for the unity and diversity of life.
   b. Organisms interact with other organisms and the physical environment.
   c. Life's processes involve the expression and transmission of genetic information.
   d. Life requires energy transfer and transformation.
   e. New properties emerge at successive levels of biological organization.

The Chemical Context of Life

Chapter Focus

This chapter considers the basic principles of chemistry that explain the behavior of atoms and molecules. You will learn how the subatomic particles—protons, neutrons, and electrons—are organized in atoms, how atoms are connected by covalent bonds, and how ions are attracted to each other in ionic bonds. The chapter also focuses on the properties of water, which emerge from the polarity and hydrogen bonding capacity of this small, essential molecule.

Chapter Review

2.1 Matter consists of chemical elements in pure form and in combinations called compounds

Elements and Compounds **Matter** is anything that takes up space and has mass. **Elements** are substances that cannot be chemically broken down to other types of matter. A **compound** is made up of two or more elements combined in a fixed ratio. The characteristics of a compound differ from those of its constituent elements, an example of *emergent properties* arising in higher levels of organization.

The Elements of Life Your body is composed of 25 different elements. The elements needed for an organism to live and reproduce are called **essential elements** (the list varies somewhat for different organisms). Carbon (C), oxygen (O), hydrogen (H), and nitrogen (N) make up 96% of living matter. Some elements, like iron (Fe) and iodine (I), may be required in very minute quantities and are called **trace elements**.

FOCUS QUESTION 2.1

Fill in the names beside the symbols of the following elements that, along with a few others, make up about 4% of an organism's mass.

Ca   K

P   S

Evolution of Tolerance to Toxic Elements Some plants exhibit evolutionary adaptations that allow them to grow in soils containing toxic elements.

2.2 An element's properties depend on the structure of its atoms

Each element has its own type of **atom**, the smallest unit of matter retaining the properties of that element.

Subatomic Particles   Three *subatomic particles* are important to your understanding of atoms. Uncharged **neutrons** and positively charged **protons** are packed tightly together to form the **atomic nucleus** of an atom. Negatively charged **electrons** form a cloud around the nucleus.

Protons and neutrons have a similar mass of about 1.7×10^{-24} g, or close to 1 dalton each. A **dalton** is the measurement unit for atomic mass. Electrons have negligible mass.

Atomic Number and Atomic Mass   So what makes the atoms of different elements different? Each element has a characteristic **atomic number**, or number of protons in the nucleus of each of its atoms. Unless otherwise indicated, an atom has a neutral electrical charge, and thus the number of protons is equal to the number of electrons. A subscript to the left of the symbol for an element indicates its atomic number; a superscript indicates its mass number. The **mass number** is equal to the number of protons and neutrons in the nucleus and approximates the mass of an atom of that element in daltons. The term **atomic mass** refers to the total mass of an atom.

FOCUS QUESTION   2.2

The difference between the mass number and the atomic number of an atom is equal to the number of _____. An atom of phosphorus, $^{31}_{15}P$, contains _____ protons, _____ electrons, and _____ neutrons. The atomic mass of phosphorus is approximately _____.

Isotopes   Although the number of protons is constant, the number of neutrons can vary among the atoms of an element, creating different **isotopes** that have slightly different masses. Some isotopes are unstable; the nuclei of **radioactive isotopes** spontaneously decay, giving off particles and energy. Radioactive isotopes are important tools in biological research and medicine. Too great an exposure to radiation from decaying isotopes poses a significant health hazard.

The Energy Levels of Electrons   **Energy** is defined as the capacity to cause change—to do work. **Potential energy** is energy stored in matter as a consequence of its position or structure. The potential energy of electrons increases as their distance from the positively charged nucleus increases. Electrons can be located in different **electron shells**, each with a characteristic energy level and distance from the nucleus.

FOCUS QUESTION   2.3

To move to a shell farther from the nucleus, an electron must _____ energy; an electron _____ energy when it moves to a closer shell.

Electron Distribution and Chemical Properties   What determines the chemical behavior of an atom? It is a function of the distribution of its electrons—in particular, the number of **valence electrons** in its outermost electron shell, or **valence shell**. A valence shell of eight electrons is complete, resulting in an unreactive or inert atom. (Remember that the first shell can hold only two electrons.) Atoms with incomplete valence shells are chemically reactive. The elements in each row, or period, of the *periodic table of the elements* have the same number of electron shells and are arranged in order of increasing number of electrons in the outer shell.

FOCUS QUESTION   2.4

Draw an electron distribution diagram for the following atoms. *(Note that electrons do not pair up in the second and third shells until after four electrons are present.)*

a. $_1H$

b. $_6C$

c. $_8O$

d. $_{11}Na$

FOCUS QUESTION   2.5

Fill in the blanks in the following concept map to help you review the atomic structure of atoms.

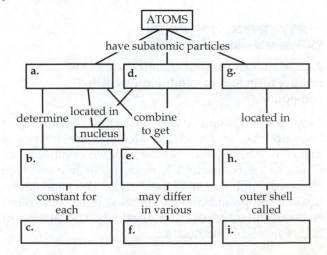

2.3   The formation and function of molecules depend on chemical bonding between atoms

Atoms with incomplete valence shells can either share electrons with or completely transfer electrons to or from other atoms such that each atom is able to complete its valence shell. These interactions usually result in attractions called **chemical bonds**, which hold the atoms close together.

Covalent Bonds   When two atoms share a pair of valence electrons, a **covalent bond** is formed. A **molecule** consists of two or more atoms held together by covalent bonds. An electron distribution diagram shows the shared electrons in a molecule. In a *structural formula*, such as H—H, the line indicates a **single bond**. A *molecular formula*, such as O_2, indicates only the kinds and numbers of atoms. In an oxygen molecule, two pairs of valence electrons are shared between oxygen atoms, forming a double covalent bond, or simply a **double bond** (O=O). The **valence**, or bonding capacity, of an atom usually equals the number of electrons required to complete its valence shell.

FOCUS QUESTION   2.6

What are the valences of the four most common elements of living matter?

Electronegativity is the attraction of a particular atom for shared electrons. If the atoms in a molecule have similar electronegativities, the electrons remain equally shared, and the bond is said to be a **nonpolar covalent bond**. If one element is more electronegative, it pulls the shared electrons closer to itself, creating a **polar covalent bond**. This unequal sharing of electrons results in a "polarity" or separation of charges, with a slight negative charge ($\delta-$) associated with the more electronegative atom and a slight positive charge ($\delta+$) associated with the atom from which the electrons are pulled.

FOCUS QUESTION   2.7

Explain whether the following molecules contain nonpolar or polar covalent bonds. (*Hint:* N and O both have high electronegativities. C and H have lower, and similar, electronegativities.)

a. nitrogen molecule $N \equiv N$ **c.** methane $H—\overset{\displaystyle H}{\underset{\displaystyle H}{|}}\overset{|}{C}—H$

b. ammonia $H\diagdown\overset{N}{\underset{H}{|}}\diagup H$ **d.** formaldehyde $\overset{H}{\underset{H}{}}\diagdown C=O$

Ionic Bonds   What happens when two atoms are very unequal in their attraction for valence electrons? The more electronegative atom may completely transfer an electron from the other atom, resulting in the formation of charged atoms called **ions**. The atom that lost the electron is a positively charged **cation**. The negatively charged atom that gained the electron is called an **anion**. An **ionic bond** may hold these ions together because of the attraction of their opposite charges.

Ionic compounds, or **salts**, often exist as three-dimensional crystalline lattice arrangements held together by electrical attractions. The number of ions present in a salt crystal is not fixed, but the atoms are present in specific ratios. Salts have strong ionic bonds when dry, but the crystal dissolves in water.

Covalent molecules that are electrically charged are also referred to as *ions*.

FOCUS QUESTION   2.8

Calcium ($_{20}$Ca) and chlorine ($_{17}$Cl) can combine to form the salt calcium chloride. Based on the number of electrons in their valence shells and their bonding capacities, what would the formula for this salt be? **a.** _____ Which atom becomes the cation? **b.** _____

Weak Chemical Bonds   Ionic bonds and other weak bonds may form temporary interactions between molecules. Weak bonds within many large molecules help to create those molecules' three-dimensional functional shapes.

A hydrogen atom that is covalently bonded to an electronegative atom has a partial positive charge and can be attracted to another nearby electronegative atom. This attraction is called a **hydrogen bond**.

All atoms and molecules are attracted to each other when in close contact by **van der Waals interactions**. Momentary uneven electron distributions produce changing positive and negative regions that create these weak attractions.

Molecular Shape and Function   A molecule's characteristic size and shape affect how it interacts with other molecules. A carbon atom bonded to four other atoms has a tetrahedral shape.

Draw the electron distribution diagram of a water molecule, showing its V shape and covalently shared electrons. Indicate the areas with slight negative and positive charges that enable a water molecule to form hydrogen bonds with other polar molecules. Then draw a second water molecule and indicate a hydrogen bond between the two.

2.4   Chemical reactions make and break chemical bonds

Chemical reactions involve the making or breaking of chemical bonds. Matter is conserved in chemical reactions; the same number and kinds of atoms are present in both **reactants** and **products**, although the rearrangement of atoms causes the properties of these molecules to be different.

Chemical reactions are reversible—the products of the forward reaction can become reactants in the reverse reaction. Increasing the concentrations of reactants can speed up the rate of a reaction. **Chemical equilibrium** is reached when the forward and reverse reactions proceed at the same rate, and the relative concentrations of reactants and products no longer change.

Fill in the missing coefficients for respiration, the conversion of glucose and oxygen to carbon dioxide and water, with the release of energy. Make sure that all atoms are conserved in the chemical reaction.

$C_6H_{12}O_6 + \underline{\quad}O_2 \rightarrow \underline{\quad}CO_2 + \underline{\quad}H_2O + $ Energy

2.5   Hydrogen bonding gives water properties that help make life possible on Earth

The V-shaped water molecule is a **polar molecule** with a slight positive charge on each hydrogen atom ($\delta+$) and a slight negative charge ($\delta-$) associated with the oxygen. Hydrogen bonds between water molecules create a structural organization that leads to the emergent properties of water.

Cohesion of Water Molecules   Liquid water is unusually cohesive due to the constant forming and reforming of hydrogen bonds that hold the molecules close together. This **cohesion** creates a more structurally organized liquid and helps water to be pulled upward in plants. The **adhesion** of water molecules to the walls of plant vessels also contributes to water transport. Hydrogen bonding between water molecules produces a high **surface tension** at the interface between water and air, making the surface unusually difficult to break.

Moderation of Temperature by Water   **Thermal energy** is a measure of the **kinetic energy** associated with the random movement of atoms and molecules. **Temperature** measures the *average* kinetic energy of the molecules in a body of matter; thermal energy reflects the *total* kinetic energy in that matter, which relates to the volume of the body of matter. The thermal energy that transfers from a warmer to a cooler body of matter is defined as **heat**.

A **calorie (cal)** is the amount of heat it takes to raise 1 g of water 1°C. A **kilocalorie (kcal)** is 1,000 calories, the amount of heat required or released to change the temperature of 1 kg of water by 1°C. A **joule (J)** equals 0.239 cal; a calorie equals 4.184 J.

Specific heat is the amount of heat absorbed or lost when 1 g of a substance changes its temperature by 1°C. Water's specific heat of 1 cal/g·°C is unusually high compared with other common substances. Why does water absorb or release a relatively large quantity of heat as its temperature changes? Heat must be absorbed to break hydrogen bonds before water molecules can move faster and the temperature can rise; conversely, heat is released when hydrogen bonds form as the temperature of water drops. The high proportion of water in the environment and within organisms keeps temperature fluctuations within limits that permit life.

Vaporization or *evaporation* occurs when molecules with sufficient kinetic energy overcome their attraction to other molecules in a liquid and escape into the air as a gas. The **heat of vaporization** is the quantity of heat that must be absorbed for 1 g of a liquid to be converted to a gas. Water's high heat of vaporization is again related to the large amount of heat needed to break the hydrogen bonds holding water molecules together. Water helps moderate Earth's climate as solar heat absorbed by tropical seas is dissipated during evaporation, and heat is released as moist tropical air moving poleward condenses to form rain.

As a liquid vaporizes, the surface left behind loses the kinetic energy of the escaping molecules and cools down. **Evaporative cooling** helps to protect terrestrial organisms from overheating and contributes to the stability of temperatures in lakes and ponds.

Floating of Ice on Liquid Water   As water cools below 4°C, it expands. By 0°C, each water molecule is hydrogen-bonded to four other molecules, creating a crystalline lattice that spaces the molecules apart. Ice is thus less dense than liquid water and so it floats.

FOCUS QUESTION 2.11

The following concept map is one way to show how the breaking and forming of hydrogen bonds are related to temperature moderation. Fill in the blanks and compare your choice of concepts to those given in the answer section. Or, even better, create your own map to help you understand how water stabilizes temperature.

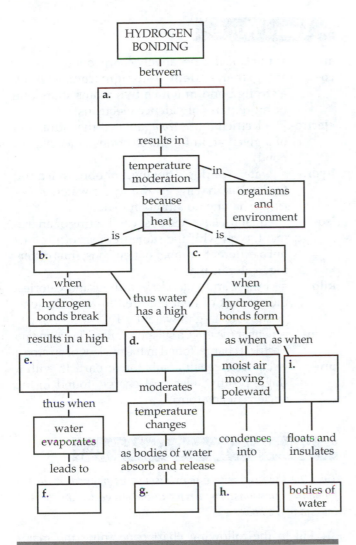

Water: The Solvent of Life A **solution** is a liquid homogeneous mixture of two or more substances; the dissolving agent is called the **solvent**, and the substance that is dissolved is the **solute**. Water is the solvent in an **aqueous solution**. The positive and negative regions of water molecules are attracted to oppositely charged ions or partially charged regions of

polar molecules. Thus, solute molecules become surrounded by water molecules (a **hydration shell**) and dissolve into solution.

Ionic and polar substances are **hydrophilic**; they have an affinity for water due to electrical attractions and hydrogen bonding. Nonpolar and nonionic substances are **hydrophobic**; they will not easily mix with or dissolve in water.

Most of the chemical reactions of life take place in water. A **mole (mol)** is the amount of a substance that has a mass in *grams* numerically equivalent to its **molecular mass** (the sum of the mass of all atoms in the molecule). A mole of any substance has exactly the same number of molecules—6.02×10^{23}, called Avogadro's number. The **molarity** of a solution (abbreviated M) refers to the number of moles of a solute dissolved in 1 liter of solution.

FOCUS QUESTION 2.12

a. How many grams of lactic acid ($C_3H_6O_3$) are in 1 liter of a 0.5 M solution of lactic acid (^{12}C, ^1H, ^{16}O)?

b. How many molecules of lactic acid are in the solution in **a**?

Acids and Bases A water molecule can dissociate into a **hydrogen ion**, H^+ (which binds to another water molecule to form a **hydronium ion**, H_3O^+), and a **hydroxide ion**, OH^-. In pure water at 25°C, the concentrations of H^+ and OH^- are the same; both are equal to $10^{-7} M$.

When acids or bases dissolve in water, the H^+ and OH^- balance shifts. An **acid** adds H^+ to a solution, whereas a **base** reduces H^+ in a solution by accepting hydrogen ions or by adding hydroxide ions (which then combine with H^+ and thus remove hydrogen ions). A strong acid or strong base dissociates completely when mixed with water. A weak acid or base reversibly dissociates, either releasing or binding H^+.

In an aqueous solution, the *product* of the $[H^+]$ and $[OH^-]$ is constant at 10^{-14}. Brackets, [ ], indicate molar concentration. If the $[H^+]$ is higher, then the $[OH^-]$ is lower, because the excess hydrogen ions combine with the hydroxide ions in solution and form water. Likewise, an increase in $[OH^-]$ causes an equivalent decrease in $[H^+]$.

The **pH** of a solution is defined as the negative log (base 10) of the $[H^+]$: pH = $-$log $[H^+]$. For a neutral aqueous solution, $[H^+]$ is $10^{-7} M$, and the pH = 7. As the $[H^+]$ increases in an acidic solution, the pH value decreases. (This inverse relationship makes sense because the exponent becomes smaller: 10^{-4} indicates a higher $[H^+]$ than 10^{-7}.) The difference between each unit of the pH scale represents a tenfold difference in the concentration of [H$^+$] and [OH$^-$].

FOCUS QUESTION 2.13

Complete the following table to review your understanding of pH.

[H$^+$]	[OH$^-$]	pH	Acidic, Basic, or Neutral?
10^{-8}			
	[10^{-7}]		
		1	

Buffers within a cell maintain a stable pH (usually close to 7) by accepting excess H$^+$ or donating H$^+$ when H$^+$ concentration decreases. Weak acid-base pairs that reversibly bind hydrogen ions are typical of most buffering systems.

FOCUS QUESTION 2.14

The carbonic acid/bicarbonate system is an important biological buffer. Label the molecules and ions in this equation, and indicate which is the H$^+$ donor and which is the acceptor.

$$H_2CO_3 \rightleftharpoons HCO_3^- + H^+$$

In which direction will this reaction proceed

a. when the pH of a solution begins to fall?

b. when the pH rises above normal level?

The increasing release of CO_2 to the atmosphere is linked to fossil fuel combustion. The oceans absorb about 25% of this CO_2, which lowers the pH of seawater. The resulting ocean acidification decreases the concentration of carbonate (CO_3^{2-}), an important ion needed for coral reef calcification.

FOCUS QUESTION 2.15

a. Add to the formula in Focus Question 2.14 to show why increasing [CO_2] dissolving in water leads to a lower pH.

b. Use this formula to explain how a lower pH would affect the [CO_3^{2-}] in the ocean.

$$HCO_3^- \rightleftharpoons CO_3^{2-} + H^+$$

c. Assuming a fairly constant [Ca^{2+}] in the ocean, how would a change in [CO_3^{2-}] affect the calcification rate—the production of calcium carbonate ($CaCO_3$)—by the coral in a reef ecosystem?

Word Roots

an-   = not (*anion:* a negatively charged ion)

co-   = together; **-valent** = strength (*covalent bond:* a strong bond in which two atoms share one or more pairs of valence electrons)

electro- = electricity (*electronegativity:* the attraction of a given atom for the electrons of a covalent bond)

hydro-   = water; **-philos** = loving; **-phobos** = fearing (*hydrophilic:* having an affinity for water; *hydrophobic:* having no affinity for water)

iso-   = equal (*isotope:* one of several forms of an element, each with the same number of protons but a different number of neutrons, thus differing in atomic mass)

kilo-   = a thousand (*kilocalorie:* a thousand calories; the amount of heat required to raise the temperature of 1 kg of water by 1°C)

neutr-   = neither (*neutron:* a subatomic particle having no electrical charge, found in the nucleus of an atom)

pro-   = before (*proton:* a subatomic particle with a single positive electrical charge, found in the nucleus of an atom)

Structure Your Knowledge

Take the time to write out or discuss your answers to the following questions. Then refer to the suggested answers at the end of the book.

1. Fill in the following chart concerning the major subatomic particles of an atom.

Particle	Charge	Mass	Location

2. Atoms can have various numbers associated with them.
   a. Define the following and show where each of them is placed relative to the symbol of an element such as C (use the most common isotope, C-12): atomic number, mass number, atomic mass.
   b. Define valence.
   c. Which of these four numbers is most related to the chemical behavior of an atom? Explain.

3. Fill in the following table, which summarizes the emergent properties of water that contribute to the fitness of the environment for life.

Property	Explanation of Property	Example of Benefit to Life
a.	Hydrogen bonds hold water molecules together and adhere them to a hydrophilic surface.	b.
High specific heat	c.	Temperature changes in environment and organisms are moderated.
d.	Hydrogen bonds must be broken for water to evaporate.	e.
f.	Water molecules with high kinetic energy evaporate; remaining molecules are cooler.	g.
Less dense as a solid	h.	i.
j.	k.	Most chemical reactions in life involve solutes dissolved in water.

Test Your Knowledge

MULTIPLE CHOICE: *Choose the one best answer.*

1. Each element has its own characteristic atom in which
   a. the atomic mass is constant.
   b. the atomic number is constant.
   c. the mass number is constant.
   d. Two of the above are correct.
   e. All of the above are correct.

2. Which of the following is *not* a trace element in the human body?
   a. iodine
   b. zinc
   c. iron
   d. calcium
   e. fluorine

3. A sodium ion (Na^+) contains 10 electrons, 11 protons, and 12 neutrons. What is the atomic number of sodium?
   a. 10   c. 12   e. 33
   b. 11   d. 23

4. Radioactive isotopes can be used in studies of metabolic pathways because
   a. their half-life allows a researcher to time an experiment.
   b. they are more reactive.
   c. the cell does not recognize the extra protons in the nucleus, so isotopes are readily used in metabolism.
   d. their location or quantity can be experimentally determined because of their radioactivity.
   e. their extra neutrons produce different colors that can be traced through the body.

5. Which of the following atomic numbers would describe the element that is least reactive?
   a. 1   c. 12   e. 18
   b. 8   d. 16

6. An atom of argon has three electron shells, all of which are full. Its atomic mass is 40. How many neutrons does it have?
   a. 8   c. 20   e. 24
   b. 16   d. 22

7. Which of the following describes what happens as a chlorophyll pigment absorbs energy from sunlight?
   a. An electron moves to a higher electron shell and the electron's potential energy increases.
   b. An electron moves to a higher electron shell and its potential energy decreases.
   c. An electron drops to a lower electron shell and releases its energy as heat.
   d. An electron drops to a lower electron shell and its potential energy increases.
   e. An electron of sunlight is transferred to chlorophyll, producing a chlorophyll ion with higher potential energy.

Use this information to answer questions 8 through 13.

The six elements most common in living organisms are

$$^{12}_{6}C \quad ^{16}_{8}O \quad ^{1}_{1}H \quad ^{14}_{7}N \quad ^{32}_{16}S \quad ^{31}_{15}P$$

8. How many electrons does phosphorus have in its valence shell?

  a. 3   **c.** 7   **e.** 16

  b. 5   **d.** 15

9. What is the atomic mass of phosphorus?

  a. 15   **c.** 31   **e.** 62

  b. 16   **d.** 46

10. A radioactive isotope of carbon has the mass number 14. How many neutrons does this isotope have?

  a. 2   **c.** 8   **e.** 14

  b. 6   **d.** 12

11. How many covalent bonds is a sulfur atom most likely to form?

  a. 1   **c.** 3   **e.** 5

  b. 2   **d.** 4

12. Based on electron configuration, which of the following elements would have chemical behavior most like that of oxygen?

  a. C   **c.** N   **e.** S

  b. H   **d.** P

13. How many of the elements listed above are found next to each other (side by side) on the periodic table?

  a. one group of two

  b. two groups of two

  c. one group of two and one group of three

  d. one group of three

  e. all of them

14. A covalent bond between two atoms is likely to be nonpolar if

  a. one of the atoms is much more electronegative than the other.

  b. the two atoms are about equally electronegative.

  c. the two atoms are of the same element.

  d. one atom is an anion and the other is a cation.

  e. Both **b** and **c** are correct.

15. A triple covalent bond would

  a. not be possible.

  b. involve the bonding of three atoms.

  c. involve the bonding of six atoms.

  d. produce a triangularly shaped molecule.

  e. involve the sharing of six electrons.

16. A cation

  a. has gained an electron.

  b. can easily form hydrogen bonds.

  c. is more likely to form in an atom with seven electrons in its valence shell.

  d. has a positive charge.

  e. Both **c** and **d** are correct.

For questions 17 through 19, choose from the following answers to identify the types of bonds in this diagram of a water molecule interacting with an ammonia molecule.

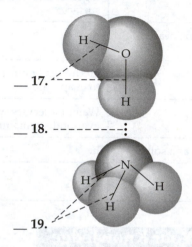

  _____ **17.**

  _____ **18.**

  _____ **19.**

  a. nonpolar covalent bond

  b. polar covalent bond

  c. ionic bond

  d. hydrogen bond

  e. cannot determine without more information

20. In what type of bond would you expect potassium ($^{39}_{19}K$) to participate?

  a. ionic; it would lose one electron and carry a positive charge

  b. ionic; it would gain one electron and carry a negative charge

  c. covalent; it would share one electron and make one covalent bond

  d. covalent; it would share two electrons and form two bonds

  e. none; potassium is an inert element

21. Which of the following may form between any closely aligned molecules?

  a. nonpolar covalent bonds

  b. polar covalent bonds

  c. ionic bonds

  d. hydrogen bonds

  e. van der Waals interactions

22. What is the molecular shape of methane (CH_4)?
    a. planar or flat, with the four H extending out from the carbon
    b. pentagonal, or a flat five-sided arrangement
    c. tetrahedral, with carbon in the center and H at each corner
    d. circular, with the four H attached in a ring around the carbon
    e. linear, since all the bonds are nonpolar covalent bonds

23. The ability of morphine to mimic the effects of the body's endorphins is due to
    a. a chemical equilibrium developing between morphine and endorphins.
    b. the one-way conversion of morphine into endorphin.
    c. molecular shape similarities that allow morphine to bind to endorphin receptors.
    d. the similarities between morphine and heroin.
    e. hydrogen bonding and other weak bonds forming between morphine and endorphins.

24. Which of the following molecules or compounds would you predict is capable of forming hydrogen bonds?
    a. CH_4
    b. CH_4O
    c. NaCl
    d. H_2
    e. a, b, and d can form hydrogen bonds.

25. Chlorine has an atomic number of 17 and a mass number of 35. How many electrons would a chloride ion have?
    a. 16
    b. 17
    c. 18
    d. 33
    e. 34

26. Taking into account the bonding capacities or valences of carbon (C) and oxygen (O), how many hydrogen (H) must be added to complete the following structural diagram of this molecule?

    $$O{\Large\diagup\mkern-18mu\diagdown}{}_{O}^{}\ C-C-C-C=C-C-C$$

    a. 9
    b. 10
    c. 11
    d. 12
    e. 13

27. What is the difference between a molecule and a compound?
    a. There is no difference; the terms are interchangeable.
    b. Molecules contain atoms of a single element, whereas compounds contain two or more elements.
    c. A molecule consists of two or more covalently bonded atoms; a compound contains two or more atoms held by ionic bonds.
    d. A compound consists of two or more elements in a fixed ratio; a molecule has two or more covalently bonded atoms of the same or different elements.
    e. Compounds always consist of molecules, but molecules are not always compounds.

28. In a reaction in chemical equilibrium,
    a. the forward and reverse reactions are occurring at the same rate.
    b. the reactants and products are in equal concentration.
    c. the forward reaction has gone further than the reverse reaction.
    d. there are equal numbers of atoms on both sides of the equation.
    e. a, b, and d are correct.

29. What would be the probable effect of adding more product to a reaction that is in equilibrium?
    a. There would be no change because the reaction is in equilibrium.
    b. The reaction would stop because excess product is present.
    c. The reaction would slow down but still continue.
    d. The forward reaction would increase and more product would be formed.
    e. The reverse reaction would increase and more reactants would be formed.

30. What coefficients must be placed in the blanks to balance the following chemical reaction?

    $$C_5H_{12} + \underline{\quad}O_2 \rightarrow \underline{\quad}CO_2 + \underline{\quad}H_2O$$

    a. 5; 5; 5
    b. 6; 5; 6
    c. 6; 6; 6
    d. 8; 4; 6
    e. 8; 5; 6

31. The polar covalent bonds of water molecules
    a. promote the formation of hydrogen bonds.
    b. help water to dissolve nonpolar solutes.
    c. lower the heat of vaporization and lead to evaporative cooling.
    d. create a crystalline structure in liquid water.
    e. do all of the above.

32. What contributes to the movement of water up the vessels of a tall tree?
    a. cohesion
    b. hydrogen bonding
    c. adhesion
    d. hydrophilic cell walls
    e. all of the above

33. You have three flasks containing 100 mL of different liquids. Each is warmed with 100 calories of heat. The temperature of the liquid in flask 1 rises 1°C; in flask 2 it rises 1.5°C; and in flask 3 it rises 2°C. Which of these liquids has the highest specific heat?
    a. the liquid in flask 1
    b. the liquid in flask 2
    c. the liquid in flask 3
    d. You cannot tell unless you know what liquid is in each flask.
    e. This type of experiment does not relate to the specific heat of a substance.

34. Climates tend to be moderate near large bodies of water because
    a. a large amount of solar heat is absorbed during the gradual rise in temperature of the water.
    b. water releases heat to the environment as it cools.
    c. the high specific heat of water helps to moderate air temperatures.
    d. a great deal of heat is absorbed and released as hydrogen bonds break or form.
    e. all of the above are true.

35. A burn from steam at 100°C is more severe than a burn from boiling water because
    a. the steam is hotter than boiling water.
    b. steam releases a great deal of heat as it condenses on the skin.
    c. steam has a higher heat of vaporization than does water.
    d. a person is more likely to come into contact with steam than with boiling water.
    e. steam stays on the skin longer than does boiling water.

36. Ice floats because
    a. air is trapped in the crystalline lattice.
    b. the formation of hydrogen bonds releases heat; warmer objects float.
    c. it has a smaller surface area than liquid water.
    d. it insulates bodies of water so they do not freeze from the bottom up.
    e. hydrogen bonding spaces the molecules farther apart, creating a less dense structure.

37. Why is water such an excellent solvent?
    a. As a polar molecule, it can surround and dissolve ionic and polar molecules.
    b. It forms ionic bonds with ions, hydrogen bonds with polar molecules, and hydrophobic interactions with nonpolar molecules.
    c. It forms hydrogen bonds with itself.
    d. It has a high specific heat and a high heat of vaporization.
    e. It is liquid and has a high surface tension.

38. The molarity of a solution is equal to
    a. Avogadro's number of molecules in 1 liter of solvent.
    b. the number of moles of a solute in 1 liter of solution.
    c. the molecular mass of a solute in 1 liter of solution.
    d. the number of solute molecules in 1 liter of solvent.
    e. 342 g if the solute is sucrose.

39. Which of the following substances would you add to enough water to yield 1 liter of solution in order to make a 0.1 M solution of glucose ($C_6H_{12}O_6$)? The mass numbers for these elements are approximately C = 12, O = 16, and H = 1.
    a. 6 g C, 12 g H, and 6 g O
    b. 72 g C, 12 g H, and 96 g O
    c. 18 g of glucose
    d. 29 g of glucose
    e. 180 g of glucose

40. How many molecules of glucose would be in 1 liter of the 0.1 M solution made in question 39?
    a. 0.1
    b. 6
    c. 60
    d. 6×10^{23}
    e. 6×10^{22}

41. Adding a base to a solution would
    a. raise the pH.
    b. lower the pH.
    c. decrease [H^+].
    d. do both a and c.
    e. do both b and c.

42. Some archaea are able to live in lakes with pH values of 11. How does pH 11 compare with the pH 7 typical of your body cells?
    a. It is four times more acidic than pH 7.
    b. It is four times more basic than pH 7.
    c. It is a thousand times more acidic than pH 7.
    d. It is a thousand times more basic than pH 7.
    e. It is ten thousand times more basic than pH 7.

43. A buffer
    a. releases excess OH^-.
    b. releases excess H^+.
    c. is often a weak acid-base pair.
    d. always maintains a neutral pH.
    e. Both **c** and **d** are correct.

44. In the past century, the average temperature of the oceans has increased by 0.74°C. Would you consider this evidence of global warming?
    a. No, the rise in temperature is too small to be significant.
    b. No, global warming affects air temperature, not water temperature.
    c. No, the change of average temperature does not reflect the quantity of thermal energy in the oceans.
    d. Yes, because of the high specific heat of water and the huge volume of water in the oceans, a small rise in temperature would reflect a large amount of heat absorbed by the oceans.
    e. Yes, the decreased rate of calcification of reef-building organisms is directly related to this temperature increase.

Carbon and the Molecular Diversity of Life

Chapter Focus

Carbon, with its ability to bond to four other atoms, is the basis for the structural and functional diversity of organic molecules. A small number of monomers or subunits are joined to form a huge variety of large molecules, which can be grouped into four classes.

Class	Monomers or Components	Functions
Carbohydrates	Monosaccharides	Energy source, raw materials, energy storage, structural compounds
Lipids	Glycerol and fatty acids → fats; phospholipids; steroids	Energy storage (fats), membrane components (phospholipids), hormones (steroids)
Proteins	Amino acids	Enzymes, transport, movement, receptors, defense, structure, storage, hormones
Nucleic acids	Nucleotides	Heredity, various functions in gene expression

Chapter Review

Organic compounds are those containing carbon and usually hydrogen. These compounds include lipids and the huge **macromolecules** of carbohydrates, proteins, and nucleic acids.

3.1 Carbon atoms can form diverse molecules by bonding to four other atoms

The Formation of Bonds with Carbon How many covalent bonds must carbon (with an atomic number of 6) form to complete its valence shell? Carbon's **valence** of four is at the center of its ability to form large and complex molecules. When a carbon atom forms four single covalent bonds, the resulting molecule or portion of a molecule is in a tetrahedral shape. When two carbons are joined by a double bond, the other atoms bonded to the carbons are in the same plane, forming a flat molecule.

Molecular Diversity Arising from Variation in Carbon Skeletons Carbon skeletons can vary in length, branching, placement of double bonds, and the presence of rings. **Hydrocarbons** consist of only carbon and hydrogen. Hydrocarbon chains are hydrophobic due to their nonpolar C—H bonds, and they release energy when broken down.

The Chemical Groups Most Important to Life The properties of organic molecules are largely determined by characteristic chemical groups attached to a carbon skeleton. The first six **functional groups** described in the following text may participate in chemical reactions. Except for the sulfhydryl group, these hydrophilic groups also increase the solubility of organic compounds in water. A seventh group, the nonpolar methyl group, alters molecular shape and may serve as a signal on organic molecules.

The **hydroxyl group** consists of an oxygen and hydrogen (—OH). Organic molecules with hydroxyl groups are called alcohols, and their names usually end in *-ol*.

A **carbonyl group** consists of a carbon double-bonded to an oxygen ($>$CO). If the carbonyl group is at the end of the carbon skeleton, the compound is called an aldehyde. Otherwise, the compound is called a ketone.

A **carboxyl group** consists of a carbon double-bonded to an oxygen and also attached to an —OH group (—COOH). Compounds with a carboxyl group are called carboxylic acids or organic acids because they tend to release H^+, becoming a carboxylate ion (—COO$^-$).

An **amino group** consists of a nitrogen atom bonded to two hydrogen atoms (—NH$_2$). Compounds with an amino group, called amines, can act as bases, picking up a hydrogen ion and becoming —NH$_3^+$. Both the amino group and carboxyl group are ionized at normal cellular pH.

The **sulfhydryl group** consists of a sulfur atom bonded to a hydrogen (—SH). Thiols are compounds containing sulfhydryl groups.

A **phosphate group** is bonded to a carbon skeleton by an oxygen attached to a phosphorus atom that is bonded to three other oxygen atoms (—OPO$_3^{2-}$). This anion contributes a negative charge to organic phosphates.

A **methyl group** is a carbon bonded to three hydrogens (—CH$_3$). Methylated compounds may have their function modified due to the addition of the methyl group.

FOCUS QUESTION 3.1

Practice recognizing the functional groups by circling and naming the groups you see in the following molecules.

ATP: An Important Source of Energy for Cellular Processes **Adenosine triphosphate**, or **ATP**, consists of the organic molecule adenosine to which three phosphate groups are attached. When ATP reacts with water, the third phosphate is split off and energy is released.

3.2 Macromolecules are polymers, built from monomers

Polymers are chainlike molecules formed from the linking together of many similar or identical small molecules, called **monomers**.

Synthesis and Breakdown of Polymers Monomers are joined by a **dehydration reaction**, in which one monomer provides a hydroxyl group (—OH) and the other contributes a hydrogen (—H) to release a water molecule. In **hydrolysis**, the bond between monomers is broken by the addition of water. The hydroxyl group of a water molecule is joined to one monomer while the hydrogen is bonded with the other. **Enzymes** catalyze both dehydration reactions and hydrolysis.

Diversity of Polymers Polymers are constructed from about 40 to 50 common monomers and a few rarer molecules. The seemingly endless variety of macromolecules arises from the essentially infinite number of possibilities in the sequencing of these basic building blocks.

FOCUS QUESTION 3.2

Monomers are linked into polymers by _____ _____, which involve the _____ of a water molecule.

Polymers are broken down to monomers by _____, which involves the _____ of a water molecule.

3.3 Carbohydrates serve as fuel and building material

Carbohydrates include sugars and their polymers.

Sugars **Monosaccharides** have the general formula of $(CH_2O)_n$. The number (n) of these units forming a sugar varies from three to seven, with hexoses ($C_6H_{12}O_6$), trioses, and pentoses being most common.

Fill in the blanks to review monosaccharides.

You can recognize a monosaccharide by its multiple (a) _____ groups and its one (b) _____ group, whose location determines whether the sugar is an (c) _____ or a (d) _____. In aqueous solutions, most five- and six-carbon sugars form (e) _____. The names for most sugars end in (f) _____.

Glucose is broken down to yield energy in cellular respiration. Monosaccharides also serve as the raw materials for the synthesis of other organic molecules. Two monosaccharides are joined by a **glycosidic linkage** to form a **disaccharide**.

Polysaccharides Polysaccharides are storage or structural macromolecules. **Starch**, a storage molecule in plants, is a polymer made of glucose molecules joined by 1–4 linkages that give starch a helical shape. Animals use **glycogen**, a highly branched polymer of glucose, as their energy storage molecule.

Cellulose, the major component of plant cell walls, is the most abundant organic compound on Earth. It differs from starch and glycogen by the configuration of the ring form of glucose (beta instead of alpha) and the resulting geometry of the glycosidic bonds. In a plant cell wall, hydrogen bonds between hydroxyl groups hold parallel cellulose molecules together to form strong microfibrils.

Enzymes that digest the α linkages of starch are unable to hydrolyze the β linkages of cellulose. Only a few organisms (some prokaryotes, protists, and fungi) have enzymes that can digest cellulose.

Chitin is a structural polysaccharide formed from glucose monomers with a nitrogen-containing group. Chitin is found in the exoskeleton of arthropods and the cell walls of many fungi.

Number the carbons in the following glucose and fructose molecules. (Each unlabeled corner of the ring represents a carbon. In glucose, carbon 1 is to the right of the O in the ring; in fructose, carbon 1 extends up from the ring on the left side.) Circle the atoms that will be removed by a dehydration reaction. Then draw the resulting sucrose molecule with its 1–2 glycosidic linkage.

Glucose   Fructose

Sucrose

FOCUS QUESTION   3.5

Fill in the following concept map that summarizes this section on carbohydrates.

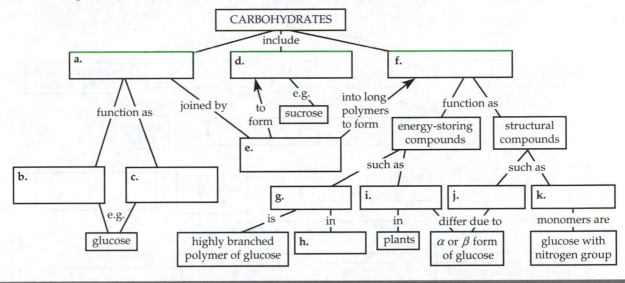

3.4   Lipids are a diverse group of hydrophobic molecules

Fats, phospholipids, and steroids are part of a diverse assemblage of biological molecules that are grouped together as **lipids** based on their hydrophobic behavior. Lipids do not form polymers.

Fats   Fats are composed of **fatty acids** attached to the three-carbon alcohol, glycerol. A fatty acid consists of a long hydrocarbon chain with a carboxyl group at one end. The nonpolar hydrocarbons make a fat hydrophobic.

A **triacylglycerol**, or fat, consists of three fatty acid molecules, each linked to glycerol by an ester linkage, a bond that forms between a hydroxyl and a carboxyl group. *Triglyceride* is another name for fats.

Fatty acids with double bonds in their carbon chain are called **unsaturated fatty acids**. The double bond creates a kink in the hydrocarbon chain and prevents fat molecules with unsaturated fatty acids from packing closely together and becoming solidified at room temperature. The fats of plants and fish are generally unsaturated and are called oils. **Saturated fatty acids**

have no double bonds in their carbon chains. Most animal fats are saturated and solid at room temperature. Diets rich in saturated fats have been linked to cardiovascular disease.

Fats are excellent energy storage molecules, containing twice the energy of carbohydrates such as starch.

Phospholipids   Phospholipids consist of a glycerol linked to two fatty acids and a negatively charged phosphate group, to which other small molecules are attached. The phosphate head of this molecule is hydrophilic and water soluble, whereas the two fatty acid chains are hydrophobic. The unique structure of phospholipids makes them ideal constituents of cell membranes. The hydrophilic heads face the aqueous environment on either side of a membrane; the hydrophobic tails associate in the center of the phospholipid bilayer, shielded from water.

Steroids   Steroids are a class of lipids distinguished by four connected carbon rings with various chemical groups attached. **Cholesterol** is a common component of animal cell membranes and a precursor for other steroids, including many hormones.

FOCUS QUESTION 3.6

Fill in this concept map to help you organize your understanding of lipids.

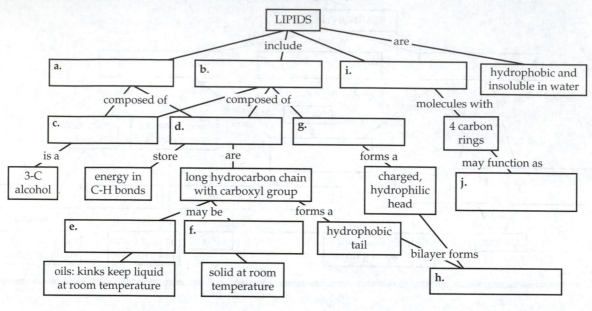

3.5   Proteins include a diversity of structures, resulting in a wide range of functions

Proteins are central to almost every function of life. Most enzymes, which function as **catalysts** that selectively speed up the chemical reactions of a cell, are proteins. A **protein** is a functional molecule that consists of one or more polypeptides, each folded into a specific three-dimensional shape. A **polypeptide** is a polymer of amino acids.

Amino Acids   **Amino acids** are composed of a central carbon, called the *alpha (α) carbon*, bonded to four partners: a hydrogen atom, a carboxyl group, an amino group, and a variable side chain called the R group. At the pH in a cell, the amino and carboxyl groups are usually ionized. The R group confers the unique physical and chemical properties of each amino acid. Side chains may be either nonpolar and hydrophobic, or polar or charged (acidic or basic) and thus hydrophilic.

Polypeptides   A **peptide bond** links the carboxyl group of one amino acid with the amino group of another. A string of amino acids making up a polypeptide has an amino end (N-terminus) and a carboxyl end (C-terminus).

FOCUS QUESTION 3.7

a. Draw the amino acids alanine (R group: —CH$_3$) and serine (R group: —CH$_2$OH) and then show how a dehydration reaction will form a peptide bond between them.

b. Which of these amino acids has a polar R group? a nonpolar R group?

c. What does the following molecular segment represent? (Note the N—C—C—N—C—C sequence.)

Protein Structure and Function A protein has a unique three-dimensional shape, or structure, created by the twisting or folding of one or more polypeptide chains. Protein structure usually arises spontaneously. Depending on the sequence of amino acids, various types of bonds form between parts of the chain as the protein is synthesized in the cell. The unique structure of a protein enables it to recognize and bind to other molecules. *Globular proteins* are roughly spherical; *fibrous proteins* are long fibers.

Primary structure is the genetically coded sequence of amino acids within a protein.

Secondary structure involves regions of coiling or folding of the polypeptide backbone, stabilized by hydrogen bonds between the oxygen (with a partial negative charge) of one peptide bond and the partially positive hydrogen attached to the nitrogen of another peptide bond. An α **helix** is a coil produced by hydrogen bonding between every fourth amino acid. A β **pleated sheet** is held by repeated hydrogen bonds along regions of the polypeptide backbone lying parallel to each other.

Tertiary structure, the three-dimensional shape of a protein, results from interactions between the various side chains (R groups) of the constituent amino acids. The following chemical interactions help produce the stable and unique shape of a protein: **hydrophobic interactions** between nonpolar side groups clumped in the center of the molecule due to their repulsion by water, van der Waals interactions among those nonpolar side chains, hydrogen bonds between polar side chains, and ionic bonds between negatively and positively charged side chains. Strong covalent bonds, called **disulfide bridges**, may occur between the sulfhydryl side groups of cysteine monomers that have been brought close together by the folding of the polypeptide.

Quaternary structure occurs in proteins that are composed of more than one polypeptide. The individual polypeptide subunits are held together in a precise structural arrangement to form a functional protein.

FOCUS QUESTION   3.8

In the following diagram of a portion of a protein, label the types of interactions that are shown. What level of structure are these interactions producing?

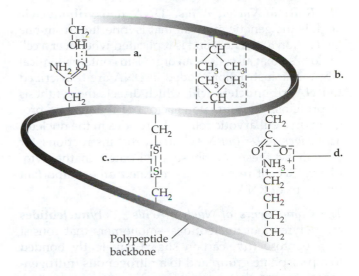

In the inherited blood disorder **sickle-cell disease**, a change in one amino acid affects the structure of a hemoglobin molecule, causing red blood cells to deform into a sickle shape that clogs tiny blood vessels.

The bonds and interactions that maintain the three-dimensional shape of a protein may be disrupted by changes in pH, salt concentration, or temperature, causing a protein to unravel. **Denaturation** also occurs if a protein is transferred to an organic solvent; in that case, its hydrophobic regions are on the outside interacting with the nonpolar solvent.

Using the technique of **X-ray crystallography** biochemists have identified the structure of thousands of proteins. These structures can then be related to the specific functions of different regions of a protein.

FOCUS QUESTION   3.9

Now that you have gained experience with concept maps, create your own map to review what you have learned about proteins. Try to include the concepts of structure and function, and look for cross-links on your map. You may want to include the functions of proteins. *One version of a protein concept map is included in the answer section, but remember that the real value is in the thinking process you must go through to create your own map.*

3.6   Nucleic acids store, transmit, and help express hereditary information

Genes are the units of inheritance that determine the primary structure of proteins. **Nucleic acids** are polymers made of nucleotide monomers.

The Roles of Nucleic Acids   **DNA, deoxyribonucleic acid**, is the genetic material that is inherited from one generation to the next and is replicated whenever a cell divides so that all cells of an organism contain identical DNA. The instructions coded in DNA are transcribed to **RNA, ribonucleic acid**, which directs the synthesis of proteins, the ultimate enactors of the genetic program. In a eukaryotic cell, DNA resides in the nucleus. *Messenger RNA (mRNA)* carries the instructions for protein synthesis to ribosomes located in the cytoplasm. Recent research has revealed other important functions of RNA.

The Components of Nucleic Acids   **Polynucleotides** are polymers of **nucleotides**—monomers that consist of a pentose (five-carbon sugar) covalently bonded to a phosphate group and to a nitrogenous (nitrogen-containing) base. A nucleotide may contain more than one phosphate group; without the phosphate group it is called a *nucleoside*.

Nitrogenous bases are either **pyrimidines** or **purines**, which consist respectively of one or two nitrogen-containing rings. Adenine, guanine, and cytosine are present in DNA and RNA. The base thymine is present only in DNA; uracil is only in RNA. In DNA, the sugar is **deoxyribose**; in RNA, it is **ribose**. In a nucleotide, the base attaches to the 1′ carbon and a phosphate group attaches to the 5′ carbon of the sugar.

Nucleotide Polymers   Nucleotides are linked together into a polynucleotide by phosphodiester linkages, which join the sugar of one nucleotide with the phosphate of the next. The polymer has two distinct ends: a 5′ end with a phosphate attached to the 5′ carbon of a sugar, and a 3′ end with a hydroxyl group on the 3′ carbon of a sugar. The nitrogenous bases extend from this backbone of repeating sugar-phosphate units. The unique sequence of bases in a gene codes for the specific amino acid sequence of a protein.

FOCUS QUESTION   3.10

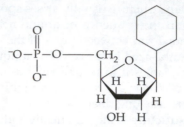

a. Label the three parts of this nucleotide. Indicate with an arrow where the phosphate group of the next nucleotide would attach to build a polynucleotide. Number the carbons of the pentose.

b. Is the base of this nucleotide a purine or a pyrimidine? How do you know?

c. Is this a DNA nucleotide or an RNA nucleotide? How do you know?

The Structure of DNA and RNA Molecules   DNA molecules consist of two polynucleotides (strands) spiraling in a **double helix**. The two sugar-phosphate backbones run in opposite 5′ to 3′ directions, an arrangement called **antiparallel**. The nitrogenous bases pair and hydrogen-bond together in the inside of the molecule. Adenine pairs only with thymine; guanine always pairs with cytosine. Thus, the sequences of nitrogenous bases on the two strands of DNA are *complementary*. Because of this specific base-pairing property, DNA can precisely replicate itself.

RNA molecules are usually single polynucleotides, although base-pairing within or between RNA molecules is common. For example, the functional shape of *transfer RNA (tRNA)*, an RNA involved in protein synthesis, involves several regions of complementary base-pairing.

DNA and Proteins as Tape Measures of Evolution   Genes form the hereditary link between generations. Closely related members of the same species share many common DNA sequences and proteins. More closely related species have a larger proportion of their DNA and proteins in common. This "molecular genealogy" provides evidence of evolutionary relationships.

FOCUS QUESTION   3.11

Take the time to create a concept map that summarizes what you have just reviewed about nucleic acids. Compare your map with that of a study partner or explain it to a friend. Refer to Figures 3.26 and 3.27 in your textbook to help you visualize polynucleotides and the three-dimensional structures of DNA and RNA.

Word Roots

carb-   = coal (*carboxyl group:* a chemical group present in organic acids, consisting of a single carbon atom double-bonded to an oxygen atom and also bonded to a hydroxyl group)

Structure Your Knowledge

1. The diversity of life is amazing. Yet the molecular logic of life is simple: Small molecules common to all organisms are ordered into unique large biological molecules. Explain why carbon is central

di-   = two; **-sacchar** = sugar (*disaccharide:* a double sugar, consisting of two monosaccharides joined through a dehydration reaction)

glyco-   = sweet (*glycogen:* an extensively branched glucose polysaccharide that stores energy in animals)

hydro-   = water; **-lyse** = break (*hydrolysis:* a chemical reaction that breaks bonds between two molecules by the addition of water; functions in disassembly of polymers to monomers)

macro-   = large (*macromolecule:* a giant molecule formed by the joining of smaller molecules, such as a polysaccharide, a protein, or a nucleic acid)

poly-   = many; **meros-** = part (*polymer:* a long molecule consisting of many similar or identical monomers linked together by covalent bonds)

sulf-   = sulfur (*sulfhydryl group:* a chemical group consisting of a sulfur atom bonded to a hydrogen atom)

tri-   = three (*triacylglycerol:* a lipid consisting of three fatty acids linked to one glycerol molecule; also called a fat or triglyceride)

to the molecular diversity of life. How do carbon skeletons, chemical groups, monomers, and polymers relate to this diversity?

2. Fill in the following table on the important chemical groups of organic compounds.

Chemical Group	Molecular Formula	Names and Characteristics of Compounds Containing Group
	—OH	
		Aldehyde or ketone; polar group
Carboxyl		
	—NH$_2$	
		Thiols; cross-links stabilize proteins
Phosphate		
	—CH$_3$	

3. Identify the type of monomer or group shown by the formulas a–g. Then match the chemical formulas with their descriptions. Answers may be used more than once.

$H_3N^+ - \overset{\overset{\displaystyle H}{|}}{\underset{\underset{\displaystyle H}{|}}{C}} - \overset{\overset{\displaystyle O}{\diagup}}{\underset{\underset{\displaystyle O^-}{\diagdown}}{C}}$ a. _____

b. _____

c. _____

d. _____

e. _____

f. _____

g. _____

_____ 1. molecules that would combine to form a fat

_____ 2. monomer that would be attached to other monomers by a peptide bond

_____ 3. molecules or groups that would combine to form a nucleotide

_____ 4. molecules that are carbohydrates

_____ 5. monomer of a protein

_____ 6. most nonpolar (hydrophobic) molecule

4. Describe the four structural levels that produce the functional shape of a protein.

Test Your Knowledge

MATCHING: *Match the molecule with its class of molecule.*

_____ **1.** glycogen
_____ **2.** cholesterol
_____ **3.** RNA
_____ **4.** collagen
_____ **5.** hemoglobin
_____ **6.** a gene
_____ **7.** triacylglycerol
_____ **8.** enzyme
_____ **9.** cellulose
_____ **10.** chitin

A. carbohydrate
B. lipid
C. protein
D. nucleic acid

MULTIPLE CHOICE: *Choose the one best answer.*

1. Carbon's valence of four most directly results from
   a. its tetrahedral shape.
   b. its very slight electronegativity.
   c. its four electrons in the valence shell that can form four covalent bonds.
   d. its ability to form single, double, and triple bonds.
   e. its ability to form chains and rings of carbon atoms.

2. Hydrocarbons are not soluble in water because
   a. they are hydrophilic.
   b. their C—H bonds are nonpolar.
   c. they do not ionize.
   d. they store energy in the many C—H bonds along the carbon backbone.
   e. they are lighter than water.

3. The chemical group that can cause an organic molecule to act as a base is
   a. —COOH.
   b. —OH.
   c. —SH.
   d. —NH$_2$.
   e. —CH$_3$.

4. The chemical group that confers acidic properties to organic molecules is
   a. —COOH.
   b. —OH.
   c. —SH.
   d. —NH$_2$.
   e. —CH$_3$.

5. Polymerization (the formation of polymers) is a process that
   a. creates bonds between glucose monomers in the formation of a polypeptide.
   b. involves the addition of a water molecule.
   c. links the nitrogenous base of one nucleotide with the phosphate of the next.
   d. involves a dehydration reaction.
   e. may involve all of the above.

6. Which of the following statements is *not* true of a hexose?
   a. It may be found in nucleic acids.
   b. It can occur in a ring structure.
   c. It has the formula $C_6H_{12}O_6$.
   d. It has one carbonyl and five hydroxyl groups.
   e. It may be an aldehyde or a ketone sugar.

7. Which of the following statements is *not* true of cellulose?
   a. It is the most abundant organic compound on Earth.
   b. It differs from starch because of the configuration of glucose and the geometry of the glycosidic linkage.
   c. It may be hydrogen-bonded to neighboring cellulose molecules to form microfibrils.
   d. Few organisms have enzymes that hydrolyze its glycosidic linkages.
   e. Its monomers are glucose with nitrogen-containing appendages.

8. Plants store most of their energy for later use as
   a. unsaturated fats.
   b. glycogen.
   c. starch.
   d. sucrose.
   e. cellulose.

9. Maltose is made from joining two glucose molecules in a dehydration reaction. What is the molecular formula for this disaccharide?
   a. $C_6H_{12}O_6$
   b. $C_{10}H_{20}O_{10}$
   c. $C_{12}H_{22}O_{11}$
   d. $C_{12}H_{24}O_{12}$
   e. $C_{12}H_{24}O_{13}$

10. A cow can derive nutrients from cellulose because
    a. it can produce the enzymes that break the β linkages between glucose molecules.
    b. it chews and rechews its cud so that cellulose fibers are finally broken down.
    c. its rumen contains prokaryotes and protists that can hydrolyze the bonds of cellulose.
    d. its intestinal tract contains termites, which harbor microbes that hydrolyze cellulose.
    e. it has enzymes that convert cellulose to starch and then hydrolyze starch to glucose.

11. Which of the following substances is the major component of the cell membrane of a fungus?
    a. cellulose
    b. chitin
    c. cholesterol
    d. phospholipids
    e. unsaturated fatty acids

12. A fatty acid that has the formula $C_{16}H_{32}O_2$ is
    a. saturated.
    b. unsaturated.
    c. branched.
    d. hydrophilic.
    e. part of a steroid molecule.

13. Three molecules of the fatty acid in question 12 are joined to a molecule of glycerol ($C_3H_8O_3$). The resulting molecule has the formula
    a. $C_{48}H_{96}O_6$.
    b. $C_{48}H_{98}O_9$.
    c. $C_{51}H_{102}O_8$.
    d. $C_{51}H_{98}O_6$.
    e. $C_{51}H_{104}O_9$.

14. Which of the following molecules is the most hydrophobic?
    a. cholesterol
    b. nucleotide
    c. chitin
    d. phospholipid
    e. glucose

15. Which of the following molecules provides the most energy (kcal/g) when eaten and digested?
    a. glucose
    b. starch
    c. glycogen
    d. fat
    e. protein

16. Which of the following is *not* one of the many functions performed by proteins?
    a. acting as signals and receptors
    b. acting as an enzymatic catalyst for metabolic reactions
    c. providing protection against disease
    d. serving as contractile components of muscle
    e. forming primary energy storage in plant seeds

17. What happens when a protein denatures?
    a. Its primary structure is disrupted.
    b. Its secondary and tertiary structures are disrupted.
    c. It always flips inside out.
    d. It hydrolyzes into component amino acids.
    e. Its hydrogen bonds, ionic bonds, hydrophobic interactions, disulfide bridges, and peptide bonds are disrupted.

18. The α helix of proteins is
    a. part of the protein's tertiary structure and is stabilized by disulfide bridges.
    b. a double helix.
    c. stabilized by hydrogen bonds and is commonly found in fibrous proteins.
    d. found in some regions of globular proteins and is stabilized by hydrophobic interactions.
    e. a complementary sequence to messenger RNA.

19. β pleated sheets are characterized by
    a. disulfide bridges between cysteine amino acids.
    b. parallel regions of the polypeptide chain held together by hydrophobic interactions.
    c. folds stabilized by hydrogen bonds between segments of the polypeptide backbone.
    d. membrane sheets composed of phospholipids.
    e. hydrogen bonds between adjacent cellulose molecules.

20. What is the *best* description of the following molecule?

    a. chitin
    b. amino acid
    c. tripeptide
    d. nucleotide
    e. protein

21. Which number(s) in the molecule in question 20 refer(s) to a peptide bond?
    a. 1
    b. 2
    c. 3
    d. 4
    e. both 2 and 4

22. What *determines* the sequence of the amino acids in a particular protein?
    a. its primary structure
    b. the sequence of nucleotides in RNA, which was determined by the sequence of nucleotides in the gene for that protein
    c. the sequence of nucleotides in DNA, which was determined by the sequence of nucleotides in RNA
    d. the sequence of RNA nucleotides making up the ribosome
    e. the three-dimensional shape of the protein

23. Both hydrophobic and hydrophilic interactions are important for which of the following types of molecules or structures?
    a. proteins
    b. cell membranes
    c. cellulose in plant cell walls
    d. a and b
    e. a, b, and c

24. How are nucleotide monomers connected to form a polynucleotide?
    a. by hydrogen bonds between complementary nitrogenous base pairs
    b. by ionic attractions between phosphate groups
    c. by disulfide bridges between cysteines
    d. by covalent bonds between the sugar of one nucleotide and the phosphate of the next
    e. by ester linkages between the carboxyl group of one nucleotide and the hydroxyl group on the ribose of the next

25. If the nucleotide sequence of one strand of a DNA helix is 5'GCCTAA3', what would be the 3'–5' sequence on the complementary strand?
    a. GCCTAA
    b. CGGAUU
    c. CGGATT
    d. ATTCGG
    e. TAAGCC

26. Monkeys and humans share many of the same DNA sequences and have similar proteins, indicating that
    a. the two groups belong to the same species.
    b. the two groups share a relatively recent common ancestor.
    c. humans evolved from monkeys.
    d. monkeys evolved from humans.
    e. the two groups evolved about the same time.

Chapter 4

A Tour of the Cell

Chapter Focus

The cell is the fundamental unit of life. The complexities in the processes of life are reflected in the complexities of the cell. It is easy to become overwhelmed by the number of new vocabulary terms for this array of cell structures. The following diagram provides you with an organizational framework for the wealth of detail found in "a tour of the cell."

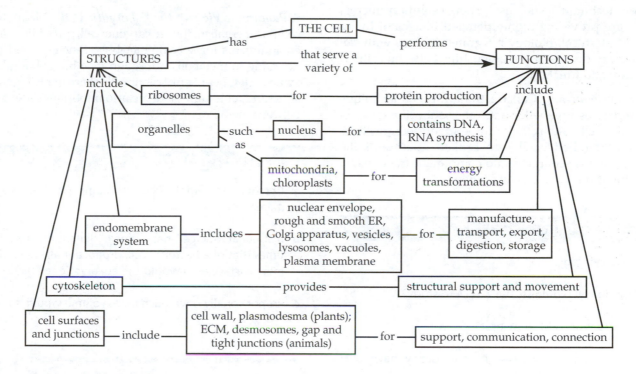

Chapter Review

The cell is the basic structural and functional unit of all organisms. In the hierarchy of biological organization, the capacity for life emerges from the structural order of the cell. All cells are related through common descent, but evolution has shaped diverse adaptations.

4.1 Biologists use microscopes and the tools of biochemistry to study cells

Microscopy   The glass lenses of a **light microscope (LM)** refract (bend) the visible light passing through a specimen such that the projected image is magnified. *Magnification* is the ratio of the size of an image to the real size of the object. *Resolution* is a measure of the clarity of an image and is determined by the minimum distance two points must be separated to be seen as separate. The resolving power of the light microscope is limited, so that details finer than 0.2 μm (micrometers = 10^{-3} mm) cannot be distinguished. Staining specimens and using techniques such as fluorescence, phase-contrast, and confocal microscopy improve visibility by increasing *contrast* between structures.

Many cellular structures, including most membrane-enclosed **organelles**, cannot be resolved by the light microscope. Their structures were first revealed when the **electron microscope (EM)** was developed in the 1950s. The electron microscope focuses a beam of electrons through a specimen or onto its surface. The short wavelength of the electron beam allows a resolution of about 2 nm (nanometers = 10^{-3} μm), about a hundred times greater than that of the light microscope.

In a **transmission electron microscope (TEM)**, a beam of electrons is passed through a thin section of a specimen stained with atoms of heavy metals. Electromagnets, acting as lenses, then focus the image onto a monitor.

In a **scanning electron microscope (SEM)**, an electron beam scans the surface of a specimen that is usually coated with a thin gold film. The beam excites electrons from the specimen, which are detected and translated into an image on a video screen. This image appears three-dimensional.

New techniques such as "super-resolution microscopy" are providing highly detailed images of living cells. Modern cell biology integrates *cytology* with *biochemistry* to understand relationships between cellular structure and function.

Cell Fractionation   **Cell fractionation** is a technique that separates organelles and other subcellular structures of a cell so that they can be identified and their functions studied. Cells are broken apart and the homogenate is separated into component fractions by centrifugation at increasing speeds.

FOCUS QUESTION 4.1

a. Define cytology.

b. What do cell biologists use a TEM to study?

c. What does an SEM show best?

d. What advantages does light microscopy have over both TEM and SEM?

4.2   Eukaryotic cells have internal membranes that compartmentalize their functions

Comparing Prokaryotic and Eukaryotic Cells   How are prokaryotic and eukaryotic cells the same? All cells are bounded by a *plasma membrane*, which encloses a semifluid substance called **cytosol**. All cells contain *chromosomes* (with genes composed of DNA) and *ribosomes*.

How are prokaryotic and eukaryotic cells different? Only members of the domains Bacteria and Archaea have **prokaryotic cells**, which are cells with no nucleus or membrane-enclosed organelles. The DNA of prokaryotic cells is concentrated in a region called the **nucleoid**. **Eukaryotic cells** have a true *nucleus* enclosed in a double membrane and numerous membrane-bounded organelles suspended in cytosol. **Cytoplasm** refers to the region between the nucleus and the plasma membrane, and also to the interior of a prokaryotic cell.

Most bacterial cells range from 1 to 5 μm in diameter, whereas eukaryotic cells range from 10 to 100 μm. The small size of cells is influenced by geometry. Area is proportional to the square of linear dimension, while volume is proportional to its cube. The **plasma membrane** surrounding every cell must provide sufficient surface area for exchange of oxygen, nutrients, and wastes relative to the volume of the cell.

A Panoramic View of the Eukaryotic Cell   Membranes compartmentalize the eukaryotic cell, providing local environments for specific metabolic functions and participating in metabolism through membrane-bound enzymes. All types of eukaryotic cells share many common structures, although there are also important differences between them.

FOCUS QUESTION 4.2

a. Describe the molecular structure of the plasma membrane.

b. If a eukaryotic cell has a linear dimension that is 10 times that of a bacterial cell, proportionally how much more surface area would the eukaryotic cell have?

c. Proportionally how much more volume would it have?

4.3   The eukaryotic cell's genetic instructions are housed in the nucleus and carried out by the ribosomes

The Nucleus: Information Central   The **nucleus** is surrounded by the **nuclear envelope**, a double membrane perforated by pores. A protein *pore complex* lining each pore regulates the movement of materials between the nucleus and the cytoplasm. The inner membrane is lined by the **nuclear lamina**, a layer of protein filaments that helps to maintain the shape of the nucleus.

Most of the cell's DNA is located in the nucleus, where it is organized into units called **chromosomes**. Chromosomes are made up of **chromatin**, a complex of DNA and proteins. Each eukaryotic species has a characteristic chromosomal number. Individual chromosomes are visible only when condensed in a dividing cell.

The **nucleolus**, a dense structure visible in the nondividing nucleus, synthesizes *ribosomal RNA (rRNA)* and combines it with protein to assemble ribosomal subunits, which then pass through nuclear pores to the cytoplasm.

Ribosomes: Protein Factories **Ribosomes** are composed of protein and ribosomal RNA. Most of the proteins produced on *free ribosomes* are used within the cytosol. *Bound ribosomes*, attached to the endoplasmic reticulum or nuclear envelope, usually make proteins that will be included within membranes, packaged into organelles, or exported from the cell.

FOCUS QUESTION 4.3

How does the nucleus control protein synthesis in the cytoplasm?

4.4   The endomembrane system regulates protein traffic and performs metabolic functions in the cell

A good approach to learning all those organelles in a eukaryotic cell is to connect the components of the **endomembrane system**—the nuclear envelope, endoplasmic reticulum, Golgi apparatus, lysosomes, vesicles, vacuoles, and the plasma membrane. Although not identical, all these membranes are related either through direct contact or by the transfer of membrane segments by membrane-bound sacs called **vesicles**.

The Endoplasmic Reticulum: Biosynthetic Factory The **endoplasmic reticulum (ER)** is a membranous system that is continuous with the nuclear envelope and encloses a network of interconnected tubules or compartments called cisternae. Ribosomes are attached to the cytoplasmic surface of **rough ER; smooth ER** lacks ribosomes.

Smooth ER serves diverse functions in different cells: Its enzymes are involved in phospholipid and steroid (including sex hormone) synthesis, carbohydrate metabolism, and detoxification of drugs and poisons. Barbiturates, alcohol, and other drugs increase a liver cell's production of smooth ER, thus leading to an increased tolerance (and thus reduced effectiveness) for these and other drugs. Smooth ER also functions in

the storage and release of calcium ions during muscle contraction.

Proteins intended for secretion are manufactured by membrane-bound ribosomes and then threaded into the lumen of the rough ER. Many are covalently bonded to small carbohydrates to form **glycoproteins**. Secretory proteins are transported from the rough ER in membrane-bound **transport vesicles**.

Rough ER also manufactures membranes. Enzymes built into the membrane assemble phospholipids, and membrane proteins formed by bound ribosomes are inserted into the ER membrane. Transport vesicles transfer ER membrane to other parts of the endomembrane system.

The Golgi Apparatus: Shipping and Receiving Center The **Golgi apparatus** consists of a stack of flattened sacs. Vesicles that bud from the ER join to the *cis* face of a Golgi stack, adding to it their contents and membrane. According to the *cisternal maturation model*, Golgi products are processed and tagged as the cisternae themselves progress from the *cis* to the *trans* face. Glycoproteins often have their attached carbohydrates modified. The Golgi apparatus of plant cells manufactures some polysaccharides, such as pectins. Golgi products are sorted into vesicles, which pinch off from the *trans* face of the Golgi apparatus. These vesicles may have surface molecules that help direct them to the plasma membrane or to other organelles.

Lysosomes: Digestive Compartments In animal cells, **lysosomes** are membrane-enclosed sacs containing hydrolytic enzymes that digest macromolecules. Lysosomes provide an acidic pH for these enzymes.

In some protists, lysosomes fuse with *food vacuoles* to digest material ingested by **phagocytosis**. Macrophages, a type of white blood cell, use lysosomes to destroy ingested bacteria. Lysosomes also recycle a cell's own macromolecules by fusing with vesicles enclosing damaged organelles or small bits of cytosol, a process known as *autophagy*.

Vacuoles: Diverse Maintenance Compartments **Vacuoles** are large vesicles. **Food vacuoles** are formed as a result of phagocytosis. **Contractile vacuoles** pump excess water out of freshwater protists. Vacuoles in plant cells may store organic compounds and inorganic ions for the cell, or contain dangerous metabolic by-products and poisonous or unpalatable compounds, which may protect the plant from predators. A large **central vacuole** is found in mature plant cells and encloses a solution called cell sap. A plant cell increases in size with a minimal addition of new cytoplasm as its vacuole absorbs water and expands.

The Endomembrane System: A Review As membranes move from the ER to the Golgi apparatus and then to other organelles, their compositions, functions, and contents are modified.

FOCUS QUESTION   4.4

FOCUS QUESTION   4.4

Name the components of the endomembrane system shown in this diagram, and list the functions of each of these membranes.

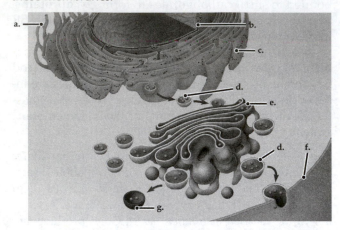

a.

b.

c.

d.

e.

f.

g.

4.5   Mitochondria and chloroplasts change energy from one form to another

Cellular respiration, the metabolic processing of fuels to produce ATP, occurs within the **mitochondria** of eukaryotic cells. Photosynthesis occurs in the **chloroplasts** of plants and algae, which produce sugars from carbon dioxide and water by absorbing solar energy.

The Evolutionary Origins of Mitochondria and Chloroplasts   According to the **endosymbiont theory**, both mitochondria and chloroplasts originated as prokaryotic cells engulfed by an ancestral eukaryotic cell. The double membranes surrounding mitochondria and chloroplasts appear to have been part of the prokaryotic *endosymbiont*. These organelles grow and reproduce independently within the cell. They contain a small amount of DNA and ribosomes (both similar to prokaryotic structures), and they synthesize some of their proteins.

Chemical Energy Conversion   Two membranes, each a phospholipid bilayer with unique embedded proteins,

enclose a mitochondrion. A narrow intermembrane space exists between the smooth outer membrane and the convoluted inner membrane. The folds of the inner membrane, called **cristae**, create a large surface area and enclose the **mitochondrial matrix**. Many respiratory enzymes, mitochondrial DNA, and ribosomes are housed in this matrix. Other respiratory enzymes and proteins are built into the inner membrane.

Chloroplasts: Capture of Light Energy   Two membranes separated by a thin intermembrane space enclose a chloroplast. Inside the inner membrane is a membranous system of connected flattened sacs called **thylakoids**, inside of which is the thylakoid space. Photosynthetic enzymes are embedded in the thylakoids, which may be stacked together to form structures called **grana**. Chloroplast DNA, ribosomes, and many enzymes are found in the **stroma**, the fluid surrounding the thylakoids.

   Plastids are plant organelles that also include *amyloplasts*, which store starch, and *chromoplasts*, which contain pigments.

FOCUS QUESTION   4.5

Sketch a mitochondrion and a chloroplast, and label their membranes and compartments.

Peroxisomes: Oxidation   **Peroxisomes** are oxidative organelles filled with enzymes that function in a variety of metabolic pathways, such as breaking down fatty acids for energy or detoxifying alcohol and other poisons. An enzyme that converts hydrogen peroxide (H_2O_2), a toxic by-product of these pathways, to water is also packaged into peroxisomes.

4.6   The cytoskeleton is a network of fibers that organizes structures and activities in the cell

Roles of the Cytoskeleton: Support and Motility   The **cytoskeleton** is a network of protein fibers that give mechanical support and function in cell motility (of both internal structures and the cell as a whole). The cytoskeleton interacts with special proteins called **motor proteins** to produce cellular movements.

Components of the Cytoskeleton Let's review the three main types of fibers found in the cytoskeleton: *microtubules, microfilaments,* and *intermediate filaments.*

All eukaryotic cells have **microtubules**, which are hollow rods constructed of columns of globular proteins called tubulins. Microtubules change length through the addition or subtraction of tubulin dimers. In addition to providing the supporting framework of the cell, microtubules separate chromosomes during cell division and serve as tracks along which organelles move with the aid of motor molecules.

In animal cells, microtubules grow out from a region near the nucleus called a **centrosome**. A pair of **centrioles**, each composed of nine sets of triplet microtubules arranged in a ring, is associated with the centrosome and replicates before cell division. Fungi and plant cells lack centrosomes with centrioles and apparently have some other microtubule-organizing center.

Cilia and **flagella** are locomotor extensions of some eukaryotic cells. Cilia are numerous and short; flagella occur one or two to a cell and are longer. Many protists use cilia or flagella to move through aqueous media. Cilia or flagella attached to stationary cells of a tissue move fluid past the cell. A signal-receiving cilium, such as the *primary cilium* found on vertebrate animal cells, transmits environmental signals to a cell's interior.

Motile cilia and flagella are composed of two single microtubules surrounded by a ring of nine doublets of microtubules (a nearly universal "9 + 2" arrangement), all of which are enclosed in an extension of the plasma membrane. The microtubule doublets slide past each other as the two "feet" of large motor proteins called **dyneins** alternately attach to adjacent doublets, pull down, release, and reattach. In conjunction with anchoring cross-linking proteins and radial spokes, this action causes the bending of the flagellum or motile cilium. A **basal body**, with a "9 + 0" pattern of microtubule triplets, anchors a cilium or flagellum in the cell.

Microfilaments are thin solid rods consisting of a twisted double chain of globular proteins called **actin**. Thus, microfilaments are also called actin filaments. They form a supporting network just inside the plasma membrane and make up the core of small cytoplasmic extensions called microvilli.

In muscle cells, thousands of actin filaments and thicker filaments made of the motor protein **myosin** produce muscle contraction. Actin and myosin also interact to cause the amoeboid movements of some cells. *Cytoplasmic streaming* in plant cells also involves actin–myosin interactions.

Intermediate filaments are intermediate in size between microtubules and microfilaments and are more diverse in their protein composition. Intermediate filaments are important in maintaining cell shape. The nucleus is securely held in a web of intermediate filaments, and the nuclear lamina lining the inside of the nuclear envelope is composed of intermediate filaments.

FOCUS QUESTION 4.6

Fill in the following table to organize what you have learned about the components of the cytoskeleton. You may wish to refer to the textbook for additional details.

Cytoskeleton Component	Structure and Monomers	Functions
Microtubules	a.	b.
Microfilaments (actin filaments)	c.	d.
Intermediate filaments	e.	f.

4.7 Extracellular components and connections between cells help coordinate cellular activities

Cell Walls of Plants Plant **cell walls** are composed of microfibrils of cellulose embedded in a matrix of polysaccharides and protein.

The **primary cell wall** secreted by a young plant cell is relatively thin and flexible. Adjacent cells are glued together by the **middle lamella**, a thin layer of polysaccharides (called pectins). When they stop growing, some cells secrete a thicker and stronger **secondary cell wall** between the plasma membrane and the primary cell wall.

Sketch two adjacent plant cells, and show the location of the primary and secondary cell walls and the middle lamella. Also include a plasmodesma (see the Cell Junctions section later in this chapter).

The Extracellular Matrix (ECM) of Animal Cells   Animal cells secrete an **extracellular matrix (ECM)** composed primarily of glycoproteins and other carbohydrate-containing molecules. **Collagen** forms strong fibers that are embedded in a network of proteoglycan complexes. These large complexes form when multiple **proteoglycans**, each consisting of a small core protein with many carbohydrate chains, link to a long polysaccharide. Cells may be attached to the ECM by **fibronectins** and other glycoproteins that bind to **integrins**, proteins that span the plasma membrane and bind, via other proteins, to microfilaments of the cytoskeleton. Thus, information about changes outside the cell can be communicated through a mechanical signaling pathway involving fibronectins, integrins, and the microfilaments of the cytoskeleton. Signals from the ECM appear to influence the activity of genes in the nucleus.

Label the indicated structures in this diagram of the ECM of an animal cell.

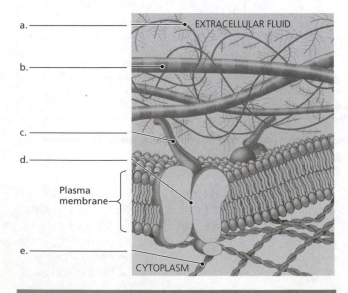

a. _____
EXTRACELLULAR FLUID
b. _____
c. _____
d. _____
Plasma membrane
e. _____
CYTOPLASM

Cell Junctions   **Plasmodesmata** are channels in plant cell walls through which the plasma membranes of bordering cells connect, thus linking most cells of a plant into a living continuum. Water, small solutes, and even some proteins and RNA molecules can move through these channels.

What types of cell junctions are found between animal cells? At **tight junctions**, proteins hold adjacent cell membranes tightly together, creating an impermeable seal across a layer of epithelial cells. **Desmosomes** (also called *anchoring junctions*) are reinforced by intermediate filaments and rivet cells into strong sheets. **Gap junctions** (also called *communicating junctions*) are cytoplasmic connections that allow for the exchange of ions and small molecules between cells through protein-lined pores.

The Cell: A Living Unit Greater Than the Sum of Its Parts   The compartmentalization and the many specialized organelles typical of cells exemplify the principle that structure correlates with function. The intricate functioning of a living cell emerges from the complex interactions of its multiple parts.

Word Roots

centro- = the center; **-soma** = a body (*centrosome:* a structure in the cytoplasm of animal cells that functions as a microtubule-organizing center)

chloro- = green (*chloroplast:* the site of photosynthesis in plants and photosynthetic protists)

cili- = hair (*cilium:* a short cellular appendage containing microtubules)

cyto- = cell (*cytosol:* the semifluid portion of the cytoplasm)

-ell = small (*organelle:* a membrane-enclosed structure with a specialized function, suspended in the cytosol of eukaryotic cells)

endo- = inner (*endomembrane system:* the collection of membranes inside and surrounding a eukaryotic cell that includes the plasma membrane, nuclear envelope, endoplasmic reticulum, Golgi apparatus, lysosomes, vesicles, and vacuoles)

eu- = true (*eukaryotic cell:* a type of cell with a membrane-enclosed nucleus and organelles)

extra- = outside (*extracellular matrix:* the meshwork surrounding animal cells, consisting of glycoproteins, polysaccharides, and proteoglycans)

flagell- = whip (*flagellum:* a long cellular appendage specialized for locomotion)

glyco- = sweet (*glycoprotein:* a protein with one or more covalently attached carbohydrates)

lamin- = sheet/layer (*nuclear lamina:* a netlike array of protein filaments lining the inner surface of the nuclear envelope)

lyso- = loosen (*lysosome:* a membrane-enclosed sac of hydrolytic enzymes found in the cytoplasm of animal cells and some protists)

micro- = small; **tubul** = a little pipe (*microtubule:* a hollow rod composed of tubulin proteins that makes up part of the cytoskeleton in all eukaryotic cells and is found in cilia and flagella)

nucle- = nucleus; **-oid** = like (*nucleoid:* a non-membrane-bounded region in a prokaryotic cell where the DNA is concentrated)

phago- = to eat; **-kytos** = vessel (*phagocytosis:* a type of endocytosis in which large particulate substances are taken up by a cell)

plasm- = molded; **-desma** = a band or bond (*plasmodesma:* an open channel in a plant cell wall, connecting the cytoplasm of adjacent cells)

pro- = before; **karyo-** = nucleus (*prokaryotic cell:* a type of cell lacking a membrane-enclosed nucleus and organelles)

thylaco- = sac or pouch (*thylakoid:* a flattened membranous sac inside a chloroplast)

vacu- = empty (*vacuole:* a relatively large vesicle with a specialized function)

Structure Your Knowledge

1. In the following table, write the organelles or structures that are associated with each of the listed functions in animal cells.

Functions	Associated Organelles and Structures
Cell division	a.
Information storage and transferral	b.
Energy conversions	c.
Manufacture of proteins, membranes, and other products	d.
Lipid synthesis, drug detoxification	e.
Digestion, recycling	f.
Oxidation, conversion of H_2O_2 to water	g.
Structural integrity	h.
Movement	i.
Exchange with the environment	j.
Cell-to-cell connections	k.

2. Write the functions of each of the following plant cell structures.

Plant Cell Structures	Functions
Cell wall	a.
Central vacuole	b.
Chloroplast	c.
Amyloplast	d.
Plasmodesmata	e.

3. Label the indicated structures in the following diagram of an animal cell.

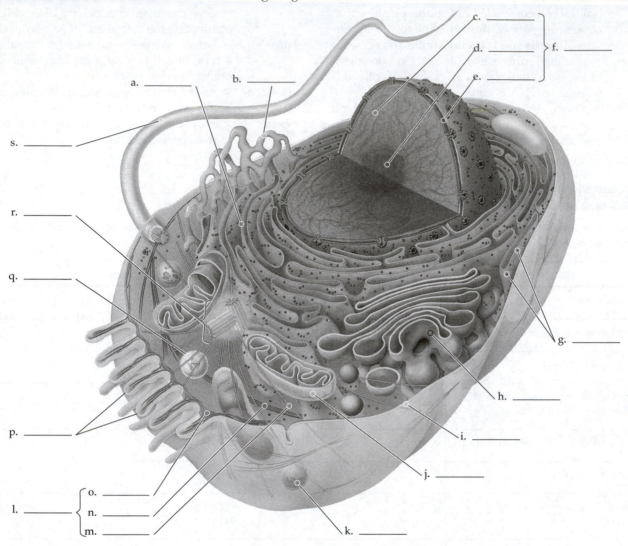

a. _____   b. _____   c. _____ ⎤
                               d. _____ ⎬ f. _____
s. _____   e. _____ ⎦

r. _____

q. _____

p. _____   g. _____

l. _____ ⎰ o. _____
           ⎱ n. _____
             m. _____

                           h. _____
                           i. _____
                           j. _____
              k. _____

4. Create a diagram or flowchart in the following space to trace the development of a secretory product (such as a digestive enzyme) from the DNA code to its export from the cell.

Test Your Knowledge

MULTIPLE CHOICE: *Choose the one best answer.*

1. Resolution of a microscope refers to
   a. the distance between two separate points.
   b. the sharpness or clarity of an image.
   c. the amount of magnification of an image.
   d. the depth of focus on a specimen's surface.
   e. the wavelength of light.

2. The detailed structure of a chloroplast can be seen with the best resolution by using
   a. transmission electron microscopy.
   b. scanning electron microscopy.
   c. phase-contrast light microscopy.
   d. differential-interference contrast microscopy.
   e. confocal fluorescence microscopy.

3. In an animal cell fractionation procedure, one of the last pellets from the series of centrifugations would most likely contain
   a. ribosomes.
   d. lysosomes.
   b. peroxisomes.
   e. nuclei.
   c. mitochondria.

4. Which of the following is/are *not* found in a prokaryotic cell?
   a. ribosomes
   b. plasma membrane
   c. mitochondria
   d. a and c
   e. a, b, and c

5. The cells of an ant and an elephant are, on average, the same size; an elephant just has more cells. What is the main advantage of small cell size?
   a. Small cells are easier to organize into tissues and organs.
   b. A small cell has a larger plasma membrane surface area than does a large cell, facilitating the exchange of sufficient materials with its environment.
   c. The cytoskeleton of a large cell would have to be so large that cells would be too heavy.
   d. Small cells require less oxygen and nutrients than do large cells.
   e. A small cell has a larger surface area relative to cytoplasmic volume, facilitating sufficient exchange of materials.

6. Proteins that function within the cytosol are generally synthesized
   a. by ribosomes bound to rough ER.
   b. by free ribosomes.
   c. by the nucleolus.
   d. within the Golgi apparatus.
   e. by mitochondria and chloroplasts.

7. The pores in the nuclear envelope provide for the movement of
   a. proteins into the nucleus.
   b. ribosomal subunits out of the nucleus.
   c. mRNA out of the nucleus.
   d. signal molecules into the nucleus.
   e. all of the above.

8. A fluorescent green tag is attached to a protein being synthesized and inserted into an area of rough ER membrane. Which of the following is a possible route for that protein to follow?
   a. rough ER → Golgi → lysosome → nuclear membrane → plasma membrane
   b. rough ER → transport vesicle → Golgi → vesicle → plasma membrane → food vacuole
   c. rough ER → nuclear envelope → Golgi → smooth ER → lysosome
   d. rough ER → transport vesicle → Golgi → smooth ER → plasma membrane
   e. rough ER → transport vesicle → Golgi → vesicle → extracellular matrix

9. A growing plant cell elongates primarily by
   a. increasing the number of vacuoles.
   b. synthesizing more cytoplasm.
   c. taking up water into its central vacuole.
   d. activating its contractile vacuoles.
   e. producing a secondary cell wall.

10. Which of the following is *not* a similarity among nuclei, chloroplasts, and mitochondria?
    a. They contain DNA.
    b. They are bounded by two phospholipid bilayer membranes.
    c. They can divide to reproduce themselves.
    d. They are derived from the endoplasmic reticulum system.
    e. Their membranes are associated with specific proteins.

11. Which of the following is a major component of the plasma membrane of a plant cell?
    a. collagen
    b. cellulose
    c. phospholipids
    d. pectins
    e. proteoglycans

12. Microtubules are components of or associated with all of the following *except*
    a. centrioles.
    b. separating chromosomes in cell division.
    c. tracks along which organelles can move using motor proteins.
    d. flagella and cilia.
    e. amoeboid movement.

13. Which of the following characteristics are shared by cristae of mitochondria, thylakoids, and microvilli?
    a. have a large surface area of membrane that enhances their particular function
    b. are bounded by a double membrane
    c. have a shape reinforced by microfilaments
    d. are not derived from the endoplasmic reticulum system
    e. arose by endosymbiosis of an aerobic bacterium, a photosynthetic bacterium, and a motile bacterium, respectively

14. The innermost portion of the cell wall of a plant cell specialized for support is the
    a. primary cell wall.
    b. secondary cell wall.
    c. middle lamella.
    d. plasma membrane.
    e. cytoskeleton.

15. Plasmodesmata in plant cells are similar in function to
    a. desmosomes.
    b. tight junctions.
    c. gap junctions.
    d. the extracellular matrix.
    e. integrins.

16. Which of the following structures is/are involved in the relay of external information into a cell?
    a. a primary cilium
    b. fibronectins
    c. integrins
    d. microfilaments just inside the plasma membrane
    e. all of the above

17. Which of the following is *incorrectly* paired with its function?
    a. peroxisome—contains enzymes that break down H_2O_2
    b. nucleolus—produces ribosomal RNA, assembles ribosome subunits
    c. Golgi apparatus—processes, tags, and ships cellular products
    d. lysosome—food sac formed by phagocytosis
    e. ECM (extracellular matrix)—supports and anchors cells, communicates information with inside of cell

Answer questions 18 through 22 using the following five choices.
    a. muscle cells in the thigh muscle of a runner
    b. pancreatic cells that manufacture digestive enzymes
    c. macrophages that engulf bacteria
    d. epithelial cells lining the digestive tract
    e. ovarian cells that produce estrogen (a steroid hormone)

18. In which cells would you expect to find the most tight junctions?

19. In which cells would you expect to find the most lysosomes?

20. In which cells would you expect to find the most smooth endoplasmic reticulum?

21. In which cells would you expect to find the most bound ribosomes?

22. In which cells would you expect to find the most mitochondria?

Chapter 5

Membrane Transport and Cell Signaling

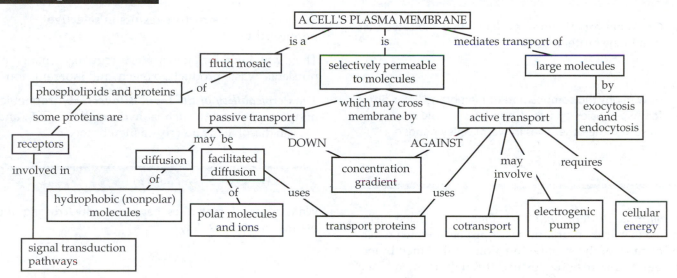

A CELL'S PLASMA MEMBRANE

is a — fluid mosaic

is — selectively permeable to molecules

mediates transport of — large molecules

of — phospholipids and proteins

some proteins are

receptors

involved in

signal transduction pathways

may be — passive transport

diffusion

of — hydrophobic (nonpolar) molecules

facilitated diffusion

of — polar molecules and ions

which may cross membrane by

DOWN — concentration gradient

uses — transport proteins

AGAINST

active transport

uses

may involve — cotransport

requires — cellular energy

electrogenic pump

by — exocytosis and endocytosis

Chapter Review

The plasma membrane is the boundary of life. Like all biological membranes, it has **selective permeability**, allowing some materials to cross it more easily than others.

5.1 Cellular membranes are fluid mosaics of lipids and proteins

According to the **fluid mosaic model**, biological membranes consist of various proteins that are embedded in a bilayer of **amphipathic** phospholipids (having both hydrophilic and hydrophobic regions). The membrane proteins are also amphipathic, with their hydrophilic regions extending into the aqueous surroundings.

FOCUS QUESTION 5.1

Label the components in the following diagram of the fluid mosaic model of membrane structure. Indicate whether regions are hydrophobic or hydrophilic.

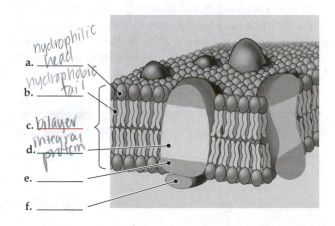

a. hydrophilic head
b. hydrophobic tail
c. bilayer
d. integral protein
e. _____
f. _____

The Fluidity of Membranes Membranes are held together primarily by weak hydrophobic interactions that allow the lipids and some of the proteins to drift laterally. Some membrane proteins seem to be held by attachments to the cytoskeleton or ECM; others appear to be directed in their movements.

Phospholipids with unsaturated hydrocarbon tails maintain membrane fluidity at lower temperatures. The steroid cholesterol, common in plasma membranes of animals, restricts movement of phospholipids and thus reduces fluidity at warmer temperatures. Cholesterol also prevents the close packing of lipids and thus enhances fluidity at lower temperatures.

Evolution of Differences in Membrane Lipid Composition Variations in membrane lipid composition and the ability to change that composition in response to changing temperatures are evolutionary adaptations.

FOCUS QUESTION 5.2

a. Cite some experimental evidence that indicates that membrane proteins drift.

b. How might the composition of membrane lipids differ for two species of fish if one lives in cold, mountain lakes and the other lives in warm valley ponds?

Membrane Proteins and Their Functions Each membrane has its own unique set of proteins, which determine most of the specific functions of that membrane. **Integral proteins** often extend through the membrane (are *transmembrane* proteins), with two hydrophilic ends. The hydrophobic midsection usually consists of one or more α helical stretches of nonpolar amino acids. **Peripheral proteins** are attached to the surface of the membrane. Attachments of membrane proteins to the cytoskeleton and to fibers of the extracellular matrix provide support for the plasma membrane.

FOCUS QUESTION 5.3

Can you list the six major kinds of functions that membrane proteins may perform?

The Role of Membrane Carbohydrates in Cell–Cell Recognition The cell's ability to distinguish other cells is based on the recognition and binding of membrane proteins to carbohydrates on other cells. The **glycolipids** and **glycoproteins** extending to the outside of plasma membranes vary from species to species, from individual to individual, and even among cell types.

Synthesis and Sidedness of Membranes Membranes have distinct faces or sides, related to the composition of the lipid layers and the directional orientation of their proteins. Carbohydrates attached to membrane proteins as they are synthesized in the ER are modified in the Golgi apparatus. Carbohydrates may also be attached to lipids in the Golgi apparatus. When transport vesicles fuse with the plasma membrane, these interior glycoproteins and glycolipids become located on the extracellular face of the membrane.

5.2 Membrane structure results in selective permeability

The plasma membrane permits a regular exchange of nutrients, waste products, oxygen, and inorganic ions.

The Permeability of the Lipid Bilayer Hydrophobic, nonpolar molecules, such as hydrocarbons, CO_2, and O_2, can dissolve in and cross a membrane.

FOCUS QUESTION 5.4

What types of molecules have difficulty crossing the plasma membrane? Why?

Transport Proteins Ions and polar molecules may move across the plasma membrane with the aid of **transport proteins**. Hydrophilic passageways through a membrane are provided for specific molecules or ions by *channel proteins*, such as **aquaporins**, which facilitate water passage. *Carrier proteins* may physically bind and transport a specific molecule across a membrane.

5.3 Passive transport is diffusion of a substance across a membrane with no energy investment

Diffusion is the movement of a substance down its **concentration gradient** due to random molecular motion (thermal energy). The diffusion of one solute is unaffected by the concentration gradients of other solutes; each solute will reach its own dynamic equilibrium. Remember that the cell does not expend energy when substances diffuse across membranes down their concentration gradients; therefore, the process is called **passive transport**.

Effects of Osmosis on Water Balance **Osmosis** is the diffusion of free water across a selectively permeable membrane. Water diffuses down its own concentration gradient, which is affected by the solute concentration. Clustering of water molecules around solute particles lowers the proportion of free water that is available to cross the membrane.

FOCUS QUESTION 5.5

A solution of 1 *M* glucose is separated by a selectively permeable membrane from a solution of 0.2 *M* fructose and 0.7 *M* sucrose. The membrane is not permeable to the sugar molecules. Indicate which side initially has more free water molecules, and which side has fewer. Show the direction of osmosis.

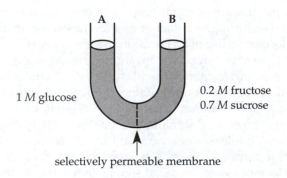

1 *M* glucose

0.2 *M* fructose
0.7 *M* sucrose

selectively permeable membrane

Tonicity, the tendency of a surrounding solution to cause a cell to gain or lose water, is affected by the relative concentrations of those solutes in the solution and in the cell that cannot cross the membrane. An animal cell will neither gain nor lose water in an **isotonic** environment. An animal cell placed in a **hypertonic** solution (which has more nonpenetrating solutes) will lose water and shrivel. If placed in a **hypotonic** solution, the cell will gain water, swell, and possibly lyse (burst). Cells without rigid walls must either live in an isotonic environment, such as salt water or isotonic body fluids, or have adaptations for **osmoregulation**, the regulation of solute concentrations and water balance.

The cell walls of plants, fungi, prokaryotes, and some protists play a role in water balance in hypotonic environments. Water moving into the cell causes the cell to swell against its cell wall, creating *turgor pressure*. **Turgid** cells provide mechanical support for nonwoody plants. Plant cells in an isotonic surrounding are **flaccid**. In a hypertonic medium, a plant cell undergoes **plasmolysis**—that is, the plasma membrane pulls away from the cell wall as water exits and the cell shrivels.

FOCUS QUESTION 5.6

a. What osmotic problems does the freshwater protist *Paramecium* face, and what adaptation enables it to osmoregulate?

b. Compare the ideal osmotic environment for animal cells and plant cells.

Facilitated Diffusion: Passive Transport Aided by Proteins **Facilitated diffusion** involves the diffusion of polar molecules and ions across a membrane with the aid of transport proteins, either channel proteins or carrier proteins. Many **ion channels** are **gated channels**, which open or close in response to electrical or chemical stimuli. The binding of a solute to a carrier protein may cause a change in the protein's shape that translocates the solute across the membrane.

FOCUS QUESTION 5.7

Why is facilitated diffusion considered a form of passive transport?

5.4 Active transport uses energy to move solutes against their gradients

The Need for Energy in Active Transport **Active transport**, which requires the expenditure of energy to transport a solute against its concentration gradient, is essential if a cell is to maintain internal concentrations of small molecules that differ from their concentrations outside the cell. The terminal phosphate group of ATP may be transferred to a carrier protein, inducing it to change its shape and translocate the bound solute across the membrane. The **sodium-potassium pump** works this way to exchange N^+ and K^+ across animal cell membranes, creating a higher concentration of potassium ions and a lower concentration of sodium ions within the cell.

How Ion Pumps Maintain Membrane Potential Cells have a **membrane potential**, a voltage across the plasma membrane due to the unequal distribution of ions on either side. This electrical potential energy results from the separation of opposite charges: The cytoplasm of a cell is negatively charged relative to the extracellular fluid. The membrane potential favors the diffusion of cations into the cell and anions out of the cell. Both the membrane potential and the concentration gradient affect the diffusion of an ion; thus, an ion diffuses down its **electrochemical gradient**.

Electrogenic pumps are membrane proteins that generate voltage across a membrane by the active transport of ions. A **proton pump** that transports H^+ out of the cell generates voltage across membranes in plants, fungi, and bacteria.

The sodium-potassium pump, the major electrogenic pump in animal cells, exchanges sodium ions for potassium ions, both of which are cations. How does this exchange generate a membrane potential?

Cotransport: Coupled Transport by a Membrane Protein   Cotransport is a mechanism through which the active transport of a solute is indirectly driven by an ATP-powered pump that transports another substance against its gradient. As that actively transported substance diffuses back down its concentration gradient through a cotransporter, the solute is carried against its concentration gradient across the membrane.

5.5   Bulk transport across the plasma membrane occurs by exocytosis and endocytosis

Bulk transport, like active transport, requires energy to transport larger biological molecules packaged in vesicles across the membrane.

Exocytosis   In **exocytosis**, the cell secretes large molecules by the fusion of vesicles with the plasma membrane.

Endocytosis   In **endocytosis**, a region of the plasma membrane sinks inward and pinches off to form a vesicle containing material that had been outside the cell. **Phagocytosis** is a form of endocytosis in which pseudopodia wrap around a food particle or another large particle, creating a vacuole that then fuses with a lysosome containing hydrolytic enzymes. In **pinocytosis**, droplets of extracellular fluid are taken into the cell in small vesicles. **Receptor-mediated endocytosis** enables a cell to acquire specific substances from extracellular fluid. Molecules bind to receptor proteins that are usually clustered in coated pits on the cell surface and are carried into the cell when a vesicle forms.

a. How is cholesterol transported into human cells?

b. Explain why cholesterol accumulates in the blood of individuals with the disease familial hypercholesterolemia.

5.6   The plasma membrane plays a key role in most cell signaling

Local and Long-Distance Signaling   Chemical signals may be communicated between cells through direct cytoplasmic connections (gap junctions or plasmodesmata) or through contact of membrane-bound surface molecules (cell–cell recognition in animal cells).

In *paracrine signaling* in animals, a signaling cell releases messenger molecules into the extracellular fluid, and these **local regulators** influence nearby cells. *Growth factors* are one class of local regulators. In another type of local signaling called *synaptic signaling*, a nerve cell releases neurotransmitter molecules, which diffuse across the narrow synapse to its target cell.

Hormones are chemical signals that travel to more distant cells. In hormonal or *endocrine signaling* in animals, the circulatory system transports hormones throughout the body to reach and bind to target cells that have appropriate receptors. Plant hormones may reach their target cells by traveling through plant vascular tissues or even through the air as a gas.

Transmission of electrical and chemical signals within the nervous system is also a type of long-distance signaling.

The Three Stages of Cell Signaling: A Preview   E. W. Sutherland's studies of epinephrine's effect on the hydrolysis of glycogen in liver cells established that cell signaling involves three stages: **reception** of a chemical signal by binding to a receptor protein either inside a target cell or on its surface; **transduction** of the signal, often by a **signal transduction pathway**—a sequence of changes in relay molecules; and the specific **response** of the cell.

5.7   Reception, the binding of a signaling molecule to a receptor protein

A signaling molecule acts as a **ligand**, which specifically binds to a receptor protein and usually induces a change in the receptor protein's shape, activating the receptor. Most ligands are large and water-soluble, and they bind to receptors in the plasma membrane.

A cell-surface transmembrane receptor that works with the aid of a G protein is called a **G protein-coupled receptor (GPCR)**. Binding of the appropriate signaling molecule to a G protein-coupled receptor activates the receptor, which then binds to and activates a specific **G protein** located on the cytoplasmic side of the membrane. The activated G protein, which carries a GTP molecule, activates a membrane-bound enzyme, which then triggers the next step in the pathway to the cell's response.

G protein-coupled receptor systems are involved in the function of many hormones and neurotransmitters and in embryological development and sensory reception. Many bacteria produce toxins that interfere with G-protein function; up to 60% of all medicines influence G-protein pathways.

The binding of a signaling molecule to a **ligand-gated ion channel** opens or closes a "gate," thereby allowing or blocking the flow of specific ions through the receptor channel. The resulting change in ion concentration inside the cell triggers a cellular response. Neurotransmitters often bind to ligand-gated ion channels in the transmission of neural signals.

Hydrophobic chemical messengers and small signaling molecules such as the gas nitric oxide may cross a cell's plasma membrane and bind to receptors in the cytoplasm or nucleus of target cells. Steroid hormones activate receptors in target cells that function as *transcription factors* that regulate gene expression.

FOCUS QUESTION 5.10

Describe the difference between signaling molecules that bind to cell membrane receptors and those that bind to intracellular receptors.

Transduction by Cascades of Molecular Interactions Multistep pathways enable a small number of extracellular signals to be amplified to produce a large cellular response. Such pathways also provide opportunities for regulation and coordination.

The relay molecules in a signal transduction pathway are usually proteins, which interact as they pass the message from the extracellular signaling molecule to the protein that produces the cellular response.

Protein kinases are enzymes that transfer phosphate groups from ATP to proteins. Relay molecules in signal transduction pathways are often protein kinases, which are sequentially phosphorylated. Phosphorylation produces a shape change that usually activates each enzyme. Hundreds of different kinds of protein kinases regulate the activity of a cell's proteins.

FOCUS QUESTION 5.11

a. What does a protein kinase do?

b. What does a **protein phosphatase** do?

c. What is a "phosphorylation cascade"?

Small, water-soluble molecules or ions often function as **second messengers**, which rapidly relay the signal from the membrane-receptor-bound "first messenger" into a cell's interior.

Binding of an extracellular signal to a G protein-coupled receptor activates a G protein that may activate adenylyl cyclase, a membrane protein that converts ATP to **cyclic AMP (cAMP)**. The cAMP often activates *protein kinase A*, which phosphorylates other proteins. A cytoplasmic enzyme converts cAMP to inactive AMP, thereby removing the second messenger in the absence of a signaling molecule.

FOCUS QUESTION 5.12

Label the components in the following diagram depicting the steps in a signal transduction pathway that uses cAMP as a second messenger.

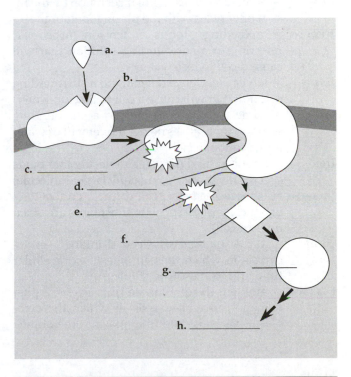

a. _____

b. _____

c. _____

d. _____

e. _____

f. _____

g. _____

h. _____

5.8   Response: Regulation of transcription or cytoplasmic activities

Signal transduction pathways may lead to the activation of transcription factors, which regulate the expression of specific genes. Signaling pathways may also activate existing cytoplasmic enzymes.

The Evolution of Cell Signaling   Similarities among cell-signaling pathways in diverse species suggest an early evolution of such mechanisms.

Word Roots

amphi-   = dual (*amphipathic molecule:* a molecule that has both hydrophobic and hydrophilic regions)

aqua-   = water; **-pori** = a small opening (*aquaporin:* a channel protein in the plasma membrane that specifically facilitates osmosis, the diffusion of free water across the membrane)

co-   = together; **trans-** = across (*cotransport:* the coupling of the "downhill" diffusion of one substance to the "uphill" transport of another against its concentration gradient)

electro-   = electricity; **-genic** = producing (*electrogenic pump:* an active transport protein that generates voltage across a membrane while pumping ions)

endo-   = inner; **cyto-** = cell (*endocytosis:* cellular uptake of matter via formation of vesicles from the plasma membrane)

exo-   = outer (*exocytosis:* cellular secretion by the fusion of vesicles with the plasma membrane)

hyper-   = exceeding; **-tonus** = tension (*hypertonic:* referring to a surrounding solution that will cause a cell to lose water)

hypo-   = lower (*hypotonic:* referring to a surrounding solution that will cause a cell to take up water)

iso-   = same (*isotonic:* referring to a surrounding solution that causes no net movement of water into or out of a cell)

liga-   = bound or tied (*ligand:* a molecule that binds specifically to another, usually larger molecule)

phago-   = eat (*phagocytosis:* "cell eating"; endocytosis in which large particulate substances are taken up by a cell)

pino-   = drink (*pinocytosis:* "cell drinking"; endocytosis in which the cell ingests extracellular fluid and its dissolved solutes)

plasm-   = molded; **-lyso** = loosen (*plasmolysis:* a phenomenon in walled cells in which the cytoplasm shrivels and the plasma membrane pulls away from the cell wall when the cell loses water to a hypertonic environment)

trans-   = across (*signal transduction pathway:* a series of steps linking a mechanical, chemical, or electrical stimulus to a specific cellular response)

Structure Your Knowledge

1. Create a concept map to illustrate your understanding of osmosis. This exercise will help you to practice using the words *hypotonic, isotonic,* and *hypertonic,* and to focus on the effect of these osmotic environments on plant and animal cells. Explain your map to a friend.

2. The following diagram illustrates passive and active transport across a plasma membrane. Use it to answer questions **a** through **d**.
   a. Which section represents facilitated diffusion? How can you tell? Does the cell expend energy in this transport? Why or why not? What types of solute molecules may be moved by this type of transport?
   b. Which section shows active transport? List two ways that you can tell.
   c. What types of solute molecules can diffuse through the membrane shown in section I?
   d. Which of these sections are considered passive transport?

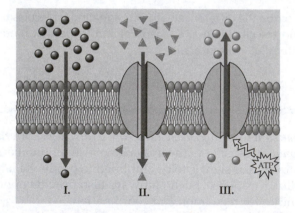

3. Briefly describe the three stages of cell signaling.

4. Some signaling pathways alter a protein's activity; others result in the production of new proteins. Explain the mechanisms for these two different responses.

Test Your Knowledge

MULTIPLE CHOICE: *Choose the one best answer.*

1. If a single layer of phospholipids coats the water in a beaker, which parts of the molecules face the air?
   a. the phosphate groups
   b. the hydrocarbon tails
   c. both heads and tails because the molecules are amphipathic and lie sideways
   d. the glycolipid regions
   e. No parts of the molecules face the air, because the phospholipids dissolve in the water and do not form a layer.

2. Glycoproteins and glycolipids are important for
   a. facilitated diffusion.
   b. active transport.
   c. cell–cell recognition.
   d. intercellular joining.
   e. signal transduction pathways.

3. Which of the following is the most probable description of an integral, transmembrane protein?
   a. amphipathic with a hydrophilic head and a hydrophobic tail region
   b. a globular protein with hydrophobic amino acids in the interior and hydrophilic amino acids arranged around the outside
   c. a fibrous protein coated with hydrophobic fatty acids
   d. a glycolipid attached to the portion of the protein facing the exterior of the cell and cytoskeletal elements attached to the interior portion
   e. a middle region composed of α helical stretches of hydrophobic amino acids, with hydrophilic regions at both ends of the protein

4. A cell is manufacturing receptor proteins for cholesterol. How would those proteins be oriented before they reach the plasma membrane?
   a. facing inside the ER lumen but outside the transport vesicle membrane
   b. facing inside the ER lumen and inside the transport vesicle
   c. attached outside the ER and outside the transport vesicle
   d. attached outside the ER but facing inside the transport vesicle
   e. embedded in the hydrophobic center of both the ER and transport vesicle membranes

5. The fluidity of membranes in a plant in cold weather may be maintained by increasing the
   a. proportion of peripheral proteins.
   b. action of an H^+ pump.
   c. concentration of cholesterol in the membrane.
   d. number of phospholipids with unsaturated hydrocarbon tails.
   e. number of phospholipids with saturated hydrocarbon tails.

Use the following U-tube setup to answer questions 6 through 8.

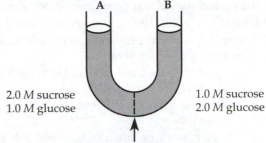

A   B

2.0 *M* sucrose   1.0 *M* sucrose
1.0 *M* glucose   2.0 *M* glucose

selectively permeable membrane

The solutions in the two arms of this U-tube are separated by a membrane that is permeable to water and glucose but not to sucrose. Side A is filled with a solution of 2.0 *M* sucrose and 1.0 *M* glucose. Side B is filled with 1.0 *M* sucrose and 2.0 *M* glucose.

6. *Initially*, the solution in side A, with respect to that in side B,
   a. has a lower solute concentration.
   b. has a higher solute concentration.
   c. has an equal solute concentration.
   d. is lower in the tube.
   e. is higher in the tube.

7. During the period *before* equilibrium is reached, which molecule(s) will show net movement through the membrane?
   a. water
   b. glucose
   c. sucrose
   d. water and sucrose
   e. water and glucose

8. *After* the system reaches equilibrium, what changes can be observed?
   a. The water level is higher in side A than in side B.
   b. The water level is higher in side B than in side A.
   c. The molarity of glucose is higher in side A than in side B.
   d. The molarity of sucrose has increased in side A.
   e. Both **a** and **c** have occurred.

9. An animal cell placed in a hypotonic environment will
   a. become flaccid.
   b. become turgid.
   c. burst (lyse).
   d. plasmolyze.
   e. shrivel.

10. You observe plant cells under a microscope as they are placed in an unknown solution. First the cells plasmolyze; after a minute, the plasmolysis reverses and the cells appear normal. What would you conclude about the unknown solution?
    a. It is hypertonic to the plant cells, and its solute cannot cross the plant cell membranes.
    b. It is hypotonic to the plant cells, and its solute cannot cross the plant cell membranes.
    c. It is isotonic to the plant cells, but its solute can cross the plant cell membranes.
    d. It is hypertonic to the plant cells, but its solute can cross the plant cell membranes.
    e. It is hypotonic to the plant cells, but its solute can cross the plant cell membranes.

11. Which of the following is *not* true about osmosis?
    a. It is a passive process in cells without walls, but an active one in cells with walls.
    b. Water moves into a cell from a hypotonic environment.
    c. Solute molecules bind to water and decrease the free water available to move.
    d. It can occur more rapidly through channel proteins known as aquaporins.
    e. There is no net osmosis when cells are in isotonic solutions.

12. Which of the following is *not* true of carrier molecules involved in facilitated diffusion?
    a. They increase the speed of transport across a membrane.
    b. They can concentrate solute molecules on one side of the membrane.
    c. They may have specific binding sites for the molecules they transport.
    d. They may undergo a change in shape upon binding of solute.
    e. They do not require an energy investment from the cell to function.

13. Facilitated diffusion of ions across a cellular membrane requires _____; and the ions move _____.
    a. energy and channel proteins; against their electrochemical gradient
    b. energy and channel proteins; against their concentration gradient
    c. cotransport proteins; against their electrochemical gradient
    d. channel proteins; down their electrochemical gradient
    e. channel proteins; down their concentration gradient

14. The membrane potential of a cell favors
    a. the movement of cations into the cell.
    b. the movement of anions into the cell.
    c. the action of an electrogenic pump.
    d. the movement of sodium out of the cell.
    e. both **b** and **d**.

15. Which of the following describes cotransport?
    a. active transport of two solutes through a cotransport protein
    b. passive transport of two solutes through a cotransport protein
    c. ion diffusion against the electrochemical gradient created by an electrogenic pump
    d. a pump such as the sodium-potassium pump that moves ions in two different directions
    e. transport of one solute against its concentration gradient in tandem with another that is diffusing down its concentration gradient

16. The proton pump in plant cells is the functional equivalent of an animal cell's
    a. cotransport mechanism.
    b. sodium-potassium pump.
    c. contractile vacuole for osmoregulation.
    d. receptor-mediated endocytosis of cholesterol.
    e. ATP pump.

17. Which of the following is an example of active transport?
    a. the transport of a solute in which a carrier protein binds the solute, changes shape, and moves the solute across a membrane
    b. the flow of K^+ through an open ion channel out of a cell
    c. the movement of water into a plant cell
    d. the movement of LDL particles into a cell
    e. the movement of O_2 into a cell

18. Exocytosis may involve all of the following *except*
    a. ligands and coated pits.
    b. the fusion of a vesicle with the plasma membrane.
    c. a mechanism to export some carbohydrates during the formation of plant cell walls.
    d. a mechanism to rejuvenate the plasma membrane.
    e. a means of exporting large molecules.

19. What is a key difference between a local regulator and a hormone?
    a. Local regulators are small, hydrophobic molecules; hormones are either larger polypeptides or steroids.
    b. Local regulators diffuse to neighboring cells; hormones usually travel throughout the plant or animal body to distant target cells.
    c. Local regulators initiate short-term responses; hormones trigger longer-lasting responses to environmental stimuli.
    d. The signal transduction pathways of local regulators do not involve second messengers; pathways triggered by hormones do involve second messengers.
    e. Local regulators often open ligand-gated channels and affect ion concentrations in a cell; hormones bind with intracellular receptors and affect gene expression.

20. A signaling molecule that binds to a plasma-membrane protein receptor functions as a
    a. ligand.
    b. second messenger.
    c. protein phosphatase.
    d. protein kinase.
    e. receptor protein.

21. A G protein is
    a. a specific type of membrane-receptor protein.
    b. a protein on the cytoplasmic side of a membrane that becomes activated by a transmembrane receptor protein.
    c. a membrane-bound enzyme that converts ATP to cAMP.
    d. an intracellular receptor protein that, once activated, functions as a transcription factor.
    e. a guanine nucleotide that converts between GDP and GTP to activate and inactivate relay proteins.

22. Which of the following compounds can activate a protein by transferring a phosphate group to it?
    a. G protein
    b. adenyl cyclase
    c. protein phosphatase
    d. protein kinase
    e. both a and c

23. Many signal transduction pathways use second messengers to
    a. transport a signaling molecule through the hydrophobic center of the plasma membrane.
    b. relay a signal from the outside to the inside of the cell.
    c. relay the message from the inside of the membrane throughout the cytoplasm.
    d. amplify the message by phosphorylating cascades of proteins.
    e. dampen the message once the signaling molecule has left the receptor.

24. When epinephrine binds to cardiac (heart) muscle cells, it speeds their contraction. When it binds to muscle cells of the small intestine, it inhibits their contraction. How can the same hormone have different effects on muscle cells?
    a. Cardiac cells have more receptors for epinephrine than do intestinal muscle cells.
    b. Epinephrine circulates to the heart first and is in higher concentration around cardiac cells.
    c. The two types of muscle cells have different signal transduction pathways for epinephrine and thus have different cellular responses.
    d. Cardiac muscle is stronger than intestinal muscle and thus has a stronger response to epinephrine.
    e. Epinephrine binds to G protein-coupled receptors in cardiac cells, and these receptors always increase a response to the signal. Epinephrine binds to intracellular receptors in intestinal muscle cells, and these receptors inhibit a response to the signal.

Chapter 6

An Introduction to Metabolism

Chapter Focus

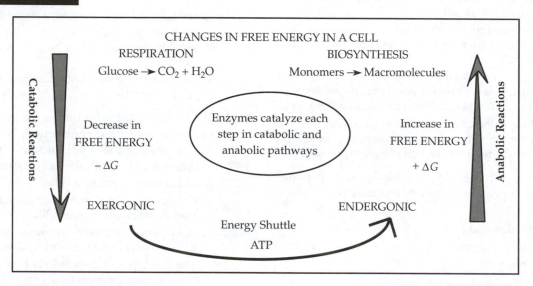

CHANGES IN FREE ENERGY IN A CELL

RESPIRATION

Glucose → CO_2 + H_2O

BIOSYNTHESIS

Monomers → Macromolecules

Catabolic Reactions

Anabolic Reactions

Decrease in
FREE ENERGY

$-\Delta G$

Enzymes catalyze each step in catabolic and anabolic pathways

Increase in
FREE ENERGY

$+\Delta G$

EXERGONIC

ENDERGONIC

Energy Shuttle

ATP

Chapter Review

6.1 An organism's metabolism transforms matter and energy

Metabolic Pathways **Metabolism** includes all the chemical reactions in an organism. These reactions are ordered into **metabolic pathways**, a sequence of steps, each controlled by an enzyme, that converts specific molecules to products.

Catabolic pathways release the energy stored in complex molecules through the breakdown of these molecules into simpler compounds. **Anabolic pathways**, sometimes called biosynthetic pathways, require energy to combine simpler molecules into more complicated ones. The energy released from catabolic pathways drives the anabolic pathways in a cell. The study of energy transformations in organisms, called **bioenergetics**, is central to understanding metabolism.

Forms of Energy **Energy** can be defined as the capacity to cause change. Some forms of energy can do

work, such as moving matter against an opposing force. **Kinetic energy** is the energy of motion, of matter that is moving. This matter does its work by transferring its motion to other matter. The kinetic energy of randomly moving atoms or molecules is **thermal energy**. Thermal energy transferring from one body of matter to another is called **heat**.

Potential energy is the capacity of matter to cause change as a consequence of its location or arrangement. **Chemical energy** is a form of potential energy that is available for release in chemical reactions.

Energy can be converted from one form to another. Plants convert light energy to the chemical energy in sugar, and cells release this potential energy to drive cellular processes.

The Laws of Energy Transformation **Thermodynamics** is the study of energy transformations in a collection of matter. In an *open system*, energy and matter may be exchanged between the *system* and its *surroundings*. (In an *isolated system*, such exchange does not occur.) Organisms are open systems.

The **first law of thermodynamics** states that energy can be neither created nor destroyed. According to this *principle of conservation of energy,* energy can be transferred and transformed from one kind to another, but the total energy of the universe is constant.

The **second law of thermodynamics** states that every energy transfer or transformation results in increasing disorder within the universe. **Entropy** is used as a measure of disorder or randomness. In every energy transfer or transformation, some of the energy is converted to thermal energy, a disordered form of energy, and released as heat.

A **spontaneous process** is one that is "energetically favorable"; in other words, it occurs without an input of energy. For a process to occur spontaneously, it must result in an increase in entropy. A nonspontaneous process will occur only if energy is added to the system.

A cell may become more ordered, but it does so with an attendant increase in the entropy of its surroundings. For example, an animal takes in and uses highly ordered organic molecules as a source of the matter and energy needed to create and maintain its own organized structure, but it returns heat and the simple molecules of carbon dioxide and water to the environment.

FOCUS QUESTION 6.1

Complete the following concept map that summarizes some of the key ideas about energy.

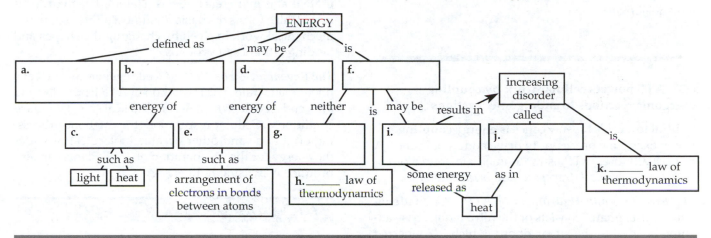

6.2 The free-energy change of a reaction tells us whether or not the reaction occurs spontaneously

Free-Energy Change (ΔG), Stability, and Equilibrium The portion of a system's energy available to perform work when the system's temperature and pressure are uniform is defined as **free energy** (symbolized by G). The *change* in free energy during a chemical reaction is represented by ΔG and is equal to $G_{\text{final state}} - G_{\text{initial state}}$. The free energy of the system must decrease ($-\Delta G$) for a reaction to be spontaneous.

When ΔG is negative, the final state has less free energy than the initial state; thus, the final state is less likely to change and is more stable. A system rich in free energy has a tendency to change spontaneously to a more stable state. This change may be harnessed to perform work. A state of maximum stability is called *equilibrium.* At equilibrium in a chemical reaction, the forward and backward reactions occur at the same rate, and the relative concentrations of products and reactants stay the same. Moving toward equilibrium is spontaneous; the ΔG of the reaction is negative. Once at equilibrium, a system is at its minimum of free energy; it will not spontaneously change, and it can do no work.

FOCUS QUESTION 6.2

Complete the following table to indicate how the free energy of a system (or a chemical reaction) relates to the system's stability, capacity to do work, tendency for spontaneous change, and equilibrium.

	Stability	Work Capacity	Spontaneous Change?	Equilibrium
System with High Free Energy				
System with Low Free Energy				

Free Energy and Metabolism   Now let's see how these energy concepts apply to metabolism. An **exergonic reaction** ($-\Delta G$) proceeds with a net release of free energy and is spontaneous (energetically favorable). The magnitude of ΔG indicates the maximum amount of work an exergonic reaction can perform. **Endergonic reactions** ($+\Delta G$) are nonspontaneous; they must absorb free energy from the surroundings. The energy released by an exergonic reaction ($-\Delta G$) is equal to the energy required by the reverse reaction ($+\Delta G$).

FOCUS QUESTION   6.3

Metabolic disequilibrium is essential to life—a cell whose metabolic reactions reached equilibrium would be dead. What mechanisms prevent a cell's reactions from reaching equilibrium?

6.3   ATP powers cellular work by coupling exergonic reactions to endergonic reactions

Central to a cell's bioenergetics is **energy coupling**, the use of exergonic processes to drive endergonic ones. A cell usually uses ATP as the immediate source of energy for its *chemical, transport,* and *mechanical work.*

The Structure and Hydrolysis of ATP   ATP **(adenosine triphosphate)** consists of the nitrogenous base adenine bonded to the sugar ribose, which is connected to a chain of three phosphate groups. ATP can be hydrolyzed to ADP (adenosine diphosphate) and an inorganic phosphate molecule (P_i), releasing 7.3 kcal (30.5 kJ) of energy per mole of ATP when measured under standard conditions. The ΔG of the reaction in the cell is estimated to be closer to -13 kcal/mol.

FOCUS QUESTION   6.4

Label the three components (**a** through **c**) of the following ATP molecule.

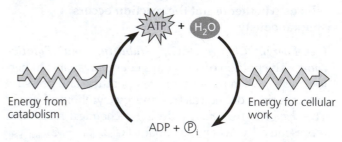

d. Indicate which bond is likely to break. By what chemical mechanism is the bond broken?

e. Explain why this reaction releases so much energy.

How the Hydrolysis of ATP Performs Work   ATP is used to couple exergonic and endergonic reactions in a cell. The free energy released from the hydrolysis of ATP is used to transfer the phosphate group to a reactant molecule, producing a **phosphorylated intermediate** that is more reactive (less stable). The hydrolysis of ATP also forms the basis for almost all transport and mechanical work in a cell by changing the shapes and binding affinities of proteins.

The Regeneration of ATP   A cell regenerates ATP at a phenomenal rate. The formation of ATP from ADP and P_i is endergonic, with a ΔG of $+7.3$ kcal/mol (standard conditions). Cellular respiration (the catabolic processing of glucose and other organic molecules) provides the energy for the regeneration of ATP. Plants can also produce ATP using light energy.

FOCUS QUESTION   6.5

Look at the following ATP cycle and explain why both the left and right sides of the cycle are examples of energy coupling.

6.4   Enzymes speed up metabolic reactions by lowering energy barriers

Enzymes are biological **catalysts**—agents that speed the rate of a reaction but are unchanged by the reaction.

The Activation Energy Barrier   Chemical reactions involve both the breaking and forming of chemical bonds. Energy must be absorbed to contort molecules to an unstable state in which bonds can break. Energy is released when new bonds form and molecules return to stable, lower energy states. **Activation energy**, or the *free energy of activation* (E_A), is the energy that must be absorbed by reactants to reach the unstable *transition state*, in which bonds are likely to break, and from which the reaction can proceed.

How Enzymes Speed Up Reactions The E_A barrier is essential to life because it prevents the energy-rich macromolecules of the cell from decomposing spontaneously. For metabolism to proceed in a cell, however, E_A must be reached. Enzymes are able to lower E_A so that specific reactions can proceed at cellular temperatures. Enzymes do not change the ΔG for a reaction.

In the following graph of an exergonic reaction with and without an enzyme catalyst, label parts **a** through **e**.

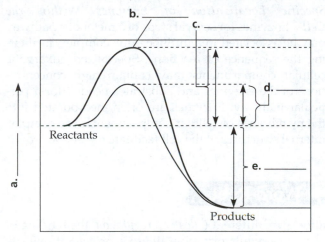

Substrate Specificity of Enzymes Protein enzymes are macromolecules with characteristic three-dimensional shapes that result in their specificity for their particular **substrate**. The substrate attaches at the enzyme's **active site**, a pocket or groove found on the surface of the enzyme that has a shape compatible to that of the substrate. The substrate is temporarily bound to its enzyme, forming an **enzyme-substrate complex**. Interactions between the substrate and active site cause the enzyme to change shape slightly, creating an **induced fit** that enhances the enzyme's ability to catalyze the chemical reaction.

Catalysis in the Enzyme's Active Site What happens in an enzyme's active site? The substrate is often held in the active site by hydrogen bonds or ionic bonds. The side chains (R groups) of some of the surrounding amino acids in the active site facilitate the conversion of substrate to product. The product then leaves the active site, and the catalytic cycle repeats, often at astonishing speed.

Whether an enzyme catalyzes the forward or backward reaction is influenced by the relative concentrations of reactants and products and the ΔG of the reactions. Enzymes catalyze reactions moving toward equilibrium.

Enzymes may catalyze reactions involving the joining of two reactants by properly orienting the substrates closely together. An induced fit can stretch critical bonds in the substrate molecule and make them easier to break. An active site may provide a microenvironment, such as a lower pH, that is necessary for a particular reaction. Enzymes may also actually participate in a reaction by forming brief covalent bonds with the substrate.

The rate of a reaction will increase with increasing substrate concentration up to the point at which all enzyme molecules are *saturated* with substrate molecules and working at full speed. At that point, only adding more enzyme molecules will increase the rate of the reaction.

In the following diagram of a catalytic cycle, sketch two appropriate substrate molecules and two products, and identify the key steps of the cycle.

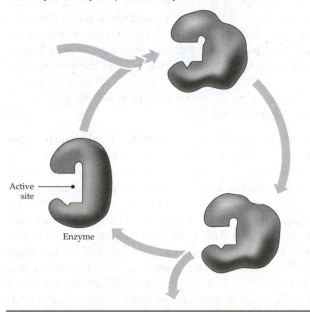

Effects of Local Conditions on Enzyme Activity The speed of an enzyme-catalyzed reaction may increase with rising temperature up to the point at which increased thermal agitation begins to disrupt the weak bonds and interactions that stabilize protein shape. Each enzyme has *optimal conditions* that include a temperature and pH that favor its most active shape.

Cofactors are small molecules that bind either permanently or reversibly with enzymes and are necessary for enzyme function. They may be inorganic, such as various metal ions, or organic molecules called **coenzymes**. Most vitamins are coenzymes or precursors of coenzymes.

Enzyme inhibitors disrupt the action of enzymes, either reversibly by binding with the enzyme with weak bonds, or irreversibly by attaching with covalent bonds. **Competitive inhibitors** compete with the substrate for the active site of the enzyme. Increasing the concentration of substrate molecules may overcome this type of inhibition. **Noncompetitive inhibitors** bind to a part of the enzyme separate from the active site and impede enzyme action by changing the shape of the enzyme.

FOCUS QUESTION 6.8

Return to the diagram in Focus Question 6.7. Draw a competitive inhibitor and a noncompetitive inhibitor, and indicate where each would bind to the enzyme molecule.

The Evolution of Enzymes *Mutations* in genes may alter the amino acid sequence and thus change the substrate specificity or activity of an enzyme. If this novel function is beneficial, natural selection would be expected to preserve the mutated gene in the population.

6.5 Regulation of enzyme activity helps control metabolism

Allosteric Regulation of Enzymes In **allosteric regulation**, molecules may inhibit or activate enzyme activity when they bind to a site separate from the active site. Enzymes made of two or more polypeptides, each with its own active site, may have regulatory sites (sometimes called allosteric sites) located where subunits join. The entire unit may oscillate between two forms. The binding of an *activator* stabilizes the catalytically active shape, whereas an *inhibitor* reinforces the inactive form of the enzyme. Allosteric enzymes may be critical regulators of both catabolic and anabolic pathways.

Through a phenomenon called **cooperativity**, the binding of a substrate molecule to one subunit changes the shape of all subunits such that their active sites are stabilized in the active form.

Metabolic pathways are commonly regulated by **feedback inhibition**, in which the end product acts as an allosteric inhibitor of an enzyme early in the pathway.

FOCUS QUESTION 6.9

Both ATP and ADP serve as regulators of enzyme activity. In catabolic pathways, which of these two molecules would you predict acts as an inhibitor?

Which would you predict would act as an activator?

Specific Localization of Enzymes Within the Cell Enzymes for several steps of a metabolic pathway may be associated in a multienzyme complex, facilitating the sequence of reactions. Specialized eukaryotic cellular compartments may contain high concentrations of the enzymes and substrates needed for a particular pathway. Enzymes are often incorporated into the membranes of cellular compartments. The complex internal structures of the cell facilitate metabolic order.

Word Roots

allo- = different (*allosteric regulation:* the binding of a regulatory molecule to a protein at one site that affects the function of the protein at a different site)

ana- = up (*anabolic pathway:* a metabolic pathway that consumes energy to synthesize a complex molecule from simpler compounds)

bio- = life (*bioenergetics:* the overall flow and transformation of energy in an organism)

cata- = down (*catabolic pathway:* a metabolic pathway that releases energy by breaking down complex molecules into simpler compounds)

endo- = within (*endergonic reaction:* a nonspontaneous chemical reaction in which free energy is absorbed from the surroundings)

ex- = out (*exergonic reaction:* a spontaneous reaction, in which there is a net release of free energy)

kinet- = movement (*kinetic energy:* the energy associated with the relative motion of objects)

therm- = heat (*thermodynamics:* the study of the energy transformations that occur in a collection of matter)

Structure Your Knowledge

This chapter introduced many new and complex concepts. See if you can step back from the details and answer the following general questions.

1. What is the relationship between the concept of free energy and metabolism?

2. What role do enzymes play in metabolism?

Test Your Knowledge

FILL IN THE BLANKS

_____ 1. the totality of an organism's chemical processes

_____ 2. pathways that use energy to synthesize complex molecules

_____ 3. the form of energy resulting from location or structure

_____ 4. the most random form of energy

_____ 5. term for the measure of disorder or randomness

_____ 6. the energy that must be absorbed by molecules to reach the transition state

_____ 7. inhibitors that decrease an enzyme's activity by binding to the active site

_____ 8. nonprotein organic molecules that bind to enzymes and are necessary for their functioning

_____ 9. regulatory device in which the product of a pathway binds to an enzyme early in the pathway

_____ 10. more reactive molecules created by the transfer of a phosphate group from ATP

MULTIPLE CHOICE: *Choose the one best answer.*

1. Catabolic and anabolic pathways are often coupled in a cell because
   a. the intermediates of a catabolic pathway are used in the anabolic pathway.
   b. both pathways use the same enzymes.
   c. the free energy released from one pathway is used to drive the other pathway.
   d. the activation energy of the catabolic pathway can be used in the anabolic pathway.
   e. their enzymes are controlled by the same activators and inhibitors.

2. According to the first law of thermodynamics,
   a. for every action there is an equal and opposite reaction.
   b. every energy transfer results in an increase in disorder or entropy.
   c. the total amount of energy in the universe is conserved or constant.
   d. energy can be transferred or transformed, but disorder always increases.
   e. potential energy is converted to kinetic energy, and kinetic energy is released as heat.

3. When a cell breaks down glucose, only about 34% of the energy is captured in ATP molecules. The remaining 66% of the energy is
   a. used to increase the order necessary for life to exist.
   b. lost as heat, in accordance with the second law of thermodynamics.
   c. used to increase the entropy of the system by converting kinetic energy into potential energy.
   d. stored in starch or glycogen for later use by the cell.
   e. released when the ATP molecules are hydrolyzed.

4. A negative ΔG means that
   a. the quantity G of energy is available to do work.
   b. the reaction is spontaneous.
   c. the reactants have more free energy than the products.
   d. the reaction is exergonic.
   e. all of the above are true.

5. One way in which a cell maintains metabolic disequilibrium is to
   a. siphon products of a reaction off to the next step in a metabolic pathway.
   b. provide a constant supply of enzymes for critical reactions.
   c. use feedback inhibition to turn off pathways.
   d. use allosteric enzymes that can bind to activators or inhibitors.
   e. use the energy from anabolic pathways to drive catabolic pathways.

6. An endergonic reaction could be described as one that
   a. proceeds spontaneously with the addition of activation energy.
   b. produces products with more free energy than the reactants.
   c. is not able to be catalyzed by enzymes.
   d. releases energy.
   e. produces ATP for energy coupling.

7. The formation of ATP from ADP and inorganic phosphate
   a. is an exergonic process.
   b. transfers the phosphate to an intermediate that becomes more reactive.
   c. produces an unstable energy compound that can drive cellular work.
   d. has a ΔG of -7.3 kcal/mol under standard conditions.
   e. involves the hydrolysis of a phosphate bond.

8. What is meant by an induced fit?
   a. The binding of the substrate is an energy-requiring process.
   b. A competitive inhibitor can outcompete the substrate for the active site.
   c. The binding of the substrate changes the shape of the active site, which can stress or bend substrate bonds.
   d. The active site creates a microenvironment ideal for the reaction.
   e. The binding of an activator to an allosteric site induces a more active form of the subunits of an enzyme.

9. In an experiment, changing the pH from 7 to 6 resulted in an increase in product formation. From this we could conclude that
   a. the enzyme became saturated at pH 6.
   b. the enzyme's optimal pH is 6.
   c. this enzyme works best at a neutral pH.
   d. the temperature must have increased when the pH was changed to 6.
   e. the enzyme was in a more active shape at pH 6.

10. When substance A was added to an enzyme reaction, product formation decreased. The addition of more substrate did not increase product formation. From this we conclude that substance A could be
    a. product molecules.
    b. a cofactor.
    c. an allosteric enzyme.
    d. a competitive inhibitor.
    e. a noncompetitive inhibitor.

11. Which of the following characteristics is most directly responsible for the specificity of a protein enzyme for its substrate?
    a. its primary structure
    b. its secondary and tertiary structures
    c. the shape and characteristics of its allosteric site
    d. its cofactors
    e. the R groups of the amino acids in its active site

12. An enzyme raises which of the following parameters?
    a. ΔG
    b. the free energy of the products
    c. the free energy of activation
    d. the speed of a reaction
    e. the equilibrium of a reaction

13. Zinc, an essential trace element, may be found bound to the active site of some enzymes. Such zinc ions most likely function as
    a. a coenzyme derived from a vitamin.
    b. a cofactor necessary for catalysis.

   c. a substrate of the enzyme.
   d. a competitive inhibitor of the enzyme.
   e. an allosteric activator of the enzyme.

Use the following diagram to answer questions 14 through 16.

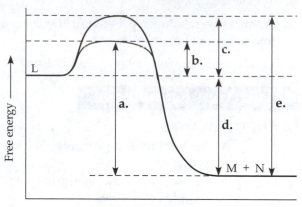

14. Which line in the diagram indicates the ΔG of the enzyme-catalyzed reaction L → M + N?

15. Which line in the diagram indicates the activation energy of the noncatalyzed reaction?

16. Which of the following terms *best* describes this reaction?
    a. nonspontaneous
    b. $-\Delta G$
    c. endergonic
    d. coupled reaction
    e. anabolic reaction

17. In cooperativity,
    a. a cellular organelle contains all the enzymes needed for a metabolic pathway.
    b. a product of a pathway serves as a competitive inhibitor of an enzyme early in the pathway.
    c. a molecule bound to the active site of one subunit of an enzyme affects the active site of other subunits.
    d. the allosteric site is filled with an activator molecule.
    e. the product of one reaction serves as the substrate for the next reaction in intricately ordered metabolic pathways.

18. In the metabolic pathway A → B → C → D → E, what effect would molecule E likely have on the enzyme that catalyzes A → B?
    a. allosteric inhibitor
    b. allosteric activator
    c. competitive inhibitor
    d. feedback activator
    e. coenzyme

Chapter 7

Cellular Respiration and Fermentation

Chapter Focus

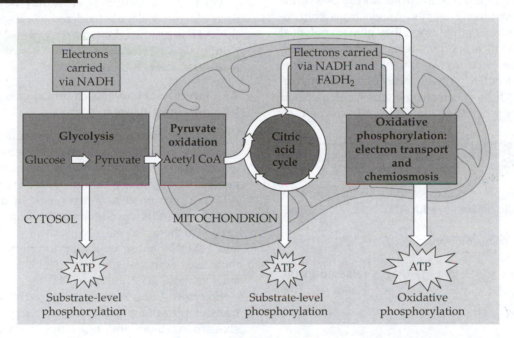

Chapter Review

7.1 Catabolic pathways yield energy by oxidizing organic fuels

Catabolic Pathways and Production of ATP The breaking down of complex organic molecules in catabolic pathways releases energy that cells can use to do work. **Fermentation** occurs without oxygen and partially degrades sugars or other fuels. **Aerobic respiration** uses oxygen in the breakdown of energy-rich organic compounds to yield carbon dioxide and water and release energy as ATP and heat. The *anaerobic respiration* of some prokaryotes does not use oxygen as a reactant but is a similar process. Aerobic respiration is often referred to as **cellular respiration.** This exergonic process has a free energy change of −686 kcal/mol of glucose.

FOCUS QUESTION 7.1

Fill in the following summary equation for cellular respiration, starting with the sugar glucose.

$$\underline{C_6H_{12}O_6} + 6O_2 \rightarrow \underline{\quad} + 6H_2O + \underline{6\,CO_2}$$

Redox Reactions: Oxidation and Reduction How does cellular respiration release energy? It involves the transfer of electrons. Oxidation–reduction reactions or **redox reactions** involve the partial or complete transfer of one or more electrons from one reactant to another. **Oxidation** is the loss of electrons from one substance; **reduction** is the addition of electrons to another substance. The substance that loses electrons becomes oxidized and acts as a **reducing agent** (electron donor). By gaining electrons, a substance acts as an **oxidizing agent** (electron acceptor) and becomes reduced.

FOCUS QUESTION 7.2

Fill in the appropriate terms in the following equation.

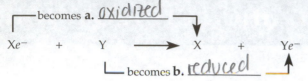

becomes a. _oxidized_

$$Xe^- \quad + \quad Y \quad \longrightarrow \quad X \quad + \quad Ye^-$$

becomes b. _reduced_

Xe^- is the reducing agent; it c. _loses_ electrons.

Y is the d. _oxidizing_ it e. _gains_ electrons.

Oxygen strongly attracts electrons and is one of the most powerful oxidizing agents. As electrons shift toward a more electronegative atom, they give up potential energy. Thus, chemical energy is released in a redox reaction that relocates electrons closer to oxygen.

Organic molecules with an abundance of hydrogen are rich in "hilltop" electrons, which release their potential energy when they "fall" closer to oxygen.

FOCUS QUESTION 7.3

a. In the conversion of glucose and O_2 to CO_2 and H_2O, which molecule becomes reduced?

_O_2_

b. Which molecule becomes oxidized?

glucose

c. What happens to the energy that is released in this redox reaction?

work

At certain steps in the oxidation of glucose, two hydrogen atoms are removed by enzymes called dehydrogenases, and the two electrons and one proton are passed to a coenzyme, **NAD$^+$** (nicotinamide adenine dinucleotide), reducing it to NADH.

Energy from respiration is slowly released in a series of redox reactions as electrons are passed from NADH down an **electron transport chain,** a group of carrier molecules located in the inner mitochondrial membrane (or in the plasma membrane of aerobic prokaryotes), to a stable location close to a highly electronegative oxygen atom, forming water.

FOCUS QUESTION 7.4

a. NAD$^+$ is called an _coenzyme_

b. Its reduced form is _NADH_.

The Stages of Cellular Respiration: A Preview It is easier to understand and learn cellular respiration when you focus on its three main stages. **Glycolysis,** which occurs in the cytosol, breaks glucose into two molecules of pyruvate. Within the mitochondrial matrix or in the cytosol of prokaryotes, pyruvate is oxidized to acetyl CoA. The **citric acid cycle** then oxidizes acetyl CoA to CO_2. In some steps, electrons are transferred to NAD$^+$. NADH passes electrons to the electron transport chain, from which they combine with H$^+$ and oxygen to form water. The energy released in this chain of redox reactions is used to synthesize ATP by **oxidative phosphorylation,** a process that includes electron transport and chemiosmosis.

Up to 32 molecules of ATP may be generated for each glucose molecule oxidized to CO_2. About 10% of this ATP is produced by **substrate-level phosphorylation,** in which an enzyme transfers a phosphate group from a substrate molecule to ADP.

7.2 Glycolysis harvests chemical energy by oxidizing glucose to pyruvate

What is the first stage of cellular respiration? Glycolysis, a ten-step process occurring in the cytosol, first breaks glucose down to pyruvate. It has an energy-investment phase and an energy-payoff phase. Two molecules of ATP are consumed as glucose is split into two three-carbon sugars (glyceraldehyde-3-phosphate). The conversion of these molecules to pyruvate produces 2 NADH and 4 ATP by substrate-level phosphorylation. For each molecule of glucose, glycolysis yields two pyruvate and a net gain of 2 ATP and 2 NADH.

Enzymes catalyze each step in glycolysis. Kinases transfer phosphate groups; other enzymes cleave the six-carbon sugar and rearrange atoms in substrate molecules, and a dehydrogenase oxidizes glyceraldehyde-3-phosphate and reduces NAD$^+$.

FOCUS QUESTION 7.5

Fill in the blanks in the following summary diagram of glycolysis.

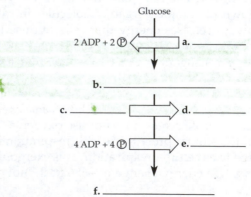

7.3 After pyruvate is oxidized, the citric acid cycle completes the energy-yielding oxidation of organic molecules

What happens to the pyruvate formed in glycolysis? After pyruvate enters a mitochondrion, enzymes catalyze a series of steps in which CO_2 is removed, the remaining two-carbon group is oxidized, with the accompanying reduction of NAD^+ to NADH; and acetyl coenzyme A, or **acetyl CoA**, is formed.

In the citric acid cycle, the acetyl group of acetyl CoA is added to oxaloacetate to form citrate, which is progressively decomposed back to oxaloacetate. For each turn of the citric acid cycle, two carbons enter from acetyl CoA; two carbons exit completely oxidized as CO_2; three NADH and one $FADH_2$ are formed; and one ATP (GTP in most animal cells) is made by substrate-level phosphorylation. It takes two turns of the citric acid cycle to oxidize the two acetyl groups derived from a single glucose molecule.

FOCUS QUESTION   7.6

Fill in the blanks in the following diagram of the citric acid cycle. Gray balls represent carbon atoms.

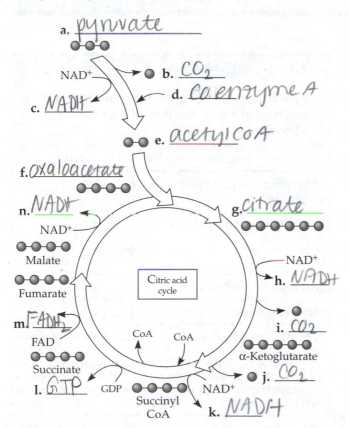

a. pyruvate
b. CO_2
c. NADH
d. coenzyme A
e. acetyl CoA
f. oxaloacetate
g. citrate
h. NADH
i. CO_2
j. CO_2
k. NADH
l. GTP
m. $FADH_2$
n. NADH

NAD⁺

Malate

Fumarate

FAD

Succinate

GDP

Succinyl CoA

α-Ketoglutarate

CoA   CoA

Citric acid cycle

7.4 During oxidative phosphorylation, chemiosmosis couples electron transport to ATP synthesis

The first two stages of respiration have produced only four ATP molecules. What happens in the final stage to get the big ATP payout?

The Pathway of Electron Transport Thousands of electron transport chains are embedded in the cristae (infoldings) of the inner mitochondrial membrane (or in the plasma membrane of prokaryotes). The components of the electron transport chain are organized into four complexes, and most are proteins to which nonprotein *prosthetic groups* are tightly bound.

The electron carriers shift between reduced and oxidized states as they accept and donate electrons. The transfer of electrons proceeds from NADH to a flavoprotein to an iron-sulfur protein to a mobile hydrophobic molecule called ubiquinone (Q or coenzyme Q). Next, electrons are passed down a series of molecules called **cytochromes,** which are proteins with an iron-containing heme group. The last cytochrome, cyt a_3, passes electrons to oxygen, which picks up two H^+, forming water.

$FADH_2$ adds its electrons to the chain at a lower energy level (within complex II); thus, less energy is provided for ATP synthesis by $FADH_2$ as compared to NADH.

Chemiosmosis: The Energy-Coupling Mechanism **ATP synthase,** a protein complex embedded in the inner mitochondrial membrane or in the prokaryotic plasma membrane, uses the energy of a proton (H^+) gradient to make ATP, an example of the process called **chemiosmosis.**

The flow of H^+ down its gradient through the stator and rotor part of the ATP synthase complex causes the rotor and attached rod to rotate, activating catalytic sites in the knob, where ADP and inorganic phosphate join to make ATP.

But where does the proton gradient that powers ATP synthase come from? When some members of the electron transport chain pass electrons, they also accept and release protons, which are deposited in the intermembrane space at three sites. The potential energy of the proton gradient is referred to as the **proton-motive force.**

In mitochondria, exergonic redox reactions produce the proton gradient that drives the production of ATP. Chloroplasts use light energy to create the proton-motive force used to make ATP by chemiosmosis. Prokaryotes use proton gradients generated across the plasma membrane to transport molecules, make ATP, and rotate flagella. These are all examples of chemiosmosis.

FOCUS QUESTION   7.7

Label the following diagram of oxidative phosphorylation in a mitochondrial membrane.

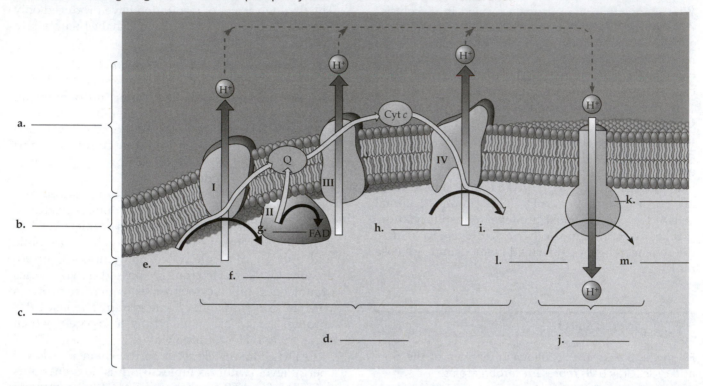

An Accounting of ATP Production by Cellular Respiration

About 30 to 32 ATPs may be produced per glucose molecule oxidized. These numbers are only estimates for three reasons: The ratio of NADH to ATP is not a whole number—experimental data indicate the production of 2.5 ATPs/NADH and of 1.5 ATPs/FADH$_2$; the electrons from NADH produced by glycolysis may be passed across the mitochondrial membrane to NAD$^+$ or FAD, depending on the type of shuttle used in the cell; and the proton-motive force generated by the electron transport chain is also used to power other work in the mitochondrion.

The efficiency of energy conversion in respiration is about 34%. The remaining 66% is lost as heat. In the brown fat cells of some mammals, an uncoupling protein allows protons to flow back across the inner mitochondrial membrane. Without the generation of ATP (which would inhibit cellular respiration), fats continue to be oxidized and heat generated during hibernation.

FOCUS QUESTION   7.8

Fill in the following tally for the maximum ATP yield from the oxidation of one molecule of glucose to six molecules of carbon dioxide.

Process	# ATP
Initial phosphorylation of glucose:	a. _____
Substrate-level phosphorylation: in glycolysis	b. _____
In c.	2
Oxidative phosphorylation:*	d. _____
Maximum Total	e. _____

*2.5 ATP for each of the f. ____ NADH from pyruvate → acteyl CoA and the g. ____ NADH from citric acid cycle; 1.5 ATP for each of the h. ____ FADH$_2$ from citric acid cycle; 2.5 or 1.5 ATP for each of the i. _____ NADH from glycolysis, depending on which shuttle passes electrons across the membrane.

7.5 Fermentation and anaerobic respiration enable cells to produce ATP without the use of oxygen

Organisms that generate ATP through anaerobic respiration have an electron transport chain that does not use oxygen as the final electron acceptor. Some bacteria use sulfate ions to accept electrons, generating H_2S instead of H_2O.

Is it possible for cells to generate ATP without using either oxygen or an electron transport chain? In fermentation, oxidation of glucose in glycolysis produces a net of 2 ATP by substrate-level phosphorylation, and NADH is recycled to NAD^+ by the transfer of electrons to pyruvate or derivatives of pyruvate.

Types of Fermentation   In **alcohol fermentation,** pyruvate is converted to acetaldehyde, and CO_2 is released. Acetaldehyde is then reduced by NADH to form ethanol (ethyl alcohol), and NAD^+ is regenerated. In **lactic acid fermentation,** pyruvate is reduced directly by NADH to lactate, recycling NAD^+. Muscle cells make ATP by lactic acid fermentation when energy demand is high and O_2 supply is low.

Comparing Fermentation with Anaerobic and Aerobic Respiration   Both fermentation and aerobic and anaerobic respiration use glycolysis, with NAD^+ as the oxidizing agent, in which glucose and other organic fuels are converted to pyruvate. To oxidize NADH back to NAD^+, fermentation uses an organic molecule such as pyruvate or acetaldehyde as the final electron acceptor. Respiration uses oxygen or another electronegative molecule (in anaerobic respiration) as the final electron acceptor after electrons are passed down an electron transport chain.

Obligate anaerobes make ATP by fermentation or anaerobic respiration and are poisoned by oxygen. **Facultative anaerobes,** such as yeasts and many bacteria, can make ATP by fermentation or respiration, depending upon the availability of oxygen.

FOCUS QUESTION 7.9

How much more ATP can be generated by respiration than by fermentation? Explain why.

The Evolutionary Significance of Glycolysis   Glycolysis is common to fermentation and respiration. This most widespread of all metabolic processes probably evolved in ancient prokaryotes. Its cytosolic location and independence of oxygen are also evidence of its antiquity.

7.6 Glycolysis and the citric acid cycle connect to many other metabolic pathways

The Versatility of Catabolism   Are carbohydrates the only type of organic molecule that can be used in cellular respiration? Proteins are digested into amino acids, which are then *deaminated* (the amino group is removed) and can enter glycolysis or the citric acid cycle at several points. The digestion of fats yields glycerol, which is fed into glycolysis, and fatty acids, which are broken down by **beta oxidation** to two-carbon fragments that enter the citric acid cycle as acetyl CoA.

Biosynthesis (Anabolic Pathways)   The organic molecules of food also provide carbon skeletons for biosynthesis. Some monomers, such as amino acids, can be directly incorporated into the cell's macromolecules. Intermediate compounds of glycolysis and of the citric acid cycle serve as precursors for anabolic pathways. Carbohydrates, fats, and proteins can be interconverted to provide for a cell's needs.

Word Roots

aero- = air (*aerobic respiration:* a catabolic pathway for organic molecules using oxygen as the final electron acceptor in an electron transport chain and ultimately producing ATP)

an- = not (*anaerobic:* refers to a chemical reaction that does not use oxygen)

chemi- = chemical (*chemiosmosis:* an energy-coupling mechanism that uses energy stored in the form of a H^+ gradient across a membrane to drive cellular work, such as the synthesis of ATP)

glyco- = sweet; **-lysis** = split (*glycolysis:* a series of reactions that splits glucose into pyruvate)

Structure Your Knowledge

1. This chapter describes how catabolic pathways release chemical energy and store it in ATP. One of the best ways to learn the three main components of cellular respiration is to teach them to someone. Find two study partners and have each person learn and explain the important concepts of glycolysis, pyruvate oxidation and the citric acid cycle, or oxidative phosphorylation. Use diagrams to illustrate the process you are explaining.

2. Create a concept map to organize your understanding of oxidative phosphorylation.

3. Fill in the following table to summarize glycolysis, the citric acid cycle, oxidative phosphorylation, and fermentation. Base the main inputs and outputs of these processes on one glucose molecule.

Process	Brief Description	Inputs	Outputs
Glycolysis			
Pyruvate to acetyl CoA and citric acid cycle			
Oxidative phosphorylation			
Fermentation			

Test Your Knowledge

MULTIPLE CHOICE: *Choose the one best answer.*

1. When electrons move closer to a more electronegative atom,
   a. energy is released.
   b. chemical energy is stored.
   c. a proton gradient is established.
   d. water is produced.
   e. ATP is synthesized.

2. In which of the following conversions is the first molecule becoming reduced to the second molecule?
   a. pyruvate → acetyl CoA
   b. glucose → pyruvate
   c. $C_6H_{12}O_6 \rightarrow 6\ CO_2$
   d. pyruvate → lactate
   e. $NADH + H^+ \rightarrow NAD^+ + 2\ H$

3. Some prokaryotes use anaerobic respiration, a process that
   a. does not involve an electron transport chain.
   b. produces ATP solely by substrate-level phosphorylation.
   c. uses a substance other than oxygen as the final electron acceptor.
   d. does not rely on chemiosmosis for ATP production.
   e. Both **a** and **b** are correct.

4. Which of the following reactions is *incorrectly* paired with its location?
   a. ATP synthesis—inner membrane of the mitochondrion, mitochondrial matrix, and cytosol
   b. fermentation—cell cytosol
   c. glycolysis—cell cytosol
   d. substrate-level phosphorylation—cytosol and mitochondrial matrix
   e. citric acid cycle—cristae of mitochondrion

5. Which of the following enzymes uses NAD^+ as a coenzyme?
   a. phosphofructokinase
   b. phosphoglucoisomerase
   c. triose phosphate dehydrogenase
   d. hexokinase
   e. phosphoglyceromutase

6. Which of the following compounds produces the most ATP when completely oxidized to CO_2?
   a. acetyl CoA
   b. glucose
   c. pyruvate
   d. fructose-1,6-bisphosphate
   e. glyceraldehyde-3-phosphate

7. When pyruvate is converted to acetyl CoA,
   a. CO_2 and ATP are released.
   b. a series of enzymatic reactions removes CO_2, transfers electrons to NAD^+, and attaches a coenzyme.
   c. one turn of the citric acid cycle is completed.
   d. NAD^+ is regenerated so that glycolysis can continue to produce ATP by substrate-level phosphorylation.
   e. phosphofructokinase is activated and glycolysis continues.

8. How many molecules of CO_2 are generated for each molecule of acetyl CoA introduced into the citric acid cycle?
   a. 1
   b. 2
   c. 3
   d. 4
   e. 6

9. Which of the following statements correctly describes the role of oxygen in cellular respiration?
   a. It is reduced in glycolysis as glucose is oxidized.
   b. It combines with H^+ diffusing through ATP synthase to produce H_2O.
   c. It provides the activation energy needed for oxidation to occur.
   d. It is the final electron acceptor for the electron transport chain.
   e. It combines with the carbon removed during the citric acid cycle to form CO_2.

10. Which of the following is an accurate description of chemiosmosis?
    a. ATP production is linked to the proton gradient established by the electron transport chain.
    b. The difference in pH between the intermembrane space and the cytosol drives the formation of ATP.
    c. The flow of H^+ through ATP synthases rotates a rotor and rod, driving the hydrolysis of ADP.
    d. The energy released by the reduction and subsequent oxidation of electron carriers transfers a phosphate to ADP.
    e. The production of water in the mitochondrial matrix by the reduction of oxygen leads to a net flow of water out of a mitochondrion.

11. When glucose is oxidized to CO_2 and water, approximately 66% of its energy is transformed to
    a. heat.
    b. ATP.
    c. a proton-motive force.
    d. potential energy.
    e. NADH.

12. Which of the following statements is *incorrect* concerning oxidative phosphorylation?
    a. It produces about 2.5 ATP for every NADH that is oxidized.
    b. It involves the redox reactions of the electron transport chain.
    c. It involves an ATP synthase located in the inner mitochondrial membrane.
    d. It uses oxygen as the final electron donor.
    e. It is an example of chemiosmosis.

13. Which of the following statements correctly describes a metabolic effect of cyanide, a poison that blocks the passage of electrons along the electron transport chain?
    a. The pH of the intermembrane space becomes much lower than normal.
    b. Electrons are passed directly to oxygen, causing cells to explode.
    c. Alcohol would build up in the mitochondria.

    d. NADH supplies would be exhausted, and ATP synthesis would cease.
    e. No proton gradient would be produced, and ATP synthesis would cease.

14. A suspension of cells is ground up and then mixed with a chemical that dissolves fats. Which of the following stages of cellular respiration would be most disrupted by this chemical?
    a. glycolysis
    b. pyruvate oxidation
    c. the citric acid cycle
    d. oxidative phosphorylation
    e. fermentation

15. Substrate-level phosphorylation
    a. involves the shifting of a phosphate group from ATP to a substrate.
    b. can use NADH or $FADH_2$.
    c. takes place only in the cytosol.
    d. accounts for 10% of the ATP formed by fermentation.
    e. is the energy source for facultative anaerobes under anaerobic conditions.

16. Fermentation produces less ATP than cellular respiration because
    a. NAD^+ is regenerated by alcohol or lactate production, without the electrons of NADH passing through the electron transport chain.
    b. pyruvate still contains most of the "hilltop" electrons that were present in glucose.
    c. its starting reactant is pyruvate and not glucose.
    d. a and b are correct.
    e. a, b, and c are correct.

17. Muscle cells in oxygen deprivation gain which of the following from the reduction of pyruvate?
    a. ATP
    b. ATP and NAD^+
    c. CO_2 and NAD^+
    d. ATP, alcohol, and NAD^+
    e. ATP and CO_2

18. Glucose made from six radioactively labeled carbon atoms is fed to yeast cells in the absence of oxygen. How many molecules of radioactive alcohol (C_2H_5OH) are formed from each molecule of glucose?
    a. 0
    b. 1
    c. 2
    d. 3
    e. 6

19. Glycolysis is considered one of the first metabolic pathways to have evolved because
    a. it relies on fermentation, which is characteristic of archaea and bacteria.
    b. it is found only in prokaryotes, whereas eukaryotes use mitochondria to produce ATP.
    c. it produces much less ATP than does the electron transport chain and chemiosmosis.
    d. it produces ATP only by substrate phosphorylation and does not involve redox reactions.
    e. it is nearly universal, is located in the cytosol, and does not involve O_2.

20. Which of the following substances produces the most ATP per gram?
    a. glucose, because it is the starting place for glycolysis
    b. glycogen or starch, because they are polymers of glucose
    c. fats, because they are highly reduced compounds
    d. proteins, because of the energy stored in their tertiary structure
    e. amino acids, because after they are deaminated, they can be fed directly into the citric acid cycle

21. Fats and proteins can be used as fuel in the cell because they
    a. can be converted to glucose by enzymes.
    b. can be converted to intermediates of glycolysis or the citric acid cycle.
    c. can pass through the mitochondrial membrane to enter the citric acid cycle.
    d. contain phosphate groups.
    e. contain more energy than glucose.

22. Brown fat, which is found in newborn infants and hibernating mammals, has uncoupling proteins that, when activated, make the inner mitochondrial membrane leaky to H^+. What function does this fat tissue serve?
    a. It produces more ATP than does regular fat and is also found in the flight muscles of ducks and geese, providing more energy for long-distance migrations.
    b. It lowers the pH of the intermembrane space, which results in the production of more ATP per gram than is produced by the oxidation of glucose or regular fat tissue.
    c. Because it dissipates the proton gradient, it generates heat through cellular respiration without producing ATP, thereby raising the body temperature of hibernating mammals or newborn infants.
    d. Its main function is insulation in the endothermic animals in which it is common.
    e. Both **a** and **b** are correct.

Chapter 8

Photosynthesis

Chapter Focus

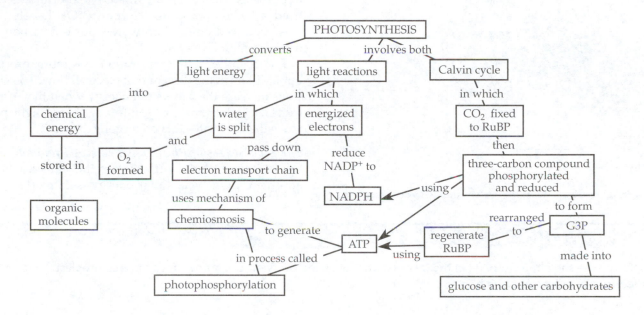

Chapter Review

In **photosynthesis,** the light energy of the sun is converted into chemical energy stored in organic molecules. **Autotrophs** "feed themselves" in the sense that they make their own organic molecules from inorganic raw materials. Autotrophs are the *producers* of the biosphere. Plants, algae, some other protists, and some prokaryotes are photoautotrophs.

Heterotrophs are *consumers.* They may eat plants or animals or decompose organic litter, but almost all ultimately depend on photoautotrophs for food and O_2.

8.1 Photosynthesis converts light energy to the chemical energy of food

The first photosynthetic organisms were likely bacteria with clusters of photosynthetic enzymes and other molecules embedded in infoldings of the plasma membrane.

Chloroplasts: The Sites of Photosynthesis in Plants Chloroplasts are located mainly in the **meso-phyll** tissue of the leaf. CO_2 enters and O_2 exits the leaf through **stomata.** Veins carry water from the roots to leaves and distribute sugar to nonphotosynthetic tissue.

A chloroplast consists of a double membrane surrounding a dense fluid called the **stroma** and an elaborate membrane system of sacs called **thylakoids,** which enclose the *thylakoid space.* Thylakoid sacs may be stacked to form *grana.* **Chlorophyll,** the green pigment that absorbs the light energy that drives photosynthesis, is embedded in the thylakoid membranes.

FOCUS QUESTION   8.1

Label the indicated parts in the following diagram of a chloroplast.

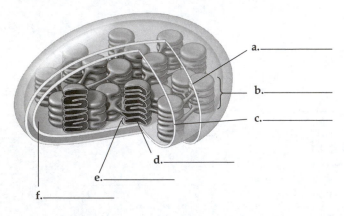

a. _____

b. _____

c. _____

d. _____

e. _____

f. _____

Tracking Atoms Through Photosynthesis: Scientific Inquiry   Taking into account only the net consumption of water and considering glucose as the product (even though the direct product is a three-carbon sugar), the equation for photosynthesis is the reverse of respiration:

$$6\ CO_2 + 6\ H_2O + \text{Light energy} \rightarrow C_6H_{12}O_6 + 6\ O_2$$

The simplest equation is $CO_2 + H_2O \rightarrow [CH_2O] + O_2$.

Using evidence from bacteria that utilize hydrogen sulfide (H_2S) for photosynthesis, C. B. van Niel hypothesized that photosynthetic organisms need a hydrogen source and that plants use H_2O as their hydrogen (and electron) source and release O_2. This hypothesis was later confirmed by experiments that used a heavy isotope of oxygen (^{18}O). Labeled O_2 was produced only when water, but not carbon dioxide, contained ^{18}O.

Photosynthesis, like respiration, is a redox process, but it differs in the direction of electron flow. The electrons increase their potential energy when they travel from water to reduce CO_2 into sugar, and light provides the energy for this endergonic process.

The Two Stages of Photosynthesis: A Preview   The two stages of photosynthesis are the light reactions and the Calvin cycle. Solar energy is converted into

FOCUS QUESTION   8.2

To help you see the big picture, fill in the blanks in the following overview of photosynthesis in a chloroplast. Indicate the locations of the processes **c** and **h**.

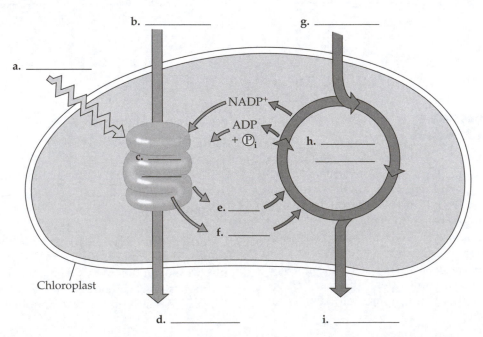

chemical energy in the **light reactions.** Light energy absorbed by chlorophyll drives the transfer of electrons and hydrogen ions from water to the electron acceptor **NADP$^+$**, which is reduced to NADPH and temporarily stores electrons. Oxygen is released when water is split. ATP is formed during the light reactions, using chemiosmosis in a process called **photophosphorylation.**

In the **Calvin cycle,** CO_2 is incorporated into existing organic compounds by **carbon fixation,** and these compounds are then reduced to form carbohydrate. NADPH and ATP from the light reactions supply the reducing power and chemical energy needed for the Calvin cycle.

8.2 The light reactions convert solar energy to the chemical energy of ATP and NADPH

The Nature of Sunlight   Electromagnetic energy, also called electromagnetic radiation, travels as rhythmic wave disturbances of electric and magnetic fields. The distance between the crests of electromagnetic waves, called their **wavelength,** ranges across the **electromagnetic spectrum,** from short gamma waves to long radio waves. The small band of radiation from about 380 nm to 750 nm is called **visible light** and is the radiation that drives photosynthesis.

Light also behaves as if it consists of discrete particles called **photons,** which have a fixed quantity of energy. The amount of energy in a photon is inversely related to its wavelength.

Photosynthetic Pigments: The Light Receptors   *Pigments* are substances that absorb light. A **spectrophotometer** measures the amount of light of different wavelengths absorbed by a pigment. The **absorption spectrum** of **chlorophyll *a*,** the pigment that participates directly in the light reactions, shows that it absorbs violet-blue and red light best. Accessory pigments such as **chlorophyll *b*** and some **carotenoids** absorb light of different wavelengths and broaden the spectrum of colors useful in photosynthesis. Some carotenoids function in *photoprotection* by absorbing excessive light energy that might damage chlorophyll or interact with oxygen to form reactive molecules.

FOCUS QUESTION   8.3

An **action spectrum** shows the relative rates of photosynthesis under different wavelengths of light. The following graph includes both an absorption spectrum and an action spectrum. Label the left *y*-axis for the absorption spectrum and the right *y*-axis for the action spectrum. Then indicate the line that represents the absorption spectrum for chlorophyll *a* and the line for the action spectrum for photosynthesis. Why are these lines different?

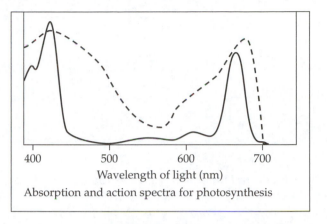

Absorption and action spectra for photosynthesis

Excitation of Chlorophyll by Light   When a pigment molecule absorbs energy from a photon, one of the molecule's electrons is elevated to an orbital where it has more potential energy. Only photons whose energy is equal to the difference between the ground state and the excited state for that molecule are absorbed.

The excited state is unstable. Energy is released as heat as the electron drops back to its ground-state orbital. Isolated chlorophyll molecules also emit photons of light, called fluorescence, as their electrons return to the ground state.

A Photosystem: A Reaction-Center Complex Associated with Light-Harvesting Complexes   Embedded in thylakoid membranes are numerous **photosystems,** each composed of several **light-harvesting complexes** and a **reaction-center complex.** A reaction-center complex contains two special chlorophyll *a* molecules and a **primary electron acceptor.** When a pigment molecule in a light-harvesting complex absorbs a photon, the energy is passed from pigment to pigment until it reaches the reaction center. What happens then? In a redox reaction, an excited electron of a reaction-center chlorophyll *a* is captured by the primary electron acceptor before it can return to the ground state.

There are two types of photosystems in the thylakoid membrane. The chlorophyll *a* molecules of the reaction center of **photosystem II (PS II)** are called P680, after the wavelength of light (680 nm) they absorb best. At the reaction center of **photosystem I (PS I)** are chlorophyll *a* molecules called P700.

FOCUS QUESTION 8.4

Describe the components of a photosystem.

Linear Electron Flow   Through a sequence called **linear electron flow,** electrons pass from water to $NADP^+$ through the two photosystems. A pigment molecule absorbs a photon of light, and the energy is relayed through other pigment molecules of the light-harvesting complex to the P680 pair of chlorophyll *a* molecules in the PS II reaction-center complex. An excited electron of P680 is trapped by the primary electron acceptor. $P680^+$ is a strong oxidizing agent, and its electron hole is filled when an enzyme removes electrons from

water, splitting it into two electrons, two H^+, and an oxygen atom that immediately combines with another oxygen atom to form O_2.

The primary electron acceptor passes the photo-excited electron to an electron transport chain made up of plastoquinone (Pq), a cytochrome complex, and plastocyanin (Pc). The energy released as electrons "fall" through the electron transport chain is used to pump protons into the thylakoid space, contributing to the proton gradient used for the synthesis of ATP.

At the bottom of the electron transport chain, the electron passes to $P700^+$ in photosystem I. It replaces the photoexcited electron that was captured by its primary electron acceptor when PS I absorbed a photon. This primary electron acceptor passes the electron down a second electron transport chain through ferredoxin (Fd), from which the enzyme $NADP^+$ reductase transfers electrons to reduce $NADP^+$ to NADPH.

A Comparison of Chemiosmosis in Chloroplasts and Mitochondria   How is chemiosmosis in mitochondria and in chloroplasts similar? Electron transport chains built into a membrane pump protons across

FOCUS QUESTION 8.5

Identify the components of linear electron flow in the following diagram. Circle the important products that will provide chemical energy and reducing power to the Calvin cycle.

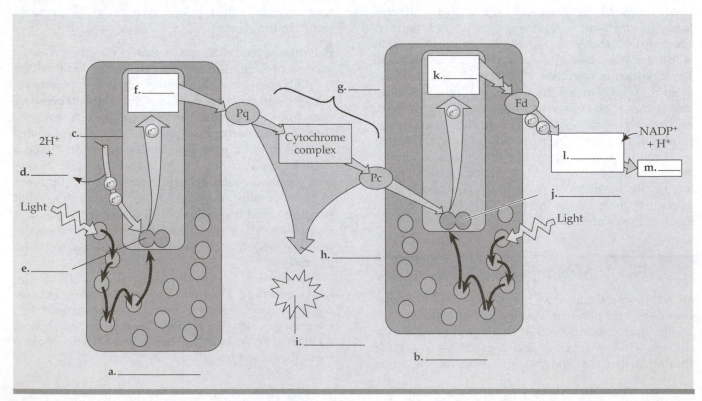

the membrane as electrons are passed down the chain in a series of redox reactions. In respiration, however, organic molecules provide the electrons, and chemical energy is transferred to ATP; in chloroplasts, by contrast, water provides the electrons, and light energy is transformed to the chemical energy of ATP.

In chloroplasts, the electron transport chain pumps protons from the stroma into the thylakoid space. As H^+ diffuses back through ATP synthase, ATP is formed on the stroma side, where it is available to the Calvin cycle.

FOCUS QUESTION 8.6

a. In the light, the proton gradient across the thylakoid membrane is as great as 3 pH units. On which side is the pH lowest?

b. What three factors contribute to the formation of this large difference in H^+ concentration between the thylakoid space and the stroma?

8.3 The Calvin cycle uses the chemical energy of ATP and NADPH to reduce CO_2 to sugar

How are the products of the light reactions used to make sugar? It takes three turns of the Calvin cycle to fix three molecules of CO_2 and produce one molecule of the three-carbon sugar **glyceraldehyde-3-phosphate (G3P)**. The cycle can be divided into three phases:

1. *Carbon fixation:* CO_2 is added to a five-carbon sugar, ribulose bisphosphate (RuBP), in a reaction catalyzed by the enzyme RuBP carboxylase (**rubisco**).

FOCUS QUESTION 8.7

Label the three phases (**a**, **b**, and **c**) and key molecules in the following diagram of the Calvin cycle. How many ATP and NADPH are needed to synthesize one G3P molecule?

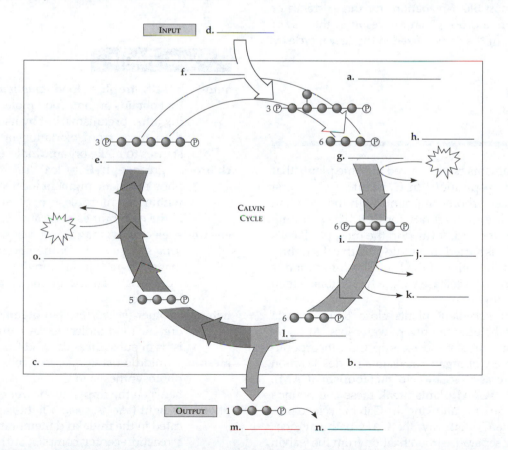

The resulting unstable six-carbon intermediate splits into two molecules of 3-phosphoglycerate.

2. *Reduction:* Each molecule of 3-phosphoglycerate is then phosphorylated by ATP to form 1,3-bisphosphoglycerate. Two electrons from NADPH reduce this compound to G3P. The cycle must turn three times to create a net gain of one molecule of G3P.

3. *Regeneration of CO_2 acceptor (RuBP):* The rearrangement of five molecules of G3P into three molecules of RuBP requires three more ATP.

Evolution of Alternative Mechanisms of Carbon Fixation in Hot, Arid Climates   In most plants, the first product of carbon fixation is a three-carbon compound, 3-phosphoglycerate, formed in the Calvin cycle. When these **C_3 plants** close their stomata on hot, dry days to limit water loss, CO_2 concentration in the leaf air spaces falls, slowing the Calvin cycle. As more O_2 than CO_2 accumulates, rubisco adds O_2 to RuBP in place of CO_2. The product splits, and a two-carbon compound exits the chloroplast and is broken down to release CO_2. This seemingly wasteful process is called **photorespiration.**

FOCUS QUESTION   8.8

What are two possible explanations for the existence of photorespiration, a process that can result in the loss of as much as 50% of the carbon fixed in the Calvin cycle?

What mechanisms have evolved in some plants that minimize photorespiration? In **C_4 plants,** CO_2 is first added to a three-carbon compound with the aid of an enzyme that has a high affinity for CO_2. The resulting four-carbon compound, formed in the mesophyll cells of the leaf, is transported to bundle-sheath cells tightly packed around the veins of the leaf. The compound is then broken down to release CO_2, which rubisco then fixes in the Calvin cycle.

Many desert succulent plants close their stomata during the day, helping to prevent water loss. At night, they open their stomata and take up CO_2, incorporating it into a variety of organic acids in a mode of carbon fixation called **crassulacean acid metabolism (CAM).** During daylight, **CAM plants** break these compounds down and release CO_2, allowing the Calvin cycle to proceed. Unlike the C_4 pathway, the CAM pathway does not structurally separate carbon fixation from the Calvin cycle; instead, the two processes are separated in time.

FOCUS QUESTION   8.9

a. Where does the Calvin cycle take place in C_4 plants?

b. How can C_4 plants successfully utilize the Calvin cycle in hot, dry conditions when C_3 plants would be undergoing photorespiration?

c. How is carbon fixation different in CAM plants?

The Importance of Photosynthesis: A Review   About 50% of the organic material produced by photosynthesis is used as fuel for cellular respiration in the mitochondria of plant cells; the rest is used as carbon skeletons for the synthesis of organic molecules (proteins, lipids, and a great deal of cellulose), stored as starch, or lost through photorespiration. Each year about 160 billion metric tons of carbohydrate are produced by photosynthesis.

Word Roots

auto-   = self; **-troph** = food (*autotroph:* an organism that obtains organic food molecules not by eating other organisms, but by using energy from the sun or from oxidation of inorganic substances to make organic molecules)

chloro-   = green; **-phyll** = leaf (*chlorophyll:* a green, photosynthetic pigment located in membranes within the chloroplasts of plants and algae and in the membranes of certain prokaryotes)

electro-   = electricity; **magnet-** = magnetic (*electromagnetic spectrum:* the entire spectrum of electromagnetic radiation, ranging in wavelength from less than a nanometer to more than a kilometer)

hetero-   = other (*heterotroph:* an organism that obtains organic food molecules by eating other organisms or substances derived from them)

meso-   = middle (*mesophyll:* leaf cells specialized for photosynthesis; in C_3 and CAM plants located between the upper and lower epidermis)

photo-   = light (*photosystem:* a light-capturing unit located in the thylakoid membrane, consisting of a reaction-center complex surrounded by numerous light-harvesting complexes)

Structure Your Knowledge

1. You have already filled in the blanks in several diagrams of photosynthesis. To really understand this process, however, try to create your own representation. In a diagrammatic form, trace the flow of electrons and the production of ATP and NADPH in the light reactions. Then outline the three major stages in the production of G3P. Indicate where these reactions occur in the chloroplast. You can compare your diagram to the sketch in the answers section. Perhaps your study group can collaboratively create a clear, concise summary of this chapter.

2. Create a concept map to confirm your understanding of the chemiosmotic synthesis of ATP in photophosphorylation.

Test Your Knowledge

MULTIPLE CHOICE: *Choose the one best answer.*

1. Which of the following processes or structures is mismatched with its location?
   a. light reactions—grana
   b. electron transport chain—thylakoid membrane
   c. Calvin cycle—stroma
   d. ATP synthase—double membrane surrounding chloroplast
   e. splitting of water—thylakoid space

2. Photosynthesis is a redox process in which
   a. CO_2 is reduced and water is oxidized.
   b. $NADP^+$ is reduced and RuBP is oxidized.
   c. CO_2, $NADP^+$, and water are reduced.
   d. O_2 acts as an oxidizing agent and water acts as a reducing agent.
   e. G3P is reduced and RuBP is oxidized.

3. Which of the following statements is *false*?
   a. When isolated chlorophyll molecules absorb photons, their electrons fall back to ground state, giving off heat and light.
   b. Accessory pigments and photorespiration may contribute to photoprotection, protecting plants from the detrimental effects of intense light.
   c. In linear electron flow in the light reactions, the electron transport chain pumps H^+ across a membrane, creating a proton-motive force used in ATP synthesis.
   d. In both photosynthetic prokaryotes and eukaryotes, ATP synthases catalyze the production of ATP within the cytosol of the cell.
   e. In sulfur bacteria, H_2S is the hydrogen (and thus electron) source for photosynthesis.

4. A spectrophotometer can be used to measure
   a. the absorption spectrum of a substance.
   b. the action spectrum of a reaction.
   c. the amount of energy in a photon.
   d. the wavelength of visible light.
   e. the efficiency of photosynthesis.

5. Accessory pigments within chloroplasts are responsible for
   a. driving the splitting of water molecules.
   b. absorbing photons of different wavelengths of light and passing that energy to P680 or P700.
   c. providing electrons to the reaction-center chlorophyll after photoexcited electrons pass to $NADP^+$.
   d. pumping H^+ across the thylakoid membrane to create a proton-motive force.
   e. anchoring chlorophyll *a* within the reaction center.

6. The following diagram is an absorption spectrum for an unknown pigment molecule. What color would this pigment appear to you?

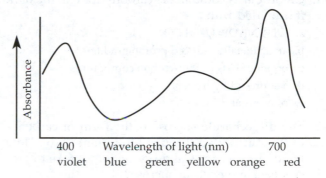

   a. violet
   b. blue
   c. green
   d. yellow
   e. red

7. In photosynthesis, the splitting of water
   a. releases oxygen, which will be used in the Calvin cycle.
   b. provides electrons that phosphorylate ADP to ATP.
   c. provides electrons that produce the gradient used in the chemiosmotic synthesis of ATP.
   d. provides electrons that reduce $NADP^+$ to NADPH.
   e. occurs in photosystem II on the stroma side of the thylakoid membrane.

8. Which of the following substances is/are the final electron acceptors for the electron transport chains in the light reactions of photosynthesis and in cellular respiration?
   a. O_2 in both
   b. CO_2 in both
   c. H_2O in the light reactions, and O_2 in respiration
   d. $NADP^+$ in the light reactions, and NAD^+ or FAD in respiration
   e. $NADP^+$ in the light reactions, and O_2 in respiration

9. Which of the following parts of an illuminated plant cell would you expect to have the lowest pH?
   a. nucleus
   b. cytosol
   c. chloroplast
   d. stroma of chloroplast
   e. thylakoid space

10. A difference between electron transport in photosynthesis and respiration is that in photosynthesis,
   a. NADPH rather than NADH passes electrons to the electron transport chain.
   b. ATP synthase releases ATP into the stroma rather than into the cytosol.
   c. light provides the energy to push electrons to the top of the electron chain, rather than energy from the oxidation of food molecules.
   d. an H^+ concentration gradient rather than a proton-motive force drives the phosphorylation of ATP.
   e. Both a and c are correct.

11. Chloroplasts could make carbohydrate in the dark if provided with
   a. ATP, NADPH, and CO_2.
   b. an artificially induced proton gradient.
   c. organic acids or four-carbon compounds.
   d. a source of hydrogen.
   e. photons and CO_2.

12. The disaccharide sucrose is the form of carbohydrate transported throughout a plant body. How many turns of the Calvin cycle does it take to produce one molecule of sucrose?
   a. 2
   b. 3
   c. 6
   d. 11
   e. 12

13. Both NADPH and ATP from the light reactions are needed
   a. in the carbon fixation stage to provide energy and reducing power to rubisco.
   b. to regenerate three RuBP from five G3P.
   c. to combine two molecules of G3P to produce glucose.
   d. to reduce 3-phosphoglycerate to G3P.
   e. to reduce the H^+ concentration in the stroma and contribute to the proton-motive force.

14. Rubisco is the enzyme that
   a. reduces CO_2 to G3P.
   b. regenerates RuBP with the aid of ATP.
   c. combines electrons and H^+ to reduce $NADP^+$ to NADPH.
   d. adds CO_2 to RuBP in the carbon fixation stage.
   e. transfers electrons from NADPH to 1,3-bisphosphoglycerate to produce G3P.

15. In C_4 plants,
   a. initial carbon fixation takes place in the mesophyll cells.
   b. photorespiration requires more energy than it does in C_3 plants.
   c. the Calvin cycle, which takes place in the bundle-sheath cells, uses an enzyme that has a greater affinity for CO_2 than does rubisco.
   d. a and b are correct.
   e. a and c are correct.

16. CAM plants avoid photorespiration by
   a. keeping their stomata closed during the day.
   b. performing the Calvin cycle at night.
   c. fixing CO_2 into four-carbon compounds in the mesophyll, which then release CO_2 in the bundle-sheath cells.
   d. storing water in their succulent stems and leaves.
   e. fixing CO_2 into organic acids during the night, which then provide CO_2 during the day.

17. In green plants, most of the ATP for synthesis of proteins, cytoplasmic streaming, and other cellular activities comes directly from
   a. photosystem I.
   b. photosystem II.
   c. the Calvin cycle.
   d. oxidative phosphorylation.
   e. photophosphorylation.

For each of the events listed in questions 18 through 22, indicate whether the event occurs during
   a. respiration.
   b. photosynthesis.
   c. both respiration and photosynthesis.
   d. neither respiration nor photosynthesis.

18. Chemiosmotic synthesis of ATP

19. Reduction of oxygen

20. Reduction of CO_2

21. Reduction of NAD^+

22. Oxidation of $NADP^+$

Chapter 9

The Cell Cycle

Chapter Focus

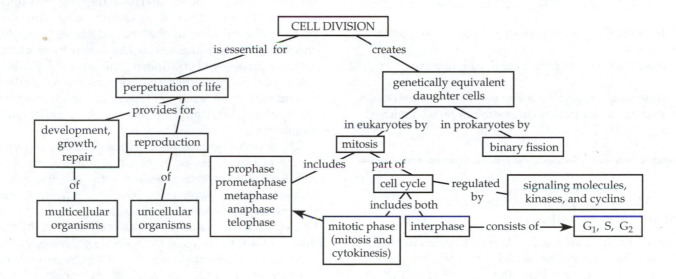

Chapter Review

Cell division, the reproduction of cells, creates duplicate offspring in unicellular organisms and provides for growth, development, and repair in multicellular organisms. The **cell cycle** extends from the creation of a new cell by the division of its parent cell to its own division into two daughter cells.

9.1 Most cell division results in genetically identical daughter cells

The process of recreating a structure as intricate as a cell necessitates the exact duplication and equal division of the DNA containing the cell's genetic program.

Cellular Organization of the Genetic Material A cell's complete complement of DNA is called its **genome,** which is organized into **chromosomes.** Each eukaryotic chromosome is a very long DNA molecule with associated proteins that help structure the chromosome and control the activity of genes, the units of inheritance. This DNA–protein complex is called **chromatin.**

Every diploid species has a characteristic number of chromosomes in each **somatic cell;** reproductive cells, or **gametes** (egg and sperm), have half that number of chromosomes.

Distribution of Chromosomes During Eukaryotic Cell Division Prior to cell division, a cell replicates its DNA. Duplicated chromosomes consist of two identical **sister chromatids,** attached along their length by proteins called *cohesins.* This attachment is called *sister chromatid cohesion.* Each sister chromatid has a **centromere,** a region where proteins bind to specific centromeric DNA sequences and hold the chromatids closely together. Regions on either side of the centromere are called the chromatid *arms.*

The two sister chromatids separate during **mitosis** (the division of the genetic material in the nucleus), and then the cytoplasm divides during **cytokinesis,** producing two separate, genetically equivalent cells.

A type of cell division called *meiosis* produces daughter cells that have half the number of chromosomes as the parent cell. When a sperm fertilizes an egg (both of which are formed by meiosis in animals), the chromosome number is restored, and the somatic cells of the new offspring again have two sets of chromosomes.

FOCUS QUESTION 9.1

a. How many chromosomes are in your somatic cells?

b. How many chromosomes are in your gametes?

c. How many chromatids are in one of your somatic cells that has duplicated its chromosomes prior to mitosis?

9.2 The mitotic phase alternates with interphase in the cell cycle

Phases of the Cell Cycle   The cell cycle consists of the **mitotic (M) phase,** which includes mitosis and cytokinesis, and the **interphase,** during which the cell grows and duplicates its chromosomes. Interphase, usually lasting for 90% of the cell cycle, includes the **G_1 phase,** the **S phase,** and the **G_2 phase.** Mitosis is conventionally divided into five stages: **prophase, prometaphase, metaphase, anaphase,** and **telophase.**

FOCUS QUESTION 9.2

a. How are the three subphases of interphase alike?

b. What key event happens during the S phase?

The Mitotic Spindle: A Closer Look   What is the cellular apparatus that moves the chromosomes during mitosis? The **mitotic spindle** consists of fibers made of microtubules and associated proteins, and its assembly in animal cells begins in the **centrosome,** or *microtubule-organizing center*. A pair of centrioles is centered in each centrosome. The single centrosome duplicates during interphase. As spindle microtubules grow out from them, the two centrosomes move to opposite poles of the cell. Radial arrays of shorter microtubules, called **asters,** extend from the centrosomes.

During prophase, the nucleoli disappear and the chromatin fibers coil and fold into visible chromosomes, consisting of sister chromatids joined at their centromeres and by sister chromatid cohesion. During prometaphase, some of the spindle microtubules attach to each chromatid's **kinetochore,** a structure of protein associated with DNA located at the centromere region. Alternate tugging by opposite kinetochore microtubules moves each chromosome to the midline of the cell. At metaphase, the chromosomes are aligned at the **metaphase plate,** a plane across the midline of the spindle. Nonkinetochore microtubules (or "polar" microtubules) extend out from each centrosome and overlap at the midline. Aster microtubules contact the plasma membrane.

The cohesins joining sister chromatids are cleaved by the enzyme *separase* to begin anaphase, and the now separate chromosomes move toward the poles. Motor proteins appear to "walk" a chromosome along the kinetochore microtubules as these shorten by depolymerizing at their kinetochore end. In addition, motor proteins at the spindle poles appear to "reel in" the chromosomes while the microtubules depolymerize at the poles. The extension of the spindle poles away from each other as an animal cell elongates is due to the overlapping nonkinetochore microtubules walking past each other, also using motor proteins.

In telophase, equivalent sets of chromosomes are at the two poles of the cell. Nuclear envelopes form, nucleoli reappear, cytokinesis begins, and the spindle microtubules depolymerize.

Cytokinesis: A Closer Look   **Cleavage** is the process that separates the two daughter cells in animals. A **cleavage furrow** forms, as a ring of actin microfilaments interacting with myosin proteins begins to contract on the cytoplasmic side of the membrane. The cleavage furrow deepens until the dividing cell is pinched in two.

In plant cells, a **cell plate** forms from the fusion of membrane vesicles derived from the Golgi apparatus. The membrane of the enlarging cell plate joins with the plasma membrane, separating the two daughter cells. A new cell wall develops between the cells from the contents of the cell plate.

FOCUS QUESTION 9.3

The following diagrams will depict the five stages of mitosis in an animal cell, after you draw in the missing chromosomes and kinetochore microtubules in stages **c, d,** and **e.** For simplicity, this cell has only four chromosomes, as you can see in stages **b** and **f.** Identify the stages and label the indicated structures.

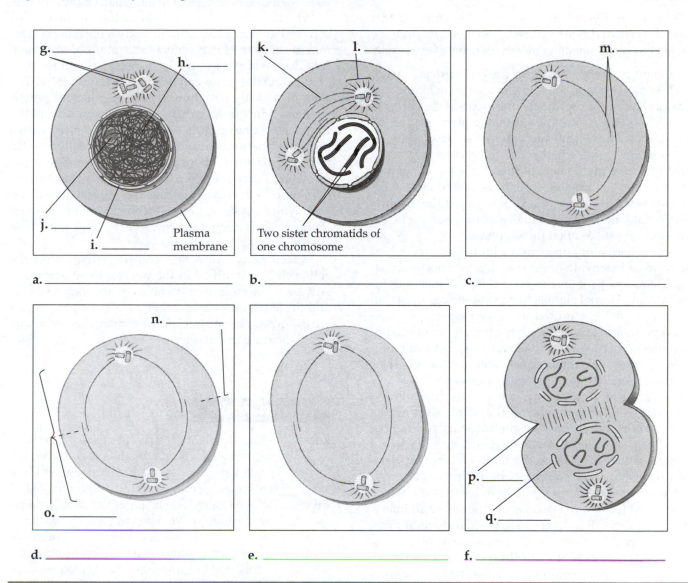

g.
h.
j.
i.
Plasma membrane

k.
l.
Two sister chromatids of one chromosome

m.

a. _____

b. _____

c. _____

n.
o.

p.
q.

d. _____

e. _____

f. _____

Binary Fission in Bacteria Single-celled eukaryotes reproduce asexually by a "dividing in half" process known as **binary fission,** which involves mitosis. The binary fission of prokaryotes does not involve mitosis. The *bacterial chromosome* is a single circular DNA molecule with associated proteins. It begins to replicate at the **origin of replication,** and one of these duplicated origins moves to the opposite pole of the cell. Replication is completed as the cell doubles in size, and the plasma membrane grows inward to divide the two identical daughter cells.

The Evolution of Mitosis Evidence for the evolution of mitosis from prokaryotic cell division includes (1) the relatedness of several proteins involved in both types of division (such as actin and tubulin) and (2) some possible intermediate stages seen in some eukaryotes, in which chromosomal separation takes place within an intact nuclear envelope.

9.3   The eukaryotic cell cycle is regulated by a molecular control system

Normal cell growth, development, and maintenance depend on proper control of the timing and rate of cell division.

Evidence for Cytoplasmic Signals   Experiments that fuse two cells at different phases of the cell cycle indicate that cytoplasmic signaling molecules drive the cell cycle.

Checkpoints of the Cell Cycle Control System   A **cell cycle control system,** consisting of a set of molecules that function cyclically, coordinates the events of the cell cycle.

What are the control points that regulate the cell cycle? Important internal and external signals are monitored to determine whether the cell cycle will proceed past the three main **checkpoints** in the G_1, G_2, and M phases. If a mammalian cell does not receive a go-ahead signal at the G_1 checkpoint, called the "restriction point," the cell will usually exit the cell cycle to a nondividing state called the **G_0 phase.**

Sets of regulatory proteins, including kinases, which activate or inactivate other proteins by phosphorylating them, and proteins called *cyclins,* interact with internal and external signals to control the cell cycle. An internal signal is required to move past the M phase checkpoint into anaphase. Only after all chromosomes are attached at their kinetochores to spindle microtubules is the enzyme separase activated and the cohesions holding sister chromatids together cleaved.

Growing animal cells in cell culture has allowed researchers to identify chemical and physical factors that affect cell division. Certain nutrients and regulatory proteins called **growth factors** have been found to be essential for cells to divide in culture. Mammalian fibroblasts have receptors on their plasma membranes for *platelet-derived growth factor (PDGF),* which is released from blood platelets at the site of an injury. Binding of PDGF initiates a signal-transduction pathway that enables cells to pass the G_1 checkpoint.

Density-dependent inhibition of cell division appears to involve the binding of cell-surface proteins of adjacent cells, which sends a cell division-inhibiting signal to both cells. Most animal cells also show **anchorage dependence** and must be attached to a substratum in order to divide.

FOCUS QUESTION   9.4

Which checkpoint is considered to be most important in the cell cycle? Why would that make sense from an energetic standpoint?

Loss of Cell Cycle Controls in Cancer Cells   When grown in cell culture, cancer cells do not exhibit density-dependent inhibition or anchorage dependence, do not depend on growth factors to divide, and may continue to divide indefinitely instead of stopping after the typical 20 to 50 divisions of normal mammalian cells in culture.

When a normal cell undergoes **transformation** or conversion to a cancer cell, the body's immune system may destroy it. If it proliferates, a mass of abnormal cells develops within a tissue. A **benign tumor** remains at its original site and can be removed by surgery. The cells of a **malignant tumor** have significant genetic changes, which enable them to invade and disrupt the functions of one or more organs. An individual with a malignant tumor is said to have cancer. Malignant tumor cells may have abnormal metabolism and unusual numbers of chromosomes. They lose their attachments to other cells and may **metastasize,** entering the blood and lymph systems and spreading to other sites. Radiation and chemicals are used to treat known or suspected metastatic tumors.

Defects in cell signaling pathways that affect the cell cycle are involved in the development of cancer. Advances in molecular techniques to analyze DNA and the levels of specific cancer-associated proteins increasingly enable physicians to match chemotherapy to a patient's particular type of tumor.

Word Roots

ana-   = up, throughout, again (*anaphase:* the mitotic stage in which the chromatids of each chromosome have separated and the daughter chromosomes are moving to the poles of the cell)

bi-   = two (*binary fission:* a method of asexual reproduction by "division in half." In prokaryotes, binary fission does not involve mitosis)

centro-   = the center; **-mere** = a part (*centromere:* in a duplicated chromosome, the region on each sister chromatid where they are most closely attached to each other by proteins that bind to specific DNA sequences)

chroma-   = colored (*chromatin:* the complex of DNA and proteins that makes up a eukaryotic chromosome. When the cell is not dividing, chromatin exists in its dispersed form, as a mass of very long, thin fibers that are not visible with a light microscope)

cyto-   = cell; **-kinet** = move (*cytokinesis:* division of the cytoplasm to form two separate daughter cells)

gamet-   = a wife or husband (*gamete:* a haploid reproductive cell, such as an egg or sperm)

gen- = produce (*genome:* the genetic material of an organism or virus)

inter- = between (*interphase:* the period in the cell cycle when the cell is not dividing)

mal- = bad or evil (*malignant tumor:* a cancerous tumor containing cells that have significant genetic and cellular changes and are capable of invading and surviving in new sites)

meta- = between (*metaphase:* the mitotic stage in which the chromosomes are attached to microtubules at their kinetochores and are aligned at the metaphase plate)

mito- = a thread (*mitosis:* a process of nuclear division that allocates replicated chromosomes equally to each of the daughter nuclei)

pro- = before (*prophase:* the first mitotic stage in which the chromatin condenses and the spindle begins to form)

-soma- = body (*centrosome:* a structure in animal cells that functions as a microtubule-organizing center and is important during cell division)

telos- = an end (*telophase:* the final stage of mitosis in which daughter nuclei are forming and cytokinesis has typically begun)

trans- = across; **-form** = shape (*transformation:* the conversion of a normal animal cell to a cancerous cell)

Structure Your Knowledge

1. In the following photomicrograph of onion root tip cells, identify the cell cycle stage for the indicated cells.

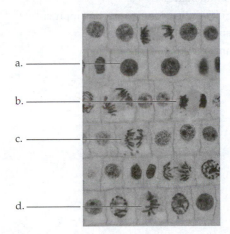

a. _____

b. _____

c. _____

d. _____

2. Describe one chromosome as it proceeds through an entire cell cycle, starting with G_1 of interphase and ending with telophase of mitosis.

Test Your Knowledge

FILL IN THE BLANK:

Identify the appropriate phase of the cell cycle.

_____ 1. most cells that will no longer divide are in this phase

_____ 2. sister chromatids separate and chromosomes move apart

_____ 3. mitotic spindle begins to form

_____ 4. cell plate forms or cleavage furrow pinches cells apart

_____ 5. chromosomes duplicate

_____ 6. chromosomes line up at central plane

_____ 7. nuclear membranes form around separated chromosomes

_____ 8. chromosomes become visible

_____ 9. kinetochore–microtubule interactions move chromosomes to midline

_____ 10. restriction point occurs in this phase

MULTIPLE CHOICE: *Choose the one best answer.*

1. One of the differences between prokaryotic binary fission and eukaryotic mitosis is that
   a. cytokinesis does not occur in prokaryotic cells.
   b. mitosis involves proteins such as tubulin and actin that function in separating chromosomes and cytokinesis in animal cells, but no such molecules are involved in prokaryotic binary fission.
   c. mitosis takes place within the nucleus but binary fission does not.
   d. the duplicated single chromosome does not separate along a mitotic spindle in prokaryotic cells.
   e. the chromosome number is reduced by half in eukaryotic cells but not in prokaryotic cells.

2. A plant cell has 12 chromosomes at the end of mitosis. How many *chromosomes* would it have in the G_2 phase of its next cell cycle?
   a. 6
   b. 9
   c. 12
   d. 24
   e. 48

3. How many *chromatids* would this plant cell have in the G_2 phase of its cell cycle?
   a. 6
   b. 9
   c. 12
   d. 24
   e. 48

4. The longest part of the cell cycle is
   a. prophase.
   b. G_1 phase.
   c. G_2 phase.
   d. mitosis.
   e. interphase.

5. In animal cells, cytokinesis involves
   a. the separation of sister chromatids.
   b. the contraction of a ring of actin microfilaments.
   c. depolymerization of kinetochore microtubules.
   d. a protein kinase that phosphorylates other enzymes.
   e. sliding of nonkinetochore microtubules past each other.

6. Humans have 46 chromosomes. That number of chromosomes will be found
   a. in a cell in anaphase.
   b. in egg and sperm cells.
   c. in somatic cells.
   d. in all the cells of the body.
   e. only in cells in G_1 of interphase.

7. Sister chromatids
   a. have one-half the amount of genetic material as does the original chromosome.
   b. start to move along kinetochore microtubules to opposite poles of a cell during telophase.
   c. each have their own centromere and kinetochore.
   d. are formed during S phase but do not join by sister chromatid cohesion until prophase.
   e. slide past each other as nonkinetochore microtubules elongate.

8. A cell in which of the following phases has the *least* amount of DNA?
   a. G_0
   b. G_2
   c. prophase
   d. metaphase
   e. anaphase

9. Which of the following is *not* true of a cell plate?
   a. It forms at the site of the metaphase plate.
   b. It results from the fusion of microtubules.
   c. It fuses with the plasma membrane.
   d. A cell wall is laid down between its membranes.
   e. It begins to form during telophase in plant cells.

10. A cell that passes the restriction point in G_1 will most likely
    a. undergo chromosome duplication.
    b. have just completed cytokinesis.
    c. continue to divide only if it is a cancer cell.
    d. move into the G_2 phase.
    e. move into the G_0 phase.

11. Which of the following would *not* be exhibited by cancer cells?
    a. loss of anchorage dependence
    b. passage through the restriction point
    c. density-dependent inhibition
    d. metastasis
    e. G_1 phase of the cell cycle

12. What initiates the separation of sister chromatids?
    a. the drop in concentration of cyclins and protein kinases
    b. the depolymerization of tubulin microtubules
    c. movement past the G_2 checkpoint
    d. a signal pathway initiated by the binding of a growth factor
    e. activation of a regulatory protein complex and the enzyme separase after all kinetochores are attached to microtubules

13. Knowledge of the cell cycle control system will be most beneficial to understanding
    a. human reproduction.
    b. plant genetics.
    c. prokaryotic growth and development.
    d. cancer.
    e. cardiovascular disease.

■ Chapter Focus

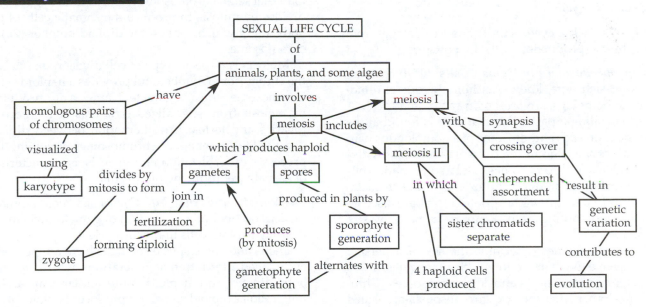

Chapter Review

Genetics is the scientific study of the transmission of traits from parents to offspring (**heredity**) and the **variation** between and within generations.

10.1 Offspring acquire genes from parents by inheriting chromosomes

Inheritance of Genes   The inheritance of traits from parents by offspring involves the transmission of discrete units of information coded in segments of DNA known as **genes**. Most genes contain instructions for synthesizing enzymes and other proteins that then guide the development of inherited traits.

When **gametes** (sperm and eggs) fuse in fertilization, genes from both parents are transmitted to offspring. The DNA of a eukaryotic cell is packaged into a species-specific number of chromosomes present in all **somatic cells**. A gene's **locus** is its location on a chromosome.

Comparison of Asexual and Sexual Reproduction   In **asexual reproduction**, a single parent passes copies of all its genes to its offspring. A **clone** is a group of genetically identical offspring. In **sexual reproduction**, an individual receives a unique combination of genes inherited from two parents.

10.2 Fertilization and meiosis alternate in sexual life cycles

An organism's **life cycle** is the sequence of stages from conception to production of its own offspring.

Sets of Chromosomes in Human Cells   Two chromosomes of each type, known as **homologous chromosomes** or homologs, are present in each somatic cell. A gene controlling a particular character is found at the same locus on each chromosome of a homologous pair.

A **karyotype** is an ordered display of an individual's condensed chromosomes. Isolated somatic cells are stimulated to undergo mitosis, arrested in metaphase, and stained. Digital photographs are used to arrange chromosomes into homologous pairs by size and shape.

Sex chromosomes determine the sex of a person: Females have two homologous X chromosomes; males have nonhomologous X and Y chromosomes. Chromosomes other than the sex chromosomes are called **autosomes**.

Where do homologous pairs of chromosomes come from? You received one set of chromosomes from each parent. The somatic cells of your body are **diploid cells**, each with a diploid number of chromosomes, abbreviated $2n$. Gametes are **haploid cells** and contain a single set of chromosomes. The number of chromosomes in a set is represented by n. The haploid number (n) of chromosomes for humans is 23.

a. How many chromosomes are there in the somatic cells of an animal in which $2n = 14$? _____ How many chromosomes are in its gametes? _____

b. If $n = 14$, how many chromosomes are there in diploid somatic cells? _____ How many sets of homologous chromosomes are in the gametes? _____

c. If $2n = 28$, how many chromatids are there in a cell after chromosome duplication has occurred prior to cell division? _____ What is the difference between sister chromatids and nonsister chromatids?

Behavior of Chromosome Sets in the Human Life Cycle   **Fertilization**, or fusion of sperm and egg, produces a **zygote** containing both a paternal and a maternal set of chromosomes. The diploid zygote then divides by mitosis to produce the somatic cells of the body, all of which contain the diploid number ($2n$) of chromosomes.

Meiosis is a special type of cell division that halves the chromosome number and provides a haploid set of chromosomes to each gamete. Gametes are produced by meiosis from specialized *germ cells* in the gonads (ovaries and testes). An alternation between diploid and haploid numbers of chromosomes, involving the processes of fertilization and meiosis, is characteristic of sexually reproducing organisms.

The Variety of Sexual Life Cycles   In most animals, meiosis occurs in the formation of gametes, which are the only haploid cells in the life cycle.

Plants and some species of algae have a type of life cycle called **alternation of generations**, which includes both diploid and haploid multicellular stages. The multicellular diploid *sporophyte* produces haploid *spores* by meiosis. These spores undergo mitosis and develop into a multicellular haploid organism, the *gametophyte*, which produces gametes by mitosis. Gametes fuse to form a diploid zygote, which develops into the next sporophyte generation.

In most fungi and some protists, the only diploid stage is the zygote. Meiosis occurs after the gametes fuse, producing haploid cells that divide by mitosis to create a unicellular or multicellular haploid organism. Gametes are produced by mitosis in these organisms.

FOCUS QUESTION 10.2

Complete the following diagrams of sexual life cycles by filling in the type of cell division, event, or cells, and indicating whether the cell labeled l. in the bottom cycle is *n* or 2*n*.

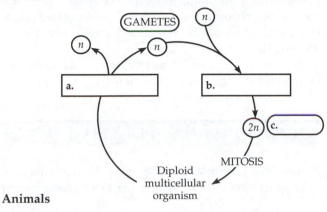

Animals

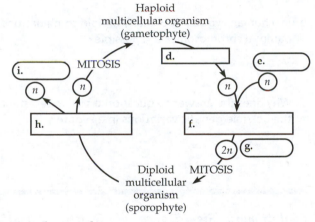

Plants and some algae

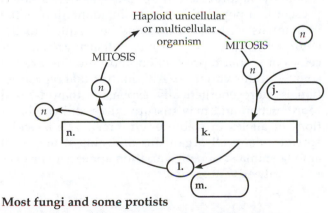

Most fungi and some protists

10.3 Meiosis reduces the number of chromosome sets from diploid to haploid

In meiosis, chromosome duplication is followed by two consecutive cell divisions—**meiosis I** and **meiosis II**—producing four haploid daughter cells, each with one set of chromosomes.

The Stages of Meiosis In interphase, DNA replication (and thus chromosome duplication) produces two genetically identical sister chromatids that remain attached at the centromere and along their length by *sister chromatid cohesion*. **Synapsis** occurs during prophase I, when homologous chromosomes pair up and are held together along their lengths by the *synaptonemal complex*. Genetic material is exchanged by **crossing over** between nonsister chromatids. After synapsis ends, crossovers remain visible as X-shaped regions called **chiasmata**, where sister chromatid cohesion continues to hold the original (but now crossed-over) sister chromatids together.

In metaphase I, homologous pairs line up at the metaphase plate with their kinetochores attached to spindle fibers from opposite poles. Each pair separates in anaphase I, with one homolog moving toward each pole. In telophase I, a haploid set of chromosomes, each composed of two sister chromatids (usually containing some regions of nonsister chromatid DNA), reaches each pole. Cytokinesis usually occurs during telophase I. There is no replication of genetic material prior to the second division of meiosis.

Meiosis II looks like a regular mitotic division, in which chromosomes line up individually on the metaphase plate. Sister chromatid cohesion at the centromere breaks down, and sister chromatids separate and move apart in anaphase II. These sister chromatids, however, are not genetically identical due to crossing over in prophase I. At the end of telophase II, there are four genetically distinct haploid daughter cells.

A Comparison of Mitosis and Meiosis Mitosis produces daughter cells that are genetically identical to the parent cell. Meiosis produces haploid cells that differ genetically from their parent cell and from each other. What are the three events in meiosis I that produce this result? In prophase I, homologous chromosomes pair and are held together (synapsis), and genetic material is exchanged by crossing over between nonsister chromatids. In metaphase I, chromosomes line up in homologous pairs, not as individual chromosomes, at the metaphase plate. And during anaphase I, the homologous pairs separate and one homolog (with sister chromatids still attached at the centromere) goes to each pole.

Homologs are held together by sister chromatid cohesion at regions of crossing over. In anaphase I, enzymes cleave the protein complexes called *cohesins* along the arms of sister chromatids, and the homologs can separate. The cohesins at the centromere hold the sister chromatids together until anaphase II.

Meiosis I is called a *reductional division* because it reduces the chromosome sets from two (diploid) to one (haploid). The sister chromatids of each homolog do not separate until meiosis II, sometimes called the *equational division*.

FOCUS QUESTION   10.3

The following diagrams represent some of the stages of meiosis (*not presented in the right order*). Label these stages. What is the diploid number for this cell? _____

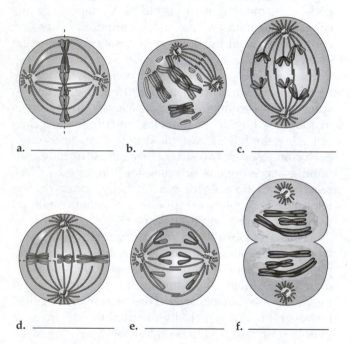

a. _____   b. _____   c. _____

d. _____   e. _____   f. _____

Now place these stages in the correct sequence.

_____ _____ _____ _____ _____ _____

10.4   Genetic variation produced in sexual life cycles contributes to evolution

Origins of Genetic Variation Among Offspring Mutations that result in different *alleles* of a gene are the original source of genetic variation. In sexual reproduction, however, three mechanisms produce the genetic variation evident in each generation: independent assortment of chromosomes, crossing over, and random fertilization.

Each homologous pair lines up independently at the metaphase plate; the orientation of the original maternal and paternal chromosomes is random. Due to this *independent assortment* of chromosomes, the number of possible combinations of maternal and paternal chromosomes in gametes is 2^n, where n is the haploid number for that species.

In prophase I, homologous segments of nonsister chromatids are exchanged by crossing over, forming **recombinant chromosomes** with new genetic combinations of maternal and paternal alleles on the same chromosome. These no-longer-equivalent sister chromatids assort independently during meiosis II.

The random nature of fertilization adds to the genetic variability established in meiosis.

FOCUS QUESTION   10.4

a. How many different assortments of maternal and paternal chromosomes are possible in a human gamete?

b. In a human zygote, how many diploid combinations of parental chromosomes are possible?

c. Why does the answer to question **b** still underestimate the possible genetic variations in a zygote?

Evolutionary Significance of Genetic Variation Within Populations In Darwin's theory of evolution by natural selection, the genetic variation present in a population results in adaptation as the individuals with the variations best suited to the environment produce the most offspring. The processes of sexual reproduction and mutation are the sources of this variation. Although sexual reproduction is more energetically expensive than asexual reproduction and may disrupt successful combinations of alleles in stable environments, it is widespread among all organisms and almost universal among animals. Genetic variation appears to be evolutionarily advantageous.

Word Roots

a- = not or without (*asexual*: generation of offspring from a single parent, occurring without the fusion of gametes)

-apsis = juncture (*synapsis*: the pairing and physical connection of duplicated homologous chromosomes during prophase I of meiosis)

auto- = self (*autosome*: a chromosome that is not directly involved in determining sex)

chiasm- = marked crosswise (*chiasma*: the X-shaped, microscopically visible region where crossing over has occurred in prophase I between homologous nonsister chromatids)

di- = two (*diploid cell*: a cell containing two sets of chromosomes [2*n*], one set inherited from each parent)

fertil- = fruitful (*fertilization*: the union of haploid gametes to produce a diploid zygote)

haplo- = single (*haploid cell*: a cell containing only one set of chromosomes [*n*])

homo- = like (*homologous chromosomes*: a pair of chromosomes of the same length, centromere position, and staining pattern that possess genes for the same characters at corresponding loci)

karyo- = nucleus (*karyotype*: a display of the chromosome pairs of a cell)

meio- = less (*meiosis*: a modified type of cell division that results in cells with half the number of chromosome sets as the original cell)

soma- = body (*somatic cell*: any cell in a multicellular organism except a gamete or its precursors)

Structure Your Knowledge

1. Label the following diagram to review the terms that describe duplicated chromosomes in a diploid cell.

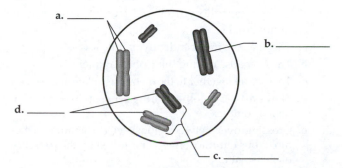

2. Describe the key events of the following stages of meiosis.

a. Interphase	
b. Prophase I	
c. Metaphase I	
d. Anaphase I	
e. Metaphase II	
f. Anaphase II	

3. Create a concept map to help you organize your understanding of the similarities and differences between mitosis and meiosis. *Compare your map with those of some classmates to see different ways of organizing the material.*

Test Your Knowledge

MULTIPLE CHOICE: *Choose the one best answer.*

1. Which of the following is *not* an accurate statement?
   a. Gametes transmit genes from two parents to offspring.
   b. You have 46 chromosomes in your somatic cells and 23 chromosomes in your gametes.
   c. Offspring differ from their parents because mutations occur during sexual reproduction.
   d. Each gene has a specific locus on a chromosome.
   e. Fertilization and meiosis alternate in sexual life cycles.

2. Asexual reproduction of a diploid organism
   a. is impossible because the chromosome number would double each generation.
   b. requires meiosis to prevent the chromosome number from doubling each generation.
   c. usually involves spores produced by mitosis.
   d. produces variation among sibling offspring.
   e. produces offspring identical to each other and the parent.

3. Homologous chromosomes
   a. have identical DNA sequences in their genes.
   b. have genes for the same characters at the same loci.
   c. are found in gametes.
   d. separate in meiosis II.
   e. have all of the characteristics listed above.

4. What are autosomes?
   a. sex chromosomes
   b. chromosomes that occur singly
   c. chromosomal abnormalities that result in genetic defects
   d. chromosomes found in mitochondria and chloroplasts
   e. none of the above

5. What is a karyotype?
   a. a genotype of an individual
   b. a pictorial display of an individual's chromosomes
   c. a blood type determination of an individual
   d. a unique combination of chromosomes found in a gamete
   e. a species-specific diploid number of chromosomes

6. The DNA content of a diploid cell is measured in the G_1 phase. After meiosis I, the DNA content of one of the two cells produced is
   a. equal to that of the G_1 cell.
   b. twice that of the G_1 cell.
   c. one-half that of the G_1 cell.
   d. one-fourth that of the G_1 cell.
   e. impossible to estimate due to independent assortment of homologous chromosomes.

7. In most fungi and some protists,
   a. the zygote is the only haploid stage.
   b. gametes are formed by meiosis.
   c. the multicellular or unicellular organism is haploid.
   d. the gametophyte generation produces gametes by mitosis.
   e. reproduction is exclusively asexual.

8. In the alternation of generations found in plants,
   a. the sporophyte generation produces spores by mitosis.
   b. the gametophyte generation produces gametes by mitosis.
   c. the zygote develops into a sporophyte generation by meiosis.
   d. spores develop into the haploid sporophyte generation.
   e. the gametophyte generation produces spores by meiosis.

9. A synaptonemal complex would be found during
   a. the early prophase of mitosis.
   b. fertilization as gametes fuse.
   c. anaphase I of meiosis.
   d. the early prophase of meiosis I.
   e. both a and d.

10. During meiosis I,
    a. homologous chromosomes separate.
    b. the chromosome number is reduced in half.
    c. crossing over between nonsister chromatids occurs.
    d. paternal and maternal chromosomes assort randomly.
    e. all of the above occur.

11. Compared with one of the four cells produced by meiosis, a cell in G_2 before meiosis has
    a. twice as much DNA and twice as many chromosomes.
    b. four times as much DNA and twice as many chromosomes.
    c. four times as much DNA and four times as many chromosomes.
    d. half as much DNA but the same number of chromosomes.
    e. half as much DNA and half as many chromosomes.

12. Meiosis II is similar to mitosis because
    a. sister chromatids separate.
    b. homologous chromosomes separate.
    c. DNA replication precedes the division.
    d. they both take the same amount of time.
    e. haploid cells are almost always produced.

13. How many *chromatids* are present in metaphase II in a cell undergoing meiosis from an organism in which $2n = 24$?
    a. 12
    b. 24
    c. 36
    d. 48
    e. 96

14. Which of the following would *not* be considered a haploid cell?
    a. a daughter cell after meiosis II
    b. a gamete
    c. a daughter cell after mitosis in the gametophyte generation of a plant
    d. a cell in prophase I
    e. a cell in prophase II

15. Which of the following is *not* true of homologous chromosomes?
    a. They behave independently in mitosis.
    b. They synapse during the S phase of meiosis.
    c. They travel together to the metaphase plate in meiosis I.
    d. They are held together during synapsis by a synaptonemal complex.
    e. Crossing over between nonsister chromatids of homologous chromosomes is indicated by the presence of chiasmata.

16. In a species with a diploid number of 6, how many different combinations of maternal and paternal chromosomes would be possible in *gametes*?
    a. 6
    b. 8
    c. 12
    d. 64
    e. 128

17. In a sexually reproducing species with a diploid number of 8, how many different diploid combinations of chromosomes would be possible in the *offspring?*
    a. 8
    b. 16
    c. 64
    d. 256
    e. 512

18. The calculation of offspring in question 17 includes only variation resulting from
    a. crossing over.
    b. random fertilization.
    c. independent assortment of chromosomes.
    d. a, b, and c.
    e. b and c.

19. Which of the following is *least* likely to be a source of genetic variation in sexually reproducing organisms?
    a. crossing over
    b. replication of DNA during S phase before meiosis I
    c. independent assortment of chromosomes
    d. random fertilization of gametes
    e. All of the above contribute equally to the genetic variation produced in sexual life cycles.

20. Which of the following statements describes why or how recombinant chromosomes add to genetic variability?
    a. They are formed as a result of random fertilization when two sets of chromosomes combine in a zygote.
    b. They are the result of mutations that change alleles.
    c. They randomly orient during metaphase II, and the nonequivalent sister chromatids separate in anaphase II.
    d. Genetic material from two parents is combined on the same chromosome.
    e. Both c and d are true.

21. What is the evolutionary significance of bdelloid rotifers, which have not engaged in sex in 40 million years?
    a. They show that asexual reproduction is more advantageous because it ensures that successful combinations of alleles are passed to offspring.
    b. Although their numerous species originated through sexual reproduction, each species has successfully cloned itself since it arose.
    c. They may pick up genetic material from other rotifers while in a state of suspended animation during dry periods, supporting the idea that genetic variation is evolutionarily advantageous.
    d. Their long history shows that sexual reproduction and genetic variation are not required for evolutionary success.
    e. Both a and d are correct.

Mendel and the Gene Idea

Chapter Focus

Through his work with garden peas in the 1860s, Gregor Mendel developed the fundamental principles of inheritance and the laws of segregation and independent assortment. This chapter describes the basic monohybrid and dihybrid crosses that Mendel performed to establish that inheritance involves particulate genetic factors (genes) that segregate independently in the formation of gametes and recombine to form offspring. The laws of probability can be applied to predict the outcome of genetic crosses.

The phenotypic expression of genotype may be affected by such factors as incomplete dominance, codominance, multiple alleles, pleiotropy, epistasis, and polygenic inheritance, as well as the environment. Genetic counseling for inherited disorders uses Mendelian principles to analyze human pedigrees.

Chapter Review

Gregor Mendel's experiments with garden peas provided evidence to replace the "blending" hypothesis of inheritance with the "particulate" hypothesis, in which discrete heritable units that retain their identities are passed from parents to offspring.

11.1 Mendel used the scientific approach to identify two laws of inheritance

Mendel's Experimental, Quantitative Approach   Why were Mendel's garden peas a good choice of study organism? They are available in many varieties, their fertilization is easily controlled, and the characteristics of their many offspring can be quantified.

Mendel studied seven **characters**, or heritable features, that occurred in alternative forms called

traits. He used **true-breeding** varieties of pea plants, which means that self-fertilizing parents always produce offspring with the parental form of the character. To follow the transmission of these well-defined traits, Mendel performed **hybridizations** in which he mated (*crossed* by cross-pollinating) contrasting true-breeding varieties, and then allowed the next generation to self-pollinate. The true-breeding parental plants are the **P generation** (parental); the offspring of the first cross are the **F₁ generation** (first filial); and the next generation, from the self- or cross-pollination of the F₁, is known as the **F₂ generation**.

The Law of Segregation   Mendel found that the F₁ offspring did not show a blending of the parental traits. Instead, only one of the parental traits of the character was found in the hybrid offspring. In the F₂ generation, however, the missing parental trait reappeared in the ratio of 3:1—three offspring with the *dominant* trait shown by the F₁ to one offspring with the reappearing *recessive* trait.

Mendel's explanation for this inheritance pattern contains four parts: (1) alternate forms of genes, called **alleles**, account for variations in characters; (2) an organism has two copies of a gene for each trait, one inherited from each parent; (3) when two different alleles occur together, one of them, called the **dominant allele**, determines the organism's appearance while the other, the **recessive allele**, has no observable effect on appearance; and (4) allele pairs separate (segregate) during the formation of gametes (Mendel's **law of segregation**), so an egg or sperm carries only one allele for each inherited character. This explanation is consistent with the behavior of chromosomes during meiosis.

A **Punnett square** can be used to predict the results of simple genetic crosses. Dominant alleles are often symbolized by a capital letter, recessive alleles by a small letter.

FOCUS QUESTION 11.1

Fill in the following diagram of a cross of round-seeded and wrinkled-seeded pea plants. The round allele (*R*) is dominant; the wrinkled allele (*r*) is recessive.

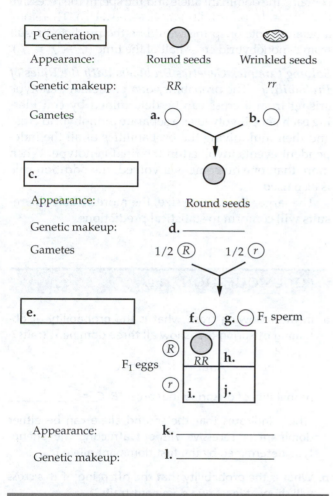

An organism that has a pair of identical alleles is said to be **homozygous** for that gene. If the organism has two different alleles, it is said to be **heterozygous**. Homozygotes are true-breeding; heterozygotes are not, since they produce gametes with one or the other allele that can combine to produce offspring that are dominant homozygotes, heterozygotes, and recessive homozygotes. **Phenotype** is an organism's observable traits; **genotype** is its genetic makeup.

How can you tell if an organism with the dominant phenotype is homozygous for the dominant allele or heterozygous? In a **testcross**, an organism of unknown genotype is crossed with a recessive homozygote to determine the genotype of the phenotypically dominant organism.

FOCUS QUESTION 11.2

A tall pea plant is crossed with a recessive dwarf pea plant. What will the phenotypic and genotypic ratio of offspring be if

a. the tall plant was *TT*?

b. the tall plant was *Tt*?

The Law of Independent Assortment   In a **monohybrid cross**, the inheritance of a single character is followed through the crossing of **monohybrids**, F_1 offspring that are heterozygous for that character. Mendel used **dihybrid crosses** between F_1 **dihybrids**, which are heterozygous for two characters, to determine whether the two characters from the parent plants were transmitted together or independently.

What should happen if the two pairs of alleles segregate independently? Gametes from an F_1 dihybrid (*AaBb*) would contain four combinations of alleles in equal quantities (*AB, Ab, aB, ab*). The random fertilization of these four types of gametes would result in 16 (4 × 4) gamete combinations that produce four phenotypic categories in a ratio of 9:3:3:1 (nine offspring showing both dominant traits, three showing one dominant trait and one recessive, three showing the opposite dominant and recessive traits, and one showing both recessive traits). Mendel obtained these ratios in the F_2 progeny of dihybrid crosses, providing evidence for the **law of independent assortment**. This principle states that each pair of alleles segregates independently from other pairs in the formation of gametes. This law applies to genes located on different chromosomes or far apart on the same chromosome. Complete the Focus Question on the law of independent assortment on the next page.

11.2   The laws of probability govern Mendelian inheritance

The probability scale goes from 0 to 1; the probabilities of all possible outcomes must add up to 1. The probability of an event occurring is the number of times that event could occur over all the possible events; for example, the probability of drawing an ace of any suit from a deck of cards is $^4/_{52}$. The outcome of independent events is not affected by previous or simultaneous trials.

FOCUS QUESTION   11.3

A true-breeding tall, purple-flowered pea plant (*TTPP*) is crossed with a true-breeding dwarf, white-flowered plant (*ttpp*).

a. What is the phenotype of the F_1 generation?

b. What is the genotype of the F_1 generation?

c. What four types of gametes are formed by F_1 plants?

_____   _____   _____   _____

d. Fill in the following Punnett square to show the offspring of the F_2 generation. Shade each phenotype a different color so you can see the ratio of offspring.

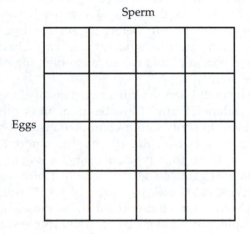

Sperm

Eggs

e. List the phenotypes and ratios found in the F_2 generation.

_____   _____

_____   _____

f. What is the ratio of tall plants to dwarf plants?

_____

Of purple-flowered plants to white-flowered plants?

_____

(Note that the alleles for each individual character segregate as in a monohybrid cross.)

The Multiplication and Addition Rules Applied to Monohybrid Crosses   The **multiplication rule** states that the probability that a certain combination of independent events will occur together is equal to the product of the separate probabilities of the independent events. The probability of a particular genotype being formed by fertilization is equal to the product of the probabilities of forming each type of gamete needed to produce that genotype.

If a genotype can be formed in more than one way, then, according to the **addition rule**, its probability is equal to the sum of the separate probabilities of the different, mutually exclusive ways it can occur. For example, a heterozygote offspring can occur if the egg contains the dominant allele and the sperm the recessive ($\frac{1}{2} \times \frac{1}{2} = \frac{1}{4}$ probability) or vice versa ($\frac{1}{4}$). Therefore, a heterozygote offspring would be the predicted result from a monohybrid cross half of the time ($\frac{1}{4} + \frac{1}{4} = \frac{1}{2}$).

Solving Complex Genetics Problems with the Rules of Probability   The probability of a particular genotype arising from a cross can be determined by considering each gene involved as a separate monohybrid cross and then multiplying the probabilities of all the independent events involved in the final genotype. When more than one outcome is involved, the addition rule is also used.

The larger the sample size, the more closely the results will conform to statistical predictions.

FOCUS QUESTION   11.4

a. In the following cross, what is the probability of obtaining offspring that show all three dominant traits?

$$AaBbcc \times AabbCC$$

probability of offspring that are $A\_B\_C\_$ = _____

(The _ indicates that the second allele can be either dominant or recessive without affecting the phenotype determined by the first dominant allele.)

b. What is the probability that the offspring of this cross will show at least two dominant traits? _____

c. What is the probability of obtaining offspring that show only one dominant trait? _____

11.3   Inheritance patterns are often more complex than predicted by simple Mendelian genetics

Extending Mendelian Genetics for a Single Gene   In the case of an allele showing **complete dominance**, the phenotype of the heterozygote is indistinguishable from that of the dominant homozygote. Intermediate phenotypes are characteristic of alleles showing **incomplete dominance**. The F_1 hybrids have a phenotype intermediate between that of the parents. The F_2 offspring show a 1:2:1 phenotypic and genotypic ratio. Alleles that exhibit **codominance** will each affect the phenotype in separate, distinguishable ways, as in the case of the MN blood group.

How might the dominant/recessive relationship of alleles vary depending on the phenotypic level one considers? **Tay-Sachs disease** is a lethal disorder in which brain cells lack a functioning enzyme to metabolize certain lipids. These lipids then accumulate and damage the brain. The Tay-Sachs allele is recessive at the organismal level—a heterozygote does not have the disease. At the biochemical level, the enzyme activity level is intermediate between that of both homozygotes (incomplete dominance); and at the molecular level, the alleles are codominant in that each produces its enzyme product, either normal or dysfunctional.

Whether or not an allele is dominant or recessive has no relation to how common it is in a population.

Most genes exist in more than two allelic forms. The gene that determines human blood groups has three alleles. The alleles I^A and I^B are codominant with each other; each codes for an enzyme that attaches a carbohydrate to the surface of red blood cells. The allele i results in no attached carbohydrate.

FOCUS QUESTION 11.5

List all the possible genotypes for the following blood groups.

a. A

b. B

c. AB

d. O

Pleiotropy is the ability of a single gene to produce multiple phenotypic effects. Certain hereditary diseases with complex sets of symptoms are caused by pleiotropic alleles.

Extending Mendelian Genetics for Two or More Genes   In **epistasis**, the expression of a gene at one locus may affect the expression of another gene. F_2 ratios that differ from the typical 9:3:3:1 often indicate epistasis.

FOCUS QUESTION 11.6

Suppose that a dominant allele M is necessary for the production of the black pigment melanin; mm individuals are white. A dominant allele B results in the deposition of a lot of pigment in an animal's hair, producing a black color. The genotype bb produces gray hair, assuming the presence of at least one M allele. Two black animals heterozygous for both genes are bred. Fill in the following table for the offspring of this cross.

Phenotype	Genotype	Ratio
Black		
	M_bb	

Quantitative characters, such as height or skin color, vary along a continuum in a population. Such phenotypic gradations are usually due to **polygenic inheritance**, in which two or more genes have an additive effect on one character. Each dominant allele contributes one "unit" to the phenotype. A polygenic character may result in a normal distribution (forming a bell-shaped curve) of the character within a population.

FOCUS QUESTION 11.7

The height of spike weed is a result of polygenic inheritance involving three genes. The dominant allele of each of these genes contributes an additional 5 cm to the base height of the plant, which is 10 cm. The tallest plant (*AABBCC*) can reach a height of 40 cm.

a. If a tall plant (*AABBCC*) is crossed with a base-height plant (*aabbcc*), what is the height of the F_1 plants?

b. How many different phenotypes will there be in the F_2?

Nature and Nurture: The Environmental Impact on Phenotype   The phenotype of an individual is the result of complex interactions between its genotype and the environment. Genotypes have a phenotypic range within which the environment influences phenotypic expression. Polygenic characters are often **multifactorial**, meaning that a combination of genetic and environmental factors affects phenotype.

Integrating a Mendelian View of Heredity and Variation   The phenotypic expression of most genes is influenced by other genes and by the environment. The theory of particulate inheritance and Mendel's principles of segregation and independent assortment, however, still form the basis for modern genetics.

11.4   Many human traits follow Mendelian patterns of inheritance

Pedigree Analysis   A family **pedigree** is a family tree with the history of a particular trait shown across the generations. By convention, circles represent females, squares are used for males, and solid symbols indicate individuals that express the trait in question. Parents are joined by a horizontal line, and offspring are listed below parents from left to right in order of birth. The genotype of individuals in the pedigree can often be deduced by following the patterns of inheritance.

Recessively Inherited Disorders   Only homozygous recessive individuals express the phenotype for the thousands of genetic disorders that are inherited as simple recessive traits. **Carriers** of the disorder are heterozygotes, who are usually phenotypically normal but may transmit the recessive allele to their offspring.

   The likelihood of two individuals carrying the same rare deleterious allele increases when they have common ancestors. Consanguineous matings, between siblings or close relatives, are indicated on pedigrees by double lines.

FOCUS QUESTION   11.8

Consider the following pedigree for the trait albinism (lack of skin pigmentation) in three generations of a family. (Solid symbols represent individuals who are albinos.) From your knowledge of Mendelian inheritance, answer the questions that follow.

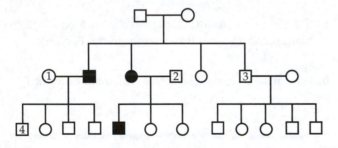

a. Is this trait caused by a dominant or recessive allele? How can you tell?

b. Determine the genotypes of the parents in the first generation. (Let *A* and *a* represent the alleles.) Genotype of father _____; genotype of mother _____. How can you tell?

c. Determine the probable genotypes of the mates of the albino offspring in the second generation and the grandson # 4 in the third generation. Genotypes: mate # 1 _____; mate # 2 _____; grandson # 4 _____.

d. Can you determine the genotype of son # 3 in the second generation? Why or why not?

Cystic fibrosis, the most common lethal genetic disease in the United States, occurs more frequently in people of European descent than in other groups. The recessive allele results in defective chloride channels in certain cell membranes. Accumulating extracellular chloride leads to the buildup of thickened mucus in various organs and a predisposition to bacterial infections.

   Sickle-cell disease is the most common inherited disease among African Americans. Due to a single amino acid substitution in the hemoglobin protein, red blood cells deform into a sickle shape when blood-oxygen concentration is low, triggering blood clumping and other pleiotropic effects. Heterozygous individuals are said to have sickle-cell trait but are usually healthy.

FOCUS QUESTION   11.9

a. What is the probability that a mating between two carriers will produce an offspring with a recessively inherited disorder? _____

b. What is the probability that a phenotypically normal child produced by a mating of two heterozygotes will be a carrier? _____

Dominantly Inherited Disorders   A few human disorders are due to dominant genes. In *achondroplasia*, dwarfism results from a single copy of a mutant allele.

   Dominant lethal alleles are rarer than recessive lethals because the harmful allele cannot be masked in the heterozygote. A late-acting lethal dominant allele can be passed on if the symptoms do not develop until after reproductive age. Molecular geneticists have developed a method to detect the lethal gene for **Huntington's disease**, a degenerative disease of the nervous system that does not develop until later in life.

Multifactorial Disorders   Many diseases have genetic (usually polygenic) and environmental components. These multifactorial disorders include heart disease, diabetes, and cancer.

Genetic Counseling Based on Mendelian Genetics   The probability of a child having a genetic defect may be determined by considering the family history of the disease.

If two prospective parents both have siblings who have a recessive genetic disorder, what is the chance that they would have a child who inherits the disorder? If this couple has two children without the disorder, what is the chance that a third child would inherit the disorder? Explain both answers.

Word Roots

co- = together (*codominance:* situation in which the phenotypes of both alleles are exhibited in the heterozygote)

di- = two (*dihybrid cross:* a cross between two organisms that are each heterozygous for both of the characters being followed [or the self-pollination of a dihybrid plant])

epi- = beside; **-stasis** = standing (*epistasis:* a type of gene interaction in which the phenotypic expression of one gene alters that of another independently inherited gene)

geno- = offspring (*genotype:* the genetic makeup, or set of alleles, of an organism)

hetero- = different (*heterozygous:* having two different alleles for a given gene)

homo- = alike (*homozygous:* having two identical alleles for a given gene)

mono- = one (*monohybrid cross:* a cross between two organisms that are heterozygous for the character being followed [or the self-pollination of a heterozygous plant])

pedi- = a child (*pedigree:* a diagram of a family tree showing the occurrence of heritable characters in parents and offspring over multiple generations)

pheno- = appear (*phenotype:* the observable physical and physiological traits of an organism, determined by its genetic makeup)

pleio- = more (*pleiotropy:* the ability of a single gene to have multiple effects)

poly- = many; **gene-** = produce (*polygenic inheritance:* an additive effect of two or more genes on a single phenotypic character)

Structure Your Knowledge

1. Relate Mendel's two laws of inheritance to the behavior of chromosomes in meiosis that you studied in Chapter 10.

2. How many different types of gametes can be formed by the following genotypes? What generalized formula describes this number?

Aa =

$AaBb$ =

$AaBbCc$ =

$AABbCc$ =

3. How can you tell whether alleles are completely dominant/recessive, incompletely dominant, or codominant?

Genetics Problems

One of the best ways to learn genetics is to work problems. You can't memorize genetic knowledge; you have to practice using it. Work through the following problems, methodically setting down the information you are given and what you are to determine. Write the symbols used for the alleles and genotypes, and the phenotypes resulting from those genotypes. Avoid using Punnett squares; while useful for learning basic concepts, they are much too laborious. Instead, break complex crosses into their monohybrid components and rely on the rules of multiplication and addition. Always look at the answers you get to see if they make logical sense. And remember that study groups are great for going over your problems and helping each other "see the light."

1. Summer squash are either white or yellow. To get white squash, at least one of the parental plants must be white. The allele for which color is dominant?

2. For the following crosses, determine the probability of obtaining an offspring with the indicated genotype.

Cross	Offspring	Probability
$AAbb \times AaBb$	$AAbb$	**a.**
$AaBB \times AaBb$	$aaBB$	**b.**
$AABbcc \times aabbCC$	$AaBbCc$	**c.**
$AaBbCc \times AaBbcc$	$aabbcc$	**d.**

3. True-breeding tall red-flowered plants are crossed with dwarf white-flowered plants. The resulting F_1 generation consists of all tall pink-flowered plants. Assuming that height and flower color are each determined by a single gene locus on different chromosomes, predict the results of an F_1 cross of dihybrid plants. Choose appropriate symbols for the alleles of the genes for height and flower color. List the phenotypes and predicted ratios for the F_2 generation.

4. Blood typing has been used as evidence in paternity cases, when the blood types of the mother and child may indicate that a man alleged to be the father could not possibly have fathered the child. For the following combinations of mother and child,

indicate the blood group(s) of men who would be exonerated (could *not* be the father of that child).

Blood Group of Mother	Blood Group of Child	Man Exonerated If He Belongs to Blood Group(s)
AB	A	a.
O	B	b.
A	AB	c.
O	O	d.
B	A	e.

5. In rabbits, *CC* is normal, *Cc* results in rabbits with deformed legs, and *cc* is lethal. For a gene for coat color, the genotype *BB* produces black, *Bb* brown, and *bb* white. Give the phenotypic ratio of offspring from a cross of a deformed-leg, brown rabbit with a deformed-leg, white rabbit.

6. Polydactyly (extra fingers and toes) results from a dominant gene. A father is polydactyl, the mother has the normal phenotype, and they have had one normal child. What is the genotype of the father? Of the mother? What is the probability that a second child will have the normal number of digits?

7. In dogs, black (*B*) is dominant to chestnut (*b*), and solid color (*S*) is dominant to spotted (*s*). What are the genotypes of the parents in a mating that produced $^3/_8$ black solid, $^3/_8$ black spotted, $^1/_8$ chestnut solid, and $^1/_8$ chestnut spotted puppies? (*Hint:* First determine what genotypes the offspring must have before you deal with the fractions.)

8. Suppose that in hamsters, a dominant gene *B* determines black coat color and *bb* produces brown. A separate gene *E/e*, however, shows epistasis over the *B/b* gene. The dominant *E* allele prevents the deposition of pigment in the hair, resulting in a golden coat color. An *ee* genotype is necessary for a black or brown hamster. A pet owner wants to know the genotypes of her three hamsters, so she breeds them and makes note of the offspring of the litters. Determine the genotypes of the three hamsters.
   a. golden female (hamster 1) × golden male (hamster 2) offspring: 7 golden, 1 black, 1 brown
   b. black female (hamster 3) × golden male (hamster 2) offspring: 8 golden, 5 black, 2 brown

9. The ability to taste phenylthiocarbamide (PTC) is controlled in humans by a single dominant allele (*T*). A nontasting woman married a tasting man, and they had three children: two tasting boys and a nontasting girl. All the grandparents could taste. Create a pedigree for this family for this trait. Solid symbols should signify nontasters (*tt*). Whenever possible, indicate whether tasters are *TT* or *Tt*.

10. Two true-breeding varieties of garden peas are crossed. One parent had red, axial flowers, and the other had white, terminal flowers. All F₁ individuals had red, terminal flowers. If 100 F₂ offspring were counted, how many of them would you expect to have red, axial flowers?

Test Your Knowledge

MULTIPLE CHOICE: *Choose the one best answer.*

1. According to Mendel's law of segregation,
   a. there is a 50% probability that a gamete will get a dominant allele.
   b. allele pairs segregate independently of other pairs of alleles in gamete formation.
   c. allele pairs separate in gamete formation.
   d. the laws of probability determine gamete formation.
   e. there is a 3:1 ratio in the F₂ generation.

2. The F₂ generation
   a. has a phenotypic ratio of 3:1.
   b. is the result of the self-fertilization or crossing of F₁ individuals.
   c. can be used to determine the genotype of individuals with the dominant phenotype.
   d. has a phenotypic ratio that equals its genotypic ratio.
   e. has 16 different genotypic possibilities.

3. A 1:1 phenotypic ratio in a testcross indicates that
   a. the alleles are dominant.
   b. one parent must have been homozygous dominant.
   c. the parent with a dominant phenotype was a heterozygote.
   d. the alleles segregated independently.
   e. the alleles are codominant.

4. Which phase of meiosis is most directly related to the law of independent assortment?
   a. prophase I
   b. prophase II
   c. metaphase I
   d. metaphase II
   e. anaphase II

5. After obtaining two heads from two tosses of a coin, the probability of obtaining a head on the next toss is
   a. $^1/_2$.
   b. $^1/_4$.
   c. $^1/_8$.
   d. $^3/_8$.
   e. $^3/_4$.

6. The probability of tossing three coins simultaneously and obtaining three heads is
   a. $1/2$.
   b. $1/4$.
   c. $1/8$.
   d. $3/8$.
   e. $3/4$.

7. The probability of tossing three coins simultaneously and obtaining two heads and one tail is
   a. $1/2$.
   b. $1/4$.
   c. $1/8$.
   d. $3/8$.
   e. $3/4$.

8. In the F_2 of a dihybrid cross involving two independently assorting genes, what proportion of the offspring will be true-breeding?
   a. $1/16$
   b. $1/8$
   c. $3/16$
   d. $1/4$
   e. $3/4$

9. In guinea pigs, the brown coat color allele (B) is dominant over red (b), and the solid color allele (S) is dominant over spotted (s). The F_1 offspring of a cross between true-breeding brown, solid-colored guinea pigs and red, spotted pigs are crossed. What proportion of their offspring (F_2) would be expected to be red and solid colored?
   a. $1/9$
   b. $1/16$
   c. $3/16$
   d. $9/16$
   e. $3/4$

10. The base height of the dingdong plant is 10 cm. Four genes contribute to the height of the plant, and each dominant allele contributes 3 cm to height. If you cross a 10-cm plant (quadruply homozygous recessive) with a 34-cm plant, how many phenotypic classes will there be in the F_2?
    a. 4
    b. 5
    c. 8
    d. 9
    e. 64

11. You think that two alleles for coat color in mice show incomplete dominance. What is the best and simplest cross to perform in order to test your hypothesis?
    a. a testcross of a homozygous recessive mouse with a mouse of unknown genotype

    b. a cross of F_1 mice to look for a 1:2:1 ratio in the offspring
    c. a reciprocal cross in which the sex of the mice of each coat color is reversed
    d. a cross of two true-breeding mice of different colors to look for an intermediate phenotype in the F_1
    e. a cross of F_1 mice to look for a 9:7 ratio in the offspring

12. The first child of a normally pigmented woman of blood group AB is a boy who is albino with type B blood. The father has normal pigmentation and has type A blood. What are the chances that the next child born to this couple will be an albino girl with type B blood?
    a. 1 chance in 2
    b. 1 chance in 4
    c. 1 chance in 8
    d. 1 chance in 16
    e. 1 chance in 32

13. A dominant allele P causes the production of purple pigment; pp individuals are white. A dominant allele C is also required for color production; cc individuals are white. What proportion of offspring will be purple from a $ppCc \times PpCc$ cross?
    a. $1/8$
    b. $3/8$
    c. $1/2$
    d. $3/4$
    e. None

14. Carriers of a genetic disorder
    a. are indicated by solid symbols on a family pedigree.
    b. are heterozygous at the genetic locus for the disorder.
    c. will produce children with the disease.
    d. are often the result of consanguineous matings.
    e. have a homozygous recessive genotype but do not phenotypically exhibit the disease.

15. Which of the following human diseases is inherited as a simple recessive trait?
    a. cystic fibrosis
    b. cancer
    c. Huntington's disease
    d. achondroplasia
    e. cardiovascular disease

16. A multifactorial disorder
    a. can usually be traced to consanguineous matings.
    b. is caused by recessively inherited lethal genes.
    c. has both genetic and environmental causes.
    d. has a collection of symptoms traceable to an epistatic gene.
    e. is usually associated with quantitative traits.

Chapter 12

The Chromosomal Basis of Inheritance

Chapter Focus

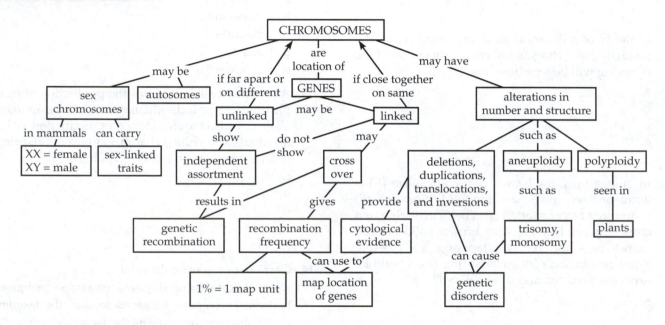

Chapter Review

12.1 Mendelian inheritance has its physical basis in the behavior of chromosomes

Mendel's laws, combined with cytological evidence of the process of meiosis, led to the **chromosome theory of inheritance:** Genes occupy specific positions (loci) on chromosomes, and it is the random alignment of pairs of homologous chromosomes at metaphase I and the separation of homologs in anaphase I that result in the independent assortment and segregation of alleles in gamete formation.

Morgan's Experimental Evidence: Scientific Inquiry T. H. Morgan worked with fruit flies, *Drosophila melanogaster,* which are prolific and rapid breeders. Fruit flies have only four pairs of chromosomes; the sex chromosomes occur as XX in female flies and as XY in male flies.

The normal phenotype found most commonly in nature for a character is called the **wild type**, whereas alternative traits, assumed to have arisen as mutations, are called *mutant phenotypes.* A mutant allele is designated with a small letter, the wild-type allele with a superscript +.

Morgan discovered a mutant white-eyed male fly, which he then mated with a wild-type red-eyed female. The F_1 offspring were all red-eyed. In the F_2, however, all female flies were red-eyed, whereas half of the males were red-eyed and half were white-eyed. How did Morgan explain these results? He deduced that the gene for eye color was located on the X chromosome. Males have only one X, so their phenotype is determined by the eye-color allele they inherit from their mother. This association of a specific gene with a chromosome provided evidence for the chromosome theory of inheritance.

FOCUS QUESTION 12.1

Complete the following summary of Morgan's crosses involving the mutant white-eyed fly by filling in the Punnett square and indicating the genotypes and phenotypes of the F_2 generation. (X^w stands for the mutant recessive white allele; X^{w^+} for the wild-type red allele.) The Y reminds you that this eye color gene is not present on the Y chromosome.

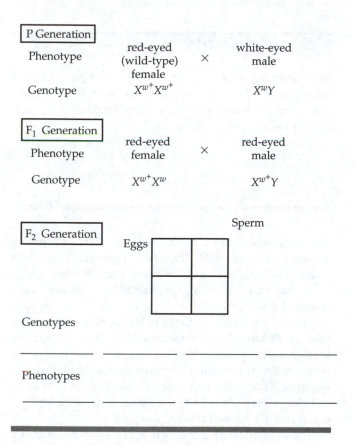

P Generation			
Phenotype	red-eyed (wild-type) female	×	white-eyed male
Genotype	$X^{w^+}X^{w^+}$		X^wY

F_1 Generation			
Phenotype	red-eyed female	×	red-eyed male
Genotype	$X^{w^+}X^w$		$X^{w^+}Y$

F_2 Generation

Sperm

Eggs

Genotypes

_____ _____ _____ _____

Phenotypes

_____ _____ _____ _____

12.2 Sex-linked genes exhibit unique patterns of inheritance

The Chromosomal Basis of Sex Sex is a phenotypic character usually determined by sex chromosomes. In humans and other mammals, females, who are XX, produce eggs that each contain an X chromosome. Males, who are XY, produce two kinds of sperm, each with either an X or a Y chromosome.

Whether the gonads of an embryo develop into testes or ovaries depends on the presence or absence of the gene *SRY*, found on the Y chromosome, whose protein product regulates many other genes. Genes located on either sex chromosome are called **sex-linked genes**. The relatively few genes located on the Y chromosome are called *Y-linked genes*. About half of these genes are expressed only in the testis. The term **X-linked genes** refers to the approximately 1,100 genes located on the X chromosome.

Inheritance of X-Linked Genes X-linked genes code for many characters that are not related to sex. Males inherit X-linked alleles from their mothers; daughters inherit X-linked alleles from both parents. Why are recessive X-linked traits seen more often in males than females? Males are *hemizygous* for X-linked genes—they have only one allele for each gene, and thus a recessive allele is always expressed.

Duchenne muscular dystrophy is an X-linked disorder resulting from the lack of a key muscle protein. **Hemophilia** is an X-linked trait characterized by excessive bleeding due to the absence of one or more blood-clotting proteins.

X Inactivation in Female Mammals Only one of the X chromosomes is fully active in most mammalian female somatic cells. The other X chromosome is contracted into a **Barr body** located inside the nuclear membrane. M. Lyon demonstrated that the selection of which X chromosome is inactivated is a random event occurring independently in embryonic cells. As a result of X inactivation, both males and females have an equal dosage of most X-linked genes. A gene called *XIST* becomes active on the X chromosome that forms the Barr body. Its RNA products appear to trigger DNA methylation and X inactivation.

FOCUS QUESTION 12.2

Two normal color-sighted individuals have two children and seven grandchildren. Fill in the probable genotype of each of the numbered individuals in the following pedigree. *Squares are males, circles are females, and solid symbols represent color blindness. Use the superscript N for the normal allele, and n for the recessive allele for color blindness.*

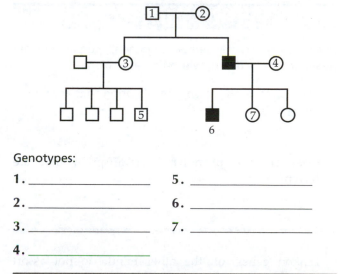

Genotypes:

1. _____   5. _____

2. _____   6. _____

3. _____   7. _____

4. _____

12.3   Linked genes tend to be inherited together because they are located near each other on the same chromosome

What are **linked genes?** They are genes located near each other on the same chromosome and tend to be inherited together.

How Linkage Affects Inheritance   Morgan performed a testcross of F_1 dihybrid wild-type flies with flies that were homozygous recessive for black bodies and vestigial wings. He found that the offspring were not in the predicted 1:1:1:1 phenotypic ratio. Rather, most of the offspring were the same phenotypes as the P generation parents—either wild type (gray, normal wings) or double mutant (black, vestigial). Morgan deduced that these traits were inherited together because their genes were located on the same chromosome.

Genetic recombination results in offspring with combinations of traits that differ from those of either P generation parent.

Genetic Recombination and Linkage   Let's first look at the recombination of genes that are not linked. Consider a cross between a dihybrid heterozygote and a doubly recessive homozygote. The dihybrid parent will produce four types of gametes; the homozygous parent only one type. Of the four types of offspring that will be produced, one-half will be **parental types** and have phenotypes like one or the other of the parental (P) generation (both dominant traits or both recessive traits). The other half of the offspring, called **recombinant types (recombinants)**, will have new combinations of the two traits. This 50% frequency of recombination is observed when two genes are located on different chromosomes, and it results from the random alignment of pairs of homologous chromosomes at metaphase I and the resulting independent assortment of the two unlinked genes.

FOCUS QUESTION   12.3

In a testcross between a heterozygote tall, purple-flowered pea plant and a dwarf, white-flowered plant,

a. what are the phenotypes of offspring that are parental types?

b. what are the phenotypes of offspring that are recombinants?

Linked genes, on the other hand, do not assort independently, and one would not expect to see recombination of parental traits in the offspring. Recombination of linked genes does occur, however, due to **crossing over**, the reciprocal trade between nonsister (a maternal and a paternal) chromatids of paired homologous chromosomes during prophase of meiosis I.

Recombinant chromosomes produced by crossing over, independent assortment, and random fertilization all contribute to the generation of new assortments of alleles, providing an abundance of genetic variation on which natural selection can work.

FOCUS QUESTION   12.4

With unlinked genes, an equal number of parental and recombinant offspring are produced. With linked genes, are more or fewer parentals than recombinants produced? Explain your answer.

Mapping the Distance Between Genes Using Recombination Data: Scientific Inquiry   A **genetic map** is an ordered list of genes on a chromosome. The percentage of recombinant offspring produced in a genetic cross is called the *recombination frequency*. A. H. Sturtevant suggested that recombination frequencies reflect the relative distance between genes; for genes that are farther apart, the probability that crossing over will occur between them is greater than for genes that are closer together. Sturtevant used recombination data to create a **linkage map**, in which one **map unit** is defined as equal to a 1% recombination frequency.

The sequence of genes on a chromosome can be determined by finding the recombination frequency between different pairs of genes. Linkage cannot be determined if genes are so far apart that crossovers between them are almost certain. They would then have the 50% recombination frequency typical of unlinked genes. Such genes are *physically linked* but *genetically unlinked*. Distant genes on the same chromosome may be mapped by adding the recombination frequencies between them and intermediate genes.

Sturtevant and his colleagues found that the genes for the various known mutations of *Drosophila* clustered into four groups of linked genes, providing additional evidence that genes are located on chromosomes—in this case, on the four *Drosophila* chromosomes.

The frequency of crossing over may vary along the length of a chromosome, and a linkage map provides the sequence but not the exact location of genes on chromosomes. **Cytogenetic maps** locate gene loci in reference to visible chromosomal features. Physical maps provide the number of DNA nucleotides between genes.

FOCUS QUESTION 12.5

The following recombination frequencies have been determined for several gene pairs. Create a linkage map for these genes, showing the map unit distance between loci.

j, k: 12%   *j, m:* 9%   *k l:* 6%   *l, m:* 15%

12.4   Alterations of chromosome number or structure cause some genetic disorders

Abnormal Chromosome Number **Nondisjunction** occurs when a pair of homologous chromosomes does not separate properly in meiosis I, or when sister chromatids do not separate in meiosis II. As a result, a gamete may receive either two copies or no copies of that chromosome. A zygote formed with one of these aberrant gametes has a chromosomal alteration known as **aneuploidy**, a nontypical number of a chromosome. The zygote will be either **trisomic** for that chromosome (chromosome number is $2n + 1$) or **monosomic** ($2n - 1$). Aneuploid organisms that survive usually have a set of symptoms caused by the abnormal dosage of genes. A mitotic nondisjunction early in embryonic development is also likely to be harmful.

Polyploidy is a chromosomal alteration in which an organism has more than two complete chromosomal sets, as in *triploidy* ($3n$) or *tetraploidy* ($4n$). Polyploidy is common in the plant kingdom and has played an important role in the evolution of plants.

FOCUS QUESTION 12.6

---

a. What is the difference between an organism with a trisomy and a triploid organism?

b. Which of these two organisms is likely to exhibit the more deleterious effects as a result of its chromosomal anomaly?

Alterations of Chromosome Structure   Chromosome breakage can result in chromosome fragments that are lost, called **deletion;** that join to a sister chromatid (or a nonsister chromatid), called **duplication;** that rejoin the original chromosome in the reverse orientation, called **inversion;** or that join a nonhomologous chromosome, called **translocation**. An unequal crossover can result in a deletion and duplication in nonsister chromatids, caused by unequal exchange between chromatids.

FOCUS QUESTION 12.7

Two nonhomologous chromosomes have gene orders, respectively, of *A-B-C-D-E-F-G-H-I-J* and *M-N-O-P-Q-R-S-T*. What types of chromosome alterations would have occurred if daughter cells were found to have a gene sequence of *A-B-C-O-P-Q-G-J-I-H* on the first chromosome?

Human Disorders Due to Chromosomal Alterations   The frequency of aneuploid zygotes may be fairly high in humans, but development is usually so disrupted that embryos spontaneously abort. Some genetic disorders, expressed as *syndromes* of characteristic traits, are the result of aneuploidy.

Down syndrome, caused by trisomy of chromosome 21, results in characteristic facial features, short stature, correctable heart defects, and developmental delays. The incidence of Down syndrome increases for older mothers.

XXY males exhibit *Klinefelter syndrome,* a condition in which the individual has abnormally small testes, is sterile, and may have subnormal intelligence. Males with an extra Y chromosome do not exhibit any well-defined syndrome.

Trisomy X results in females who are healthy and distinguishable only by karyotype, although they are slightly taller than average and at risk of learning disabilities. Monosomy X individuals (X0) exhibit *Turner syndrome* and are phenotypically female, sterile individuals with usually normal intelligence.

Structural alterations of chromosomes, such as deletions or translocations, may be associated with specific human disorders, such as *cri du chat* syndrome and *chronic myelogenous leukemia (CML)*. The reciprocal translocation involved in CML produces a short, recognizable *Philadelphia chromosome*.

Why do most sex chromosome aneuploidies have less deleterious effects than do autosomal aneuploidies?

Word Roots

aneu- = without (*aneuploidy*: a chromosomal aberration in which one or more chromosomes are present in extra copies or are deficient in number)

cyto- = cell (*cytogenetic map*: a map of a chromosome that locates genes with respect to chromosomal features distinguishable in a microscope)

hemo- = blood (*hemophilia*: a human genetic disease that is caused by an X-linked recessive allele and characterized by excessive bleeding following injury)

mono- = one (*monosomic*: referring to a diploid cell that has only one copy of a particular chromosome instead of the normal two)

non- = not; **dis-** = separate (*nondisjunction*: an error in meiosis or mitosis in which members of a pair of homologous chromosomes or a pair of sister chromatids fail to separate properly from each other)

poly- = many (*polyploidy*: a chromosomal alteration in which the organism possesses more than two complete chromosome sets; result of an accident of cell division)

re- = again; **com-** = together; **bin-** = two at a time (*recombinant*: an offspring whose phenotype differs from that of the true-breeding P generation parents; also refers to the phenotype itself)

trans- = across (*translocation*: an aberration in chromosome structure resulting from attachment of a chromosomal fragment to a nonhomologous chromosome)

tri- = three; **soma-** = body (*trisomic*: referring to a diploid cell that has three copies of a particular chromosome)

Structure Your Knowledge

1. Mendel's law of independent assortment applies to genes that are on different chromosomes. However, at least two of the genes Mendel studied were actually located on the same chromosome. Explain why genes located more than 50 map units apart behave as though they are not linked. How can one determine whether these genes are linked, and what the relative distance is between them?

2. You have found a new mutant phenotype in fruit flies that you suspect is recessive and X-linked. What is the single, best cross you could make to confirm your predictions?

3. Various human disorders or syndromes are related to chromosomal abnormalities. What explanation can you give for the adverse phenotypic effects associated with these chromosomal alterations?

Genetics Problems

Again, one of the best ways to learn genetics is to do problems.

1. The following pedigree traces the inheritance of a genetic trait.

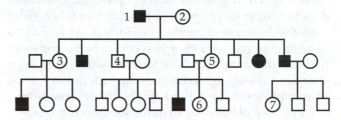

   a. What type of inheritance does this trait show?
   b. Choose an appropriate allele labeling system, and give the predicted genotype for the following individuals:

   1. _____   5. _____

   2. _____   6. _____

   3. _____   7. _____

   4. _____

   c. What is the probability that a child of individual # 6 and a phenotypically normal male will have this trait?

2. Use the following recombination frequencies to determine the order of the genes on the chromosome.

   a, c: 10%   a, d: 30%   b, c: 24%   b, d: 16%

3. In guinea pigs, black (*B*) is dominant to brown (*b*), and solid color (*S*) is dominant to spotted (*s*). A heterozygous black, solid-colored pig is mated with a brown, spotted pig. The offspring from several litters are as follows: black solid: 16; black spotted: 5; brown solid: 5; and brown spotted: 14. Are these genes linked or unlinked? If they are linked, how many map units are they apart?

4. A woman is a carrier for an X-linked lethal allele that causes an embryo with the allele to spontaneously abort. She has nine children. How many of these children do you expect to be boys?

5. A dominant sex-linked allele *B* produces white bars on black chickens, as seen in the Barred Plymouth Rock breed. A clutch of chicks has equal numbers of black and barred chicks. (Remember that sex is determined by the Z-W system in birds: ZZ are males, ZW are females.)
    a. If only the females are found to be black, what were the genotypes of the parents?
    b. If males and females are evenly represented in the black and barred chicks, what were the genotypes of the parents?

Test Your Knowledge

MULTIPLE CHOICE: *Choose the one best answer.*

1. Which of the following statements is part of the chromosome theory of inheritance?
    a. Genes are located on chromosomes.
    b. Homologous chromosomes and their associated genes undergo segregation during meiosis.
    c. Chromosomes and their associated genes undergo independent assortment in gamete formation.
    d. Mendel's laws of inheritance relate to the behavior of chromosomes in meiosis.
    e. all of the above

2. A wild type is
    a. the phenotype found most commonly in nature.
    b. the dominant allele.
    c. designated by a small letter if it is recessive or a capital letter if it is dominant.
    d. a trait found on the X chromosome.
    e. your basic party animal.

3. Sex-linked traits
    a. are carried on an autosome but expressed only in males.
    b. are coded for by genes located on a sex chromosome.
    c. are found in only one sex, depending on the sex-determination system of the species.
    d. are always inherited from the mother in mammals and fruit flies.
    e. depend on whether the gene was inherited from the mother or the father.

4. In which of the following structures would you expect to find a Barr body?
    a. an egg
    b. a sperm
    c. a liver cell of a man
    d. a liver cell of a woman
    e. a mitochondrion

5. A cross of a wild-type red-eyed female *Drosophila* with a violet-eyed male produces all red-eyed offspring. If the gene is X-linked, which of the following should the reciprocal cross (violet-eyed female × red-eyed male) produce? (Assume that the red allele is dominant to the violet allele.)
    a. all violet-eyed flies
    b. 3 red-eyed flies to 1 violet-eyed fly
    c. a 1:1 ratio of red and violet eyes in both males and females
    d. red-eyed females and violet-eyed males
    e. all red-eyed flies

6. Linkage and cytogenetic maps for the same chromosome
    a. are both based on mutant phenotypes and recombination data.
    b. may have different orders of genes.
    c. have both the same order of genes and intergenic distances.
    d. have the same order of genes but different intergenic distances.
    e. are created using chromosomal abnormalities.

7. A 1:1:1:1 ratio of offspring from a dihybrid testcross indicates that
    a. the genes are linked.
    b. the dominant organism was homozygous.
    c. crossing over has occurred.
    d. the genes are 25 map units apart.
    e. the genes are not linked or are more than 50 map units apart.

8. Genes *A* and *B* are linked and 12 map units apart. A heterozygous individual, whose parents were *AAbb* and *aaBB*, would be expected to produce gametes in which of the following frequencies?
    a. 44% *AB*   6% *Ab*   6% *aB*   44% *ab*
    b. 6% *AB*   44% *Ab*   44% *aB*   6% *ab*
    c. 12% *AB*   38% *Ab*   38% *aB*   12% *ab*
    d. 6% *AB*   6% *Ab*   44% *aB*   44% *ab*
    e. 38% *AB*   12% *Ab*   12% *aB*   38% *ab*

9. Which of the following chromosomal alterations does not alter genic balance but may affect gene expression and thus phenotype?
    a. deletion
    b. inversion
    c. duplication
    d. nonidentical duplication
    e. nondisjunction

10. A female tortoiseshell cat is heterozygous for the gene that determines black or orange coat color, which is located on the X chromosome. A male tortoiseshell cat
    a. cannot occur.
    b. is hemizygous at this locus.
    c. must have resulted from a nondisjunction and has a Barr body in each of his cells.
    d. must have two alleles for coat color in each of his cells, one from his father and one from his mother.
    e. would be hermaphroditic.

11. When we say that a few of the genes for Mendel's pea characters were physically linked but genetically unlinked, we mean that
    a. the genes are on the same chromosome, but they are more than 50 map units apart.
    b. the genes assort independently even though the chromosomes they are on travel to the metaphase plate together.
    c. their alleles segregate in anaphase I, and each gamete receives a single allele for all of these genes.
    d. dihybrid crosses with these genes produce more than 50% recombinant offspring even though they are on the same chromosome.
    e. Mendel could not determine that the genes were on the same chromosome because he did not perform crosses with these gene pairs.

12. Two true-breeding *Drosophila* are crossed: a normal-winged, red-eyed female and a miniature-winged, vermilion-eyed male. The F_1s all have normal wings and red eyes. When F_1 offspring are crossed with miniature-winged, vermilion-eyed flies, the following offspring resulted:

    233 normal wing, red eye
    247 miniature wing, vermilion eye
    7 normal wing, vermilion eye
    13 miniature wing, red eye

    From these results, you could conclude that the alleles for miniature wings and vermilion eyes are
    a. both X-linked and dominant.
    b. located on autosomes and dominant.
    c. recessive, and that these genes are located 4 map units apart.
    d. recessive, and that these genes are located 20 map units apart.
    e. recessive, and that the deviation from the expected 9:3:3:1 ratio is due to epistasis.

13. Which of the following statements is *not* true about genetic recombination?
    a. Recombination of linked genes occurs by crossing over.
    b. Recombination of unlinked genes occurs by independent assortment of chromosomes.

    c. Genetic recombination results in offspring with combinations of traits that differ from the phenotypes of both parents.
    d. Recombinant offspring outnumber parental type offspring when two genes are 50 map units apart on a chromosome.
    e. The number of recombinant offspring is proportional to the distance between two gene loci on a chromosome.

14. Suppose that alleles for an X-linked character for wing shape in flies show incomplete dominance. The X^+ allele codes for pointed wings, the X^r for round wings, and X^+X^r individuals have oval wings. In a cross between an oval-winged female and a round-winged male, the following offspring were observed: oval-winged females, round-winged females, pointed-winged males, and round-winged males. A rare pointed-winged female was noted. Cytological study revealed that she had two X chromosomes. Which of the following events could account for this unusual offspring?
    a. a crossover between the two X chromosomes
    b. a crossover between the X and Y chromosomes
    c. a nondisjunction in meiosis II between two X^+ chromatids
    d. a nondisjunction between the X and Y chromosomes, producing some sperm with no sex chromosome
    e. Both c and d together could produce an X^+X^+ female when an XX egg was fertilized by a sperm in which there was no sex chromosome.

15. Some girls who fail to undergo puberty are found to have Swyer syndrome, a condition in which they are externally female but have an XY genotype. Which of the following statements may explain the origin of this syndrome?
    a. A mutation in the *XIST* gene, which codes for RNA molecules that coat the X chromosome and initiate X-inactivation, must have occurred.
    b. A nondisjunction in the egg from the mother resulted in both sex chromosomes coming from the father.
    c. These individuals are actually XXY; the second X is not seen because it is condensed into a Barr body. They have small testes and are sterile but otherwise appear female.
    d. A mutation or deletion of the *SRY* gene on the Y chromosome prevented development of testes and production of the male sex hormones required for a male phenotype.
    e. A translocation of part of an X chromosome to the Y chromosome resulted in a double dose of female-determining genes.

Chapter 13

The Molecular Basis of Inheritance

Chapter Focus

This chapter outlines the key evidence that was gathered to establish DNA as the molecule of inheritance. Watson and Crick's double helix, with its rungs of specifically paired nitrogenous bases and twisting phosphate-sugar side ropes, provided the three-dimensional model that explained DNA's ability to encode a great variety of information and produce exact copies of itself through semiconservative replication. The replication of DNA is an extremely fast and accurate process involving many enzymes and proteins. Eukaryotic chromosomes are complexes of DNA and protein that exhibit varying levels of packing during the cell cycle. Understanding the structure and replication of DNA has enabled the development of the various techniques of genetic engineering.

Chapter Review

Deoxyribonucleic acid, or DNA, is the genetic material that is transmitted from one generation to the next and encodes the blueprints that direct and control the biochemical, anatomical, physiological, and behavioral traits of organisms. DNA is precisely copied in the process of **DNA replication.**

13.1 DNA is the genetic material

The Search for the Genetic Material: Scientific Inquiry Chromosomes were shown to carry hereditary information and to consist of proteins and DNA. Until the 1940s, proteins seemed the most likely candidate to be the genetic material. The role of DNA was established through work with microorganisms—bacteria and viruses.

In 1928, F. Griffith was working with two strains of *Streptococcus pneumoniae*. When he mixed the remains of heat-killed pathogenic bacteria with harmless bacteria, some bacteria were changed into disease-causing bacteria. These bacteria incorporated external genetic material in a process called **transformation**, which results in a change in genotype and phenotype.

O. Avery worked to identify the transforming agent by testing the DNA, RNA, and proteins from heat-killed pathogenic cells. In 1944, Avery, McCarty, and MacLeod announced that DNA was the molecule that transformed bacteria.

Viruses consist of DNA (or sometimes RNA) contained in a protein coat. They reproduce by infecting a cell and commandeering that cell's metabolic machinery. **Bacteriophages**, or **phages**, are viruses that infect bacteria. In 1952, A. Hershey and M. Chase showed that DNA was the genetic material of a phage known as T2 that infects the bacterium *E. coli*.

In 1950, E. Chargaff noted that the percentages of the four nitrogenous bases in DNA were species specific. Chargaff also determined that the number of adenines and thymines was approximately equal, and the number of guanines and cytosines was also equal. The variation in base composition among species and the A = T and G = C properties of DNA became known as *Chargaff's rules.*

FOCUS QUESTION   13.1

Hershey and Chase devised an experiment using radioactive isotopes to determine whether it was a phage's DNA or protein that entered the bacteria and served as the genetic material of the T2 phage.

a. How did they label phage protein?

b. How did they label phage DNA?

After infecting separate samples of *E. coli* with the differently labeled T2 cells, they blended and centrifuged the samples to isolate the bacterial cells from the lighter viral particles.

c. Where was the radioactivity found in the samples with labeled phage protein?

d. Where was the radioactivity found in the samples with labeled phage DNA?

e. What did Hershey and Chase conclude from these results?

Building a Structural Model of DNA: Scientific Inquiry
By the early 1950s, the arrangement of covalent bonds in a nucleic acid single polymer was established, but the three-dimensional structure of DNA was yet to be determined.

In X-ray crystallography, an X-ray beam passed through a substance produces an X-ray diffraction photo with a pattern of spots that a crystallographer interprets as information about three-dimensional molecular structure. J. Watson saw an X-ray photo produced by R. Franklin that indicated the helical shape of DNA and that it consisted of two strands—thus the term **double helix**.

Watson and F. Crick constructed wire models of a double helix that had the paired nitrogenous bases on the inside of the helix and two sugar-phosphate chains running in opposite directions (**antiparallel**) on the outside. The helix makes one full turn every 3.4 nm; ten layers of nucleotide pairs, stacked 0.34 nm apart, are present in each turn of the helix.

To produce the molecule's uniform 2-nm width, a purine base must pair with a pyrimidine base. The side groups of the bases permit two hydrogen bonds to form between adenine and thymine, and three hydrogen bonds between guanine and cytosine. This complementary pairing explains Chargaff's rules. Van der Waals attractions between the closely stacked bases help hold the molecule together.

In 1953, Watson and Crick published a paper in *Nature* reporting the double helix as the molecular model for DNA.

FOCUS QUESTION   13.2

Review the structure of DNA by labeling the following diagrams.

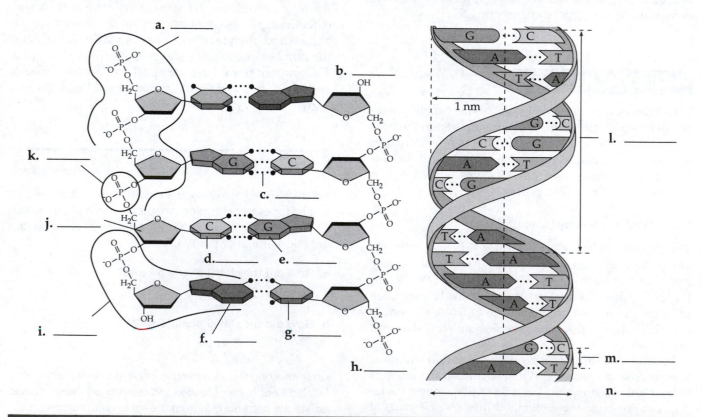

13.2 Many proteins work together in DNA replication and repair

The Basic Principle: Base Pairing to a Template Strand Watson and Crick noted that the base-pairing rule of DNA sets up a mechanism for its replication. Each side of the double helix is an exact complement to the other. When the two sides separate, each strand serves as a template for rebuilding a double-stranded molecule identical to the "parental" molecule.

The **semiconservative model** of DNA replication predicts that the two daughter DNA molecules each have one parental strand and one newly formed strand. In contrast, a conservative model predicts that the parental double helix reforms and the duplicated molecule is totally new, whereas a dispersive model predicts that all four strands of the two DNA molecules are a mixture of parental and new DNA.

M. Meselson and F. Stahl tested these models by growing *E. coli* in a medium with ^{15}N, a heavy isotope that the bacteria incorporated into their nitrogenous bases. Cells with labeled DNA were transferred to a medium with a lighter isotope, ^{14}N. Samples were removed after one and two generations of bacterial growth, and DNA was extracted and centrifuged. The locations of the density bands in the centrifuge tubes confirmed the semiconservative model of DNA replication.

FOCUS QUESTION 13.3

Using different colors for heavy (parental) and light (new) strands of DNA, sketch the DNA molecules formed in two replication cycles after *E. coli* were moved from medium containing ^{15}N to medium containing ^{14}N. Show the resulting density bands in the centrifuge tubes.

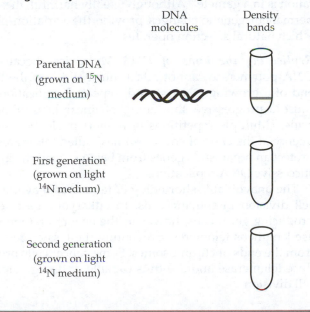

DNA molecules   Density bands

Parental DNA (grown on ^{15}N medium)

First generation (grown on light ^{14}N medium)

Second generation (grown on light ^{14}N medium)

DNA Replication: A Closer Look Replication of most bacterial chromosomes begins at a single **origin of replication**, where proteins that initiate replication bind to a specific sequence of nucleotides and separate the two strands to form a replication "bubble." Replication proceeds in both directions in the two Y-shaped **replication forks**. Eukaryotic chromosomes have many origins of replication.

An enzyme called **helicase** unwinds the helix and separates the parental strands at each replication fork. **Single-strand binding proteins** keep the separated strands apart while they serve as templates. **Topoisomerase** helps relieve the strain from the tighter twisting of DNA strands in front of helicase. An enzyme called **primase** joins about 5 to 10 RNA nucleotides base-paired to the parental strand to form the **primer** needed to start the new DNA strand.

Enzymes called **DNA polymerases** connect nucleotides to the growing end of a new DNA strand. DNA polymerase III and I are involved in replication in *E. coli*; at least 11 different DNA polymerases have been discovered so far in eukaryotes. A nucleotide lines up with its complementary base on the template strand; it loses two phosphate groups, and the hydrolysis of this pyrophosphate to two inorganic phosphates ($\circled{P}_i$) provides the energy for polymerization.

What do we mean when we say the two strands of a DNA molecule are antiparallel? Each strand has a polarity, and they run in opposite directions. The deoxyribose sugar of each nucleotide is connected to its own phosphate group at its 5' carbon and connects to the phosphate group of the adjacent nucleotide by its 3' carbon. Thus, the polarity of a strand of DNA runs from the 5' end where the first nucleotide's phosphate group is exposed to the 3' end where the nucleotide at the other end has a hydroxyl group attached to its 3' carbon.

FOCUS QUESTION 13.4

Look back to Focus Question 13.2 and label the 5' and 3' ends of the left strand of the DNA molecule.

DNA polymerases add nucleotides only to the free 3' end of a primer or growing DNA strand; DNA is replicated in a 5' → 3' direction. The **leading strand** is the new continuous strand being formed along one template strand as DNA polymerase III (DNA pol III) remains in the progressing replication fork.

The **lagging strand** is created as a series of short segments, called **Okazaki fragments**, which are elongated in the 5' → 3' direction away from the replication fork.

Each fragment on the lagging strand requires a primer. A continuous strand of DNA is produced after both DNA polymerase I (DNA pol I) replaces the RNA primer with DNA nucleotides and an enzyme called **DNA ligase** joins the sugar-phosphate backbones of the fragments.

The various proteins that function in replication form a large DNA replication complex. In eukaryotic cells, many such complexes may anchor to the nuclear matrix, and the DNA polymerase molecules may pull the parental DNA strands through them.

FOCUS QUESTION   13.5

In this diagram of bacterial DNA replication, label the following items: leading and lagging strands, Okazaki fragment, DNA pol III, DNA pol I, DNA ligase, helicase, single-strand binding proteins, primase, RNA primer, and 5' and 3' ends of parental DNA.

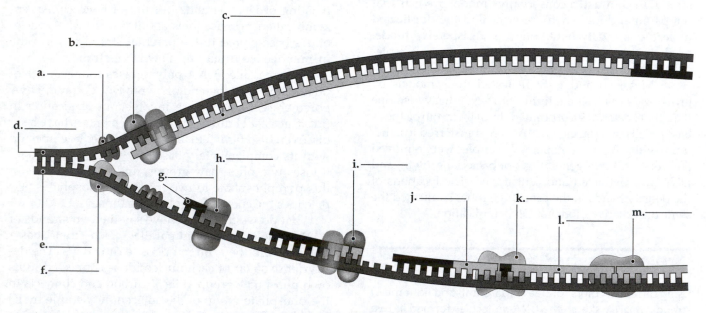

Proofreading and Repairing DNA   Initial pairing errors in nucleotide placement may occur as often as 1 per 100,000 base pairs. The amazing accuracy of DNA replication (one error in ten billion nucleotides) is achieved because DNA polymerases check each newly added nucleotide against its template and remove incorrect nucleotides. Other enzymes also fix incorrectly paired nucleotides, a process called **mismatch repair**.

A large number of different DNA repair enzymes monitor and repair damaged DNA. In **nucleotide excision repair**, the damaged strand is cut out by a **nuclease** and the gap is correctly filled through the action of a DNA polymerase and ligase. In skin cells, nucleotide excision repair frequently corrects *thymine dimers* caused by ultraviolet rays in sunlight.

Evolutionary Significance of Altered DNA Nucleotides   Mutations result when uncorrected mismatched nucleotides are replicated and the change is passed on to daughter cells (or to offspring, if the mutation is in a gamete). Although usually harmful, these permanent genetic changes provide the variation on which natural selection operates.

Replicating the Ends of DNA Molecules   Because DNA polymerase cannot attach nucleotides to the 5' end of a growing DNA strand, repeated replications cause a progressive shortening of linear DNA molecules. Multiple repetitions of a short nucleotide sequence at the ends of chromosomes, called **telomeres**, protect an organism's genes from being eroded during successive DNA replications.

The unavoidable shortening of telomeres may limit cell division in somatic cells. In eukaryotic gamete-producing germ cells, however, the enzyme **telomerase** lengthens telomeres, preventing the loss of genes from the ends of chromosomes. Some cancer cells produce telomerase and are thus capable of unregulated cell division.

13.3 A chromosome consists of a DNA molecule packed together with proteins

A bacterial cell's double-stranded, circular DNA molecule (associated with a small amount of protein) is supercoiled and tightly packed into a region of the cell called the **nucleoid**.

In eukaryotes, each chromosome consists of a single, extremely long DNA double helix precisely associated with a large amount of protein, forming a complex called **chromatin**. This DNA fits into the nucleus through a multilevel system of coiling and folding.

Histones are small, positively charged proteins that bind tightly to the negatively charged DNA. Unfolded chromatin appears as a string of beads, each bead a **nucleosome** consisting of the DNA helix wound around a protein core of four pairs of the main histone types. The "string" between beads is called *linker DNA*. The amino end (*histone tail*) of each of the eight histones extends outward. Also called the *10-nm fiber* because of its diameter, nucleosomes are the basic unit of DNA packing.

The *30-nm fiber* is a tightly coiled fiber of nucleosomes organized with the aid of a fifth histone attached near each nucleosome, the histone tails, and linker DNA. The *looped domain* is a loop of the 30-nm chromatin fiber attached to a protein scaffold, forming a *300-nm fiber*. In a metaphase chromosome, looped domains coil and fold, further compacting the chromatin into 700-nm wide chromatids, which are visible with a light microscope.

Some DNA packing is evident in interphase chromosomes. Certain regions of chromatin, called **heterochromatin**, are in a highly condensed state during interphase. The more open form of interphase chromatin, called **euchromatin**, is available for the transcription of genes.

FOCUS QUESTION 13.6

List the multiple levels of packing in a metaphase chromosome in order of increasing complexity.

13.4 Understanding DNA structure and replication makes genetic engineering possible

The complementary nature of the two strands of a DNA molecule is the basis of **nucleic acid hybridization**.

This base pairing of one strand of nucleic acid to complementary sequences of another strand is central to most techniques used in **genetic engineering**, the direct manipulation of genetic material for practical purposes.

DNA Cloning: Making Multiple Copies of a Gene or Other DNA Segment One of the approaches used in *DNA cloning* makes use of the **plasmids** of bacterial cells. **Recombinant DNA**, in which nucleotide sequences from different organisms or species are combined into the same DNA molecule *in vitro*, may be made by inserting "foreign" DNA into plasmids. These plasmids are put back into bacterial cells, where they will replicate as the *recombinant bacteria* reproduce to form clones of identical cells. Such **gene cloning** *amplifies* or provides multiple copies of the gene and may also be used to produce protein coded for by the foreign DNA.

Using Restriction Enzymes to Make Recombinant DNA Restriction endonucleases, or **restriction enzymes**, protect bacteria from phages or other organisms by cutting up foreign DNA. Most restriction enzymes recognize short nucleotide sequences, called **restriction sites**, and cut both strands of DNA at specific points within them. The cell protects its own DNA from restriction by methylating nucleotide bases within its own restriction sequences.

A restriction enzyme may cut a length of DNA at several places, and the resulting **restriction fragments** can be visualized using **gel electrophoresis**. Due to the negative charge of their phosphate groups, DNA molecules migrate through the electric field produced in a thin slab of agarose gel toward the positive electrode. How does this gel separate the different fragments? Linear molecules of DNA move at a rate inversely proportional to their length, producing band patterns in the gel containing fragments of decreasing size.

A restriction site is usually symmetrical; the same sequence, often of four to eight nucleotides, runs in opposite directions on the two strands. The most useful restriction enzymes cut the sugar-phosphate backbone in a staggered way, leaving **sticky ends** of short single-stranded sequences on both sides of the resulting restriction fragment.

DNA from different sources can be combined in the laboratory when the various DNA molecules are cut by the same restriction enzyme, and the complementary bases on the resulting sticky ends of the restriction fragments form hydrogen-bonded base pairs. DNA ligase is used to seal the strands together.

FOCUS QUESTION 13.7

Which of the following DNA sequences would most likely function as a restriction site for a restriction enzyme? Why?

··CAGCAG··   ··GTGCTG··   ··GAATTC··
··GTCGTC··   ··CACGAC··   ··CTTAAG··

Cloning vectors are DNA molecules that can move foreign DNA into a cell and replicate there. Recombinant plasmids returned to bacterial cells will replicate the foreign DNA as the bacteria reproduce.

FOCUS QUESTION 13.8

The following schematic diagram depicts an overview of gene cloning. Identify components **a–g**. Briefly describe what is happening in the three numbered steps of the process. For what two purposes might the bacteria produced in step 3 be used?

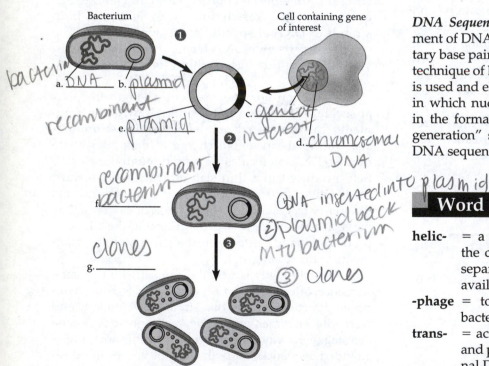

Amplifying DNA in Vitro: The Polymerase Chain Reaction (PCR) and Its Use in Cloning The **polymerase chain reaction (PCR)** can produce billions of copies of a section of DNA in only a few hours. DNA containing the region of interest is incubated with the four nucleotides, a special heat-resistant type of DNA polymerase, and specially synthesized primers that bind upstream from the target sequence on both DNA strands. The solution is heated to denature (separate) the DNA strands, then cooled so the primers can anneal (hydrogen-bond) to complementary sequences. DNA polymerase then adds nucleotides to the 3′ ends of the primers. The solution is heated again and the process repeated. The desired DNA segment need not be purified from the starting material, and very small samples can be used.

FOCUS QUESTION 13.9

a. Why is PCR often used prior to cloning a gene in cells?

Need multiple genes

b. Why even bother cloning genes in cells, given that PCR produces so many copies so fast?

PCR isn't the most accurate

DNA Sequencing The nucleotide sequence of a segment of DNA can be determined based on complementary base pairing. In the so-called *sequencing by synthesis* technique of **DNA sequencing**, a single template strand is used and electronic monitors determine the sequence in which nucleotides are added by DNA polymerase in the formation of a complementary strand. "Third-generation" sequencing techniques continue to make DNA sequencing faster and less expensive.

Word Roots

helic- = a spiral (*helicase:* an enzyme that untwists the double helix of DNA at replication forks, separating the two strands and making them available as template strands)

-phage = to eat (*bacteriophage:* a virus that infects bacteria)

trans- = across (*transformation:* a change in genotype and phenotype due to the assimilation of external DNA by a cell)

Structure Your Knowledge

1. Summarize the evidence and techniques Watson and Crick used to deduce the double-helix structure of DNA.

2. Review your understanding of DNA replication by describing the key enzymes and proteins (in the order of their functioning) that direct replication.

3. Describe how the complementary nature of DNA strands forms the basis for many of the techniques of genetic engineering.

Test Your Knowledge

MULTIPLE CHOICE: *Choose the one best answer.*

1. One of the reasons most scientists believed proteins were the carriers of genetic information was that
   a. proteins are more heat stable than nucleic acids.
   b. the protein content of duplicating cells always doubles prior to division.
   c. proteins are much more complex and heterogeneous than nucleic acids.
   d. early experimental evidence pointed to proteins as the hereditary material.
   e. proteins are found in DNA.

2. Transformation involves
   a. the uptake of external genetic material, often from one bacterial strain to another.
   b. the creation of a strand of RNA from a DNA molecule.
   c. the infection of bacterial cells by phage.
   d. the type of semiconservative replication shown by DNA.
   e. the replication of DNA along the lagging strand.

3. The DNA of an organism has thymine as 20% of its bases. What percentage of its bases would be guanine?
   a. 20%
   b. 30%
   c. 40%
   d. 60%
   e. 80%

4. In his work with pneumonia-causing bacteria, Griffith found that
   a. DNA was the transforming agent.
   b. the pathogenic and harmless strains mated.
   c. heat-killed harmless cells could cause pneumonia when mixed with heat-killed pathogenic cells.
   d. a substance was transferred to harmless cells to transform them into pathogenic cells.
   e. a T2 phage transformed harmless cells to pathogenic cells.

5. T2 phage is grown in *E. coli* with radioactive phosphorus and then allowed to infect other *E. coli*. The culture is blended to separate the viral coats from the bacterial cells and then centrifuged. Which of the following statements *best* describes the expected results of such an experiment?
   a. Both viral and bacterial DNA molecules are labeled; radioactivity is found in the liquid above the pellet.
   b. Viral DNA is labeled; radioactivity is found in the pellet.
   c. Viral proteins are labeled; radioactivity is found in the liquid but not in the pellet.
   d. Both viral and bacterial proteins are labeled; radioactivity is present in both the liquid and the pellet.
   e. The virus destroyed the bacteria; no pellet is formed.

6. Watson and Crick concluded that each base could not pair with itself because
   a. there would not be room for the helix to make a full turn every 3.4 nm.
   b. the uniform width of 2 nm would not permit two purines or two pyrimidines to pair together.
   c. the bases could not be stacked 0.34 nm apart.
   d. identical bases could not hydrogen-bond together.
   e. they would be on antiparallel strands.

7. In their classic experiment, Meselson and Stahl
   a. provided evidence for the semiconservative model of DNA replication.
   b. were able to separate phage protein coats from *E. coli* by using a blender.
   c. found that DNA labeled with ^{15}N was of intermediate density.
   d. grew *E. coli* on labeled phosphorus and sulfur.
   e. found that DNA composition was species specific.

8. The continuous elongation of a new DNA strand along one of the template strands of DNA
   a. requires the action of DNA ligase as well as polymerase.
   b. occurs because DNA ligase can only elongate in the $5' \rightarrow 3'$ direction.
   c. occurs on the leading strand.
   d. occurs on the lagging strand.
   e. a, b, and c are correct.

9. Which of the following statements about DNA polymerase is *incorrect*?
   a. It forms the bonds between complementary base pairs.
   b. It is able to proofread and correct errors in base pairing.
   c. It is unable to initiate synthesis; it requires an RNA primer.
   d. It only works in the $5' \rightarrow 3'$ direction.
   e. It is found in eukaryotes and prokaryotes.

10. Thymine dimers—covalent links between adjacent thymine bases in DNA—may be induced by UV light. When these dimers occur, they are repaired by
   a. excision enzymes (nucleases).
   b. DNA polymerase.
   c. ligase.
   d. primase.
   e. a, b, and c are all needed.

11. How does DNA synthesis along the lagging strand differ from that on the leading strand?
   a. Nucleotides are added to the 5′ end instead of the 3′ end.
   b. Ligase is the enzyme that polymerizes DNA on the lagging strand.
   c. An RNA primer is needed on the lagging strand but not on the leading strand.
   d. Okazaki fragments, which each grow 5′ → 3′, must be joined along the lagging strand.
   e. Helicase synthesizes Okazaki fragments, which are then joined by ligase.

12. Which of the following enzymes or proteins is paired with an incorrect or inaccurate function?
   a. helicase—unwinds and separates parental double helix
   b. telomerase—adds telomere repetitions to ends of chromosomes
   c. single-strand binding protein—holds strands of unwound DNA apart and straight
   d. nuclease—cuts out (excises) damaged DNA strands
   e. primase—forms DNA primer to start replication

Use the following diagram to answer questions 13 through 16.

13. Which letter indicates the 5′ end of this single DNA strand?
   a.
   b.
   c.
   d.
   e.

14. At which letter would the next nucleotide be added?
   a.
   b.
   c.
   d.
   e.

15. Which letter indicates a phosphodiester bond formed by DNA polymerase?
   a.
   b.
   c.
   d.
   e.

16. The base sequence of the DNA strand made from this template would be (from top to bottom)
   a. A T C.
   b. C G A.
   c. T A C.
   d. U A C.
   e. A T G.

17. Which of the following statements about telomeres is correct?
   a. They are ever-shortening tips of chromosomes that may signal cells to stop dividing at maturity.
   b. They are highly repetitive sequences at the tips of chromosomes that protect the lagging strand during replication.
   c. They are repetitive sequences of nucleotides at the centromere region of a chromosome.
   d. They are enzymes in germ cells that allow these cells to undergo repeated divisions.
   e. Both a and b are correct.

18. You are trying to test your hypothesis that DNA replication is *conservative*—that is, the parental strands separate, newly made complementary strands join together to make a new DNA molecule, and the parental strands then rejoin. You take a sample of *E. coli* grown in a medium containing only heavy nitrogen (^{15}N) and transfer it to a medium containing light nitrogen (^{14}N). After allowing time for only one DNA replication, you centrifuge a sample and compare the density

band(s) formed to the bands formed from bacteria grown on either normal ^{14}N or ^{15}N medium. Which band location would support your hypothesis of *conservative* DNA replication?

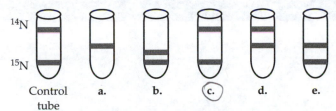

Control tube   a.   b.   c.   d.   e.

Bands after one replication on ^{14}N

19. Given the experimental procedure explained in question 18, which centrifuge tube (obtained after one DNA replication) would represent the band distribution indicating that DNA replication is *semiconservative*? a

20. If the following structures were put in order from smallest to largest, which structure would be in the middle of that size range?
    a. looped domain
    b. histone
    c. nucleosome
    d. 30-nm fiber
    e. metaphase chromosome

21. The role of restriction enzymes in DNA technology is to
    a. provide a vector for the transfer of recombinant DNA.
    b. provide template strands for the sequencing by synthesis technique of DNA sequencing.
    c. produce a cut (usually staggered) at specific restriction sites on DNA.
    d. reseal "sticky ends" after base pairing of complementary bases.
    e. denature DNA into single strands that can hybridize with complementary sequences.

22. Which of the following DNA sequences would most likely be a restriction site?
    a. AACCGG
       TTGGCC
    b. GGTTGG
       CCAACC
    c. AAGG
       TTCC
    d. AATTCCGG
       TTAAGGCC
    e. GAATTC
       CTTAAG

23. If the first three nucleotides in a six-nucleotide restriction site are CTG, what would the next three nucleotides most likely be?
    a. AGG
    b. GTC
    c. CTG
    d. CAG
    e. GAC

24. The following segment of DNA has restriction sites I and II, which create restriction fragments **a, b,** and **c.** Which of the following gels produced by electrophoresis would represent the separation and identity of these fragments?

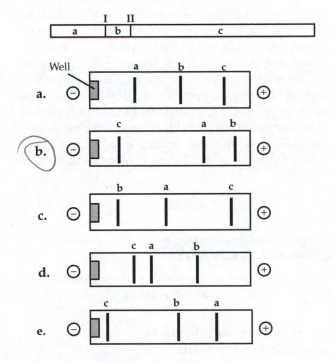

Use the following choices to answer questions 25 through 28.
    a. restriction enzyme
    b. primase
    c. DNA ligase
    d. DNA polymerase
    e. RNA polymerase

25. Which is the first enzyme used in making recombinant plasmids? a

26. Which is the last enzyme involved in making recombinant plasmids? C

27. Which enzyme is used in the polymerase chain reaction? d

28. Which enzyme is used in DNA sequencing? d

Chapter 14

Gene Expression: From Gene to Protein

Chapter Focus

This chapter examines the pathway from DNA to RNA to proteins. The instructions of DNA are transcribed to a sequence of codons in mRNA. In eukaryotes, mRNA is processed before it leaves the nucleus. Complexed with ribosomes, mRNA is translated as a sequence of amino acids in a polypeptide as tRNAs match their anticodons to the mRNA codons. Mutations, which alter the nucleotide pairs in DNA, may alter the protein product. A gene may be defined as a DNA sequence whose final product is either a polypeptide or an RNA molecule.

Chapter Review

The DNA-directed synthesis of proteins (or sometimes just RNA) is called **gene expression**.

14.1 Genes specify proteins via transcription and translation

Evidence from the Study of Metabolic Defects In 1909, A. Garrod first suggested that genes determine phenotype through the action of enzymes, reasoning that inherited diseases were caused by an inability to make certain enzymes.

G. Beadle and E. Tatum treated cells of the haploid bread mold, *Neurospora crassa*, with X-rays to produce "nutritional mutants" that were unable to grow on *minimal media*. By transferring samples of these mutants growing on *complete growth medium* to combinations of minimal medium and one added nutrient, they were able to identify the specific metabolic defect for each mutant. Experiments that supplemented different precursors of a particular metabolic pathway indicated that different mutants were blocked at a different enzymatic step in that pathway. Such research supported Beadle and Tatum's *one gene–one enzyme hypothesis*.

Biologists revised this hypothesis to one gene–one protein. Because many proteins consist of more than one polypeptide chain, the axiom changed to *one gene–one polypeptide*. (Exceptions include genes that code for more than one protein due to alternative splicing and genes that code for RNA molecules.)

Basic Principles of Transcription and Translation RNA is the link between a gene and the protein for which it codes.

Transcription is the transfer of information from DNA to **messenger RNA (mRNA)**. **Translation** transfers information from mRNA to a polypeptide, changing from the language of nucleotides to that of amino acids. **Ribosomes** are the sites of translation—the sites of the synthesis of a polypeptide.

In prokaryotes, which lack a nucleus, transcription and translation can occur simultaneously. In eukaryotes, the mRNA, called *pre-mRNA*, is processed before it exits the nucleus and enters the cytoplasm, where translation occurs. The initial RNA transcript of any gene is called the **primary transcript**.

FOCUS QUESTION   14.1

a. In what three ways does RNA differ from DNA?

b. Fill in the following sequence in the flow of genetic information, often called the *central dogma*. Above each arrow, write the name of the process involved.

$$\underline{\hspace{3cm}} \rightarrow \underline{\hspace{3cm}} \rightarrow \underline{\hspace{3cm}}$$

The Genetic Code A sequence of three nucleotides provides 4^3 (64) possible unique sequences of nucleotides, more than enough to code for the 20 amino acids. The translation of nucleotides into amino acids uses a **triplet code** to specify each amino acid.

The nucleotide base triplets along the **template strand** of a gene are transcribed into complementary mRNA **codons**. One strand of a DNA molecule can serve as the template strand for one gene and the nontemplate strand for another. The mRNA is complementary to the DNA template because its nucleotides follow the same base-pairing rules, with the exception that uracil substitutes for thymine in RNA. The term *codon* can also refer to the DNA nucleotide triplets on the *nontemplate* strand, which are identical to the mRNA, except that they have T instead of U.

During translation, the sequence of codons, read in the 5′ → 3′ direction, determines the sequence of amino acids in the polypeptide.

How did molecular biologists crack the genetic code? In the early 1960s, M. Nirenberg added artificial "poly U" mRNA to a test tube containing all the biochemical ingredients necessary for protein synthesis and obtained a polypeptide containing a single amino acid. By the mid-1960s, molecular biologists had deciphered all 64 codons. Three codons function as stop signals, or termination codons. The codon AUG both codes for methionine and functions as an initiation codon, a start signal for translation.

The code is often redundant, meaning that more than one codon may specify a single amino acid. The code is never ambiguous: No codon specifies two different amino acids.

The nucleotide sequence on mRNA is read in the correct **reading frame**, starting at a start codon and reading each triplet sequentially.

FOCUS QUESTION 14.2

Practice using the codon table in your textbook. Determine the amino acid sequence for a polypeptide coded for by the following mRNA transcript (written 5′ → 3′):

AUGCCUGACUUUAAGUAG

The genetic code of codons and their corresponding amino acids is almost universal. A bacterial cell can translate the genetic messages of human cells. The near universality of a common genetic language provides compelling evidence of the antiquity of the code and the evolutionary connection of all living organisms.

14.2 Transcription is the DNA-directed synthesis of RNA: *a closer look*

Molecular Components of Transcription The **promoter** is the DNA sequence where **RNA polymerase** attaches and initiates transcription. RNA polymerase joins RNA nucleotides that are complementary to the DNA template strand in a 5′ → 3′ direction. In bacteria, the **terminator** is the sequence that signals the end of transcription. A **transcription unit** is the sequence of DNA that is transcribed into one RNA molecule.

Bacteria have one type of RNA polymerase. Eukaryotes have three types; the one that synthesizes mRNA is called RNA polymerase II.

Synthesis of an RNA Transcript The specific binding of RNA polymerase to the promoter determines where transcription starts and which DNA strand is used as the template. The promoter includes the transcription **start point** and recognition sequences, such as the **TATA box** common in eukaryotes, upstream from the start point. In eukaryotes, **transcription factors** must first recognize and bind to the promoter before RNA polymerase II can attach, at which point the assembly is called the **transcription initiation complex**.

RNA polymerase untwists the double helix, exposing DNA nucleotides for base pairing with RNA nucleotides, and joins the nucleotides to the 3′ end of the growing polymer. The new RNA peels away from the DNA template, and the DNA double helix re-forms. Several molecules of RNA polymerase may be transcribing simultaneously along a gene, enabling a cell to produce large quantities of mRNA.

In prokaryotes, transcription ends after RNA polymerase transcribes the terminator sequence. In eukaryotes, polymerase continues past a polyadenylation signal sequence (coding for AAUAAA), and proteins cut loose the pre-mRNA.

FOCUS QUESTION 14.3

Describe the key steps of transcription as they occur in eukaryotes:

a.

b.

c.

14.3   Eukaryotic cells modify RNA after transcription

In eukaryotes, pre-mRNA is modified by **RNA processing** before it leaves the nucleus.

Alteration of mRNA Ends   A modified guanine nucleotide is attached to the 5′ end of a pre-mRNA, and a string of adenine nucleotides, called a **poly-A tail**, is added to the 3′ end. The **5′ cap** and poly-A tail may facilitate transport of mRNA from the nucleus, aid ribosome attachment, and protect the ends of mRNA from hydrolytic enzymes. The cap and tail are attached to the untranslated regions (UTRs) at the 5′ and 3′ ends.

Split Genes and RNA Splicing   Long segments of noncoding nucleotide sequences, known as **introns** or intervening sequences, occur within the boundaries of eukaryotic genes. The remaining coding regions are called **exons**, since they are expressed in protein synthesis (with the exception of the 5′ and 3′ UTRs, which are not translated). After a primary transcript is made of a gene, introns are removed and exons joined together before the mRNA leaves the nucleus—a process called **RNA splicing**.

Alternative RNA splicing allows different polypeptides to be produced from a single gene, depending on which exons are combined during RNA processing.

The signals for RNA splicing are sets of a few nucleotides at either end of each intron. Small RNAs within a large protein complex called a **spliceosome** snip an intron out of the RNA transcript and connect the adjoining exons.

Such RNA molecules that act as enzymes are called **ribozymes**. In some cases of RNA splicing, intron RNA catalyzes its own removal. What three properties of RNA relate to its ability to function as an enzyme? RNA is single-stranded and can base-pair with itself, forming a specific three-dimensional structure; some of its bases contain functional groups that can participate in catalysis; and it can hydrogen-bond with other nucleic acid molecules, allowing it to precisely locate splicing regions.

FOCUS QUESTION   14.4

How does the mRNA that leaves the nucleus differ from pre-mRNA?

14.4   Translation is the RNA-directed synthesis of a polypeptide: a closer look

Molecular Components of Translation   **Transfer RNA (tRNA)** molecules carry amino acids to ribosomes, where they are added to a growing polypeptide. Each tRNA carries a specific amino acid and has a nucleotide triplet, called an **anticodon**, that base-pairs with a complementary codon on mRNA, thus assuring that amino acids are arranged in the sequence prescribed by the transcription from DNA.

Transfer RNA is transcribed in the nucleus of a eukaryote and moves into the cytoplasm, where it can be used repeatedly. Due to hydrogen bonding between complementary nucleotide base sequences, these single-stranded, short RNA molecules fold into a three-dimensional, roughly L-shaped structure. The anticodon is at one end of the L; the 3′ end is the attachment site for its specific amino acid.

Each amino acid has a specific **aminoacyl-tRNA synthetase** that attaches it to its appropriate tRNA molecules to create an aminoacyl tRNA, or charged tRNA. The hydrolysis of ATP drives this process.

Sixty-one codons for amino acids can be read from mRNA, but there are only about 45 different tRNA molecules. A phenomenon known as **wobble** enables the third nucleotide of some tRNA anticodons to pair with more than one kind of nucleotide in the codon. Thus, a tRNA may recognize more than one mRNA codon, all of which code for the same amino acid carried by that tRNA.

FOCUS QUESTION   14.5

Using some of the codons and the amino acids you identified in Focus Question 14.2, fill in the following table.

DNA Triplet 3′→ 5′	mRNA Codon 5′→ 3′	Anticodon 3′→ 5′	Amino Acid
			methionine
		GCA	
TTC			
	UAG		

Ribosomes facilitate the specific pairing of tRNA anticodons with mRNA codons during protein synthesis. Ribosomes consist of a large and a small subunit, each composed of proteins and a form of RNA called **ribosomal RNA (rRNA)**. Subunits are constructed in the nucleolus in eukaryotes. Prokaryotic ribosomes differ enough in molecular composition that some antibiotics can inhibit them without affecting eukaryotic ribosomes.

A large and a small subunit join to form a ribosome when they attach to an mRNA molecule. Ribosomes have a binding site for mRNA and three tRNA binding sites: a **P site** (*p*eptidyl tRNA-binding site) that holds the tRNA carrying the growing polypeptide chain, an **A site** (*a*minoacyl tRNA-binding site) that holds the tRNA carrying the amino acid to be added next, and an **E site** (*e*xit site) from which discharged tRNAs leave the ribosome. A ribosome also has an *exit tunnel* through which the growing polypeptide passes out of the ribosome.

Building a Polypeptide Each of the three stages of protein synthesis—initiation, elongation, and termination—requires the aid of protein "factors." The first two stages also require energy provided by the hydrolysis of GTP (guanosine triphosphate).

The initiation stage begins as the small subunit of the ribosome binds to an mRNA and an initiator tRNA carrying methionine, which attaches to the start codon AUG on the mRNA. With the aid of proteins called *initiation factors* and the hydrolysis of GTP, the large subunit of the ribosome attaches to the small one, forming a *translation initiation complex*. The initiator tRNA fits into the P site.

The addition of amino acids in the elongation stage involves several proteins called *elongation factors* and occurs in a three-step cycle.

In codon recognition, an aminoacyl-tRNA base-pairs with the mRNA codon in the A binding site. This step requires energy from the hydrolysis of GTP.

In the peptide bond formation step, an RNA molecule of the large subunit catalyzes the formation of a peptide bond between the carboxyl end of the polypeptide held in the P site and the amino group of the new amino acid in the A site. The polypeptide is now held by the tRNA in the A site.

In translocation, the tRNA carrying the growing polypeptide is translocated to the P site, a process requiring energy from the hydrolysis of GTP. The empty tRNA from the P site is moved to the E site and released. The next mRNA codon moves into the A site as the mRNA moves through the ribosome.

Termination occurs when a stop codon—UAA, UAG, or UGA—reaches the A site of the ribosome. A *release factor* binds to the stop codon and hydrolyzes the bond between the polypeptide and the tRNA in the P site. The completed polypeptide leaves through the exit tunnel of the large subunit. With the hydrolysis of GTP and the aid of other protein factors, the two ribosomal subunits and other components dissociate. Review translation by completing Focus Question 14.6.

FOCUS QUESTION 14.6

In the following diagrams of polypeptide synthesis, name the stages (1–4), identify the components (a–l), and then briefly describe what happens in each stage. (This diagram does not include the initiation stage.)

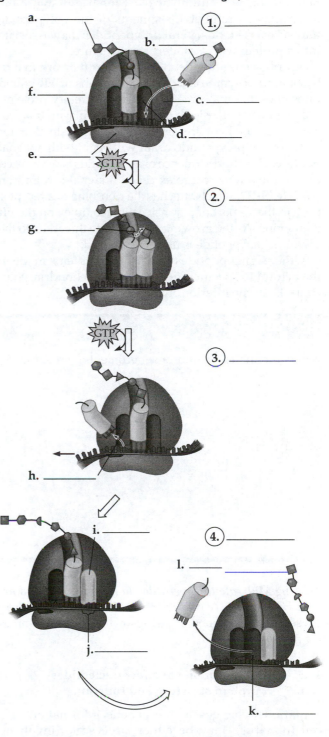

Completing and Targeting the Functional Protein

During and following translation, a polypeptide folds spontaneously into its secondary and tertiary structures. Chaperone proteins often facilitate the correct folding.

The protein may need to undergo *post-translational modifications:* Amino acids may be chemically modified; one or more amino acids at the beginning of the chain may be removed; segments of the polypeptide may be excised; or several polypeptides may associate into a protein with a quaternary structure.

All ribosomes are identical, whether they are free ribosomes that synthesize cytosolic proteins or ER-bound ribosomes that make membrane and secretory proteins. Polypeptide synthesis begins in the cytoplasm. If a protein is destined for the endomembrane system or for secretion, its polypeptide chain will begin with a **signal peptide**. This short sequence of amino acids is recognized by a protein-RNA complex called a **signal-recognition particle (SRP)**, which attaches the ribosome to a receptor protein that is part of a translocation complex on the ER membrane. As the growing polypeptide threads into the ER, the signal peptide is usually removed.

Other signal peptides direct some proteins made in the cytosol to specific sites such as mitochondria, chloroplasts, or the interior of the nucleus.

FOCUS QUESTION   14.7

What determines if a ribosome becomes bound to the ER?

Making Multiple Polypeptides in Bacteria and Eukaryotes

An mRNA may be translated simultaneously by several ribosomes in strings called polyribosomes (or *polysomes*).

14.5   Mutations of one or a few nucleotides can affect protein structure and function

Mutations, changes in the genetic information of a cell (or virus), may be either large scale (involving long segments of a chromosome) or small scale (such as **point mutations**, which affect just one nucleotide pair). If the mutation occurs in a cell that gives rise to a gamete, it may be passed on to offspring.

Types of Small-Scale Mutations

This first type of small-scale mutation is a **nucleotide-pair substitution**, in which one nucleotide and its complementary partner is replaced with another pair of nucleotides. Due to the redundancy of the genetic code, some substitutions do not change the amino acid translation and are called **silent mutations**. A **missense mutation**, which results in the insertion of a different amino acid, may not alter the character of the protein if the new amino acid has similar properties or is not located in a region crucial to that protein's function.

A nucleotide-pair substitution that results in a different amino acid in a critical portion of a protein, such as the active site of an enzyme, may significantly impair protein function. Occasionally such a change proves beneficial.

Nonsense mutations occur when a point mutation changes an amino acid codon into a stop codon, prematurely halting the translation of the polypeptide chain and usually creating a nonfunctional protein.

The second type of small-scale mutations includes nucleotide-pair **insertions** or **deletions**. If these are not in multiples of three, they will alter the reading frame. All nucleotides downstream from the mutation will be improperly grouped into codons, creating extensive missense mutations and usually ending in nonsense. Such **frameshift mutations** almost always produce nonfunctional proteins.

Mutagens   *Spontaneous mutations* include nucleotide-pair substitutions, insertions, deletions, and longer mutations that occur during DNA replication, repair, or recombination. Physical agents (such as X-rays and UV light) and a variety of chemical agents can cause mutations and are called **mutagens**. Tests can measure the mutagenic effects of chemicals and thus their potential carcinogenic risk.

FOCUS QUESTION   14.8

Define the following terms, and explain what type of small-scale mutation could cause each of these types of mutations.

a. silent mutation

b. missense mutation

c. nonsense mutation

d. frameshift mutation

What Is a Gene? Revisiting the Question Our definition of a gene has evolved from Mendel's heritable factors, to Morgan's loci along chromosomes, to one gene–one polypeptide. Research continually refines our understanding of the structural and functional aspects of genes, which now include introns, promoters, and other regulatory regions. Currently, the best working definition of a gene is that it is a region of DNA whose final product is either a polypeptide or an RNA molecule.

Word Roots

anti- = opposite (*anticodon:* a nucleotide triplet at one end of a tRNA molecule that base-pairs with a particular complementary codon on an mRNA molecule)

exo- = out, outside, without (*exon:* a sequence within a primary transcript that remains in the RNA after RNA processing; also the region of DNA from which this sequence was transcribed)

intro- = within (*intron:* a noncoding, intervening sequence within a primary transcript that is removed during RNA processing; also the region of DNA from which this sequence was transcribed)

muta- = change; **-gen** = producing (*mutagen:* a chemical or physical agent that interacts with DNA and can cause a mutation)

trans- = across; **-script** = write (*transcription:* the synthesis of RNA using a DNA template)

Structure Your Knowledge

1. You have been introduced to several types of RNA in this chapter. List three of these types and their functions. (You may also recall a fourth type that functions in spliceosomes.) What explains the functional versatility of RNA molecules?

2. Make sure you understand and can explain the processes of transcription and translation. To help you review these processes, fill in the following table describing various aspects of eukaryotic gene expression (either by yourself or in a study group).

	Transcription	Translation
Template		
Location		
Molecules involved		
Enzymes involved		
Control—start and stop		
Product		
Product processing		

3. What is the genetic code? Explain redundancy and the wobble phenomenon. What is the significance of the fact that the genetic code is nearly universal?

4. Prepare a concept map showing the types and consequences of small-scale mutations.

Test Your Knowledge

MULTIPLE CHOICE: *Choose the one best answer.*

1. A series of studies on mutants of *Neurospora* identified three classes of mutants that needed arginine added to minimal media in order to grow. The production of arginine includes the following steps: precursor → ornithine → citrulline → arginine. What nutrient(s) have to be supplied to the mutants that had a defective enzyme for the ornithine → citrulline step in order for them to grow?
   a. the precursor
   b. ornithine
   c. citrulline
   d. either ornithine or citrulline
   e. the precursor, ornithine, and citrulline

2. Transcription involves the transfer of information from
   a. DNA to RNA.
   b. RNA to DNA.
   c. mRNA to an amino acid sequence.
   d. DNA to an amino acid sequence.
   e. the nucleus to the cytoplasm.

3. If the 5' → 3' nucleotide sequence on the nontemplate DNA strand is CAT, what is the corresponding codon on mRNA?
   a. UAC   d. GTA
   b. CAU   e. CAT
   c. GUA

4. A bacterial gene 600 nucleotides long can code at most for a polypeptide of how many amino acids?
   a. 100   d. 600
   b. 200   e. 1,800
   c. 300

5. RNA polymerase
   a. is the protein responsible for the production of ribonucleotides.
   b. is the enzyme that creates hydrogen bonds between nucleotides on the DNA template strand and their complementary RNA nucleotides.
   c. is the enzyme that transcribes exons but does not transcribe introns.
   d. is a ribozyme.
   e. moves along the template strand of DNA, elongating an RNA molecule in a $5' \rightarrow 3'$ direction.

6. How is the template strand for a particular gene determined?
   a. It is the DNA strand that runs from the $5' \rightarrow 3'$ direction.
   b. It is the DNA strand that runs from the $3' \rightarrow 5'$ direction.
   c. It is established by the promoter.
   d. It doesn't matter which strand is the template because they are complementary and will produce the same mRNA.
   e. It is signaled by a polyadenylation signal sequence.

7. Which enzyme synthesizes tRNA?
   a. DNA polymerase
   b. RNA polymerase
   c. reverse transcriptase
   d. aminoacyl-tRNA synthetase
   e. ribosomal RNA

8. Which of the following is *not* involved in the formation of a eukaryotic transcription initiation complex?
   a. TATA box
   b. transcription factors
   c. small RNA molecules
   d. RNA polymerase II
   e. promoter

9. Which of the following is *true* of RNA processing?
   a. Exons are excised before the mRNA is translated.
   b. The RNA transcript that leaves the nucleus may be much longer than the original transcript.
   c. Assemblies of protein and small RNAs, called spliceosomes, may catalyze splicing.
   d. Large quantities of rRNA are assembled into ribosomes.
   e. Signal peptides are added to the 5' end of the transcript.

10. All of the following are transcribed from DNA *except*
    a. exons.
    b. introns.
    c. tRNA.

d. 3' and 5' UTRs.
e. promoter.

11. A ribozyme is
    a. an exception to the one gene–one RNA molecule axiom.
    b. an enzyme that adds the 5' cap and poly-A tail to mRNA.
    c. an example of rearrangement of exons caused by alternative RNA splicing.
    d. an RNA molecule that functions as an enzyme.
    e. an enzyme that produces both small and large ribosomal subunits.

12. Which of the following would *not* be found in a bacterial cell?
    a. mRNA
    b. rRNA
    c. small RNAs in spliceosome
    d. RNA polymerase
    e. simultaneous transcription and translation

13. Which of the following is transcribed and then translated to form a protein product?
    a. a gene for tRNA
    b. an intron
    c. a gene for a transcription factor
    d. 5' and 3' UTRs
    e. a gene for rRNA

14. Transfer RNA
    a. translocates a growing polypeptide destined for export to the endoplasmic reticulum.
    b. binds to its specific amino acid in the active site of an aminoacyl-tRNA synthetase.
    c. has catalytic activity and is thus a ribozyme.
    d. is translated from mRNA.
    e. is produced in the nucleolus.

15. Place the following events in the synthesis of a polypeptide in the proper order.
    1. A peptide bond forms.
    2. An aminoacyl tRNA matches its anticodon to the codon in the A site.
    3. A tRNA translocates from the A site to the P site, and an unattached tRNA exits from the E site.
    4. The large subunit attaches to the small subunit, with the initiator tRNA in the P site.
    5. A small subunit binds to an mRNA and an initiator tRNA.
       a. 4-5-3-2-1
       b. 4-5-2-1-3
       c. 5-4-3-2-1
       d. 5-4-1-2-3
       e. 5-4-2-1-3

16. Translocation in the process of translation involves
    a. the hydrolysis of GTP.
    b. movement of the tRNA in the A site to the P site.
    c. movement along the mRNA a distance of one triplet.
    d. the release of the unattached tRNA from the E site.
    e. all of the above.

17. Which of the following types of molecules catalyzes the formation of a peptide bond?
    a. RNA polymerase
    b. rRNA
    c. mRNA
    d. aminoacyl-tRNA synthetase
    e. tRNA

18. Which of the following is *not* true of an anticodon?
    a. It consists of three nucleotides.
    b. It lines up in the 5′ → 3′ direction along the 5′ → 3′ mRNA strand.
    c. It extends from one loop of a tRNA molecule.
    d. It may pair with more than one codon.
    e. Its base uracil base-pairs with adenine.

19. Changes in a polypeptide following translation may involve
    a. the addition of sugars or lipids to certain amino acids.
    b. the enzymatic addition of amino acids at the beginning of the chain.
    c. the removal of poly-A from the end of the chain.
    d. the addition of a 5′ cap of a modified guanosine.
    e. all of the above.

20. Several proteins may be produced at the same time from a single mRNA by
    a. the action of several ribosomes in a string, called a polyribosome.
    b. several RNA polymerase molecules working sequentially.
    c. signal peptides that associate ribosomes with rough ER.
    d. the action of several promoter regions.
    e. the involvement of multiple spliceosomes.

21. A signal peptide
    a. is most likely to be found on cytosolic proteins produced by bacterial cells.
    b. directs an mRNA molecule into the lumen of the ER.
    c. is a sign to bind the small ribosomal unit at the initiation codon.
    d. would be the first 20 or so amino acids of a protein destined for a membrane location or for secretion from the cell.
    e. is part of the UTR following the 5′ cap.

22. A nucleotide deletion early in the coding sequence of a gene would most likely result in
    a. a nonsense mutation.
    b. a frameshift mutation.
    c. multiple missense mutations.
    d. a nonfunctional protein.
    e. all of the above.

23. The type of mutation responsible for sickle-cell anemia is
    a. a silent mutation.
    b. a nucleotide-pair insertion.
    c. a point mutation.
    d. a nucleotide-pair substitution.
    e. Both c and d describe the type of mutation.

Use the following diagram of coupled transcription and translation in bacteria to answer questions 24 through 28.

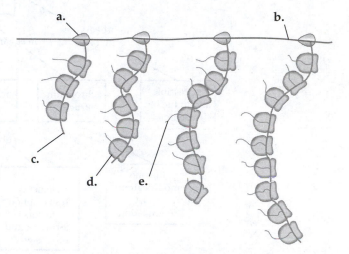

24. Which letter refers to an mRNA molecule?

25. Which letter refers to a forming polypeptide?

26. Which letter refers to RNA polymerase?

27. Which letter refers to a ribosome?

28. Which letters indicate structures or molecules containing nucleotides?
    a. a and b
    b. a, b, and d
    c. b, c, and d
    d. b, d, and e
    e. c and d

Chapter 15

Regulation of Gene Expression

Chapter Focus

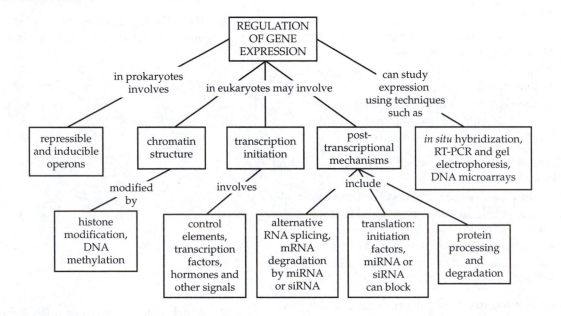

Chapter Review

15.1 Bacteria often respond to environmental change by regulating transcription

Bacteria can respond to short-term environmental fluctuations by regulating enzyme activity through *feedback inhibition*, usually of the first enzyme in a metabolic pathway. A second type of response involves regulation of enzyme production through the control of gene transcription in response to changing metabolic needs. F. Jacob and J. Monod described the *operon model* for gene regulation in 1961.

Operons: The Basic Concept   In bacteria, genes for the different enzymes of a single metabolic pathway may be grouped together into one transcription unit and served by a single promoter. Thus, these genes are *coordinately controlled*. An **operator** is a segment of DNA within or near the promoter that controls the access of RNA polymerase to the genes. An **operon** is the DNA segment that includes the clustered genes, the promoter, and the operator.

How is the transcription of the genes in an operon controlled? A **repressor** is a protein that binds to a specific operator, blocking attachment of RNA polymerase and thus turning an operon "off." **Regulatory genes** code for repressor proteins. These allosteric proteins, which may assume active or inactive shapes, are usually produced at a slow but continuous rate. The activity of the repressor protein may be determined by the presence or absence of a **corepressor**.

In the *trp* operon, tryptophan is the corepressor that binds to the *trp* repressor, changing it into its active shape, which has a high affinity for the *trp* operator and switches off the *trp* operon. What happens when the tryptophan concentration of the cell falls? Repressor proteins are no longer bound with tryptophan; the operator is no longer repressed; RNA polymerase attaches to the promoter; and mRNA for the enzymes needed for tryptophan synthesis is produced.

Repressible and Inducible Operons: Two Types of Negative Gene Regulation   The transcription of a *repressible operon,* such as the *trp* operon, is inhibited when a specific small molecule binds to and activates

a repressor. The transcription of an *inducible operon* is stimulated (induced) when a specific small molecule binds to and inactivates a repressor.

The *lac* operon, controlling lactose metabolism in *E. coli,* is an inducible operon that contains three genes. The *lac* repressor protein, coded for by the regulatory gene *lacI,* is innately active, binding to the *lac* operator and switching off the operon. Allolactose, an isomer of lactose, acts as an **inducer,** binding to and inactivating the repressor protein, so that the operon can be transcribed.

FOCUS QUESTION   15.1

In the following diagram of the *lac* operon, an inducible operon, identify components **a** through **i**.

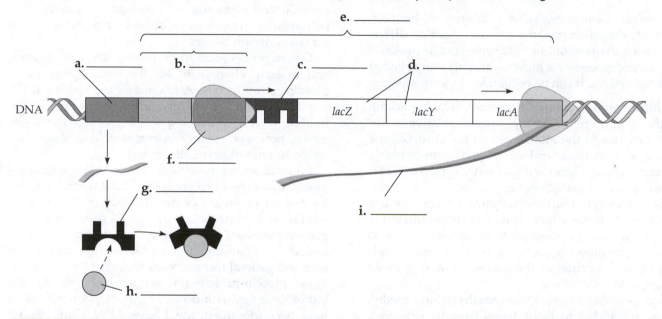

Both repressible and inducible operons involve *negative* control of gene expression involving the action of repressor proteins. Complete Focus Question 15.2 to help you distinguish between these two types of negative gene regulation.

FOCUS QUESTION   15.2

a. *Repressible enzymes* usually function in _____ pathways. The pathway's product serves as a _____ to activate the repressor and turn off enzyme synthesis, thus preventing overproduction of the product of the pathway. Genes for repressible enzymes are usually switched _____ and the repressor is synthesized in an _____ form.

b. *Inducible enzymes* usually function in _____ pathways. Nutrient molecules serve as _____ to stimulate production of the enzymes necessary for their breakdown. Genes for inducible enzymes are usually switched _____ and the repressor is synthesized in an _____ form.

Positive Gene Regulation   *E. coli* cells preferentially use glucose. Should glucose levels fall, transcription of operons for other catabolic pathways can be increased through the action of the allosteric regulatory protein *catabolite activator protein (CAP),* which is an **activator.** When glucose is scarce, **cyclic AMP (cAMP)** accumulates in the cell and binds with

CAP, changing it to its active shape. The active CAP attaches at the upstream end of the promoter region and stimulates transcription by facilitating the binding of RNA polymerase.

The regulation of the *lac* operon includes both negative control by the repressor protein that is inactivated by the presence of lactose, and positive control by CAP when complexed with cAMP.

15.2 Eukaryotic gene expression is regulated at many stages

All cells control the expression of their genes in response to their external and internal environments. Eukaryotic cells in multicellular organisms also regulate gene expression to produce specialized cells.

Differential Gene Expression Differences between cells with the same genome are the result of **differential gene expression**. In eukaryotes (as in prokaryotes), gene expression is most commonly controlled at transcription, but it can be regulated at various stages.

Regulation of Chromatin Structure DNA packing and the location of a gene's promoter (relative to nucleosomes and to the attachment to the chromosome scaffold) may help control which genes are available for transcription. Genes within highly condensed heterochromatin are usually not expressed.

Histone acetylation, the attachment of acetyl groups ($—COCH_3$) to histone tails, opens up chromatin structure, giving transcription proteins greater access to genes. The addition of methyl groups to histone tails leads to condensation of the chromatin and reduced transcription.

In a process called **DNA methylation**, methyl groups are added to DNA bases (usually cytosine). Genes are generally more heavily methylated in cells in which they are not expressed. Methylation records are usually passed on during subsequent DNA replications.

Epigenetic inheritance is the transmission of traits by mechanisms that do not involve changes in DNA sequences, such as chromatin modifications that affect gene expression in future generations of cells.

FOCUS QUESTION 15.3

a. Give an example of highly methylated and inactive DNA that is common in female mammalian cells.

b. Would the removal of acetyl groups from histone tails in a nucleosome increase or decrease the transcription of a gene?

Regulation of Transcription Initiation A typical eukaryotic gene consists of a promoter sequence, to which a *transcription initiation complex* that includes RNA polymerase II attaches, and a sequence of introns interspersed among the coding exons. After transcription, RNA processing of the pre-mRNA removes the introns and adds a 5' cap and a poly-A tail at the 3' end. **Control elements** are noncoding sequences that help regulate transcription by serving as binding sites for transcription factors.

General transcription factors bind to RNA polymerase and to each other to initiate the transcription of all protein-coding genes. A few may bind specific DNA sequences, such as the TATA box in the promoter. The binding of *specific transcription factors* to control elements, however, greatly increases or decreases transcription rates of particular genes.

Proximal control elements are located close to the promoter. **Enhancers** are groups of *distal control elements* located far upstream or downstream from a gene or within an intron. According to a current model, a protein-mediated bend in the DNA brings activators bound to enhancers into contact with *mediator proteins* and general transcription factors at the promoter. These protein-protein interactions assemble the initiation complex. Hundreds of transcription activators have been identified. Most have a DNA-binding domain and one or more activation domains that bind other regulatory proteins.

Some specific transcription factors act as repressors in various ways, such as by binding to specific control elements in an enhancer. Some activators and repressors influence chromatin structure by recruiting proteins that either acetylate histones or remove acetyl groups from histones. Most eukaryotic gene repression may occur by this *silencing* at the level of chromatin modification.

Only a dozen or so unique nucleotide sequences have been found in control elements, but each enhancer contains about ten different control elements. The *combination* of these elements ensures that a specific collection of activator proteins must be present to activate transcription. Thus, the specific activators and repressors made in a cell determine which genes are expressed.

Most eukaryotic genes coding for enzymes in the same metabolic pathway are scattered on the chromosomes. The coordinated control of these genes may involve a specific collection of control elements that are associated with each related gene and require the same activators.

For example, steroid hormones may activate multiple genes when they bind to specific receptor proteins within a cell that function as transcription activators for several genes. The same chemical signal will activate genes with the same control elements and thus coordinate gene expression.

FOCUS QUESTION 15.4

Label the components of the following diagram that illustrates how enhancers, activators, mediator proteins, and transcription factors facilitate the formation of a transcription initiation complex.

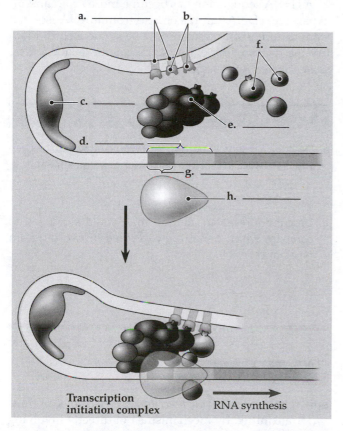

a. _____   b. _____

f. _____

c. _____

e. _____

d. _____

g. _____

h. _____

Transcription
initiation complex   RNA synthesis

Mechanisms of Post-Transcriptional Regulation Gene expression, measured by the amount of functional protein that is made, can be regulated at any post-transcriptional step. At the level of RNA processing, **alternative RNA splicing** may produce different mRNA molecules when regulatory proteins control

which segments of the primary transcript are chosen as introns and which are chosen as exons. More than 90% of human protein-coding genes probably undergo alternative splicing.

In contrast to prokaryotic mRNA, which is degraded after a few minutes, eukaryotic mRNA can last hours or even weeks. The untranslated region (UTR) at the 3' end of mRNA may contain nucleotide sequences that affect how long an mRNA lasts.

The translation of eukaryotic mRNA can be delayed by regulatory proteins that bind to the 5' or 3' UTR and prevent ribosome attachment. A great deal of mRNA is synthesized and stored in egg cells; translation may begin when enzymes add more A nucleotides to poly-A tails or when translation initiation factors are activated following fertilization.

Following translation, polypeptides are often cleaved or chemical groups added to yield an active protein. Some proteins must be transported to target locations.

Selective degradation of proteins may serve as a control mechanism in a cell. Molecules of the small protein ubiquitin are added to mark proteins for destruction.

FOCUS QUESTION 15.5

a. How is the process of alternate RNA splicing related to the control of gene expression?

b. The untranslated regions (UTR) at both the 5' and 3' ends of an mRNA may contribute to regulation of gene expression. Describe their effects.

15.3 Noncoding RNAs play multiple roles in controlling gene expression

Only 1.5% of the human genome is protein-coding DNA; an additional small percentage codes for genes for ribosomal and transfer RNA (rRNA and tRNA). Mounting evidence indicates that a large amount of the genome may be transcribed into non-protein-coding RNAs (*noncoding RNAs* or *ncRNAs*). A large population of small RNAs and longer RNA transcripts may be involved in regulating gene expression.

Effects on mRNAs by MicroRNAs and Small Interfering RNAs Small, single-stranded RNA molecules called **microRNAs (miRNAs)** are processed from longer RNA precursors. They associate with a complex of proteins and then base-pair with any mRNA that has seven to eight nucleotides of a complementary sequence. This miRNA-protein complex either blocks translation or degrades the mRNA. An estimated one-half of all human genes may be regulated by miRNAs.

Small interfering RNAs (siRNAs) appear to inhibit the expression of genes by the same mechanism as mi-RNAs. They differ from miRNAs in that many siRNAs may be produced from one long double-stranded RNA precursor. The experimental injection of double-stranded RNA into a cell can inhibit the expression genes with matching sequences and is called **RNA interference (RNAi)**. The cell's machinery cuts these longer double-stranded RNA molecules into many short siRNAs.

Chromatin Remodeling and Effects on Transcription by ncRNAs   In some yeasts, siRNAs are involved in the formation of heterochromatin. Evidence indicates that an RNA transcript produced from DNA in the centromeric region is copied into double-stranded RNA and associated with proteins. This complex then targets the centromere region, where it initiates chromatin remodeling into heterochromatin.

Newly discovered *piwi-associated RNAs (piRNAs)* also induce heterochromatin formation, thus blocking transcription of some transposons. They also appear to help re-establish proper methylation patterns in the genome during gamete formation.

FOCUS QUESTION   15.6

a. Describe how miRNAs regulate gene expression.

b. Do siRNAs regulate gene expression by affecting translation or transcription? Explain.

15.4   Researchers can monitor expression of specific genes

Base-pairing between complementary nucleotide sequences forms the basis for techniques that study the differential expression of genes.

Studying the Expression of Single Genes   Using the technique of **nucleic acid hybridization**, a **nucleic acid probe** (either RNA or DNA) complementary to the mRNA transcribed from a specific gene allows researchers to determine in which cells that particular gene is being expressed. The technique of *in situ* **hybridization** uses labeled probes to detect the locations of specific mRNAs within an intact organism.

Sample sets of mRNA can be studied using the technique called **reverse transcriptase-polymerase chain reaction (RT-PCR)**. The enzyme *reverse transcriptase* is used to make complementary copies of DNA from mRNAs—a type of *reverse transcript*. The resulting **complementary DNA (cDNA)** is amplified by PCR using primers specific for that gene, and gel electrophoresis produces bands of amplified DNA only in samples in which the gene has been transcribed to mRNA.

Studying Expression of Groups of Genes   In order to study patterns of gene expression, researchers isolate the mRNA made in different tissues or at different developmental stages, create cDNA using reverse transcriptase, and then use the cDNA as probes to explore collections of genomic DNA. Using this cDNA in **DNA microarray assays**, scientists can test all the genes expressed in a tissue for hybridization with short, single-stranded DNA fragments from thousands of genes arrayed on a grid (called a *DNA chip*).

As DNA sequencing has become more rapid and inexpensive, cDNA samples made from mRNA isolated from different sources can now be sequenced to identify the genes being expressed (a method called *RNA sequencing* or *RNA-seq*).

FOCUS QUESTION   15.7

a. Describe a research method that would allow you to identify the tissues in which a particular gene is expressed.

b. Describe a research method that would allow you to compare which genes are expressed in two different types of cells.

Word Roots

epi- = beside (*epigenetic inheritance*: inheritance of traits transmitted by mechanisms not directly involving the nucleotide sequence of a genome)

Structure Your Knowledge

1. This concept map may look complex, but working through it will help you review the mechanisms by which bacteria regulate gene expression in response to varying metabolic needs.

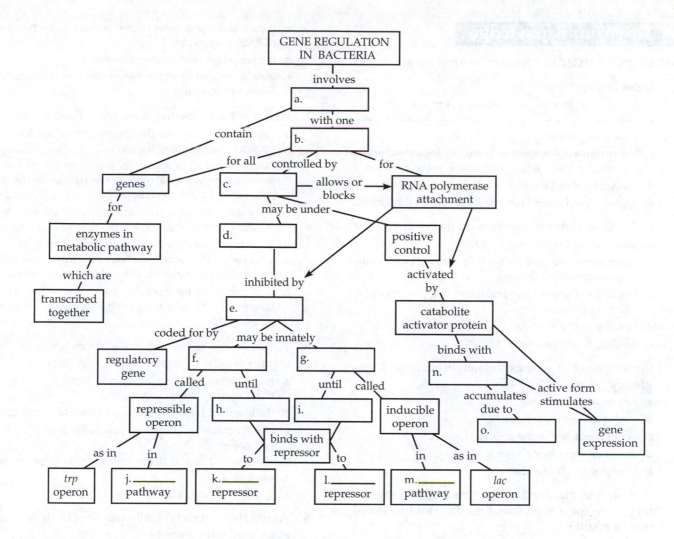

2. Fill in the following table to help you organize the major mechanisms that can regulate the expression of eukaryotic genes.

Level of Control	Examples
Chromatin modification	a.
Transcriptional regulation	b.
RNA processing	c.
RNA degradation	d.
Translational regulation	e.
Protein processing; degradation	f.

Test Your Knowledge

MULTIPLE CHOICE: *Choose the one best answer.*

1. Inducible enzymes
   a. are usually involved in anabolic pathways.
   b. are produced when a small molecule inactivates the repressor protein.
   c. are produced when an activator molecule enhances the attachment of RNA polymerase to the operator.
   d. are regulated by inherently inactive repressor molecules.
   e. are regulated almost entirely by feedback inhibition.

2. In *E. coli*, tryptophan switches off the *trp* operon by
   a. inactivating the repressor protein.
   b. inactivating the gene for the first enzyme in the pathway (feedback inhibition).
   c. binding to the repressor and increasing the latter's affinity for the operator.
   d. binding to the operator.
   e. binding to the promoter.

3. In the control of gene expression in bacteria, a regulatory gene
   a. has its own promoter.
   b. is transcribed continuously.
   c. is not contained in the operon it controls.
   d. codes for repressor proteins.
   e. is or does all of the above.

4. A mutation that renders the product of a regulatory gene for a repressible operon nonfunctional would result in
   a. continuous transcription of the genes of the operon.
   b. complete blocking of the attachment of RNA polymerase to the promoter.
   c. irreversible binding of the repressor to the operator.
   d. no difference in transcription rate when an activator protein was present.
   e. negative control of transcription.

5. The control of gene expression is more complex in eukaryotic cells because
   a. DNA is associated with protein.
   b. gene expression differentiates specialized cells.
   c. the chromosomes are linear and more numerous.
   d. operons are controlled by more than one promoter region.
   e. inhibitory or activating molecules may help regulate transcription.

6. DNA methylation of cytosine residues
   a. initiates the acetylation of histones.
   b. may be a mechanism of epigenetic inheritance when methylation patterns are repeated in daughter cells.
   c. occurs in the promoter region and enhances binding of RNA polymerase.
   d. is a signal for protein degradation.
   e. may be related to the transformation of proto-oncogenes to oncogenes.

7. Which of the following is *not* true of enhancers?
   a. They may be located thousands of nucleotides upstream from the genes they affect.
   b. When bound with activators, they interact with the promoter region and other transcription factors to produce an initiation complex.
   c. They may complex with steroid-activated receptor proteins, which selectively activate specific genes.
   d. They may coordinate the transcription of enzymes involved in the same metabolic pathway when they contain the same combination of control elements.
   e. Each gene may have several enhancers, and each enhancer may be associated with and regulate several genes.

8. Which of the following is *not* an example of the control of gene expression after transcription?
   a. mRNA stored in the cytoplasm needing activation of translation initiation factors
   b. the length of time mRNA lasts before it is degraded
   c. rRNA genes amplified in multiple copies in the genome
   d. alternative RNA splicing before mRNA leaves the nucleus
   e. splicing or modification of a polypeptide

9. A eukaryotic gene typically has all of the following associated with it *except*
   a. a promoter.
   b. an operator.
   c. enhancers.
   d. introns and exons.
   e. control elements.

10. Which of the following would you expect to find as part of a receptor protein that binds with a steroid hormone?
    a. a TATA box
    b. a domain that binds to DNA and protein-binding domains
    c. an activated operator region that allows attachment of RNA polymerase
    d. an enhancer sequence located at some distance upstream or downstream from the promoter
    e. transmembrane domains that facilitate the protein's localization in a plasma membrane

11. Which of the following statements explains why a larger portion of the DNA in a human cell is transcribed than would be predicted by the 1.5% of the genome that is protein-coding DNA?

    a. Multiple enhancer regions are being transcribed to amplify the transcription of protein-coding genes.

    b. Much of this non-protein-coding RNA functions to regulate the translation or degradation of mRNAs.

    c. Many of these transcriptions produce double-stranded siRNAs that regulate the transcription of other genes.

    d. The additional DNA that is transcribed represents introns that are excised from the primary transcript in the production of mRNA.

    e. These transcriptions are of noncoding "junk" DNA that is a remnant of mutated protein-coding segments, and the transcripts are degraded by nuclear enzymes.

12. How is the coordinated transcription of eukaryotic genes involved in the same pathway regulated?

    a. The genes are transcribed in one transcription unit, although each gene has its own promoter.

    b. The genes are located in the same region of the chromosome, and enzymes acetylate the entire region so that transcription may begin.

    c. All the genes respond to the same general transcription factors, although they may respond to different specific transcription factors.

    d. A steroid hormone selectively binds to the promoters for all the genes.

    e. The genes have the same combination of control elements in their enhancers.

13. You have affixed the chromosomes from a cell onto a microscope slide. Which of the following would *not* make a good radioactively labeled probe to help map a particular gene to one of those chromosomes? (Assume that the DNA of chromosomes and probes is single stranded.)

    a. cDNA made from the mRNA transcribed from the gene

    b. a portion of the amino acid sequence of that protein

    c. mRNA transcribed from the gene

    d. a piece of the chromosome on which the gene is located

    e. a sequence of nucleotides determined from a known sequence of amino acids in the protein product of the gene

14. Could you use cDNA produced from a fertilized egg to produce a DNA microarray "chip" of all the genes in that organism? Explain.

    a. Yes, because a fertilized egg is the only cell of an organism that has a complete set of genes.

    b. Yes, but you would have to produce the cDNA from the chromosomal DNA to make sure all genes were included in the grid.

    c. Yes, but only if the cell has been activated to begin protein synthesis.

    d. No, because cDNA is produced using the mRNA that a cell is producing, and no cell expresses all its genes at the same time.

    e. No, because the chromosomes of a fertilized egg are highly methylated and condensed, so no mRNA is being made.

Chapter 16

Development, Stem Cells, and Cancer

Chapter Focus

Much of what has been learned about the genetic basis of development has come from experiments using **model organisms**, species that are easy to study and grow in the lab.

Chapter Review

16.1 A program of differential gene expression leads to the different cell types in a multicellular organism

A Genetic Program for Embryonic Development   The three key processes of embryonic development are (1) cell division, the production of large numbers of cells; (2) cell **differentiation**, the formation of cells specialized in structure and function; and (3) **morphogenesis**, the physical processes that produce body structures and shape. All three processes are based on differential gene expression resulting from differences in gene regulation in cells.

Cytoplasmic Determinants and Inductive Signals   The cytoplasm of an unfertilized egg cell contains maternal mRNA, proteins, and other substances that are unevenly distributed, and the first few mitotic divisions separate these components and expose the nuclei in these new cells to different environments. These maternal components of the egg that influence early development by regulating gene expression are called **cytoplasmic determinants**.

   The other important source of developmental control is signals from other embryonic cells in the form of contact with cell-surface molecules or secreted molecules. Change in the gene expression of target cells resulting from signals from other cells is called **induction**.

Sequential Regulation of Gene Expression During Cellular Differentiation   A cell's developmental history leads to its eventual differentiation as a cell with a specific structure and function. The term **determination** is used to describe the condition in which a cell is irreversibly committed to its fate, even if it has not yet developed its final structure. When a cell becomes differentiated, it expresses genes for *tissue-specific proteins*, and that expression is usually controlled at the level of transcription.

   How do muscle cells become determined and differentiated? The embryonic precursor cells from which muscle cells arise have the potential to develop into a number of different cell types. Once they become committed to becoming muscle cells, they are called *myoblasts*. Researchers have identified several "master regulatory genes" that cause myoblast determination. One of these is *myoD*, which codes for MyoD protein, a transcription factor that binds to specific control elements and initiates transcription of other muscle-specific transcription factors. These secondary transcription factors then activate muscle-protein genes. MyoD protein also turns on genes that block the cell cycle and stop cell division, and it activates its own transcription, thereby maintaining the cell's differentiated state.

FOCUS QUESTION   16.1

What is the difference between determination and differentiation?

Apoptosis: A Type of Programmed Cell Death   In the best understood type of "programmed cell death," called **apoptosis**, cellular components are chopped up and packaged into vesicles that are released as "blebs" and then engulfed by scavenger cells.

   Apoptosis occurs 131 times during normal development in *C. elegans*, a nematode that is a model organism for studies of embryonic development and genetics. Signal transduction pathways activate a cascade of "suicide" proteins. There are similarities in genes encoding apoptotic proteins in all animals, and this process is essential to development as well as maintenance.

Pattern Formation: Setting Up the Body Plan **Pattern formation** is the spatial ordering of cells and tissues into their characteristic structures and locations. An animal's three major body axes are laid out early in development. Cytoplasmic determinants and inductive signals provide the molecular cues, called **positional information**, that tell a cell where it is located relative to the body axes and neighboring cells and determine how the cell and its progeny will develop.

Anatomical, genetic, and biochemical studies of *Drosophila* development have led to the discovery of some common developmental principles. A fruit fly's body consists of a series of segments grouped into the head, thorax, and abdomen. The anterior-posterior and dorsal-ventral axes are determined by positional information provided by cytoplasmic determinants localized in the unfertilized egg. Nurse cells and follicle cells surrounding each egg supply mRNAs and nutrients needed for development. A fertilized egg develops into a segmented larva. The third larval stage forms a cocoon, in which the larva metamorphoses into an adult fly.

In the 1940s, E. B. Lewis studied developmental mutants and was able to map certain mutations that control pattern formation to specific genes, called **homeotic genes**. In the 1970s, C. Nüsslein-Volhard and E. Wieschaus undertook a search for the genes that control segment formation. They studied mutations that were **embryonic lethals**, which prevented the development of viable larvae. They exposed flies to a chemical mutagen and then performed many thousands of crosses to detect recessive mutations that caused the death of embryos or resulted in larvae with abnormal segmentation. They identified 120 genes involved in pattern formation leading to normal segmentation.

Maternal effect genes are genes of the mother that, when mutant, result in a mutant offspring, regardless of the offspring's genotype. They code for proteins or mRNA that are deposited in the unfertilized egg. These genes are also called **egg-polarity genes** because they determine the anterior-posterior and dorsal-ventral axes of the egg and consequently of the embryo.

One egg-polarity gene is *bicoid*. The product of the *bicoid* gene is concentrated at one end of the embryo and responsible for determining its anterior end. Offspring of a mother with two mutant alleles for this gene have two tail regions and lack the front half of the body. Researchers located bicoid mRNA concentrated in the most anterior end of egg cells. Following

fertilization, the mRNA is translated into Bicoid protein, which diffuses posteriorly, forming a gradient in the early embryo. Gradients of such substances, which are called **morphogens**, establish an embryo's axes or other features—an example of the *morphogen gradient hypothesis*. Gradients of proteins transcribed from maternal mRNAs also determine the posterior end and establish the dorsal-ventral axis. Positional information encoded by the embryo's genes then establishes the proper number of segments and finally triggers the formation of each segment's characteristic structures.

FOCUS QUESTION   16.2

What type of evidence established that Bicoid protein is a morphogen that determines the anterior end of a fruit fly?

16.2 Cloning of organisms showed that differentiated cells could be "reprogrammed" and ultimately led to the production of stem cells

Research established the *genomic equivalence* of all the cells of an organism, but is each cell able to express all of its genes? *Organismal cloning* involves producing genetically identical individuals (clones) from a single cell of a multicellular organism.

Cloning Plants and Animals F. C. Steward demonstrated that differentiation does not necessarily irreversibly change the DNA of a cell by growing new carrot plants from root cells. Most plant cells remain **totipotent**, retaining the ability to give rise to a complete new organism.

Early evidence of the totipotency of differentiated animal cells was provided by the work of Briggs, King, and Gurdon, who transplanted nuclei from embryonic and tadpole cells into enucleated frog egg cells, a method called *nuclear transplantation*. The ability of the transplanted nucleus to direct normal development was inversely related to its developmental age.

In 1997, Scottish researchers reported cloning an adult sheep by transplanting a nucleus from a fully differentiated mammary cell into an unfertilized enucleated egg cell, and then implanting the resulting early embryo into a surrogate mother. The mammary cell was induced to dedifferentiate by culturing it in a nutrient-poor medium.

The *reproductive cloning* of numerous mammals has shown that cloned animals do not always look and behave identically. Environmental influences and random events play a role in development.

Although numerous mammals have now been cloned successfully, most cloned embryos fail to develop normally, and many cloned animals have various defects. What is a likely cause of these developmental failures?

***Stem Cells of Animals*   Stem cells** are relatively unspecialized cells that continue to reproduce themselves and can, under proper conditions, differentiate into one or more types of cells. *Embryonic stem (ES) cells* taken from early embryos can be cultured indefinitely and can differentiate into all cell types. Thus, ES cells are **pluripotent**. *Adult stem cells* have been isolated from various tissues and grown in culture. Such cells are capable of producing multiple (but not all) types of cells.

Stem cell research has the potential to provide cells to repair organs that are damaged or diseased. *Therapeutic cloning* of embryonic stem cells, although different from reproductive cloning of humans, still raises ethical and political issues. The transformation of adult stem cells into *induced pluripotent stem (iPS) cells* may provide sources of model cells for studying diseases and potential treatments, and may someday provide a patient's own iPS cells for regenerative treatments.

16.3   Abnormal regulation of genes that affect the cell cycle can lead to cancer

***Types of Genes Associated with Cancer*   Chemical carcinogens, X-rays, or certain viruses most often cause the changes in the genes that regulate cell growth and division that lead to cancer. **Oncogenes**, or cancer-causing genes, were first found in certain types of viruses. Similar genes were later recognized in the genomes of humans and other animals. Cellular **proto-oncogenes**, which code for proteins that stimulate cell growth and division, may become oncogenes by several mechanisms, resulting in an overproduction or increased activity of growth-stimulating proteins.

Mutations in **tumor-suppressor genes** can contribute to the onset of cancer when they result in a decrease in the activity of proteins that prevent uncontrolled cell growth.

a. Describe three genetic changes that can convert a proto-oncogene into an oncogene.

b. List three possible functions of tumor-suppressor proteins.

***Interference with Cell-Signaling Pathways*   In about 30% of human cancers, the *ras* proto-oncogene is mutated. The ***ras* gene** codes for a G protein that connects a growth-factor receptor on the plasma membrane to a cascade of protein kinases that leads to the production of a cell cycle stimulating protein. A mutation may create a hyperactive version of the Ras protein that relays a signal without the binding of a growth factor.

The ***p53* gene** is mutated in about 50% of human cancers. It codes for a tumor-suppressor protein that is a specific transcription factor for several genes. It often activates the *p21* gene, whose product binds to cyclin-dependent kinases, halting the cell cycle and allowing time for the cell to repair damaged DNA. It also activates several miRNAs that inhibit the cell cycle. The p53 protein can also activate genes involved in DNA repair. Should DNA damage be irreparable, p53 activates "suicide genes" that initiate apoptosis.

***The Multistep Model of Cancer Development*   More than one mutation appears to be needed to produce a cancerous cell. Mutation of a single proto-oncogene can stimulate cell division, but usually both alleles for several tumor-suppressor genes must be defective to allow uncontrolled cell growth.

***Inherited Predisposition and Other Factors Contributing to Cancer*   A genetic predisposition to certain cancers may involve the inheritance of an oncogene or a recessive mutant allele for a tumor-suppressor gene. Approximately 15% of colorectal cancers involve inherited mutations, often in the tumor-suppressor gene *APC*, which regulates cell migration and adhesion.

About 5–10% of breast cancer cases are linked to an inherited mutant allele for either *BRCA1* or *BRCA2*, both of which appear to be tumor-suppressor genes involved in a cell's DNA damage repair pathway.

The ultraviolet radiation in sunlight and the chemicals in cigarette smoke may contribute to cancer through their DNA-damaging effects.

Tumor viruses appear to be involved in about 15% of human cancers. Viruses can interfere with gene regulation when the insertion of their genetic material into a cell's DNA introduces an oncogene or affects a proto-oncogene or tumor-suppressor gene. Viral proteins may also inactivate p53 or other tumor-suppressor proteins.

Word Roots

morph- = form; **-gen** = produce (*morphogen:* a substance that provides positional information in the form of a concentration gradient along an embryonic axis)

proto- = first, original; **onco-** = tumor (*proto-oncogene:* a normal cellular gene that has the potential to become an oncogene, which is involved in triggering molecular events that lead to cancer)

Structure Your Knowledge

1. How might the mechanism of transcriptional regulation differ for cytoplasmic determinants and for the cell-cell signaling involved in induction?

2. What are stem cells? What are the differences between embryonic stem cells, adult stem cells, and induced pluripotent stem cells? Do plants have stem cells? Explain.

Test Your Knowledge

MULTIPLE CHOICE: *Choose the one best answer.*

1. Cytoplasmic determinants are
    a. unevenly distributed cytoplasmic components of an unfertilized egg.
    b. often involved in transcriptional regulation.
    c. usually separated in the first few mitotic divisions following fertilization.
    d. maternal contributions that help to direct the initial stages of development.
    e. all of the above.

2. Pattern formation in animals is based on
    a. positional information a cell receives from gradients of morphogens.
    b. the induction of cells by the nurse cells in the mother's ovary.
    c. the packing of chromatin in the nucleus.

    d. the differentiation of cells that then migrate together to form tissues and organs.
    e. the first few mitotic divisions.

3. What would be the fate of a *Drosophila* larva that inherits two copies of a mutant *bicoid* gene (one mutant allele from each heterozygous parent)?
    a. It develops two heads, one at each end of the larva.
    b. It develops two tails, one at each end of the larva.
    c. It develops normally but, if female, produces mutant larvae that have two tail regions.
    d. It develops into an adult with legs growing out of its head.
    e. It receives no *bicoid* mRNA from the nurse cells of its mother.

4. In the following hypothetical embryo, a high concentration of a morphogen called morpho is needed to activate gene *P*; gene *Q* is active at or above medium concentrations of morpho; and gene *R* is expressed so long as any quantity of morpho is present. A different morphogen, called phogen, activates gene *S* and inactivates gene *Q* when at medium to high concentrations. If morpho and phogen are diffusing from their sites of production at opposite ends of this embryo, which genes will be expressed in region 2? (Assume a gradient of morphogen concentrations in the three regions, from high at the source, to medium in the middle, and to low at the opposite end.)

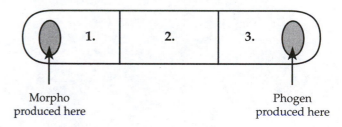

Morpho produced here   Phogen produced here

    a. genes *P, Q, R,* and *S*
    b. genes *P, Q,* and *R*
    c. genes *Q* and *R*
    d. genes *R* and *S*
    e. gene *R*

5. Apoptosis is
    a. a cell suicide program that may be initiated by p53 protein in response to DNA damage.
    b. metastasis, or the spread of cancer cells to a new location in the body.
    c. a type of programmed cell death that is a normal part of development.
    d. the transformation of a proto-oncogene to an oncogene by a point mutation.
    e. both **a** and **c**.

6. Which of the following is *not* true of adult stem cells?

   a. They have been found not only in bone marrow, but also in other tissues, including the adult brain.

   b. They come from skin cells that have been induced to become pluripotent by the introduction of cloned "stem cell" master regulatory genes.

   c. They are capable of developing into several (but not all) types of cells.

   d. These relatively unspecialized cells continually reproduce themselves in the body.

   e. They have been successfully grown in culture and made to differentiate into specialized cells.

7. Which of the following might a proto-oncogene code for?

   a. DNA polymerase

   b. RNA polymerase

   c. receptor protein for growth factors

   d. an enhancer

   e. transcription factors that inhibit cell division genes

8. A gene can develop into an oncogene when

   a. it is present in more copies than normal.

   b. it undergoes a translocation that removes it from its normal control region.

   c. a mutation results in a more active or resistant protein.

   d. a mutation in a control element increases expression.

   e. any of the above occur.

9. A tumor-suppressor gene could cause the onset of cancer if

   a. both alleles have mutations that decrease the activity of the gene product.

   b. only one allele has a mutation that alters the gene product.

   c. it is inherited from a parent in mutated form.

   d. a proto-oncogene has also become an oncogene.

   e. both **a** and **d** have occurred.

10. Which of the following would most likely account for a family history of colorectal cancer?

    a. a diet that is low in fats and high in fiber

    b. inheritance of one mutated *APC* allele that regulates cell adhesion and migration

    c. a family history of breast cancer

    d. inheritance of the *ras* oncogene, which locks the G protein in an active configuration

    e. inheritance of a proto-oncogene

Chapter 17

Viruses

█ Chapter Focus

A virus is an infectious particle consisting of a genome of single-stranded or double-stranded DNA or RNA enclosed in a protein capsid, and sometimes within a membrane envelope derived from the host. Viruses replicate using the metabolic machinery of their bacterial, animal, or plant host. Viral infections may destroy the host cell and cause diseases within the host organism. Viruses may have evolved from plasmids or transposons.

█ Chapter Review

17.1 A virus consists of a nucleic acid surrounded by a protein coat

A **virus** is an infectious particle consisting of genes inside a protein coat and, in some viruses, a membranous envelope.

Viral Genomes Viral genomes may be single-stranded or double-stranded DNA or RNA. Viral genes are usually contained on a single linear or circular nucleic acid molecule.

Capsids and Envelopes The **capsid**, or protein shell, is built from a large number of often identical protein subunits *(capsomeres)* and may be rod-shaped *(helical viruses)*, polyhedral *(icosahedral viruses)*, or more complex in shape. **Viral envelopes**, which are derived from membranes of the host cell but also include viral proteins and glycoproteins, may cloak the capsids of viruses that infect animals. Some viruses also contain a few viral enzymes.

Complex capsids are found among **bacteriophages**, or **phages**, viruses that infect bacteria. Of the phages that infect the bacterium *E. coli*, T2, T4, and T6 have similar capsid structures consisting of an icosahedral head and a protein tail piece with tail fibers for attaching to a bacterium.

17.2 Viruses replicate only in host cells

Viruses are obligate intracellular parasites that lack metabolic enzymes and other equipment needed to replicate. Each virus type has a limited **host range** due to proteins on the outside of the virus that recognize only specific receptor molecules on the host cell surface.

General Features of Viral Replicative Cycles Once the viral genome enters the host cell, the cell's enzymes, nucleotides, amino acids, ribosomes, ATP, and other resources are used to replicate the viral genome and produce capsid proteins. Many DNA viruses use host DNA polymerases to copy their genome, whereas RNA viruses use virus-encoded RNA polymerases for replicating their RNA genome.

After replication, viral nucleic acid and capsid proteins spontaneously self-assemble to form new viruses within the host cell. Hundreds or thousands of newly formed virus particles are released, often destroying the host cell in the process.

Replicative Cycles of Phages A **lytic cycle** culminates in lysis of the host cell and release of newly produced phages. **Virulent phages** replicate only by a lytic cycle.

The T4 phage uses its tail fibers to stick to a receptor site on the surface of an *E. coli* cell. The sheath of the tail contracts and thrusts its viral DNA into the cell, leaving the empty capsid behind. The *E. coli* cell begins to transcribe and translate phage genes, one of which codes for an enzyme that chops up host cell DNA. Nucleotides from the degraded bacterial DNA are used to produce viral DNA. Capsid proteins are assembled into phage tails, tail fibers, and heads. The viral components assemble into phage particles. These are released after the manufacture of an enzyme that damages the bacterial cell wall, which causes the cell to swell and burst.

Mutations that change their receptor sites and **restriction enzymes** that chop up viral DNA once it enters the cell help bacteria defend against viral infection.

In a **lysogenic cycle**, a virus replicates its genome without killing its host. **Temperate phages** can replicate by lytic and lysogenic cycles.

When the phage lambda (λ) injects its DNA into an *E. coli* cell, it can begin a lytic cycle, or its DNA may be incorporated into the host cell's chromosome and begin a lysogenic cycle as a **prophage**. Most of the genes of the inserted phage genome are repressed by a protein coded for by a prophage gene. Reproduction of the host cell replicates the phage DNA along with the bacterial DNA. The prophage may exit the bacterial chromosome, usually in response to environmental stimuli, and start a lytic cycle.

Several disease-causing bacteria would be harmless except for the expression of prophage genes that code for toxins.

FOCUS QUESTION   17.1

In the following diagram of lytic and lysogenic cycles, describe steps numbered **1–8** and label structures **a–e**.

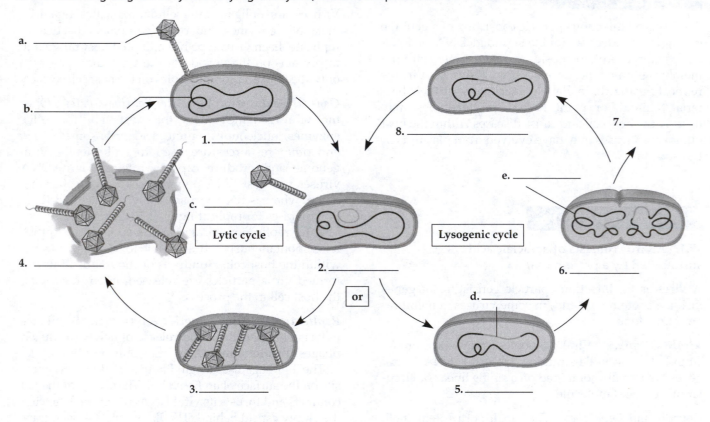

a. _____

b. _____

1. _____

c. _____

Lytic cycle

4. _____

3. _____

2. _____

or

8. _____

e. _____

7. _____

Lysogenic cycle

6. _____

d. _____

5. _____

Replicative Cycles of Animal Viruses   The genomes of animal viruses may be double- or single-stranded DNA or RNA. A viral envelope surrounds the capsid of almost all animal viruses that have RNA genomes.

What is the replicative cycle of an enveloped RNA virus? Glycoproteins extending from the viral membrane attach to receptor sites on a host cell plasma membrane. The capsid is transported into the cell. The viral genome replicates and directs the synthesis of proteins. Viral glycoproteins are produced and embedded in the ER membrane, processed by the Golgi apparatus, and then transported to the plasma membrane. New viruses bud off within an envelope that is derived from the host's plasma membrane and bears viral glycoproteins.

Herpesviruses replicate within the host cell nucleus and are temporarily cloaked in host cell nuclear membrane. The herpesvirus's double-stranded DNA can remain latent as a mini-chromosome in the nucleus of certain nerve cells until it initiates herpes infections in times of stress.

The single-stranded RNA of some animal viruses can serve directly as mRNA. The RNA genome of other viruses must first be transcribed into a strand of complementary RNA (using a viral enzyme packaged inside the capsid) that then serves as mRNA and as a template for making genome RNA.

In the complicated replicative cycle of **retroviruses**, the viral RNA genome is transcribed into double-stranded DNA by a viral enzyme, **reverse transcriptase**. This viral DNA is then integrated into a chromosome, where it is transcribed by the host cell into viral RNA, which acts both as new viral genome and as mRNA for viral proteins.

HIV (human immunodeficiency virus) is a retrovirus that causes **AIDS (acquired immunodeficiency syndrome)**. The integrated viral DNA remains as a **provirus** within the host cell DNA. New viruses, assembled with two copies of the RNA genome and two molecules of reverse transcriptase within a capsid, bud off covered in host cell plasma membrane studded with viral glycoproteins.

FOCUS QUESTION 17.2

Summarize the flow of genetic information during replication of a retrovirus. Indicate the enzymes that catalyze this flow.

_____ → _____ → _____

Enzymes:

Evolution of Viruses Viruses may have evolved from fragments of cellular nucleic acids that moved from one cell to another and eventually evolved special packaging. Sources of viral genomes may have been *plasmids*, self-replicating circles of DNA found in bacteria and yeast, and *transposons*, segments of DNA that can change locations within a cell's genome. Thus, viruses, plasmids, and transposons are all *mobile genetic elements*.

17.3 Viruses are formidable pathogens in animals and plants

Viral Diseases in Animals The symptoms of a viral infection may be caused by viral-programmed toxins produced by infected cells, cells killed or damaged by the virus, or the body's defense mechanisms fighting the infection.

Vaccines are harmless variants or derivatives of pathogens that induce the immune system to react against the actual disease agent. Vaccinations have greatly reduced the incidence of many viral diseases.

Unlike bacteria, viruses use the host's cellular machinery to replicate, and few drugs have been found to treat or cure viral infections. Some antiviral drugs resemble nucleosides and interfere with viral nucleic acid synthesis.

Emerging Viruses Examples of *emerging viruses* include HIV, Ebola virus, and West Nile virus. A general outbreak of a disease is called an **epidemic**. The flu epidemic of 2009 spread rapidly, becoming a global epidemic or **pandemic**.

The sudden emergence of viral diseases may be linked to the mutation of an existing virus (more common in RNA viruses, which have higher mutation rates), to the dissemination of an existing virus to a more widespread population (as in HIV), or to the spread from one host species to another (as in the 2009 flu pandemic, which likely passed to humans from pigs). Many new human diseases are thought to originate by the third mechanism.

Influenza types B and C infect only humans; type A infects other animals as well as humans. If different strains of influenza A undergo genetic recombination within an animal's cells and accumulate mutations that allow the virus to infect human cells, the recombinant virus may be highly pathogenic. The H1N1 virus (named for the form of the viral surface proteins hemagglutinin and neuraminidase) caused both the 1918 and 2009 flu pandemics. The H5N1 virus is carried by wild and domestic birds. The human mortality rate for avian flu infections is greater than 50%, but thus far the virus is not easily transmitted from person to person.

Viral Diseases in Plants Most plant viruses are RNA viruses. Plant viral diseases may spread through *vertical transmission* from a parent plant via infected seeds or cutting, or through *horizontal transmission* from an external source. Plant injuries increase susceptibility to viral infections, and insects can act as carriers of viruses.

FOCUS QUESTION 17.3

How does a virus spread throughout a plant? Are there cures for viral plant diseases?

Word Roots

capsa- = a box (*capsid:* the protein shell that encloses a viral genome)

lyto- = loosen (*lytic cycle:* a type of phage replicative cycle resulting in the release of new phages by lysis (and death) of the host cell)

-phage = to eat (*bacteriophage:* a virus that infects bacteria)

pro- = before (*provirus:* a viral genome that is permanently inserted into a host genome)

retro- = backward (*retrovirus:* an RNA virus that replicates by transcribing its RNA into DNA and then inserting the DNA into a cellular chromosome)

virul- = poisonous (*virulent phage:* a phage that replicates only by a lytic cycle)

Structure Your Knowledge

1. Create a concept map that describes the lytic and lysogenic cycles of a phage.

2. Complete the following concept map to help organize your understanding of viruses.

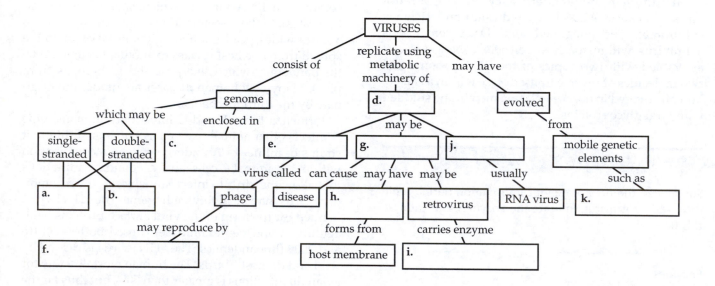

Test Your Knowledge

MULTIPLE CHOICE: *Choose the one best answer.*

1. The study of viruses has provided information on all of the following topics *except*
   a. the molecular biology of all organisms.
   b. the sexual replicative cycles of viruses.
   c. new techniques for manipulating genes.
   d. the causes of diseases.
   e. the role of mutation in the relationship between host and virus.

2. Viral genomes may be any of the following *except*
   a. single-stranded DNA.
   b. double-stranded RNA.
   c. misfolded infectious proteins.
   d. a linear single-stranded RNA molecule.
   e. a circular double-stranded DNA molecule.

3. The reverse transcriptase carried by retroviruses
   a. uses viral RNA as a template for making complementary RNA strands.
   b. protects viral DNA from degradation by restriction enzymes.
   c. destroys the host cell DNA.
   d. translates RNA into proteins.
   e. uses viral RNA as a template for DNA synthesis.

4. Virus particles are formed from capsid proteins and nucleic acid molecules
   a. by spontaneous self-assembly.
   b. at the direction of viral enzymes.
   c. using host cell enzymes.
   d. using ATP stored in the tail piece.
   e. by both **b** and **d**.

5. A virus has a base ratio of (A + G)/(U + C) = 1. What type of virus is this?

    a. a single-stranded DNA virus

    b. a single-stranded RNA virus

    c. a double-stranded DNA virus

    d. a double-stranded RNA virus

    e. a retrovirus

6. Vertical transmission of a plant virus involves

    a. movement of viral particles through plasmodesmata.

    b. inheritance of an infection from a parent.

    c. a bacteriophage transmitting viral particles.

    d. insects carrying viral particles between plants.

    e. entry through damaged cells.

7. Bacteria defend against viral infection

    a. with antibiotics they produce.

    b. with restriction enzymes that chop up foreign DNA.

    c. through the transfer of R plasmids.

    d. with reverse transcriptase.

    e. through the incorporation of viral DNA into the bacterial chromosome.

8. Drugs that are effective in treating viral infections

    a. induce the body to produce antibodies.

    b. inhibit the action of viral ribosomes.

    c. interfere with the synthesis of viral nucleic acid.

    d. change the cell-recognition sites on the host cell.

    e. are vaccines that stimulate the immune system to create immunity.

9. An RNA viral genome may be replicated by

    a. DNA polymerase from the host.

    b. RNA polymerase coded by viral genes and carried in the viral capsid.

    c. reverse transcriptase that synthesizes RNA.

    d. RNA polymerase from the host.

    e. restriction enzymes from the host.

Chapter 18

Genomes and Their Evolution

Chapter Focus

This chapter introduces genomics and bioinformatics, new approaches to analyzing and comparing the genomes of life's diverse organisms. Researchers in these fields address questions about genome organization, gene expression, growth and development, and evolution. The various types of noncoding DNA in the human genome are described. The chapter also covers the processes that contribute to genome evolution.

Chapter Review

The complete genome sequences of humans, chimpanzees, and numerous other eukaryotes and prokaryotes are enabling the study of whole sets of genes and their interactions, called **genomics**. The new field of **bioinformatics** is applying computational methods to the analysis of the ever-growing volume of biological data.

18.1 The Human Genome Project fostered development of faster, less expensive sequencing techniques

The international effort to sequence the human genome, called the **Human Genome Project**, was begun in 1990 and declared "virtually completed" in 2006. Technological advances during this project have tremendously accelerated DNA sequencing. Today's sequencing machines are "high-throughput" devices, able to analyze biological materials rapidly and produce enormous volumes of data.

J. C. Venter, founder of the company Celera Genomics, developed a **whole-genome shotgun approach** that relies on powerful computer programs to order the large number of sequenced, overlapping short fragments cut from a chromosome. This approach is now widely used. Newer sequencing techniques, called *sequencing by synthesis,* have both increased the speed and decreased the cost of sequencing genomes.

Metagenomics is an approach in which an environmental sample that includes DNA from many species is sequenced, and then computers sort the sequences into specific genomes.

FOCUS QUESTION 18.1

What types of organisms have been best studied using metagenomics?

18.2 Scientists use bioinformatics to analyze genomes and their functions

Centralized Resources for Analyzing Genome Sequences The National Center for Biotechnology Information (NCBI) and other genome centers maintain websites with databases of DNA sequences and protein sequences and structures, as well as software for analyzing and comparing data. The ever-expanding NCBI database of sequences, called GenBank, makes the resources of bioinformatics available to researchers worldwide. A Protein Data Bank contains all the three-dimensional protein structures determined thus far.

Understanding the Functions of Protein-Coding Genes The sequences of newly identified genes are compared with sequences of known genes of other species to look for similarities that might indicate the gene's function. A combination of biochemical techniques (identifying the three-dimensional structure and potential binding sites of the protein) and functional studies (blocking or disabling the gene to determine the effect on phenotype) helps to reveal protein function.

Understanding Genes and Gene Expression at the Systems Level In addition to comparing genomes of different species, genomics also considers the interactions of genes in a genome. A research project called ENCODE (Encyclopedia of DNA Elements) analyzed 1% of the human genome to identify protein-coding and noncoding RNA genes, regulatory sequences, and chromatin modifications. Researchers found that only 2% of the region they studied codes for proteins, but over 90% of the region was transcribed into RNA. This approach has now been extended to the entire human genome and to those of the nematode *C. elegans* and the fruit fly *D. melangaster*.

Proteomics is the identification and study of entire protein sets (*proteomes*) coded for by a genome. With compiled lists of DNA sequences and proteins now available, researchers are studying the functional integration of these components in biological systems. This **systems biology** approach seeks to model the dynamic behavior of whole systems, using bioinformatics to process and integrate huge amounts of data.

The Cancer Genome Atlas is taking a systems biology approach to analyzing changes in genes and patterns of gene expression in cancer cells. The project catalogues the mutations found in several common cancers and identifies genes of known and unknown functions that may provide new targets for therapies. Silicon "chips" holding arrays of most known human genes are used to analyze gene expression in patients with various diseases. Such approaches may help to match treatments with a person's unique genetic makeup.

FOCUS QUESTION 18.2

The software program BLAST, available on the NCBI website, allows a researcher to compare a DNA sequence to every sequence in GenBank to identify similar regions. What are some uses of this function?

18.3 Genomes vary in size, number of genes, and gene density

Genome Size Of the 3,700 genomes that had been sequenced as of August 2012, most are of bacteria, whose genomes have between 1 and 6 million base pairs (Mb). Archaeal genomes appear to be of a similar size, whereas most animals and plants have genomes of at least 100 Mb. Among eukaryotes, there does not appear to be a correlation between genome size and an organism's phenotype.

Number of Genes Bacteria and archaea have from 1,500 to 7,500 genes; eukaryotes range from about 5,000 genes for unicellular fungi to 40,000 for some multicellular organisms. The human genome contains fewer than 21,000 genes. Alternative splicing of the ten or so exons in most human genes can yield many different proteins for each gene. Small RNAs (such as miRNAs) that regulate gene expression may contribute to greater organismal complexity.

Gene Density and Noncoding DNA Eukaryotes have fewer genes per million base pairs than bacteria or archaea. Mammals appear to have the lowest gene density. Most of the DNA of eukaryotic genomes is noncoding DNA that is located within and between genes (such as introns and complex regulatory sequences) as well as non-protein-coding DNA between genes.

FOCUS QUESTION 18.3

Refer to the organisms listed in Table 18.1 in your text to answer the following questions.

a. Which organism has the highest gene density? _____ the lowest gene density? _____

b. Which organism has the largest number of genes? _____ the smallest number? _____

c. Which organism has the largest haploid genome size? _____ the smallest genome size? _____

d. What is the estimated number of genes in the human genome? _____ Explain the fact that there are many more different polypeptides than genes.

18.4 Multicellular eukaryotes have much noncoding DNA and many multigene families

About 1.5% of the human genome consists of exons that code for proteins, rRNA, or tRNA. The rest includes gene-related regulatory sequences (5%), introns (20%), gene fragments and nonfunctional former genes called **pseudogenes** (15%), and sequences present in many copies, called **repetitive DNA**. Much of this repetitive DNA (44% of the human genome) is either made up of or related to transposable elements.

Noncoding DNA, which was previously referred to as "junk DNA," may turn out to have important functions. Almost 500 identical regions of noncoding DNA have been identified in humans, rats, and mice, a higher level of sequence conservation than for protein-coding regions in these species.

Transposable Elements and Related Sequences

Stretches of DNA that can move about within a genome through a process called *transposition* are called *transposable genetic elements* or **transposable elements**.

What are the two types of eukaryotic transposable elements? **Transposons** move about a genome as a DNA intermediate, either by a "cut-and-paste" mechanism or a "copy-and-paste" mechanism. The enzyme *transposase*, generally encoded by the transposon, is required for both mechanisms. **Retrotransposons**, which make up the majority of transposable elements, are first transcribed into an RNA intermediate. This RNA transcript is converted back to DNA by reverse transcriptase, which is coded for by the retrotransposon itself.

Transposable elements may be represented as multiple (although not identical) copies of transposons or as related sequences that have lost the ability to move. In humans, about 10% of the genome is made up of *Alu elements*. Many of these 300-nucleotide-long sequences are transcribed into RNA, which is of unknown function.

About 17% of the human genome consists of *LINE-1*, or *L1*, retrotransposons. The introns of about 80% of analyzed human genes contain L1 sequences, suggesting that L1 may help regulate gene expression.

FOCUS QUESTION 18.4

Why do retrotransposons always move by the "copy-and-paste" mechanism?

Other Repetitive DNA, Including Simple Sequence DNA

About 14% of the human genome is repetitive DNA that appears to have arisen from mistakes in DNA replication. Scattered large-segment duplications account for 5–6% of the human genome. **Simple-sequence DNA**, by contrast, makes up 3% of the human genome and consists of multiple copies of tandemly repeated sequences. When the repeat consists of two to five nucleotides, the unit is called a **short tandem repeat**, or **STR**. The variation in repeat numbers between genomes is the basis for determining **genetic profiles**, which are used by forensic scientists. Much of a genome's simple sequence DNA is located at centromeres, where it functions in cell division and chromatin organization, and at telomeres, which protect the tips of chromosomes.

Genes and Multigene Families

More than half of the gene-related DNA occurs in **multigene families**, collections of similar or identical genes.

With the exception of the genes for histone proteins, *identical* multigene families code for RNA products.

The genes coding for the three largest rRNA molecules are arranged in a single transcription unit repeated in huge tandem arrays, enabling cells to produce the millions of ribosomes needed for protein synthesis.

Examples of multigene families of *nonidentical* genes are the two families of genes that code for globins, including the α and β polypeptide subunits of hemoglobin. Different versions of each globin subunit are clustered together on two different chromosomes and are expressed at the appropriate times during development. The families also include several pseudogenes.

FOCUS QUESTION 18.5

For each of the following types of DNA sequences found in the human genome, write the letter of the correct description and the percentage of the genome (listed beneath the descriptions) in the blanks provided.

Types of DNA	Description	%
1. Exons or rRNA/tRNA-coding	_____	_____
2. Introns	_____	_____
3. Regulatory sequences	_____	_____
4. Transposable elements and related sequences	_____	_____
5. *Alu* elements	_____	_____
6. L1 sequences	_____	_____
7. Unique noncoding DNA	_____	_____
8. Large-segment duplications	_____	_____
9. Simple sequence DNA	_____	_____

Descriptions

A. DNA in centromeres and telomeres, also STRs

B. multiple copies of mostly movable sequences

C. gene fragments and pseudogenes

D. protein- and RNA-coding sequences

E. family of short sequences related to transposable elements

F. multiple copies of large sequences

G. retrotransposons found in introns of most genes

H. enhancers, promoters, and other such sequences

I. noncoding sequences within genes

Choices of percentages: 1.5, 3, 5, 5–6, 10, 15, 17, 20, and 44. (These percentages do not add up to 100 because some of these types of DNA are subsets of other categories, and some types are not listed.)

18.5 Duplication, rearrangement, and mutation of DNA contribute to genome evolution

Duplication of Entire Chromosome Sets Extra sets of chromosomes may arise by accidents in meiosis. The resulting extra genes might diverge through mutation, leading to genes with novel functions. Polyploidy is fairly common in plants.

Alterations of Chromosome Structure Using genomic sequence information, researchers can compare the locations of DNA sequences on chromosomes among different species and reconstruct the evolutionary history of chromosomal rearrangements. Duplications and inversions of chromosomes are thought to contribute to speciation, in that matings between individuals from populations with differing chromosomal rearrangements would be less successful.

Duplication and Divergence of Gene-Sized Regions of DNA Errors such as unequal crossing over during meiosis (as may occur between copies of a transposable element on misaligned nonsister chromatids) and slippage of template strands during DNA replication might lead to the duplication of genes.

The α-globin and β-globin gene families appear to have evolved from a common ancestral globin gene, which was duplicated and then diverged. Multiple duplications and mutations within each family have led to the current family of genes with related functions along with several intervening pseudogenes.

In other cases, mutation of a duplicated gene may lead to a protein product with a new function.

FOCUS QUESTION 18.6

Lysozyme and α-lactalbumin have similar amino acid sequences but different functions. The genes for both proteins are found in mammals, but birds have only the gene for lysozyme. What does this observation suggest about the evolution of these genes?

Rearrangements of Parts of Genes: Exon Duplication and Exon Shuffling Unequal crossing over can lead to a gene with a duplicated exon. Exons often code for structural or functional regions called **domains**, and their duplication could provide a protein with enhanced properties. Errors in meiotic recombination could also lead to *exon shuffling* within a gene or between nonallelic genes.

How Transposable Elements Contribute to Genome Evolution Recombination events can take place between homologous transposable element sequences that are scattered throughout the genome, causing chromosomal mutations that may occasionally be beneficial to the organism. Transposable elements that insert within a gene may disrupt its functioning; those that insert within regulatory sequences may increase or decrease gene expression. A transposable element can also move a copy of a gene or an exon to a new location. The increased genetic diversity provided by these mechanisms provides raw material for natural selection.

FOCUS QUESTION 18.7

a. Explain two ways in which exon shuffling could occur.

b. What is a potential benefit of exon shuffling?

18.6 Comparing genome sequences provides clues to evolution and development

Comparing Genomes Comparisons of *highly conserved* genes illuminate the evolutionary relationships among species that are distantly related. Such analyses support the theory that bacteria, archaea, and eukaryotes represent the three domains of life, and also demonstrate the advantages of using model organisms to study both basic biological processes and human biology.

The similarity between genomes of two closely related species allows researchers to use one genome sequence as a framework for mapping the other genome. Also, the identified small differences between genomes can be correlated with the phenotypic divergence of the species. The human and chimpanzee genomes differ in single nucleotide substitutions by only 1.2%. Insertions or deletions of larger regions in the genome result in an additional 2.7% difference. There are more *Alu* elements in the human genome, and a third of the human duplications are not present in the chimpanzee genome.

Comparisons of genetic changes since species diverged show that some genes are changing faster in humans than in the chimpanzee or mouse. Many of these more-quickly evolving genes code for transcription factors; one example is the *FOXP2* gene, which appears to function in vocalization in vertebrates and in speech and language in humans. Mutations in this gene cause verbal impairment in humans; it is expressed in the brains of songbirds during the period they are learning their songs; and knock-out experiments with mice have shown that homozygous mutant mice had malformed brains and they, along with heterozygous mice, did not produce their normal vocalizations.

Comparisons of human genomes have revealed several million **single nucleotide polymorphisms (SNPs)**, single base-pair sites where variation is found in at least 1% of the population. Comparisons have also revealed inversions, deletions, duplications, and a surprisingly high number of *copy-number variants* (*CNVs*) in which some individuals have one or multiple copies of a gene or genetic region rather than the normal two. Such CNVs likely have phenotypic effects. Genetic markers such as SNPs, CNVs, and variations in repetitive DNA (such as STRs) will contribute to the study of human evolution.

Comparing Developmental Processes   Biologists in the field of evolutionary developmental biology (**evo-devo**) compare developmental processes to understand how they have evolved and how minor changes in gene sequence or regulation may lead to diverse forms of life.

A sequence of 180 nucleotides called a **homeobox**, which codes for a *homeodomain*, has been found in *Drosophila* homeotic genes. The same or very similar homeobox nucleotide sequences have been identified in homeotic genes of many animals. Homeotic genes in the fruit fly and mouse are found in the same linear sequence on chromosomes. Related sequences are found in regulatory genes of yeast and plants. These similarities indicate that the homeobox sequence must have arisen early and been conserved through evolution as part of the genes involved in the regulation of gene expression and development.

Homeotic genes are often called *Hox* genes in animals. Proteins with homeodomains probably coordinate the transcription of groups of developmental genes.

Many other genes involved in development, such as those coding for components of signaling pathways, are highly conserved. The differing patterns of expression of these genes in different body areas may explain the development of animals with different body plans.

FOCUS QUESTION   18.8

If all *Hox* genes contain the same or very similar homeobox, how can they control different developmental sequences?

Word Roots

pseudo- = false (*pseudogene:* a DNA segment that is very similar to a real gene but does not yield a functional product)

retro- = backward (*retrotransposon:* a transposable element that moves within a genome by means of an RNA intermediate, a transcript of the retrotransposon DNA)

Structure Your Knowledge

1. About 25% of the human genome relates to the production of proteins or RNA products (exons, introns, or regulatory sequences). Is the remaining 75% just "junk"? Describe the following types of noncoding DNA, including some of their possible functions.
   a. transposable elements
   b. *Alu* elements
   c. L1 sequences
   d. simple sequence DNA
   e. pseudogenes

2. Describe some of the processes that contribute to genome evolution.

Test Your Knowledge

MULTIPLE CHOICE: *Choose the one best answer.*

1. Why is the whole-genome shotgun approach now widely used to sequence genomes?
   a. It uses only one, very efficient restriction enzyme to create fragments.
   b. It makes use of linkage and physical maps to order sections of a chromosome.
   c. Newer sequencing techniques, such as sequencing by synthesis, and enhanced computer software can rapidly assemble overlapping fragments into complete sequences.
   d. It uses multiple research labs, which share their results on the Internet.
   e. All of the above contribute to its widespread use.

2. Metagenomics is a new approach that
   a. identifies proteomes and protein interaction networks.
   b. analyzes genomes for all functionally important elements.
   c. provides sequence data and software programs on Internet websites.

d. sequences all the DNA in an environmental sample and uses computer software to assemble the sequences into specific genomes.

e. applies genome-wide association studies to the identification of human genes of medical importance.

3. Why is proteomics important in the systems biology approach?

a. The interactions of networks of proteins are central to the functioning of cells and organisms.

b. This bioinformatics field allows for the mathematical modeling of biological systems.

c. Determining the proteins expressed in a cell identifies the genes more accurately than can be done through genomics.

d. The three-dimensional structure of a protein can be used to predict its function.

e. Comparing the proteins produced by a normal allele and the allele associated with a disease can facilitate improved treatments.

4. Bacterial genes have an average length of 1,000 base pairs; human genes average about 27,000 base pairs. Which of the following statements is the best explanation for that difference?

a. Prokaryotes have smaller, but many more, individual genes.

b. Prokaryotes are more ancient organisms; longer genes arose later in evolution.

c. Prokaryotes are unicellular; humans have many types of differentiated cells.

d. Prokaryotic genes do not have introns; human genes have multiple introns.

e. Prokaryotic proteins are not as large and complex as human proteins.

5. Which of the following statements best explains the discovery that a complex human has roughly the same number of genes as the simple nematode *C. elegans*?

a. The unusually long introns in human genes are involved in regulation of gene expression.

b. More than one polypeptide can be produced from a human gene by alternative splicing.

c. Human genes code for many more types of domains.

d. The human genome has a high proportion of noncoding DNA.

e. The large number of SNPs (single nucleotide polymorphisms) in the human genome provides a great deal of genetic variability.

6. Which of the following statements *best* describes what pseudogenes and introns have in common?

a. They do not result in a functional product.

b. They are DNA segments that lack a promoter but have other control regions.

c. They are transcribed but their translation is blocked by miRNAs.

d. They code for RNA products, not proteins.

e. They appear to have arisen from retrotransposons.

7. Which of the following techniques can be used to determine the function of a newly identified gene?

a. comparisons with genes of known functions that have similar sequences

b. blockage of gene function to see the effect on the phenotype

c. searches for similar sequences for domains of known function in other proteins.

d. both **a** and **b**

e. **a, b,** and **c**

8. Which of the following statements is *not* descriptive of transposable elements?

a. Barbara McClintock's work with maize provided the first evidence of such DNA segments.

b. Transposable elements or related sequences make up 85% of the corn genome.

c. Retrotransposons called *LINE-1* are found within the introns of many human genes and may help regulate gene expression.

d. Transposable elements often encode the enzymes, such as transposase or reverse transcriptase, necessary for their movement.

e. Each transposable element is present as multiple identical copies, often clustered in the centromere or telomere regions of a chromosome.

9. The protein tissue plasminogen activator (TPA) has three types of domains. Which of the following statements *best* explains why one of each of these types of domains is found in three different proteins (epidermal growth factor, fibronectin, and plasminogen)?

a. The genes for all four proteins are members of a multigene family involved in cell signaling.

b. The gene for TPA was the first gene to evolve; the other three genes each lost two of the domains from the ancestral TPA gene.

c. The gene for TPA arose by exon shuffling involving the other three genes.

d. The gene for TPA has many *Alu* elements that provide alternative splice sites to incorporate these exons.

e. Several duplication events led to the evolution of the TPA gene.

10. Genes that are highly conserved are useful for
    a. identifying genes that led to new species.
    b. determining the function of newly discovered genes.
    c. establishing the sequence of divergence of closely related species.
    d. tracing the relationships of groups that diverged early in the evolution of life.
    e. both **a** and **c**.

11. A highly conserved nucleotide sequence that has been found in developmental regulatory genes in many diverse organisms is called
    a. a homeodomain.
    b. a homeobox.
    c. a retrotransposon.
    d. a homeotic gene.
    e. an L1 sequence.

12. Which of the following approaches would be most useful in tracing human evolution?
    a. evo-devo and the comparison of developmental genes in plants and animals
    b. metagenomics and proteomics
    c. systems biology and the use of "knock-out" experiments
    d. analysis of single nucleotide polymorphisms and copy-number variants across individuals from the same and different populations
    e. All of the above make important contributions to studying the evolution of human populations.

Chapter Focus

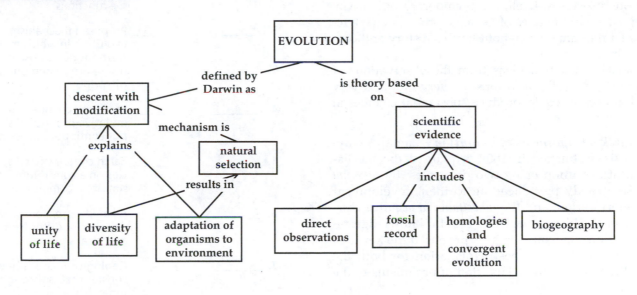

Chapter Review

Charles Darwin presented his scientific explanation for the adaptations of organisms to their environments, the diversity of life, and the unity of life in his book *On the Origin of Species.* **Evolution** may be defined in Darwin's terms as *descent with modification*, or more narrowly as the changes in a population's genetic composition over time.

Evolution can be viewed as both the *pattern* of evolutionary change observable in the natural world and the *process* or mechanisms underlying those changes.

141

19.1 The Darwinian revolution challenged traditional views of a young Earth inhabited by unchanging species

Scala Naturae and Classification of Species The Greek philosopher Aristotle proposed that all forms of life were permanent and perfect and could be arranged on a "scale of nature" of increasing complexity.

The Old Testament account of creation asserts that species are perfect and fixed. In the 1700s, Linnaeus developed both a *binomial* system for naming organisms according to their genus and species, and a hierarchy of classifications grouping his named species.

Ideas About Change over Time **Fossils** are remnants or traces of past organisms, usually found in sedimentary rocks formed through the compression of layers of sand and mud into superimposed layers called **strata**. **Paleontology**, the study of fossils, was developed by Cuvier. He maintained that the extinctions and differences in the fossils found in different strata are the result of local sudden catastrophic events and are not indicative of evolution.

Hutton proposed that immense changes in Earth's geology are the cumulative result of slow but continuous processes. Lyell, a contemporary of Darwin, proposed that the rates of geologic processes have remained the same throughout Earth's history and continue in the present.

Darwin took two ideas from the observations of Hutton and Lyell: Earth must be very old, and very slow processes could produce substantial changes in species.

Lamarck's Hypothesis of Evolution Lamarck's hypothesis, published in 1809, explained the mechanism of evolution using two principles: the *use and disuse* of body parts, leading to their development or deterioration, and the *inheritance of acquired characteristics*. Although current genetic knowledge rejects his mechanism, Lamarck proposed the key idea that evolution is the best explanation for both the fossil record and the adaptation of organisms to the environment.

FOCUS QUESTION 19.1

a. Write the capital letter representing the theory or philosophy, and the lowercase letter(s) representing its proponent(s), in the blanks preceding the following seven descriptions.

A. extinctions result of catastrophic events   a. Aristotle

B. inheritance of acquired characteristics   b. Cuvier

C. gradual geologic changes   c. Darwin

D. descent with modification   d. Hutton

E. classification system   e. Lamarck

F. scale of nature   f. Linnaeus

G. uniform rate of geologic processes   g. Lyell

Theory	Proponent(s)	
1. _____	_____	Describes the diversity of God's creations by naming and classifying species
2. _____	_____	The history of Earth is marked by sudden floods or droughts that resulted in extinctions
3. _____	_____	Proposed mechanism of evolution in which modifications due to use or disuse are passed on to offspring
4. _____	_____	Profound change is the cumulative product of slow but continuous processes
5. _____	_____	Earth contains fixed species on a continuum from simple to complex
6. _____	_____	All of life is related; present-day species differ from ancestral species
7. _____	_____	Geologic processes have constant rates throughout time

b. Now place the men listed in **a** through **g** in chronological order.

_____ _____ _____ _____ _____ _____ _____

19.2 Descent with modification by natural selection explains the adaptations of organisms and the unity and diversity of life

Darwin's Research   Darwin was 22 years old when he sailed from Great Britain on the HMS *Beagle*. He spent the voyage collecting thousands of specimens of the fauna and flora of South America, observing the various adaptations of organisms living in very diverse habitats, and making special note of the geographic distribution of the distinctly South American species. Darwin also read and was influenced by Lyell's *Principles of Geology*.

   Adaptations are inherited characteristics that contribute to an organism's survival and reproduction in a specific environment. Darwin proposed that adaptations arise through **natural selection**, a process in which individuals with beneficial characteristics produce more offspring than others because of those characteristics. In 1844 he wrote an essay on the origin of species and natural selection but did not publish it. In 1858 Darwin received Wallace's manuscript describing a nearly identical theory. Darwin then published *On the Origin of Species by Means of Natural Selection* in 1859.

Ideas from The Origin of Species   Darwin's book developed two main points: Descent with modification is the basis of life's unity and diversity, and natural selection is the mechanism that matches organisms with their environment.

   Darwin's concept of descent with modification explains that all organisms are related through descent from some unknown ancestor and develop increasing modifications as they adapt to various habitats. The history of life is analogous to a tree, with a common ancestor at the fork of each new branch and with present-day species at the tips of the youngest twigs. Many branches (probably 99% of all species that have ever lived) end in extinction, accounting for the morphological gaps sometimes present between related groups.

   Darwin used the example of **artificial selection** in the breeding of domesticated plants and animals as evidence that selection among the variations present in a population can lead to substantial changes. Malthus's essay on human population growth influenced Darwin's idea of the overproduction of offspring.

Inference 1: Individuals whose inherited characteristics give them a better chance of surviving and reproducing in a given environment tend to leave more offspring.

Inference 2: This unequal survival and reproduction leads to an accumulation of favorable traits in a population over generations.

Remember these three points about evolution by natural selection: (1) Natural selection results in the evolution of populations, not individuals; (2) natural selection affects only those traits that are heritable and that differ in a population; and (3) natural selection depends on the specific environmental factors present in a region at a given time. If the environment changes, different adaptations will be favored.

19.3 Evolution is supported by an overwhelming amount of scientific evidence

Direct Observations of Evolutionary Change   One of the four types of data that document the pattern and process of evolution involves scientific studies. As an example, the beak length of soapberry bugs has been shown to correlate with the size of the fruits that house the seeds on which they feed. Populations of soapberry bugs have become adapted to introduced plant species, often in relatively short amounts of time.

   The development of new antibiotics is usually followed rapidly by the evolution of resistance to them. Current strains of MRSA—methicillin-resistant *Staphylococcus aureus*—are resistant to multiple antibiotics and can cause potentially lethal infections.

   The evolution of drug-resistant bacteria and changes in soapberry bug beak length illustrate two facets of natural selection: It is an editing, not a creative, mechanism that selects for variations already present in a population. And it is regional and temporal, selecting for traits that are beneficial in the local environment at that current time.

FOCUS QUESTION   19.3

a. Explain how the rapid evolution of drug resistance in bacteria is an example of natural selection.

b. How might multidrug-resistant strains of MRSA have evolved?

FOCUS QUESTION   19.2

List the two observations from which Darwin drew the two inferences that explain natural selection.

   Observation 1:

   Observation 2:

Homology What is a second type of evidence for evolution? **Homology** is similarity resulting from common ancestry. The forelimbs of all mammals are **homologous structures**, containing the same skeletal elements regardless of function or external shape. Comparative anatomy and embryology illustrate that evolution is a remodeling process in which ancestral structures become modified for new functions.

Vestigial structures, which may be of little or no value to the organism, are historical remnants of ancestral structures.

Homologies can be seen on a molecular level. DNA, RNA, and an essentially universal genetic code, which have been passed along through all branches of evolution, are important evidence that all forms of life descended from the earliest organisms and are thus related. Homologous genes are found across the wide diversity of living organisms.

Distantly related organisms may appear similar as a result of **convergent evolution**, the independent evolution of similar characteristics. These **analogous** structures arise as a result of evolutionary adaptation to similar environments.

The Fossil Record What does the fossil record tell us about evolution? It documents that present and past organisms differ, and that many species have become extinct. Fossils also trace the evolution of new groups, as in the origin of whales from land mammals.

The major branches of evolutionary descent established with evidence from anatomy and molecular data are supported by the sequence of fossil forms found in the fossil record.

Biogeography The study of the geographic distributions of species, or **biogeography**, provides a fourth type of evidence for evolution. The distribution of organisms has been influenced by *continental drift*, the slow movements of Earth's continents. The single land mass called **Pangaea** formed 250 million years ago and began to break apart 200 million years ago. The gradual separation of continents helps to explain and predict where fossils of different groups and their descendants are found.

Islands often have **endemic** species, found nowhere else and usually closely related to species on the nearest island or mainland. Widely separated areas having similar environments are not likely to be populated by closely related species. Rather, each area is more likely to have species that are taxonomically related to those of their region, regardless of environment.

FOCUS QUESTION 19.4

Complete the following concept map that summarizes the main sources of evidence for evolution.

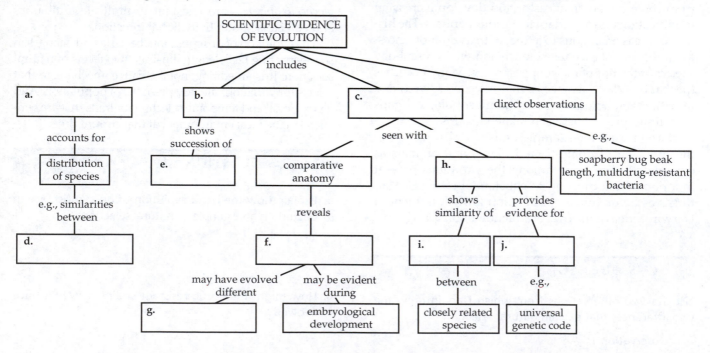

What Is Theoretical About the Darwinian View of Life? A scientific *theory* is a unifying concept with broad explanatory power and with predictions that have been and continue to be tested by experiments and observations. Evolution is the underlying theory of life—its unity and diversity.

Word Roots

bio- = life; **geo-** = the Earth (*biogeography*: the scientific study of the past and present distribution of species)

end- = within (*endemic*: referring to a species that is confined to a specific geographic area)

homo- = like, resembling (*homology*: similarity in characteristics resulting from a shared ancestry)

paleo- = ancient (*paleontology*: the scientific study of fossils)

vestigi- = trace (*vestigial structure*: a feature of an organism that is an historical remnant of a structure that served a function in the organism's ancestors)

Structure Your Knowledge

1. Explain in your own words the main components of Darwin's theory of evolution.

Test Your Knowledge

MULTIPLE CHOICE: *Choose the one best answer.*

1. The best description of natural selection is
   a. the survival of the fittest.
   b. the struggle for existence.
   c. the reproductive success of the members of a population best adapted to the environment.
   d. the overproduction of offspring in environments with limited natural resources.
   e. a change in allele frequencies in a population.

2. To Cuvier, the differences in fossils from different strata were evidence for
   a. changes occurring as a result of cumulative but gradual processes.
   b. divine creation.
   c. evolution by natural selection.
   d. continental drift.
   e. local catastrophic events such as droughts or floods.

3. Darwin proposed that new species evolve from ancestral forms by
   a. the gradual accumulation of adaptations to changing or different environments.
   b. the inheritance of acquired adaptations to the environment.
   c. the struggle for limited resources.
   d. the accumulation of mutations.
   e. the excessive production of offspring.

4. All of the following influenced Darwin as he synthesized the theory of evolution by natural selection *except*
   a. the biogeographic distribution of species such as the mockingbirds on the Galápagos Islands.
   b. Lyell's book, *Principles of Geology*, on the gradualness of geologic changes.
   c. Linnaeus's hierarchical classification of species, which could be interpreted as evidence of evolutionary relationships.
   d. examples of artificial selection that produce rapid changes in domesticated species.
   e. Mendel's paper in which he described his "laws of inheritance."

5. The smallest unit that can evolve is
   a. an individual.
   b. a mating pair.
   c. a species.
   d. a population.
   e. a community.

6. Which of the following statements is *not* considered part of the process of natural selection?
   a. Many of the variations among individuals in a population are heritable.
   b. More offspring are produced than are able to survive and reproduce.
   c. Individuals with traits best adapted to the environment are likely to leave more offspring.
   d. Many adaptive traits may be acquired during an individual's lifetime, contributing to that individual's reproductive success.
   e. Unequal reproductive success leads to gradual change in a population.

7. What might you conclude from the observation that the bones in your arm and hand are similar to the bones that make up a bat's wing?
   a. The bones in the bat's wing are vestigial structures, no longer useful as "arm" bones.
   b. The bones in a bat's wing are homologous to your arm and hand bones.
   c. Bats and humans evolved in the same geographic area.
   d. Bats lost their opposable digits during the course of evolution.
   e. Our ancestors could fly.

8. The remnants of pelvic and leg bones in a snake
   a. are vestigial structures.
   b. show that lizards evolved from snakes.
   c. are homologous structures.
   d. provide evidence for inheritance of acquired characteristics.
   e. resulted from artificial selection.

9. The hypothesis that whales evolved from land-dwelling ancestors is supported by
   a. evidence from the biogeographic distribution of whales.
   b. molecular comparisons of whales, fish, and reptiles.
   c. historical accounts of walking whales.
   d. the ability of captive whales to be trained to walk.
   e. fossils of extinct whales that had increasingly reduced hind limbs.

10. Which of the following sources of evidence provides the best support for Darwin's claim that all of life is descended from a common ancestor?
    a. the fossil record
    b. comparative embryology
    c. classification system of species, genus, family, etc.
    d. DNA comparisons and the common genetic code
    e. comparative anatomy

11. When cytochrome *c* molecules are compared, yeasts and molds are found to differ by approximately 46 amino acids per 100 residues (amino acids in the protein); insects and vertebrates are found to differ by 29 amino acids per 100 residues. What can one conclude from these data?
    a. Very little, unless the DNA sequences for the cytochrome *c* genes are compared.
    b. Yeasts evolved from molds, but vertebrates did not evolve from insects.
    c. Insects and vertebrates diverged from a common ancestor more recently than did yeasts and molds.
    d. Yeasts and molds diverged from a common ancestor more recently than did insects and vertebrates.
    e. The evolution of cytochrome *c* occurred more rapidly in yeasts and molds than in insects and vertebrates.

Use the following evolutionary tree representing the relationships among a group of vertebrates to answer questions 12 and 13. The letters at each branch point indicate the common ancestor for groups beyond that point.

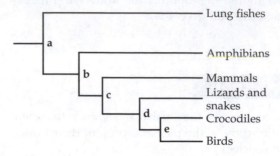

12. Which was the last common ancestor of crocodiles and the lineage of lizards and snakes?

13. Which of the following can be concluded from this evolutionary tree?
    a. Mammals are more closely related to amphibians than to birds.
    b. Mammals are more closely related to lizards and snakes than to birds.
    c. Birds are more closely related to lizards and snakes than to mammals.
    d. Birds and mammals are more closely related than are birds to lizards and snakes.
    e. Lungfishes are not related to any of these groups.

14. Which of the following is an example of convergent evolution?
    a. the evolution of multiple-drug resistance in MRSA
    b. similarities between the marsupial Tasmanian wolf and the eutherian North American wolf
    c. two very different plants that are found in different habitats but evolved from a fairly recent common ancestor
    d. the remodeling of the vertebrate forelimb in the evolution of a bird wing
    e. the many different bill sizes and shapes of finches on the Galápagos Islands

Chapter 20

Phylogeny

Chapter Focus

A goal of systematics is to reconstruct the phylogenetic history of species. Cladistics uses the distribution of shared derived characters to identify monophyletic taxa, or clades. The best phylogenetic hypotheses are parsimonious trees based on morphological and molecular homologies as well as on fossil evidence. Current hypotheses for the basic tree of life suggest that horizontal gene transfers were common in the early history of life.

Chapter Review

Phylogeny is the evolutionary history of a species or group. **Systematics** focuses on classifying organisms and determining evolutionary relationships.

20.1 Phylogenies show evolutionary relationships

Binomial Nomenclature **Taxonomy** is the scientific discipline that names and classifies species. As instituted by Linnaeus, each species is assigned a two-part Latinized name—a **binomial**—consisting of the name of the **genus** and the specific epithet, designating one species in that genus.

Hierarchical Classification In the Linnaean taxonomic system, genera are grouped into **families**, which are then placed into the increasingly broader categories of **orders**, **classes**, **phyla**, and **kingdoms**, and more recently into **domains**. A taxonomic unit at any level is called a **taxon**.

Linking Classification and Phylogeny A **phylogenetic tree** represents hypotheses about evolutionary relationships among groups. Its branching patterns may correspond to hierarchical Linnaean classification.

How can you read a phylogenetic tree? Each dichotomous **branch point** represents the divergence of two taxa from a common ancestor. **Sister taxa**, sharing an immediate common ancestor, are each other's

closest relatives. Trees are **rooted**, with the common ancestor to all the taxa in the tree usually on the left. A **basal taxon** is one that diverged early in the evolutionary history of a group. A **polytomy** is a branch point involving more than two descendants, indicating that these evolutionary relationships are not yet clear.

What We Can and Cannot Learn from Phylogenetic Trees Phylogenetic trees show patterns of evolutionary descent, which do not always correspond to morphological similarity. Unless otherwise indicated, branching sequences do not indicate when species evolved or the amount of genetic change in each lineage. Also, species located next to each other on the tree share a common ancestor; one did not evolve from the other.

Applying Phylogenies The information contained in phylogenetic trees has practical applications in such areas as molecular forensics or crop improvement.

FOCUS QUESTION 20.1

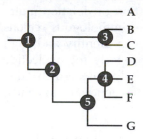

a. In this hypothetical phylogenetic tree, which number represents the common ancestor of all the taxa?

b. Which letter represents the basal taxon?

c. Which branch point is a polytomy?

d. Which taxa are sister taxa?

20.2 Phylogenies are inferred from morphological and molecular data

Morphological and Molecular Homologies   Homologies are similarities due to shared ancestry. The more similar organisms' morphologies or DNA sequences are, the more likely those organisms are closely related.

Sorting Homology from Analogy   **Analogy** is similarity due to convergent evolution, in which unrelated species develop similar features because natural selection has led to similar adaptations. Analogous structures are also called **homoplasies**. In general, when two complex structures share many similar features, it is more likely that those structures were inherited from a common ancestor. And when genes from different organisms share many nucleotide sequences, it is more likely that the genes are homologous.

FOCUS QUESTION   20.2

What two complications may make it difficult to determine phylogenetic relationships based on morphological similarities between species? Give examples.

Evaluating Molecular Homologies   How are molecular homologies identified? Molecular comparisons are often complicated by insertion or deletion mutations that change the lengths of homologous regions of DNA. Computer programs can identify and align homologous DNA segments properly for nucleotide comparisons. Statistical tools help distinguish "distant" homologies in divergent sequences from coincidental molecular homoplasies.

20.3 Shared characters are used to construct phylogenetic trees

Cladistics   What is the most common methodology used to reconstruct phylogeny? **Cladistics** uses common ancestry to classify organisms. A **clade** consists of an ancestral species and all of its descendant species. Such a clade is **monophyletic**. A **paraphyletic** group excludes some species that share a common ancestor with other species in the group, and a **polyphyletic** group includes several groups with different ancestors.

How are clades identified? **Shared ancestral characters** are found in a particular clade but originated in an ancestor that is not a member of that clade. **Shared derived characters** are unique to a particular clade.

To determine the branching sequence of a group of related species, the group is compared to an **outgroup**, a species or group of species that diverged before the group being studied (the **ingroup**). A comparison of the characters that are present in each taxon of the ingroup indicates the sequence in which shared derived characters evolved and determines the branch points used to produce a phylogenetic tree.

FOCUS QUESTION   20.3

Place the taxa (outgroup, A, B, C, and D) on the following phylogenetic tree based on the presence or absence of the characters 1–4 as shown in the table. Indicate before each branch point the number for the shared derived character that evolved in the ancestor of the clade.

	Outgroup	Taxa			
	O	A	B	C	D
1	0	1	1	1	1
2	0	0	1	0	1
3	0	1	1	0	1
4	0	0	1	0	0

(Characters)

Phylogenetic Trees with Proportional Branch Lengths   The branching pattern of most phylogenetic trees is relative rather than absolute, indicating only the order in which members of each clade last shared a common ancestor. The branch length of some trees can be scaled to reflect rates of evolutionary change (for instance, number of changes in DNA sequences) or time (using the dates of branch points as indicated in the fossil record).

Maximum Parsimony   Systematists use morphological characters or molecular comparisons to choose among many possible phylogenetic trees using the principle of **maximum parsimony**—that the smallest number of evolutionary changes is the simplest explanation and thus the best hypothesis to consider first. Computer programs search for the most parsimonious trees.

Phylogenetic Trees as Hypotheses A phylogenetic tree represents the best hypothesis of the relationships among a set of species; the more data that can be compared, the more reliable the tree becomes.

Phylogenetic hypotheses can be used to make and test predictions. *Phylogenetic bracketing* predicts that features shared by two closely related organisms will be present in their common ancestor and all its descendants. Fossil discoveries, for example, support the prediction that dinosaurs, as descendants of the common ancestor of birds and crocodiles, built nests and brooded their eggs.

FOCUS QUESTION 20.4

According to the principle of parsimony, the evolution of the four-chambered heart should place birds and mammals in the same clade. Why does the most accepted evolutionary tree show them as separate branches from the reptilian line?

20.4 Molecular clocks help track evolutionary time

Molecular Clocks Some regions of DNA appear to evolve at constant rates, and comparisons of the number of nucleotide substitutions in related genes can serve as **molecular clocks** to estimate the time since two species branched from their common ancestor.

Some genes appear to have a reliable average rate of evolution. Graphs that plot nucleotide or amino acid differences against the times for known evolutionary branch points can be used to estimate phylogenetic branchings that are not evident from the fossil record.

FOCUS QUESTION 20.5

Assuming that harmful mutations are removed quickly from the gene pool, but neutral mutational changes that have little effect on fitness should occur at a constant rate, explain why different genes might have a different molecular clock rate.

Natural selection, which favors some DNA changes over others, may disrupt the smooth running of the molecular clock. Molecular clocks may be less reliable when used to date evolutionary divergences that occurred billions of years ago. When molecular clocks are calibrated using many genes, fluctuations due to natural selection may average out. For example, molecular and fossil-based estimates of divergence times in vertebrate evolution agree closely

Applying a Molecular Clock: The Origin of HIV By comparing nucleotide sequences of HIV from samples taken at various times during the epidemic, researchers have observed a remarkably consistent rate of evolution. They estimate that the HIV-1 M strain first infected humans in the 1930s.

20.5 New information continues to revise our understanding of evolutionary history

From Two Kingdoms to Three Domains Historically, taxonomists divided the diversity of life into two kingdoms—plants and animals. In the five-kingdom system, the prokaryotes were set apart from the eukaryotes and placed in kingdom Monera. The Protista contained mostly unicellular eukaryotes while kingdoms Plantae, Fungi, and Animalia consisted of multicellular eukaryotes.

The current three-domain system creates a taxon above the kingdom level. The domains Bacteria and Archaea, although both consisting of single-celled prokaryotes, differ in many characteristics. The domain Eukarya includes all the eukaryotes: plants, fungi, animals, and many groups of mostly single-celled organisms.

The Important Role of Horizontal Gene Transfer The rRNA genes evolve so slowly that they have been used as the basis for constructing a tree of life. The tree's first major split represents the divergence of the bacteria from the other two domains. Analysis of other genes, however, suggests that eukaryotes share a more recent common ancestor with bacteria than with archaea.

Genome comparisons from the three domains indicate that **horizontal gene transfer**, perhaps through transposable elements, plasmid exchange, and viral infection, or even through fusions of different organisms, occurred during the early history of life. As a result, some scientists suggest that the early history of life is best represented as a tangled network of branches.

FOCUS QUESTION   20.6

This tangled web of life represents a current hypothesis on the origin of the three domains. Identify the domains **a**, **b**, and **c**, and the two major episodes of horizontal gene transfer labeled **d** and **e**.

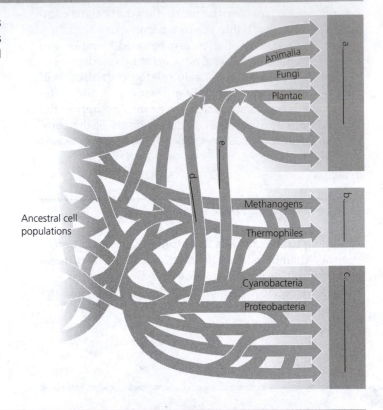

Word Roots

analog- = proportion (*analogy:* similarity between two species due to convergent evolution)

bi- = two; **nom-** = name (*binomial:* two-part latinized format for naming of a species, consisting of the genus and specific epithet)

clad- = branch (*clade:* a group of species that includes an ancestral species and all its descendants)

homo- = like, resembling (*homoplasy:* a similar structure or molecular sequence that has evolved independently in two species)

mono- = one (*monophyletic:* pertaining to a group of taxa that consists of a common ancestor and all its descendants)

parsi- = few (*principle of parsimony:* the premise that a theory about nature should be the simplest explanation that is consistent with the facts)

phylo- = tribe; **-geny** = origin (*phylogeny:* the evolutionary history of a species or group of related species)

Structure Your Knowledge

1. Draw a phylogenetic tree that best represents the relationships among taxa A–E described as follows: Taxon A is the basal taxon. B and C are sister taxa. They share a more recent common ancestor with D than with E.

2. Describe some of the tools that systematists use for constructing phylogenetic trees.

3. What is a molecular clock and how can it be used?

Test Your Knowledge

MULTIPLE CHOICE: *Choose the one best answer.*

1. Related families are grouped into the next-highest taxon, which is called a(n)
   a. class.
   b. phylum.
   c. order.
   d. genus.
   e. kingdom.

2. Four of the following trees describe the same phylogenetic relationships among taxa A, B, C, D, E, and F. Which tree shows a different phylogeny?

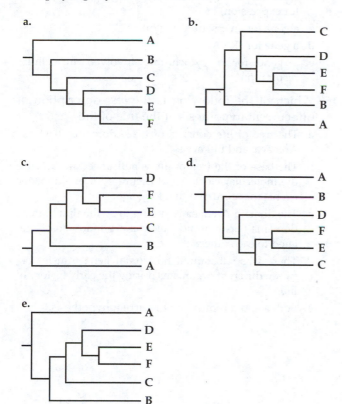

3. Convergent evolution may occur
   a. when ancestral structures are co-opted for new functions.
   b. when homologous structures are adapted for different functions.
   c. from adaptive radiation.
   d. when species are widely separated geographically.
   e. when species have similar ecological roles.

4. Which of the following provides the best example of analogous structures?
   a. forelimbs of bat and mole
   b. treelike and shrublike silversword plants of Hawaii
   c. wings of bee and hummingbird
   d. skulls of apes and humans
   e. hindlegs of Australian and North American moles

5. Which of the following is a shared derived character for monotreme, marsupial, and eutherian (placental) mammals?
   a. parental care
   b. internal fertilization
   c. amnion
   d. production of milk for young
   e. complete embryonic development inside a uterus

6. Analysis of which of the following data sets would produce the most reliable phylogenetic tree?
   a. DNA sequences of many homologous genes
   b. the fossil record
   c. morphology
   d. homoplasies
   e. Using all of these would produce the best supported phylogeny.

7. A taxon such as the class Reptilia—which does not include its relatives, the birds—is
   a. really an order.   d. paraphyletic.
   b. a clade.   e. polyphyletic.
   c. monophyletic.

The following phylogenetic tree includes the dates of divergence for taxa A through E as determined from the fossil record. Use this tree to answer questions 8 through 12.

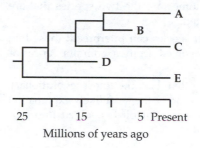

Millions of years ago

8. Which letter represents the basal taxon?

9. Which letter(s) refer to extinct groups(s)?
   a. A   d. E and D
   b. B   e. B and D
   c. C

10. Which two extant taxa are most closely related?
    a. A and B   d. C and D
    b. A and C   e. C and E
    c. A and E

11. How many million years ago did taxa A and D last share a common ancestor?
    a. 25   d. 10
    b. 20   e. cannot tell
    c. 15

12. The technique that enables biologists to make predictions about certain characteristics of taxon B by comparing shared derived characters of taxa A and C is called
    a. cladistics.
    b. systematics.
    c. phylogenetic bracketing.
    d. maximum parsimony.
    e. taxonomy.

13. The greatest number of shared derived characters should be found in two organisms that were traditionally placed in the same
    a. order.
    b. domain.
    c. family.
    d. class.
    e. phylum.

14. Which of the following approaches would allow a biologist studying the evolution of four similar species of birds to choose the best phylogenetic tree from all possible phylogenies?
    a. Draw the simplest tree and choose that one.
    b. From a comparison of DNA sequences, determine the number of evolutionary events required for each tree and then choose the most parsimonious tree.
    c. Compare the entire genomes of all species; the two most similar genomes are the two species that are most closely related.
    d. Determine which species can interbreed; those that can interbreed evolved from a common ancestor most recently.
    e. Choose the tree that has the most evolutionary changes, as this would be the most likely explanation for how these very similar birds evolved into four distinct species.

15. Which of the following segments of DNA would likely have the fastest molecular clock rate?
    a. noncoding DNA that has a regulatory function
    b. a pseudogene (gene that has lost sequences needed for expression)
    c. a gene for an essential enzyme
    d. a gene for rRNA
    e. a gene for a cytochrome involved in cellular respiration

16. Which of the following is the *best* description of our current hypothesis of the tree of life?
    a. The tree of life consists of three domains: Bacteria, Archaea, and Eukarya.
    b. The base of the tree of life is still uncertain because the molecular clock is not accurate for evolutionary events that occurred that long ago.
    c. The domain Archaea is known to be the first branch; domains Bacteria and Eukarya are more closely related to each other.
    d. There was substantial horizontal gene transfer between different organisms during the early history of life.
    e. Both **a** and **d** represent our current hypothesis.

Chapter 21

The Evolution of Populations

Chapter Focus

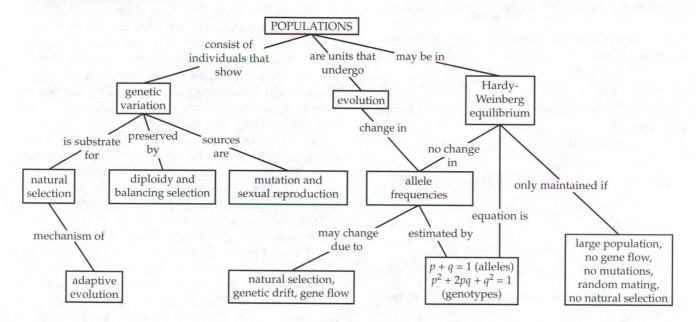

Chapter Review

Although individuals are selected for or against by natural selection, it is populations that evolve. **Microevolution** is defined as changes in the allele frequencies of a population from generation to generation.

21.1 Genetic variation makes evolution possible

Genetic Variation Genetic variation—differences in genes or other DNA sequences among individuals—is the raw material for natural selection. Some phenotypic differences vary as distinct, either-or phenotypes and usually are determined by a single gene locus. Other characters vary along a continuum and are usually affected by two or more gene loci.

Measures of genetic variation in a population include *gene variability*, the average percent of loci that

are heterozygous, and *nucleotide variability*, differences in nucleotide base-pair sites between individuals of a population.

Sources of Genetic Variation New alleles can originate by *mutation*, a change in the sequence of nucleotides in DNA. Mutations that occur in somatic cells in animals cannot be passed on to the next generation. Point mutations that occur in noncoding DNA or do not change the amino acid sequence of a protein are often harmless. Mutations that do alter the phenotype are usually harmful. In rare cases, however, a mutant allele may actually enhance an individual's reproductive success.

Chromosomal mutations are most often deleterious. Duplication of small DNA segments, introduced by transposable elements or errors in cell division, may provide extra loci that could eventually take on new functions by mutation.

Mutation rates in animals and plants average about one in every 100,000 genes per generation. Mutation produces genetic variation very rapidly in prokaryotes and viruses due to their short generation spans.

In a sexually reproducing population, most of the genetic variation comes from the reshuffling of alleles into new combinations in each individual. How do those alleles get shuffled? The processes of crossing over and independent assortment in meiosis and random combination of gametes in fertilization create unique genetic combinations in offspring.

FOCUS QUESTION   21.1

a. What is a major source of genetic variation for prokaryotes and viruses?

b. What is the major source of genetic variation for plants and animals?

c. Explain why your answers to **a** and **b** are different.

21.2   The Hardy-Weinberg equation can be used to test whether a population is evolving

Gene Pools and Allele Frequencies   A **population** is a localized, interbreeding group of individuals of a species. The **gene pool** is the term for all the alleles at all the loci present in a population. If all individuals are homozygous for the same allele, the allele is said to be *fixed*. More often, two or more alleles are present in the gene pool in some relative proportion or frequency. In a case with two alleles at a particular gene locus, the letters p and q represent the frequencies of the two alleles within the population, and their combined frequencies must equal 1: $p + q = 1$.

FOCUS QUESTION   21.2

In a population of 200 mice, 98 are homozygous dominant for brown fur (*BB*), 84 are heterozygous (*Bb*), and 18 are homozygous recessive for white fur (*bb*).

a. The genotype frequencies of this population are

   _____*BB* _____*Bb* _____*bb*.

b. The allele frequencies of this population are

   _____B allele _____b allele.

The Hardy-Weinberg Principle   How do allele and genotype frequencies change from generation to generation? If only Mendelian segregation and recombination of alleles in sexual reproduction are involved, the frequencies of alleles and genotypes in a population will remain constant, as described by the **Hardy-Weinberg principle**. Such a nonevolving gene pool is said to be in *Hardy-Weinberg equilibrium*.

The allele frequency within a population determines the proportion of gametes that will contain that allele. The random combination of gametes will yield offspring with genotypes that reflect and reconstitute the allele frequencies of the previous generation.

With the equation for Hardy-Weinberg equilibrium, the frequencies of genotypes in the next generation can be calculated from the probability of each combination of alleles. According to the rule of multiplication, the probability that two gametes containing the same allele will come together is equal to ($p \times p$) or p^2, or to ($q \times q$) or q^2. A p allele and a q allele can combine in two different ways, depending on which parent contributes which allele; therefore, the frequency of a heterozygous offspring is equal to $2pq$. The sum of the frequencies of all possible genotypes in the population adds up to 1: $p^2 + 2pq + q^2 = 1$.

FOCUS QUESTION   21.3

Use the allele frequencies you determined in Focus Question 21.2 to predict the genotype frequencies of the next generation.

Frequencies of

B (*p*) = _____ b (*q*) = _____

$BB = p^2 =;$ _____ $Bb = 2pq =$ _____ $bb = q^2 =$ _____

Hardy-Weinberg equilibrium is maintained only if all of the following five conditions are met: (1) no mutations, (2) random mating (because nonrandom mating changes genotype frequencies), (3) no natural selection (no unequal survival and reproductive success), (4) an extremely large population to offset chance fluctuations (called genetic drift), and (5) no gene flow or movement of alleles into or out of the population.

If the frequency of homozygous recessive individuals is known (q^2), then the frequency of q may be estimated as the square root of q^2 (assuming the population is in Hardy-Weinberg equilibrium for that gene).

FOCUS QUESTION 21.4

Practice using the Hardy-Weinberg equation so that you can easily determine genotype frequencies from allele frequencies, and vice versa.

a. The allele frequencies in a population are $A = 0.6$ and $a = 0.4$. Predict the genotype frequencies for the next generation.

   AA _____ *Aa* _____ *aa* _____

b. What would the allele frequencies be for the generation you predicted in part **a**?

   A _____ *a* _____

c. Suppose you are able to determine the actual genotype frequencies in the population and find that these frequencies differ significantly from what you predicted in part **a**. What would such results indicate?

21.3 Natural selection, genetic drift, and gene flow can alter allele frequencies in a population

Natural Selection   Individuals with traits that are better suited to their environment tend to be more successful in producing viable, fertile offspring, and they pass their alleles to the next generation in disproportionate numbers, resulting in *adaptive evolution*.

Genetic Drift   Chance deviations from expected results are more likely to occur in a small sample. Fluctuation in a population's allele frequencies from one generation to the next as a result of chance events is called **genetic drift**.

Genetic drift that occurs when only a few individuals colonize a new area is known as the **founder effect**. Allele frequencies in the small sample are unlikely to be representative of the parent population.

The **bottleneck effect** occurs when some disaster or other factor reduces population size dramatically, and the few surviving individuals are unlikely to represent the genetic makeup of the original population. Genetic drift will remain a factor until the population grows large enough for chance events to have less of an impact.

How can we summarize genetic drift? Genetic drift has a larger effect on small populations, causes gene frequencies to randomly fluctuate over time, can lead to the loss of genetic variation within populations, and can cause fixation of harmful alleles.

Gene Flow   **Gene flow**, the migration of individuals or the transfer of gametes between populations, may change allele frequencies. Differences in allele frequencies between populations tend to be reduced by gene flow. Gene flow may transfer alleles that either improve or reduce the ability of populations to adapt to local conditions.

FOCUS QUESTION 21.5

Fill in the following concept map that summarizes three causes of microevolution. Even better, create your own concept map to help you review the ways in which a population's genetic composition may be altered.

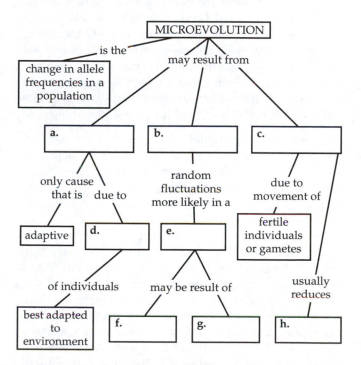

21.4   Natural selection is the only mechanism that consistently causes adaptive evolution

By increasing the frequencies of alleles that enhance survival and reproduction, natural selection leads to **adaptive evolution**.

Natural Selection: A Closer Look   **Relative fitness** is a measure of an individual's contribution to the gene pool of the next generation, relative to the contributions of others.

The frequency distribution of a trait may be affected by three modes of selection. **Directional selection** occurs most frequently during periods of environmental change, when individuals on one end of a phenotypic range may be favored. **Disruptive selection** occurs when the environment favors individuals on both extremes of a phenotypic range. **Stabilizing selection** acts against extreme phenotypes and favors more intermediate forms, tending to reduce phenotypic variation.

The Key Role of Natural Selection in Adaptive Evolution   As natural selection increases the frequencies of alleles that enhance survival and reproduction, the match between organisms and their environment increases.

Sexual Selection   **Sexual selection** is the selection for characteristics that enhance an individual's chances of obtaining mates. Sexual selection can lead to **sexual dimorphism**, the distinction between males and females on the basis of secondary sexual characteristics. Sexual selection may involve *intrasexual selection*, in which individuals of the same sex compete for mates, or *intersexual selection*, in which individuals of one sex (usually female) discriminate in choosing a mate. Also called *mate choice*, intersexual selection may be based on showy traits that reflect the general health of the male, and thus the quality of his genes.

The Preservation of Genetic Variation   How is it that genetic variation is not eliminated as natural selection continually selects for the most favorable alleles? Some of the genetic variation seen in populations may be **neutral variation** that does not confer a selective advantage or disadvantage.

The diploidy of most eukaryotes maintains genetic variation by hiding recessive alleles in heterozygotes, enabling those alleles to persist in the population and be selected for, should the environment change.

Balancing selection can maintain stable frequencies of two or more phenotypes in a population. **Heterozygote advantage** tends to maintain two or more alleles at a locus when heterozygotes have survival and reproductive advantages. In **frequency-dependent selection**, a phenotype's reproductive success declines if it becomes too common in the population.

FOCUS QUESTION   21.6

a. Why hasn't the highly deleterious sickle-cell allele been selected against and eliminated from the gene pool of the U.S. population?

b. Why is this allele at such a relatively high frequency in the gene pool of some African populations?

Why Natural Selection Cannot Fashion Perfect Organisms   Natural selection can act only on variations that are available; new alleles do not arise as they are needed. Each species has evolved from a long line of ancestral forms, many of whose structures have been co-opted for new situations. Adaptations are often compromises between the need to do different things, such as swim and walk. And finally, chance events affect a population's evolutionary history.

Word Roots

micro- = small (*microevolution:* a change in the allele frequencies of a population over generations)

Structure Your Knowledge

1. a. What is the Hardy-Weinberg principle?
    b. Define the variables of the equation for Hardy-Weinberg equilibrium. Make sure you can use this equation to determine allele frequencies and predict genotype frequencies.

2. It seems that natural selection would work toward genetic unity; the genotypes that are most fit produce the most offspring, increasing the frequency of adaptive alleles and eliminating less beneficial alleles from the population. Yet there remains a great deal of variability within populations of a species. Describe some of the factors that contribute to this genetic variability.

Test Your Knowledge

MULTIPLE CHOICE: *Choose the one best answer.*

1. Mutations are rarely a direct source for microevolution in eukaryotes because
   a. they are most often harmful and do not get passed on.
   b. they do not directly produce most of the genetic variation present in a diploid population.
   c. they occur very rarely.
   d. they are passed on only when they occur in gametes.
   e. all of the above are true.

2. The average heterozygosity of *Drosophila* is estimated to be about 14%, which means that
   a. 86% of fruit fly genes are identical.
   b. on average, 14% of a fruit fly's gene loci are heterozygous.
   c. 14% of nucleotide sites differ between individuals.
   d. nucleotide variability must be very great between individuals.
   e. the fruit fly population never experienced a bottleneck effect.

3. A scientist observes that the height of a certain species of asters decreases as the altitude on a mountainside increases. She gathers seeds from samples at various altitudes, plants them in a uniform environment, and measures the height of the new plants. All of her experimental asters grow to approximately the same height. From this she concludes that
   a. height is not a quantitative trait.
   b. the trend in height that she observed was due to genetic variations.
   c. the differences in the parent plants' heights were due to directional selection.
   d. the height variation she initially observed was an example of environmental influence.
   e. stabilizing selection was responsible for height differences in the parent plants.

4. Humans have an estimated 1,000 olfactory receptor genes. This is most likely a result of
   a. gene flow.
   b. gene duplication.
   c. frequency-dependent selection.
   d. neutral variation.
   e. disruptive selection.

5. Which of the following provides most of the genetic variation found in plant and animal populations?
   a. mutations
   b. sexual reproduction
   c. sexual selection
   d. geographic variation
   e. recessive masking in heterozygotes

6. According to the Hardy-Weinberg principle,
   a. the allele frequencies of a population should remain constant from one generation to the next if the population is large and only sexual reproduction is involved.
   b. only natural selection, resulting in unequal reproductive success, will cause evolution.
   c. the square root of the frequency of individuals showing the dominant trait will equal the frequency of p.
   d. p and q can only be determined for a population that is not evolving.
   e. all of the above are correct.

7. If a population has the following genotype frequencies—$AA = 0.42$, $Aa = 0.46$, and $aa = 0.12$—what are the allele frequencies?
   a. $A = 0.42$; $a = 0.12$
   b. $A = 0.6$; $a = 0.4$
   c. $A = 0.65$; $a = 0.35$
   d. $A = 0.76$; $a = 0.24$
   e. $A = 0.88$; $a = 0.12$

8. In a population with two alleles, B and b, the allele frequency of b is 0.4. What would be the frequency of heterozygotes if the population is in Hardy-Weinberg equilibrium?
   a. 0.16
   b. 0.24
   c. 0.48
   d. 0.6
   e. You cannot tell from this information.

9. In a population that is in Hardy-Weinberg equilibrium for two alleles, C and c, 16% of the population show a recessive trait. Assuming C is dominant to c, what percent show the dominant trait?
   a. 36%
   b. 48%
   c. 60%
   d. 84%
   e. 96%

10. In a study of a population of field mice, you find that 48% of the mice have a coat color that indicates that they are heterozygous for a particular gene. What would be the frequency of the dominant allele in this population?
   a. 0.24
   b. 0.48
   c. 0.50
   d. 0.60
   e. You cannot estimate allele frequency from this information.

11. In a random sample of a population of shorthorn cattle, 73 animals were red ($C^R C^R$); 63 were roan, a mixture of red and white ($C^R C^r$); and 13 were white ($C^r C^r$). Estimate the allele frequencies of C^R and C^r, and explain whether or not the population is in Hardy-Weinberg equilibrium.

   a. $C^R = 0.64$, $C^r = 0.36$; because the population is large and a random sample was chosen, the population is in equilibrium.

   b. $C^R = 0.7$, $C^r = 0.3$; the genotype ratio is not what would be predicted from these frequencies, and the population is not in equilibrium.

   c. $C^R = 0.7$, $C^r = 0.3$; the genotype ratio is close to what would be predicted from these frequencies, and the population is in equilibrium.

   d. $C^R = 1.04$, $C^r = 0.44$; the allele frequencies add up to greater than 1, and the population is not in equilibrium.

   e. You cannot estimate allele frequency from this information.

12. Genetic drift is likely to be seen in a population
   a. that has a high migration rate.
   b. that has a low mutation rate.
   c. in which natural selection is occurring.
   d. that is very small.
   e. for which environmental conditions are changing.

13. Which of the following is a likely result of gene flow?
   a. a decrease in the adaptive evolution of neighboring populations that inhabit quite different environments
   b. an increased migration of individuals to a favorable environment
   c. a reduction of allele frequency differences between populations
   d. Both a and c may result from gene flow.
   e. All three (a, b, and c) may result from gene flow.

14. Genetic analysis of a large population of mink inhabiting an island in Michigan revealed an unusual number of loci where one allele was fixed. Which of the following is the most probable explanation for this genetic homogeneity?
   a. The population exhibited nonrandom mating, producing a high proportion of homozygous genotypes.
   b. A very small number of mink may have colonized this island, and this founder effect and subsequent genetic drift could have fixed many alleles.
   c. The gene pool of this population never experienced gene flow.
   d. Natural selection has selected for and fixed the best-adapted alleles at these loci.
   e. The colonizing population may have had much more genetic diversity, but very recent genetic drift may have fixed these alleles by chance.

15. All of the following tend to maintain two or more alleles at a particular locus in a population *except*
   a. balancing selection.
   b. disruptive selection.
   c. heterozygote advantage.
   d. directional selection.
   e. frequency-dependent selection.

16. Sexual selection
   a. selects for traits that enhance an individual's chance of mating.
   b. increases the size of individuals.
   c. results in individuals better adapted to the environment.
   d. produces more offspring.
   e. selects for traits that increase fertility.

17. A plant population is found in an area that is becoming more arid. The average surface area of leaves has been decreasing over the generations. This trend is an example of
   a. stabilizing selection.
   b. directional selection.
   c. disruptive selection.
   d. gene flow.
   e. genetic drift.

18. Mice that are homozygous for a lethal recessive allele die shortly after birth. In a large breeding colony of mice, you find that a surprising 5% of all newborns die from this trait. In checking lab records, you discover that the same proportion of offspring have been dying from this trait in this colony for the past three years. (Mice breed several times a year and have large litters.) How might you explain the persistence of this lethal allele at such a high frequency?
   a. Homozygous recessive mice have a reproductive advantage.
   b. A large mutation rate keeps producing this lethal allele.
   c. There is some sort of heterozygote advantage and perhaps selection against the homozygous dominant trait.
   d. Genetic drift has kept the recessive allele at this high frequency in the population.
   e. Since this is a diploid species, the recessive allele cannot be selected against when it is in the heterozygote.

19. If an allele is recessive and lethal in homozygotes shortly after birth, then
    a. the allele is present in the population at a frequency of 0.001.
    b. the allele will be removed from the population by natural selection in approximately 1,000 years.
    c. the relative fitness of the homozygous recessive genotype is 0.
    d. the allele will most likely remain in the population at a low frequency because it cannot be selected against when in a heterozygote.
    e. Both **c** and **d** are correct.

20. Which of these types of selection is *mismatched* with its example?
    a. disruptive—a population of black-bellied seedcrackers consists of birds with either small bills (more effective at eating soft seeds) or large bills (able to crack hard seeds)
    b. intrasexual—elephant seal males are more than four times larger than females; males fight over areas of beach where females congregate during breeding season
    c. intersexual—female gray tree frogs choose mates that give long mating calls
    d. stabilizing—the frequencies of A, B, AB, and O blood groups remain constant in a population
    e. frequency-dependent—as fish of one coloration become more numerous, predators form a "search image" for that coloration and preferentially feed on them

21. In an area of erratic rainfall, a biologist found that, in a certain wildflower population, homozygous plants with alleles for curled leaves reproduced better in dry years, whereas homozygous plants with alleles for flat leaves reproduced better in wet years. This situation would tend to
    a. result in genetic drift in the wildflower population.
    b. result in frequency-dependent selection against the most common leaf phenotype in these wildflowers.
    c. lead to balancing selection as a result of the heterozygote advantage of wildflowers with both alleles.
    d. preserve genetic variation in this wildflower population.
    e. result in disruptive selection during years of moderate rainfall.

Chapter 22

The Origin of Species

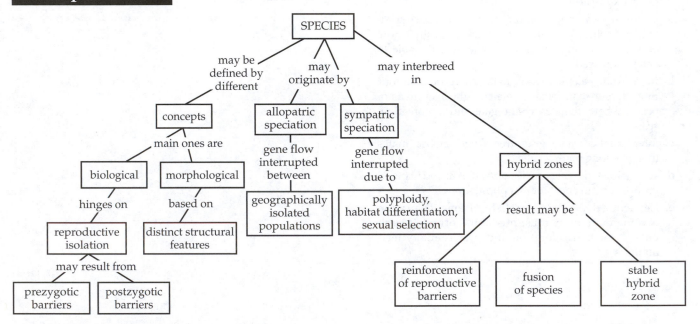

Chapter Review

Speciation, the splitting of one species into two or more species, creates the diversity of life while maintaining the relatedness or unity of life. **Microevolution** involves evolutionary changes in allele frequencies within the gene pool of a population. **Macroevolution** considers changes above the species level, such as the origin of new taxonomic groups.

22.1 The biological species concept emphasizes reproductive isolation

Different kinds of organisms are most often characterized by their physical form or morphology, although differences in physiology, biochemistry, and DNA sequences also support the existence of distinct species.

The Biological Species Concept According to the **biological species concept**, a **species** is a group of populations whose members have the potential to interbreed in nature and produce viable, fertile offspring, but do not successfully interbreed with members of other species. *Gene flow* between populations tends to hold a species together genetically, and thus morphologically.

The **reproductive isolation** that is necessary for the formation of a new species results from barriers that prevent individuals of different species from producing viable, fertile offspring, or **hybrids**. **Prezygotic barriers** function before the formation of a zygote by preventing mating between members of different species or successful fertilization should gametes meet. Should a hybrid zygote form, **postzygotic barriers** may prevent it from developing into a viable, fertile adult.

FOCUS QUESTION 22.1

For the following examples, indicate whether it is a prezygotic barrier or a postzygotic barrier and the type of reproductive isolation.

Pre- or Post-	Type of Isolation	Example
a.		Two species of frogs mate in a laboratory setup and produce viable but sterile offspring.
b.		Two species of sea urchins release gametes at the same time, but the sperm fail to fuse with eggs of a different species.
c.		The genital openings of two species of land snails cannot line up because their shells spiral in opposite directions.
d.		Two species of short-lived mayflies emerge during different weeks in spring.
e.		Two species of salamanders mate and produce offspring, but the hybrid's offspring are sterile.
f.		Two similar species of birds have different mating rituals.
g.		Embryos of two species of mice bred in the lab usually abort.
h.		Peepers breed in woodland ponds; leopard frogs breed in swamps.

What are some limitations of the biological species concept? It does not work for species that are asexual, and reproductive isolation cannot be determined for extinct species. Also, gene flow does occur between some species that remain morphologically and ecologically different.

Other Definitions of Species   Most species have been identified on the basis of physical characteristics, an approach called the **morphological species concept**. The **ecological species concept** defines species on the basis of their ecological niche, the role they play, and the resources they use in the specific environments in which they are found. In the **phylogenetic species concept**, a species is an evolutionary lineage that represents one branch on the tree of life—the smallest group of individuals that share a common ancestor.

FOCUS QUESTION 22.2

Fill in the following table to review four of the approaches that biologists have proposed for conceptualizing a species.

Concept	Emphasis
biological	a.
b.	anatomical differences, most commonly used
c.	unique roles in specific environments
phylogenetic	d.

22.2   Speciation can take place with or without geographic separation

Allopatric ("Other Country") Speciation   **Allopatric speciation** may occur when geographic isolation interrupts gene flow between two subpopulations. Geographic separation alone is not a reproductive barrier in the biological sense. Intrinsic reproductive barriers may arise coincidentally as allopatric populations go down separate evolutionary paths due to mutation, genetic drift, and natural selection. Evidence of allopatric speciation has been found in both field and laboratory studies.

Sympatric ("Same Country") Speciation   **Sympatric speciation** occurs when reproductive barriers prevent gene flow between populations that share the same area. How can this happen?

Mistakes during cell division may lead to **polyploidy**, the presence of extra sets of chromosomes. An **autopolyploid** has more than two sets of chromosomes that have all come from the same species. Failures in cell division can produce tetraploids ($4n$), which can fertilize themselves or other tetraploids but cannot reproduce with diploids from the parent population, resulting in reproductive isolation in just one generation.

Polyploid species may also arise when two different species interbreed. The resulting hybrids may propagate asexually but are usually sterile due to difficulties in the meiotic production of gametes. Future cell division mistakes, however, can result in the production of a fertile **allopolyploid**.

Polyploid speciation has been frequent and important in plant evolution. Researchers now hybridize plants by inducing meiotic and mitotic errors to create new species.

FOCUS QUESTION 22.3

a. A new plant species B forms by autopolyploidy from species A, which has a chromosome number of $2n = 10$. How many chromosomes would species B have?

b. If species A were to hybridize with species C ($2n = 14$) and produce a new allopolyploid species D, how many chromosomes would species D have?

Sympatric speciation in animals may involve isolation within the geographic range of the parent population based on different resource usage. Sexual selection may also lead to sympatric speciation. In a laboratory study of two sympatric species of cichlids, mate choice based on coloration was shown to be the reproductive barrier that normally separates the two species.

FOCUS QUESTION 22.4

a. Differentiate between allopatric and sympatric speciation.

b. How might reproductive barriers arise in each type of speciation?

22.3 Hybrid zones reveal factors that cause reproductive isolation

What is a **hybrid zone?** It is an area in which members of different species come into contact and successfully interbreed.

Patterns Within Hybrid Zones In the hybrid zone found between the yellow-bellied toad and the fire-bellied toad, the frequencies of alleles specific to one species range from almost 100% near that species' edge of the hybrid zone to 0% at the edge near the other species. What keeps these hybrids from introducing alleles of each other's species into the parent populations? Their poor survival and reproduction mean that these hybrids rarely produce viable offspring with the parent species.

Hybrid Zones over Time In a process called *reinforcement*, natural selection may strengthen prezygotic barriers to reproduction when hybrids are less fit than

members of the parent species. In other cases, weak reproductive barriers may allow sufficient gene flow that barriers break down even more, resulting in the *fusion* of the two hybridizing species into a single species. Many hybrid zones appear to exhibit *stability*, with hybrids continuing to be formed but with the two parent gene pools remaining separate.

FOCUS QUESTION 22.5

Use the following diagrams to explain the three possible outcomes for a hybrid zone over time.

a.

b.

c.

22.4 Speciation can occur rapidly or slowly and can result from changes in few or many genes

The Time Course of Speciation In the fossil record, new forms often appear rather suddenly, persist unchanged for a long time, and then disappear. **Punctuated equilibria** is the term used to describe these long periods of stasis punctuated by episodes of relatively rapid speciation and change. For other species, the fossil record indicates a much more gradual divergence.

Various studies suggest that once divergence begins, the process of speciation may be relatively rapid. For example, the chromosomes of experimentally produced hybrid sunflowers came to resemble those of the wild sunflower hybrid species in only five generations of breeding in the lab.

FOCUS QUESTION 22.6

If the process of speciation appears to occur relatively rapidly, why don't we see new species evolving all the time?

Studying the Genetics of Speciation Researchers have been able to identify the genes that play a key role in speciation in some organisms. For example, reproductive isolation in two species of monkey flowers has been linked to a few genes, including a flower-color gene that affects pollinator preference.

From Speciation to Macroevolution As species diverge and diversify, some differences may accumulate that lead to the formation of whole new groups of organisms, becoming part of the large-scale evolutionary changes of macroevolution.

Word Roots

allo- = other; **-patri** = father (*allopatric speciation:* the formation of new species in populations that are geographically isolated from one another)

auto- = self; **poly-** = many (*autopolyploid:* an individual that has more than two chromosome sets that are all derived from a single species)

macro- = large (*macroevolution:* evolutionary change above the species level)

post- = after (*postzygotic barrier:* a reproductive barrier that prevents hybrid zygotes produced by two different species from developing into viable, fertile adults)

sym- = together (*sympatric speciation:* the formation of new species in populations that live in the same geographic area)

Structure Your Knowledge

1. How are speciation and microevolution different?

2. In which type of speciation, sympatric or allopatric, would you expect reproductive barriers to arise more easily? Explain your answer.

3. What does the term *punctuated equilibria* describe?

Test Your Knowledge

MULTIPLE CHOICE: *Choose the one best answer.*

1. Which of the following is *not* a type of *intrinsic* reproductive isolation?
   a. mechanical isolation
   b. behavioral isolation
   c. geographic isolation
   d. gametic isolation
   e. temporal isolation

2. Which concept of species applies to the work of both Linnaeus, who first named organisms according to genus and species, and Darwin?
   a. biological
   b. ecological
   c. phylogenetic
   d. morphological
   e. both **c** and **d**

3. For which of the following is the biological species concept least appropriate?
   a. plants
   b. animals
   c. prokaryotes
   d. fossils
   e. both **c** and **d**

4. A horse ($2n = 64$) and a donkey ($2n = 62$) can mate and produce a mule. How many chromosomes would there be in a mule's cells?
   a. 31   d. 64
   b. 62   e. 126
   c. 63

5. Allopatric speciation is more likely to occur when an isolated population
   a. is large and thus has more genetic variation.
   b. is reintroduced to its original homeland.
   c. is small and exposed to different selection pressures in its new habitat.
   d. inhabits an island close to the mainland.
   e. All of the above contribute to allopatric speciation.

6. A tetraploid plant species (with four identical sets of chromosomes) is probably the result of
   a. allopolyploidy.
   b. autopolyploidy.
   c. hybridization and nondisjunction.
   d. allopatric speciation.
   e. **a** and **c**.

7. A botanist identifies a new species of plant that has 32 chromosomes. It grows in the same habitat with three similar species: species A ($2n = 14$), species B ($2n = 16$), and species C ($2n = 18$). Which of the following is a possible speciation mechanism for the new species?
   a. allopatric divergence by development of a reproductive isolating mechanism
   b. change in a key developmental gene that causes the plants to flower at different times
   c. autopolyploidy, perhaps due to a nondisjunction in the formation of gametes of species B
   d. allopolyploidy, a hybrid of species A and C
   e. Either answer **c** or **d** could account for the formation of this new plant species.

8. Which of the following would *not* contribute to allopatric speciation?
   a. geographic separation
   b. genetic drift
   c. gene flow
   d. different selection pressures
   e. founder effect

9. Morphological and genetic comparisons group 30 species of snapping shrimp (genus *Alpheus*) into 15 pairs of closely related species. What is the best explanation for the fact that one member of each pair lives on the Atlantic side of the Isthmus of Panama, whereas the other member of each pair lives on the Pacific side?
   a. Different predator pressures in the Atlantic and Pacific selected for differences that resulted in the reproductive isolation of these species.
   b. The pairs of species arose by allopatric speciation when the Isthmus of Panama separated their ancestral species.
   c. When these pairs meet in a hybrid zone, the reinforcement of reproductive barriers maintains each as a separate species.
   d. The 30 species of snapping shrimp evolved by sympatric speciation before the Isthmus of Panama formed.
   e. These 15 pairs of species illustrate the pattern of punctuated equilibrium, in which most speciation events take place in a short period of time.

10. When hybrids in a hybrid zone can breed with each other and with both parent species, and also have equal fitness to the parent species, one would predict that
   a. the hybrid zone would be stable.
   b. allopatric speciation would occur.
   c. reinforcement of reproductive barriers would keep the parent species separate.
   d. reproductive barriers would lessen and the two parent species would fuse.
   e. a new hybrid species would form by sympatric speciation.

11. Which of the following is the best description of punctuated equilibria?
   a. Long periods of stasis are punctuated by episodes of relatively rapid speciation and change.
   b. The equilibrium of separate species may be punctuated by gene flow within hybrid zones.

c. Most rapid speciation events involve polyploidy in plants; speciation in animals is a much more gradual process.
   d. Rapid environmental changes produce rapid speciation events; gradual environmental changes result in gradual speciation events. Speciation does not occur if the environment remains constant.
   e. In the framework of geologic time periods, speciation events occur rapidly, and the equilibrium of species is punctuated by frequent extinctions.

12. This chapter introduced several research studies that illustrate various aspects of speciation. Which of the following is *not* an accurate description of a conclusion based on one of these studies?
   a. After several generations of being raised on different food sources, both "maltose flies" and "starch flies" tended to mate with like-adapted partners. These mating preferences appear to be the beginning of reproductive isolation by way of a behavioral prezygotic barrier between the two populations.
   b. Two closely related species of monkey flowers have different pollinators (bumblebees and hummingbirds). Transferring an allele for flower color between the two species resulted in both types of pollinators visiting flowers with the transferred flower allele. Thus, this reproductive barrier based on pollinator choice may have been influenced by a change in a single gene locus.
   c. The apparent stability of the hybrid zone between two species of *Bombina* toads may relate to the narrow width of the zone. Even though hybrid toads have increased rates of embryonic mortality and a variety of morphological abnormalities, members of both parent species continue to migrate into the zone, leading to the production of hybrids.
   d. Populations of mosquitofish inhabit ponds in the Bahamas. In ponds that contain predatory fishes, natural selection has favored a body shape that facilitates rapid bursts of speed. In mate choice experiments, female mosquitofish choose males with body shapes similar to their own. The reproductive isolation between populations is forming as a by-product of natural selection for predator avoidance.
   e. In the many species of cichlids found in Lake Victoria, the reproductive barriers between species as a result of female mate choice based on male coloration appear to be breaking down because the introduced Nile perch is a strong predator of cichlids.

Chapter 23

Broad Patterns of Evolution

Chapter Focus

The fossil record has been used to establish a geologic record, in which divisions between eras and periods are often marked by major biological transitions. These transitions have been influenced by plate tectonics, mass extinctions, and adaptive radiations. The mechanisms underlying the major changes seen in the fossil record involve changes in and control of developmental genes.

Chapter Review

The history of life on Earth as chronicled in the fossil record illustrates **macroevolution**, evolutionary changes above the species level.

23.1 The fossil record documents life's history

The Fossil Record The fossil record, based on the sequence of fossils found in the *strata* of sedimentary rocks, provides evidence of great changes in the organisms that have lived on Earth. The record is incomplete, however, because (1) large numbers of species that lived probably left no fossils, (2) geologic processes destroy many fossils, and (3) only a fraction of existing fossils have been found.

How Rocks and Fossils Are Dated The order in which fossils appear in the strata of sedimentary rocks indicates their relative age.

Radiometric dating is used to determine the actual ages of rocks and fossils. Each radioactive isotope has a fixed rate of decay—its **half-life**, which is the number of years it takes for 50% of the "parent" isotope to decay to its "daughter" isotope. During an organism's lifetime, it accumulates isotopes in proportions equal to their relative abundance in the environment. After the organism dies, its radioactive isotopes decay at a fixed rate.

Carbon-14, with its half-life of 5,730 years, can date fossils up to about 75,000 years old. Isotopes that decay more slowly can be used to date older fossils. The age of volcanic rock can be used to infer the age of fossils associated with such layers of rock.

A fossil has one-eighth of the atmospheric ratio of carbon-14 to carbon-12. Estimate the age of this fossil.

The Geologic Record Geologists have established a **geologic record** of Earth's history. The Hadean, Archaean, and Proterozoic eons encompass the first 4 billion years. The Phanerozoic eon covers the last half billion years and is divided into three eras: the Paleozoic, Mesozoic, and Cenozoic. These eras are delineated by major extinction events.

The oldest known fossils are 3.5-billion-year-old fossils of **stromatolites**, which are rocklike layers of prokaryotes and sediment.

The accumulation of atmospheric oxygen resulting from the photosynthesis of early prokaryotes began around 2.4 billion years ago. Eukaryotes originated about 1.8 billion years ago, and multicellular organisms date from 1.5 billion years ago.

The Origin of New Groups of Organisms The fossil record can shed light on the origin of new groups of organisms. The gradual evolution of mammals can be traced from a group of tetrapods called synapsids through a sequence of ancestors in the group cynodonts.

23.2 The rise and fall of groups of organisms reflect differences in speciation and extinction rates

What large-scale processes have influenced rates of speciation and extinction and thus the rise and fall of major groups of organisms?

Plate Tectonics The theory of **plate tectonics** describes how the continents are part of great plates that float on Earth's molten mantle. Earthquakes occur and islands and mountains are formed in regions where these shifting plates slide past or collide with each other. In the past 1.1 billion years, *continental drift* has brought most of the landmasses together to form a supercontinent three separate times.

The formation of the supercontinent **Pangaea** about 250 million years ago destroyed and altered habitats, causing many species to become extinct. The shifting of landmasses north or south greatly changes their climate. The separation and drifting apart of continents creates huge geographic isolation events, promoting allopatric speciation. Continental drift helps explain the past and current distributions of organisms.

FOCUS QUESTION 23.2

Which mountain range began to form when India collided with Eurasia about 45 million years ago? Looking at Figure 23.8 in your text, can you find where the landmass that became India originated?

Mass Extinctions A species may become extinct due to a change in its physical or biological environment. **Mass extinctions** have occurred during periods of major worldwide environmental change. There have been five mass extinctions over the past 500 million years.

The Permian mass extinction, occurring at the boundary between the Paleozoic and Mesozoic eras 251 million years ago (mya), claimed about 96% of marine animal species and many terrestrial organisms. These extinctions may be related to massive volcanic eruptions in Siberia, which probably warmed the global climate and slowed the mixing of oceans, thus reducing the oxygen supply for marine organisms. This *ocean anoxia* would have favored anaerobic bacteria that emit H_2S. In addition to killing land organisms, this poisonous gas could have contributed to the destruction of the protective ozone layer.

The Cretaceous mass extinction, which marks the boundary between the Mesozoic and Cenozoic eras about 65.5 mya, claimed more than one-half of the marine species and many families of land plants and animals, including all dinosaurs (except birds). The thin layer of clay rich in iridium (an element common in meteorites) may have accumulated from the fallout from a huge cloud of dust created when an asteroid or comet collided with Earth. This cloud

would have blocked sunlight and severely affected global climate.

The current rate of extinction may be 100 to 1,000 times the typical rate seen in the fossil record. A sixth mass extinction caused by human destruction of habitat may occur in the next few centuries or millennia.

The fossil record indicates it takes 5 to 10 million years for the diversity of life to recover following a mass extinction. Once an evolutionary lineage becomes extinct, it cannot reappear. Mass extinctions may alter the types of organisms making up an ecological community.

Adaptive Radiations **Adaptive radiations** are periods in which many new species evolve and fill various ecological niches. Adaptive radiations have occurred following mass extinctions, the evolution of major adaptations that allow organisms to exploit new ecological roles, and the colonization of new regions.

FOCUS QUESTION 23.3

a. Mammals originated 180 million years ago but did not change much until their adaptive radiation around 65.5 mya. Explain this observation.

b. Give some examples of adaptive radiations following major evolutionary innovations that opened new ecological roles.

c. What factors have contributed to the adaptive radiation of the thousands of endemic species of the Hawaiian Archipelago?

23.3 Major changes in body form can result from changes in the sequences and regulation of developmental genes

Effects of Developmental Genes The combination of evolutionary and developmental biology, called evo-devo, explores how slight changes in developmental genes can result in major morphological differences between species.

Heterochrony is an evolutionary change in the rate or timing of development. A minor genetic alteration that affects the relative growth rates of body parts can produce a very differently proportioned adult form.

Paedomorphosis is the retention in the adult of juvenile traits of an ancestral species and can result from changes in the timing of reproductive development.

Homeotic genes control the spatial arrangement of body parts. Changes in *Hox* genes (whose products provide positional information in animal embryos) or in how or where these genes are expressed can drastically alter body form.

The Evolution of Development   Changes in the number and sequences of developmental genes as well as changes in gene regulation may produce new body forms.

FOCUS QUESTION   23.4

a. Researchers identified the key difference in the *Hox* gene *Ubx* of *Artemia* (brine shrimp, a crustacean) and *Drosophila* (fruit fly, an insect) that is responsible for the suppression of legs in *Drosophila*. What was the significance of this study?

b. Marine populations of threespine stickleback fish have protective spines on their ventral surface; lake populations of these fish do not. How did researchers determine that the loss or reduction of these spines was caused by a difference in the regulation of a developmental gene *(Pitx1)* and not in the gene itself?

23.4   Evolution is not goal oriented

Evolutionary Novelties   How can the evolution of novel and complex structures be explained? Often very complex organs, such as the eyes of vertebrates and some molluscs, have evolved gradually from simpler structures that served similar needs in ancestral species.

Evolutionary novelties may also evolve by the gradual modification of existing structures for new functions. *Exaptation* is the term for structures that evolved and functioned in one setting and were then co-opted for a new function.

Evolutionary Trends   The fossil record documents that *Equus*, the modern horse, descended from its much smaller, browsing, multitoed ancestor, *Hyracotherium*, through a series of speciation episodes that produced many different species and diverging trends.

FOCUS QUESTION   23.5

a. According to the model of *species selection*, "differential speciation success" plays a role in macroevolution similar to the role of differential reproductive success in microevolution. Explain.

b. Do evolutionary trends indicate that evolution is goal directed?

Word Roots

hetero-   = different (*heterochrony*: evolutionary change in the timing or rate of an organism's development)

macro-   = large (*macroevolution*: evolutionary change above the species level, such as the emergence of a new group of organisms through a series of speciation events)

paedo-   = child (*paedomorphosis*: the retention in the adult organism of the juvenile features of its evolutionary ancestors)

stromato-   = something spread out; **-lite** = a stone (*stromatolite*: layered rock that results from the activities of prokaryotes that bind thin films of sediment together)

Structure Your Knowledge

1. Describe three major processes that have influenced the changing diversity of life on Earth.

2. What genetic changes would most likely lead to the evolution of new morphological forms?

Test Your Knowledge

MULTIPLE CHOICE: *Choose the one best answer.*

1. The half-life of carbon-14 is 5,730 years. A fossil that is 22,920 years old would have what amount of the normal proportion of carbon-14 to carbon-12?

   a. ½   c. ⅙   e. 1⁄16
   b. ¼   d. ⅛

2. How many half-lives should have elapsed if a sample contains 12.5% of the normal quantity of a parent isotope?
   a. 1
   b. 2
   c. 3
   d. 4
   e. 5

3. Which of the following characteristics would increase the likelihood that organisms would be well represented in the fossil record?
   a. abundant and widespread
   b. exist over a long time period
   c. have hard shells or skeletons
   d. live in regions where geology favors fossil formation
   e. All of the above contribute.

4. Look back at the Geologic Record in Table 23.1. Which eon was the longest; in which eon did the oldest fossils of prokaryotes appear?
   a. Hadean; Archaean
   b. Archaean; Proterozoic
   c. Archaean; Archaean
   d. Proterozoic; Archaean
   e. Proterozoic; Proterozoic

5. Stromatolites are
   a. early prokaryotic fossils found in sediments around hydrothermal vents.
   b. the earliest protocells that formed when lipids assembled into a bilayer surrounding organic molecules.
   c. layers of rusted terrestrial rocks formed when O_2 produced by early photosynthetic prokaryotes entered the atmosphere.
   d. fossils appearing to be eukaryotes that are about twice the size of prokaryotes.
   e. layered communities of prokaryotes, fossils of which represent the oldest known prokaryotes.

6. The Permian mass extinction between the Paleozoic and Mesozoic eras
   a. may have been caused by extreme volcanism, which produced multiple environmental effects.
   b. coincided with the breaking apart of Pangaea.
   c. made way for the adaptive radiation of mammals, birds, and pollinating insects.
   d. appears to have been caused by a large asteroid striking Earth.
   e. involved all of the above.

7. The more than 500 species of fruit fly on the various Hawaiian Islands, all apparently descended from a single ancestral species, are an excellent example of
   a. plate tectonics.
   b. adaptive radiation.
   c. biogeography.
   d. exaptation.
   e. an evolutionary trend.

8. Which of the following is *not* an accurate statement about ocean anoxia?
   a. Ocean anoxia is considered to be a major cause of the Permian mass extinction.
   b. About 50% of marine animal species were lost in the Permian mass extinction, but many were able to withstand the reduced O_2 concentration by gulping atmospheric oxygen.
   c. As global temperatures warm, reduced temperature differences between the poles and equator slow the mixing of ocean water, which causes a drop in oxygen concentrations.
   d. Anaerobic bacteria could thrive in such reduced oxygen environments and emit H_2S, a poisonous metabolic by-product.
   e. As H_2S bubbles into the atmosphere, it can kill land organisms and initiate chemical reactions that destroy the protective ozone layer.

9. The evolution of major changes in animal body form is most likely to involve
   a. duplication of *Hox* genes.
   b. changes in regulatory sequences that determine where developmental genes are expressed.
   c. mutations that alter the products of *Hox* genes.
   d. genetic changes that affect the growth rates of different body parts during development.
   e. All of the above can contribute to major morphological differences between species.

10. The evolution of the middle ear bones from bones of the jaw hinge of cynodonts is an example of
    a. heterochrony.
    b. exaptation.
    c. paedomorphosis.
    d. the gradual refinement of a structure with the same function.
    e. an evolutionary trend.

11. Pterosaurs are extinct flying reptiles whose membranous wings were supported by one greatly elongated finger. A bird's wing includes feathers, which are outgrowths of the skin along the whole length of the forelimb (arm). Four elongated fingers support most of the membrane that makes up a bat's wing. Which of the following phrases *best* describes the relationship of the "finger" wings of pterosaurs, the "arm" wings of birds, and the "hand" wings of bats?

    a. the gradual refinement of a structure that served the same function in all these species and their ancestors

    b. three different exaptations of the basic skeletal structure of the tetrapod forelimb to the function of flight

    c. changes in the regulation of the same developmental genes in all three groups

    d. the independent evolution of the skeletal support, membrane or feather coverings, and aerodynamic shape of wings in these three unrelated groups

    e. an evolutionary trend toward powered flight

12. What is meant by the concept of species selection?

    a. Reproductive isolating mechanisms are responsible for maintaining the integrity of individual species.

    b. Characteristics that increase the probability that a species will separate into two or more species will accumulate in the most successful species.

    c. Natural selection can act on the species level as well as on the population level.

    d. The species that last the longest and speciate the most often influence the direction of evolutionary trends.

    e. The colonization of new and diverse habitats can lead to the adaptive radiation of numerous species.

4

The Evolutionary History of Life

Chapter 24

Early Life and the Diversification of Prokaryotes

Chapter Focus

This chapter begins with an hypothesis of how life may have originated on Earth. It then presents the morphology, reproduction, phylogeny, and ecology of prokaryotes. The collective impact of these generally single-celled organisms is huge.

Chapter Review

Earth formed about 4.6 billion years ago and was bombarded by huge rocks and ice until about 4 billion years ago. The first fossil evidence of life dates from 3.5 billion years ago. **Prokaryotes** were the first and only inhabitants of Earth until eukaryotes, who (according to the fossil record) appeared 1.8 billion years ago. These most abundant of all organisms flourish in all habitats, including those that are too harsh for any other forms of life.

24.1 Conditions on early Earth made the origin of life possible

Scientists have proposed a four-stage hypothesis for the origin of life on early Earth: (1) The abiotic synthesis of small organic molecules; (2) The joining of these small molecules into macromolecules; (3) The packaging of these molecules into **protocells**; and (4) The origin of self-replicating molecules.

Synthesis of Organic Compounds on Early Earth As Earth cooled, water vapor condensed into oceans, and the atmosphere probably contained nitrogen and its oxides, carbon dioxide, methane, ammonia, and hydrogen.

How might the first stage in the origin of life have occurred? In the 1920s, A. Oparin and J. Haldane independently hypothesized that conditions on early Earth—in particular, the reducing atmosphere, lightning, and intense ultraviolet radiation—favored the synthesis of organic compounds from simpler molecules.

In 1953, S. Miller experimentally supported the Oparin-Haldane hypothesis using an apparatus that simulated the hypothetical conditions of early Earth. Numerous laboratory replications using various combinations of atmospheric gases have also produced organic compounds.

The openings of volcanoes and deep-sea vents may have provided the environments in which the first abiotic synthesis of organic molecules occurred. Meteorites have been found to contain carbon compounds, including many amino acids, lipids, sugars, and nitrogenous bases. Thus, meteorites may have been a second source of organic compounds.

Abiotic Synthesis of Macromolecules   How might the second stage have occurred? Researchers have created polymers by dripping solutions of amino acids or RNA nucleotides onto hot sand, clay, or rock. Such polymers may have served as weak catalysts on early Earth.

Protocells   How might the first protocells have formed? Life requires accurate replication of genetic information and a metabolism to provide the building blocks and carry out this replication. Laboratory experiments have shown that membrane-bound *vesicles* can form spontaneously from mixtures of organic ingredients. Vesicle formation is speeded by the presence of *montmorillonite* clay, thought to have been common on early Earth. These vesicles can maintain a unique internal chemistry, undergo growth and reproduction, and perform some metabolic reactions.

Self-Replicating RNA   RNA probably functioned as the first hereditary material—the fourth necessary stage in the origin of life. **Ribozymes** are enzyme-like RNA catalysts. Some laboratory-produced ribozymes can make complementary copies of their sequences. Vesicles with self-replicating, catalytic RNA could grow, split, and pass some of their RNA to daughter protocells. As the more successful offspring continued to reproduce and be acted on by natural selection, metabolic and hereditary improvements may have accumulated. DNA, a more stable genetic molecule, eventually replaced RNA as the carrier of genetic information.

FOCUS QUESTION   24.1

Why do we say that, for life to have begun, the ability to carry out reproduction and metabolism must have evolved?

Fossil Evidence of Early Life   Fossil *stromatolites*, layered rocks of prokaryotes and sediment, date from 3.5 billion years ago. Fossils of individual prokaryotic cells date from 3.4 billion years ago. By 2.5 billion years ago, diverse cyanobacteria lived in the oceans. The release of oxygen during the photosynthesis of early cyanobacteria greatly altered Earth's atmosphere.

24.2   Diverse structural and metabolic adaptations have evolved in prokaryotes

Let's explore some of the diverse adaptations that have evolved in these prokaryotic, mostly unicellular organisms over their long evolutionary history. Prokaryotic cells are usually 0.5–5 μm in diameter. The three most common cell shapes are spheres (cocci), rods (bacilli), and spirals.

Cell-Surface Structures   Prokaryotic cell walls maintain cell shape and protect the cell from bursting in a hypotonic surrounding. Most prokaryotes plasmolyze in a hypertonic environment. The cell walls of bacteria (but not those of archaea) contain **peptidoglycan**, a matrix composed of modified sugars cross-linked by short polypeptides.

The Gram stain is an important tool for identifying bacteria as **gram-positive** (bacteria with walls containing a thicker layer of peptidoglycan) or **gram-negative** (bacteria with more complex walls including an outer lipopolysaccharide membrane). Gram-negative bacteria are often more toxic and more resistant to antibiotics.

Many prokaryotes secrete a dense, sticky **capsule** or a thinner *slime layer* outside the cell wall that may serve as protection from a host's immune system and as glue for adhering to a substrate or other cells. Some bacteria produce **endospores**, which are tough-walled, resistant, dormant cells formed in response to a lack of nutrients.

Some prokaryotes may also attach by means of surface hairlike appendages called **fimbriae**. Fimbriae are shorter and more numerous than **pili**, sometimes called *sex pili*, which are specialized for pulling cells together prior to the exchange of DNA between cells.

Motility   About half of all prokaryotes exhibit **taxis**, an oriented movement in response to chemical, light, or other stimuli. Many prokaryotes are equipped with flagella, either scattered over the cell surface or concentrated at one or both ends of the cell. These flagella differ from eukaryotic flagella in their lack of a plasma membrane covering, and in their size, structure, and function. Though similar in size and rotational mechanism, the flagella of bacteria and archaea are composed of different and unrelated proteins, indicating an independent origin.

ATP-driven pumps and the diffusion of H^+ back into the cell power the motor, which turns a hook and rotates the filament of the flagellum. Of the 21 proteins shown to be universally required in bacterial flagella, 19 are homologous to cellular proteins with other functions. Thus, flagella may have evolved as a series of steps from simpler structures, as proteins took on new functions (*exaptation*).

Internal Organization and DNA Extensive internal compartmentalization is not found in prokaryotic cells, although some may have membranes, usually infoldings of the plasma membrane, that function in respiration or photosynthesis.

The circular DNA chromosome, which has relatively little protein associated with it, is concentrated in a region called the **nucleoid**. Smaller rings of independently replicating DNA, called **plasmids**, may carry a few genes.

The ribosomes of prokaryotes are smaller than eukaryotic ribosomes and differ in their protein and RNA content. Some antibiotics work by binding to prokaryotic ribosomes and blocking protein synthesis.

Nutritional and Metabolic Adaptations Nutrition refers to how an organism obtains both the energy (*photo-* or *chemotroph*) and the carbon it needs for synthesizing organic compounds (*auto-* or *heterotroph*).

Thus, there are four major nutritional categories: (1) photoautotrophs, which use light energy and CO_2 (or HCO_3^-) to synthesize organic compounds; (2) chemoautotrophs, which obtain energy by oxidizing inorganic substances (such as H_2S, NH_3, or Fe^{2+}) and need only an inorganic carbon source; (3) photoheterotrophs, which use light energy but must obtain carbon in organic form; and (4) chemoheterotrophs, which use organic molecules as both an energy and a carbon source.

Obligate aerobes need O_2 for cellular respiration; *obligate anaerobes* are poisoned by O_2. Some obligate anaerobes use fermentation, others use **anaerobic respiration** to break down nutrients, with substances other than O_2 serving as the final electron acceptor. *Facultative anaerobes* can use O_2 but also can grow in anaerobic conditions using fermentation or anaerobic respiration.

Nitrogen is an essential component of proteins and nucleic acids. Some cyanobacteria and some methanogens obtain nitrogen through **nitrogen fixation**, the conversion of atmospheric N_2 to ammonia (NH_3).

The cyanobacterium *Anabaena* forms filaments in which most cells carry out photosynthesis, while a few cells called **heterocysts** perform nitrogen fixation. Members of the filament share nutrients through intercellular connections. **Biofilms** are surface-coating colonies of one or more species. Biofilms are characterized by intercellular signaling, proteins that adhere cells to each other and to the substrate, and channels in the colony for movement of nutrients and wastes.

FOCUS QUESTION   24.2

The following micrograph shows the cyanobacterium *Anabaena*. What mode of nutrition does this bacterium use? What is the function of the spherical cell indicated by the arrow?

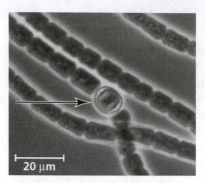

20 μm

Reproduction Prokaryotes reproduce rapidly by *binary fission* and have a huge reproductive potential. Their population growth, however, is regulated by the supply of nutrients, accumulation of toxic wastes, competition, or predation.

Adaptation of Prokaryotes: A Summary Although simpler than eukaryote cells, prokaryotic cells have structural adaptations that enhance success. And their metabolic diversity allows them to thrive in a range of physical and chemical conditions.

FOCUS QUESTION   24.3

Fill in the following table with brief descriptions of the characteristics of prokaryotic cells.

Property	Description
Cell shape	a.
Cell size	b.
Cell surface	c.
Motility	d.
Internal membranes	e.
Genome	f.
Reproduction and growth	g.

24.3 Rapid reproduction, mutation, and genetic recombination promote genetic diversity in prokaryotes

Rapid Reproduction and Mutation Mutations, although statistically rare, produce a great deal of genetic diversity because prokaryotes have such short generation times and large population sizes. In *E. coli* populations, scientists have documented rapid adaptation—occurring in 20,000 generations—to a challenging new, low-glucose environment.

Genetic Recombination The mechanisms of *genetic recombination* in prokaryotes are different from the eukaryotic mechanisms of meiosis and fertilization. Three mechanisms can transfer DNA within species of prokaryotes and between different species (called *horizontal gene transfer*).

Prokaryotes can take up DNA from the environment in a process called **transformation**. The foreign DNA is integrated into the chromosome by an exchange of homologous DNA segments. Many bacteria have surface proteins specialized for uptake of DNA from closely related species.

Bacteriophages can transfer genes from one prokaryote to another by a process called **transduction**. A random piece of DNA may be accidentally packaged within a phage and introduced into another prokaryotic cell. Through crossing over, the newly introduced DNA (from the donor cell) can be incorporated into the recipient cell's chromosome.

In **conjugation**, DNA is transferred while two prokaryotic cells are temporarily joined. In *E. coli*, a donor cell attaches to another cell by a sex pilus and transfers DNA to the recipient. The ability to form pili and donate DNA usually results from the presence of an **F factor**, a DNA segment that is either part of the chromosome or a plasmid.

Bacterial cells containing the F factor on the **F plasmid** are called F⁺ cells. During conjugation, a single strand of the plasmid DNA is transferred through a mating bridge to the recipient cell. The single parental strands in both cells replicate, and the recipient cell is now a recombinant F⁺ cell.

In cells in which the F factor is integrated into the bacterial chromosome, chromosomal genes can be transferred during conjugation. Crossovers between homologous regions of the transferred DNA and the recipient cell's chromosome produce new genetic combinations in this recombinant bacterium.

R plasmids carry genes that code for various mechanisms of resistance to antibiotics. Many R plasmids also have genes coding for pili and are transferred to nonresistant cells during conjugation, contributing to the medical problem of antibiotic-resistant pathogens.

FOCUS QUESTION 24.4

Complete the following concept map that summarizes the genetic characteristics of prokaryotes.

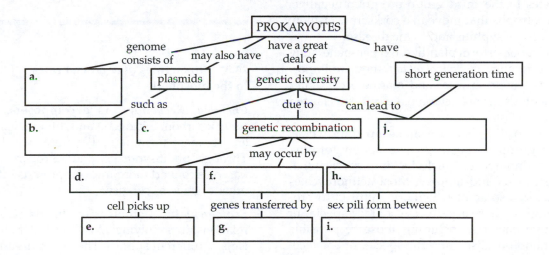

24.4 Prokaryotes have radiated into a diverse set of lineages

An Overview of Prokaryotic Diversity Molecular comparisons of base sequences of SSU-rRNA indicated that prokaryotes should be divided into two domains: Bacteria and Archaea. Analyses of prokaryotic genomes have expanded with the field of *metagenomics,* which enables researchers to sample directly from the environment. This ongoing research has shown that the genetic diversity of prokaryotes is huge and that horizontal gene transfer has been extensive in the evolution of prokaryotes.

Bacteria Bacteria show great diversity in their modes of nutrition and metabolism. Five major groups are described in the text.

Proteobacteria are a nutritionally diverse clade of aerobic and anaerobic gram-negative bacteria. Its five subgroups are alpha proteobacteria, which includes the nitrogen-fixing *Rhizobium* species and the *Agrobacterium* species used in the genetic engineering of plants; beta proteobacteria, which includes the nitrogen-recycling soil bacterium *Nitrosomonas;* gamma proteobacteria, which includes autotrophic sulfur bacteria and some serious pathogens and intestinal inhabitants such as *E. coli;* delta proteobacteria, including the colony-forming myxobacteria; and epsilon proteobacteria, the mostly pathogenic group that includes the stomach ulcer-causing *Helicobacter pylori.*

The second major group of bacteria includes chlamydias, which are obligate intracellular parasites of animals. One of these gram-negative species is the most common cause of blindness and nongonococcal urethritis, a sexually transmitted disease.

Spirochetes in the third group are gram-negative, helical heterotrophs that move in a corkscrew fashion. Spirochetes cause syphilis and Lyme disease.

Cyanobacteria perform plantlike, oxygen-generating photosynthesis, and, as *phytoplankton,* they are important producers in freshwater and marine ecosystems. Some filamentous cyanobacteria have cells specialized for nitrogen fixation.

The gram-positive bacteria make up the diverse, fifth major group of bacteria. The colony-forming subgroup actinomycetes includes the species that cause tuberculosis and leprosy. Most actinomycetes are soil bacteria, some of which are cultured to produce antibiotics. There are diverse solitary species of gram-positive bacteria, including those responsible for anthrax and botulism, and the species of *Staphylococcus* and *Streptococcus.* Mycoplasmas, the smallest of all cells, are the only bacteria that lack cell walls.

Archaea Archaea have some characteristics in common with bacteria, some in common with eukaryotes, and some unique to themselves. Many archaea are **extremophiles,** species that live in extreme habitats. **Extreme halophiles** (salt-lovers) live in extremely saline waters. **Extreme thermophiles** may be found in hot sulfur springs and near deep-sea hydrothermal vents.

Methanogens have a unique energy metabolism in which CO_2 is used to oxidize H_2, producing the waste product methane (CH_4). These strict anaerobes often live in swamps and marshes, are important decomposers in sewage treatment, and are gut inhabitants that contribute to the nutrition of cattle and other herbivores.

The methanogens and most extreme halophiles are placed in the clade Euryarchaeota. Most thermophiles fit into the Crenarchaeota. Not all euryarchaeotes and crenarchaeotes are extremophiles; numerous species have been found in more moderate habitats.

Prokaryote phylogeny continues to change as new groups are discovered, including such clades as the Korarchaeota and Nanoarchaeota.

FOCUS QUESTION 24.5

a. Evidence indicates that mitochondria evolved from an aerobic alpha proteobacterium through endosymbiosis. From which bacterial group did chloroplasts likely evolve?

b. What is metagenomics? How does it contribute to prokaryotic phylogeny?

24.5 Prokaryotes play crucial roles in the biosphere

Chemical Recycling As **decomposers,** prokaryotes return carbon, nitrogen, and other elements to the environment for assimilation into new living forms. Through photosynthesis and nitrogen fixation, prokaryotes convert inorganic compounds into forms that other organisms can use.

Ecological Interactions **Symbiosis** is an ecological relationship involving close contact between organisms of different species. The larger organism is called the **host,** and the smaller is known as the **symbiont.** In **mutualism,** both species benefit. In **commensalism,** one organism benefits while the other is neither harmed nor helped. In **parasitism**, a **parasite** eats from its host. **Pathogens** are parasites that cause disease.

FOCUS QUESTION 24.6

Explain the important role of chemoautotrophic bacteria in hydrothermal vent communities.

Impacts on Humans   Many of the 500–1,000 species of bacteria living in the human intestines are mutualists that aid in food digestion and synthesize important nutrients that are used by the host.

About one-half of all human diseases are caused by pathogenic bacteria. Pathogens most commonly cause disease by producing toxins. **Exotoxins** are secreted proteins that cause such diseases as botulism and cholera. **Endotoxins**, which are lipopolysaccharides released from the outer membrane of gram-negative bacteria that have died, cause such diseases as typhoid fever and *Salmonella* food poisoning.

Improved sanitation and the development of antibiotics have decreased the incidence of bacterial disease in developed countries. The evolution of antibiotic-resistant strains of pathogenic bacteria, however, poses a serious health threat. Horizontal gene transfer also spreads genes connected with virulence, as in the emergence of the dangerous *E. coli* strain O157:H7.

Bioremediation is the use of organisms to remove environmental pollutants. Prokaryotes are used to treat sewage and clean up oil spills. They can be used to produce biodegradable plastics. Prokaryotes have been engineered to make vitamins, antibiotics, ethanol, and other products.

Word Roots

an-   = without, not; **aero-** = the air (*anaerobic respiration*: a catabolic pathway in which inorganic molecules other than oxygen accept electrons at the end of electron transport chains)

endo-   = inner, within (*endotoxin*: a toxic component of the outer membrane of certain gram-negative bacteria that is released only when the bacteria die)

exo-   = outside (*exotoxin*: a toxic protein secreted by a prokaryote or other pathogen that produces specific symptoms, even if the pathogen is no longer present)

-gen   = produce (*methanogen*: organism that produces methane as a waste product of the way it obtains energy; all known methanogens are in domain Archaea)

halo-   = salt; **-philos** = loving (*extreme halophile*: an organism that lives in a highly saline environment)

mutu-   = reciprocal (*mutualism*: a symbiotic relationship in which both participants benefit)

-oid   = like, form (*nucleoid*: a non-membrane-bounded region in a prokaryotic cell where its chromosome is located)

proto-   = first (*protocell*: an abiotic precursor of a living cell that had a membrane-like structure and that maintained an internal chemistry different from that of its surroundings)

sym-   = with, together; **-bios** = life (*symbiosis*: an ecological relationship between organisms of two different species that live together in direct contact)

taxis   = to arrange (*taxis*: an oriented movement toward or away from a stimulus)

thermo-   = temperature (*extreme thermophile*: an organism that thrives in hot environments (often 60–80°C or hotter))

Structure Your Knowledge

1. List the sequence of four stages that could have led to the origin of the first living cells.

2. One might think of prokaryotes as primitive, simple single cells. This chapter, however, provides many examples of how diverse, highly adaptable, metabolically versatile, cooperative, and communal they can be. List some examples of these prokaryotic characteristics.

3. Describe four positive ways in which prokaryotes have an impact on our lives and on the world around us.

Test Your Knowledge

FILL IN THE BLANKS

_____ 1. the name for spherical prokaryotes

_____ 2. region in which the prokaryotic chromosome is found

_____ 3. common laboratory technique that identifies two groups of bacteria

_____ 4. surface appendages of prokaryotes used for adherence to substrate

_____ 5. an oriented movement in response to light or chemical stimuli

_____ 6. resistant cell that can survive harsh conditions

_____ 7. surface-coating cooperative colonies of prokaryotes

_____ 8. proteins that are secreted by pathogens and are potent poisons

_____ 9. use of organisms to remove environmental pollutants

_____ 10. type of respiration that uses molecules other than O_2 as final electron acceptors

MULTIPLE CHOICE: *Choose the one best answer.*

1. The primitive atmosphere of Earth may have favored the abiotic synthesis of organic molecules because it
   a. was highly oxidative.
   b. was reducing (near openings of volcanoes) and had energy sources in the form of lightning and UV radiation.
   c. had a great deal of methane and organic fuels.
   d. had plenty of water vapor, carbon, oxygen, and nitrogen, providing the C, H, O, and N needed for organic molecules.
   e. consisted almost entirely of hydrogen gas, creating a very reducing environment.

2. Which of the following is a proposed hypothesis for the origin of genetic information?
   a. Early DNA molecules coded for RNA, which then catalyzed the production of proteins.
   b. Early polypeptides were the first "genes," followed by RNA and then DNA.
   c. Short RNA strands were capable of self-replication and evolved by natural selection.
   d. As protocells grew and split, they distributed copies of their molecules to their offspring.
   e. Early RNA molecules coded for the order of amino acids in a polypeptide.

3. All of the following types of evidence support the hypothesis that life on Earth formed spontaneously *except* for
   a. the discovery of ribozymes, showing that RNA molecules may be autocatalytic.
   b. laboratory studies involving the formation and reproduction of vesicles.
   c. the fossil record.
   d. the abiotic synthesis of polymers from monomers dripped onto hot rocks.
   e. the production of organic compounds in S. Miller's experimental apparatus.

4. Stromatolites are
   a. early prokaryotic fossils found in sediments around hydrothermal vents.
   b. protocells that form when lipids assemble into a bilayer surrounding organic molecules.
   c. a surface-coating colony of one or more species of prokaryotes.
   d. nitrifying archaeans that are extremely abundant in the oceans.
   e. layered communities of prokaryotes, fossils of which represent the oldest known prokaryotes.

5. Many prokaryotes secrete a sticky capsule outside the cell wall that
   a. protects them from plasmolyzing in a hypertonic environment.
   b. serves as protection from host defenses and as glue for adherence.
   c. reacts with the Gram stain.
   d. is used for attaching cells during conjugation.
   e. is composed of peptidoglycan, modified sugars cross-linked by short polypeptides.

6. Some prokaryotes have specialized internal membranes that
   a. have attached ribosomes and function in protein synthesis.
   b. arose from endosymbiosis of smaller prokaryotes, evolving into mitochondria and chloroplasts.
   c. form from infoldings of the plasma membrane and may function in cellular respiration or photosynthesis.
   d. are produced by the endoplasmic reticulum but differ in composition from eukaryotic membranes.
   e. enclose the nucleoid and separate plasmids from the single prokaryotic chromosome.

7. Genetic variation in prokaryotes may result from
   a. horizontal gene transfer.
   b. mutation.
   c. conjugation.
   d. transformation and transduction.
   e. all of the above.

8. Chemoautotrophs
   a. are photosynthetic.
   b. use organic molecules for both energy and carbon sources.
   c. oxidize inorganic substances for energy and use CO_2 as a carbon source.
   d. use light to generate ATP but need organic molecules for a carbon source.
   e. use light energy to extract electrons from H_2S.

9. Facultative anaerobes
   a. can survive with or without oxygen.
   b. are poisoned by oxygen.

c. can carry out anaerobic respiration or fermentation but not aerobic cellular respiration.

d. are able to fix atmospheric nitrogen to make NH_3.

e. include the methanogens, which oxidize H_2 and reduce CO_2 to CH_4.

10. Which of the following statements is *not* true of members of clade Archaea?

   a. They appear to be more closely related to eukaryotes than to bacteria.

   b. They have cell walls that lack peptidoglycan.

   c. They are often found in harsh habitats.

   d. They lack membrane-enclosed organelles but do have a nuclear envelope.

   e. They include the methanogens, extreme halophiles, and extreme thermophiles, as well as groups found in more moderate habitats.

11. The placement of prokaryotes into major clades within the domains Archaea and Bacteria has been based on

   a. fossil records of prokaryotes that have been recently discovered.

   b. molecular comparisons.

   c. Gram stain results and colony characteristics when grown on solid media.

   d. shape and motility.

   e. nutritional modes and ecology.

12. Which of the following groups is *incorrectly* described?

   a. Proteobacteria—diverse gram-negative bacteria, including pathogens such as *Salmonella* and *Helicobacter pylori* and beneficial species such as *Rhizobium*

   b. Chlamydias—intracellular parasites, including a species that causes blindness and nongonococcal urethritis

   c. Spirochetes—helical heterotrophs, including pathogens that cause syphilis and Lyme disease

   d. Cyanobacteria—photosynthetic filaments of cells; many species use the pigment bacteriorhodopsin and inhabit very saline waters

   e. Gram-positive bacteria—diverse group that includes actinomycetes, mycoplasmas, and pathogens that cause anthrax, botulism, and tuberculosis

13. Which of the following statements accurately describes a role of prokaryotes in chemical recycling?

   a. Chemoheterotrophic prokaryotes may function as decomposers to return chemical elements to an ecosystem.

   b. Photoheterotrophs convert atmospheric CO_2 into organic compounds.

   c. Chemoautotrophs such as the soil bacterium *Burkholderia glathei* increase the availability of nitrogen for growing seedlings.

   d. Photoautotrophic bacteria living on hydrothermal vents support communities of eukaryotic species.

   e. All of the above are correct.

14. Some cyanobacteria are capable of nitrogen fixation, a process that

   a. oxidizes nitrogen-containing compounds to produce ATP.

   b. allows these bacteria to live in anaerobic environments.

   c. removes soil nitrogen and returns N_2 gas to the atmosphere.

   d. converts N_2 to ammonia, making nitrogen available for incorporation into proteins and nucleic acids.

   e. is an essential part of the nitrogen cycle that extracts nitrogen trapped in mineral deposits and makes it available to organisms for their nitrogen metabolism.

15. O157:H7, a pathogenic strain of *E. coli*, has over 1,000 genes that are not present in the harmless strain K-12. Many of these genes are associated with phage DNA in the bacterial chromosome. These observations suggest that

   a. this virulent strain has many more R plasmids than does the harmless strain and is resistant to almost all antibiotics.

   b. at least some of these genes were acquired through phage-mediated horizontal gene transfer.

   c. these genes were not acquired through conjugation with K-12.

   d. most of these novel genes must be a result of mutations and short generation times.

   e. both **b** and **c** are correct.

Chapter 25

The Origin and Diversification of Eukaryotes

Chapter Focus

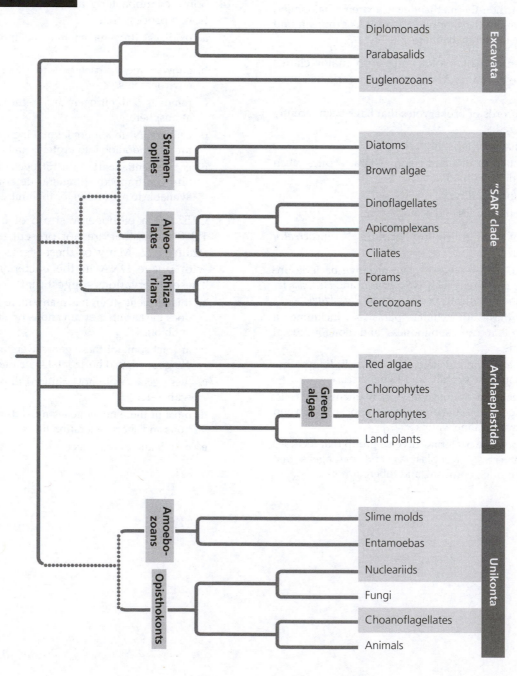

Chapter Review

Eukaryotic cells are structurally more complex than prokaryotic cells, with a cytoskeleton whose structural support enables changes in cell shape and movement. Diverse groups of eukaryotes that are mostly unicellular are informally known as **protists**.

25.1 Eukaryotes arose by endosymbiosis more than 1.8 billion years ago

The Fossil Record of Early Eukaryotes   Although chemical evidence of eukaryotes dates back 2.7 billion years, the first fossil evidence dates from 1.8 billion years, with a period of initial diversification ranging from then to 1.3 billion years ago. The next stage in eukaryotic evolution dates from 1.3 billion to 635 million years ago, and includes the appearance of novel features such as multicellularity, sexual life cycles, and eukaryotic photosynthesis. The earliest taxonomically identifiable fossils are of red algae. **Algae** refers to all groups of photosynthetic protists. Large, multicellular eukaryotes, referred to as the **Ediacaran biota**, lived from 635 to 535 million years ago. Body size, as well as taxonomic and morphological diversity, increased in these organisms.

Endosymbiosis in Eukaryotic Evolution   What is **endosymbiosis?** It is a symbiotic relationship in which one organism lives within another, and it may explain the fact that eukaryotes exhibit both archael and bacterial features. Molecular data indicate that eukaryotes originated when an archaeal cell (or a cell of archaeal ancestry) engulfed an aerobic, heterotrophic bacterium. According to the **endosymbiont theory**, mitochondria originated from endosymbiotic bacteria within the earliest eukaryotes, and plastids evolved from a photosynthetic bacterium that became an endosymbiont of a heterotrophic eukaryote.

The lineage of cells that acquired a photosynthetic cyanobacterium eventually gave rise to two lineages of photosynthetic protists: red algae and green algae. In several instances of **secondary endosymbiosis**, a red or green alga was engulfed by a heterotrophic eukaryote, leading to new protist lineages.

FOCUS QUESTION   25.1

a. List some of the evidence that supports the endosymbiont theory.

b. What is the **serial endosymbiosis** hypothesis?

c. The plastids of chlorarachniophytes are surrounded by four membranes and enclose a vestigial nucleus called a *nucleomorph*. What do these facts suggest?

25.2 Multicellularity has originated several times in eukaryotes

Multicellular Colonies   Colonies are collections of cells with little or no cellular differentiation. They may arise when dividing cells remain attached by their shared cell wall or by proteins that connect adjacent cells together.

Independent Origins of Complex Multicellularity   Lineages of complex multicellular organisms (with differentiated cells) arose independently in red, green, and brown algae, as well as plants, fungi, and animals. The proteins holding together colonies of the *Volvox* lineage are homologous to proteins of the unicellular *Chlamydomonas* cell wall.

Steps in the Origin of Multicellular Animals   Morphological and molecular similarities indicate that choanoflagellates are the closest relative of animals. Many of their shared protein domains function in cell adhesion or cell signaling, which are both important for complex multicellularity. Animal cadherin proteins that attach animal cells together include many choanoflagellate domains, along with a conserved, novel domain—the CCD region.

25.3 Four "supergroups" of eukaryotes have been proposed based on morphological and molecular data

Four Supergroups of Eukaryotes   A current hypothesis of the phylogeny of eukaryotes includes four supergroups. As you can see in the phylogenetic tree on the previous page, all four supergroups have members that were formerly classified in the kingdom Protista. The sequence of their divergence from a common ancestor is not known. New data continue to revise our understanding of eukaryotic evolution.

Excavates   The supergroup **Excavata** is based on morphological studies of the cytoskeleton and an "excavated" feeding groove found in some members. Molecular evidence indicates that each of the three groups of excavates—the diplomonads, parabasalids, and euglenozoans—is monophyletic, and recent genomic studies support the monophyly of the supergroup.

Most diplomonads and parabasalids are found in anaerobic environments. They have modified mitochondria and lack plastids. The *mitosomes* of **diplomonads** lack functional electron transport chains. The reduced mitochondria of **parabasalids** are called *hydrogenosomes* and harvest energy anaerobically, releasing H_2. *Giardia intestinalis* is a diplomonad intestinal parasite, and *Trichomonas vaginalis* is a sexually transmitted parabasalid parasite.

The diverse clade of **euglenozoans** includes predatory heterotrophs, autotrophs, and parasites, all of which have a spiral or crystalline rod inside their flagella. The *euglenids* are characterized by one or two flagella that emerge from the anterior end.

A single large mitochondrion containing a mass of DNA called a kinetoplast is characteristic of the group called *kinetoplastids*. Kinetoplastids include a number of parasites of plants, animals, and other protists.

The "SAR" Clade   The large, diverse supergroup known as the **"SAR" clade** is proposed based on recent metagenomic studies and includes the stramenopiles, alveolates, and rhizarians.

The **stramenopiles** originated by secondary endosymbiosis and include important photosynthetic protists. **Diatoms** are a major component of marine and freshwater phytoplankton; unique boxlike silica walls protect these unicellular protists. The mostly marine **brown algae** are the largest and most complex algae. Some brown algal "seaweeds" have tissues and organs that are analogous to those found in plants. They may have a rootlike **holdfast** and a stemlike **stipe**, which supports leaflike **blades**. Some brown algae have floats, which keep the photosynthesizing blades near the surface. The giant, fast-growing kelps live in deeper waters.

Membrane-bounded alveoli under the plasma membrane are characteristic of the diverse subgroup of the "SAR" clade, the **alveolates**. Alveolates include three clades: the dinoflagellates, ciliates, and apicomplexans.

Dinoflagellates originated by secondary endosymbiosis, although many are now purely heterotrophic. Others are **mixotrophs**, which combine photosynthesis and heterotrophic nutrition. The beating of two flagella in grooves between the cell's cellulose plates produces a characteristic spinning movement. *Blooms*, or population explosions, of dinoflagellates are responsible for the often-harmful "red tides."

Ciliates are characterized by the cilia they use to move and feed. Their numerous cilia may be widespread or clumped.

Rhizarians are the third major subgroup of the "SAR" clade. **Amoebas**, which move and feed with their cellular extensions called **pseudopodia**, are

FOCUS QUESTION   25.2

Label the indicated structures in the following diagram of a *Paramecium*. To which clade, subgroup, and supergroup does this organism belong?

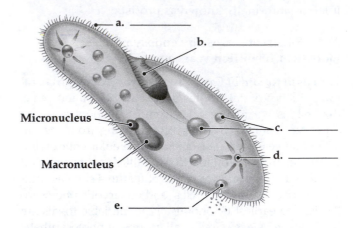

Micronucleus

Macronucleus

spread across many eukaryotic taxa. The rhizarian amoebas have threadlike pseudopodia.

Foraminiferans, or **forams**, are known for their porous shells, called **tests**, made of organic material and calcium carbonate. Pseudopodia extending through the pores function in swimming, test formation, and feeding. Many forams obtain nourishment from symbiotic algae.

Cercozoans are a large group of amoeboid and flagellated protists that feed with threadlike pseudopodia. Most cercozoans are parasites or predators, although one group is mixotrophic and at least one cercozoan is autotrophic.

FOCUS QUESTION   25.3

The cercozoan *Paulinella chromatophora* has a unique photosynthetic structure called a chromatophore. What does genetic and morphological evidence suggest about the origin of the chromatophore?

Archaeplastida   Descendants of the heterotrophic protists that acquired a cyanobacterium endosymbiont evolved into red algae and green algae. Land plants arose from the green algal lineage. The supergroup **Archaeplastida** includes red algae, green algae, and land plants.

The photosynthetic pigment phycoerythrin produces the color of **red algae** and allows them to absorb those wavelengths of light that penetrate into deep water. Most red algae species are marine and multicellular, the largest of which are called seaweeds. Red algae reproduce sexually, but have no flagellated gametes.

The chloroplasts of **green algae** resemble those of land plants, and molecular and cytological evidence indicate that green algae and plants are closely related. Some systematists favor including green algae in an extended plant kingdom, Viridiplantae. The two main green algal groups are chlorophytes and charophytes. The latter group is very closely related to land plants.

Most chlorophytes live in fresh water. Nearly all reproduce sexually with biflagellated gametes. Unicellular forms may live independently or be symbionts in other eukaryotes. Multicellular chlorophytes include colonies and multicellular organisms.

Unikonts The fourth supergroup, **Unikonta**, includes animals, fungi, and some protists. Molecular systematics supports two major clades: the amoebozoans and the opisthokonts (animals, fungi, and closely related protists). Some studies support the close relationship of these two clades; other studies do not.

Research has yet to establish the root of the eukaryotic tree. Two researchers have proposed that the unikonts were the first eukaryotes to diverge. They base this hypothesis on the DHFR and TS genes, which in bacteria and the unikont groups remain separate (the ancestral character), and which are fused (the derived character) in members of the Excavata, "SAR" clade, and Archaeplastida.

Amoebozoans form a clade that includes many species of amoeba with lobe- or tube-shaped pseudopodia. The resemblance of the slime molds to fungi is the result of convergent evolution. Molecular evidence shows them to be descended from unicellular amoebozoan ancestors. Cellular slime molds consist of haploid solitary amoeboid cells during the feeding stage. Two amoebas may fuse, and the resulting zygote undergoes meiosis and then mitosis, releasing new haploid amoebas.

FOCUS QUESTION 25.4

Describe the asexual reproductive stage of a cellular slime mold, during which the organism is multicellular.

Of the protists in the diverse clade **opisthokonts**, nucleariids are more closely related to fungi, and choanoflagellates are more closely related to animals.

25.4 Single-celled eukaryotes play key roles in ecological communities and affect human health

Structural and Functional Diversity in Protists Protists (organisms that aren't plants, animals, or fungi) are mostly unicellular, aquatic organisms, many with structurally complex cells. The three modes of nutrition—photoautotrophy, heterotrophy, and mixotrophy—are spread throughout the diverse protist lineages.

Photosynthetic Protists Many protists are **producers**. Most aquatic food webs are based on photosynthetic protists and prokaryotes. Protists are estimated to perform 30% of the world's photosynthesis. Diatom blooms can function as a carbon "pump" that sequesters carbon dioxide in ocean deposits.

FOCUS QUESTION 25.5

Satellite data have shown a negative correlation in most ocean regions between increasing sea surface temperatures and the biomass of photosynthetic protists and prokaryotes. What may account for this decrease in the growth of producers? What effects might these changes have on marine ecosystems and the global carbon cycle?

Symbiotic Protists Dinoflagellates are important photosynthetic symbionts in coral. Many termites harbor wood-digesting protists in their guts. Other protists are destructive parasites of plants and animals.

Effects of Human Health Protists that cause disease include the excavate *Trypanosoma*, which causes sleeping sickness; an amoebozoan in the genus *Entamoeba*, which causes amebic dysentery; and the alveolate **apicomplexans**, which cause diseases such as malaria. Nearly all apicomplexans are animal parasites, and most have complex life cycles that often include sexual and asexual stages and several host species. The control of malaria, caused by *Plasmodium*, is complicated by development of insecticide resistance in *Anopheles* mosquitoes, which spread the disease, and drug resistance in *Plasmodium*.

FOCUS QUESTION   25.6

a. Why is it so difficult for the human immune system to defend against *Trypanosoma*?

b. Apicomplexans have a nonphotosynthetic plastid called an apicoplast. Why might this plastid provide an effective target for the treatment of malaria?

Word Roots

archae-   = ancient (*archaeplastida:* one of four supergroups of eukaryotes. This monophyletic group, which includes red algae, green algae, and land plants, descended from an ancient protist ancestor that engulfed a cyanobacterium)

opisthios-   = posterior; **-kont** = flagellum (*opisthokont:* member of the diverse clade of organisms that descended from an ancestor with a posterior flagellum, including fungi, animals, and certain protists)

pseudo-   = false; **-podium** = foot (*pseudopodium:* a cellular extension of amoeboid cells used in moving and feeding)

stramen-   = straw; **-pilos** = hair (*stramenopiles:* one of the three major subgroups for which the "SAR" eukaryotic supergroup is named; includes diatoms and brown algae. Named for their characteristic "hairy" flagellum covered with fine, hairlike projections)

uni-   = one (*Unikonta:* one of four supergroups of eukaryotes, which includes cells with a single flagellum or amoebae with no flagella. This clade consists of amoebozoans and opisthokonts.)

Structure Your Knowledge

1. Evidence indicates that all plastids (except for one recent example) evolved from a cyanobacterium that was engulfed by an ancestral heterotrophic eukaryote (primary endosymbiosis). Fill in the blanks in the following diagram, which depicts the diversification of this ancestral eukaryote into red algae and green algae and then to various protist groups.

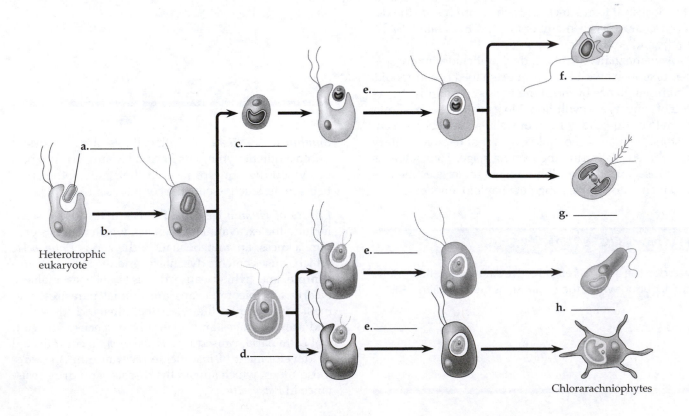

Heterotrophic eukaryote

Chlorarachniophytes

Test Your Knowledge

MATCHING: *Match the protist clades (A–H) with their characteristics and examples (1–8).*

_____ 1. modified mitochondria; *Giardia* and *Trichomonas*

_____ 2. plant-type chloroplasts; *Chlamydomonas* and *Ulva*

_____ 3. lobe-shaped pseudopodia; free-living amoeba, parasites, slime molds

_____ 4. phycoerythrin, no flagellated stage; *Porphyra*

_____ 5. secondary endosymbiosis of a red alga; diatoms and brown algae

_____ 6. amoebas with threadlike pseudopodia; forams and cercozoans

_____ 7. subsurface sacs; dinoflagellates, apicomplexans, ciliates

_____ 8. flagella with spiral or crystalline rod; *Trypanosoma* and *Euglena*

A. Alveolates

B. Amoebozoans

C. Diplomonads and parabasalids

D. Euglenozoans

E. Rhizarians

F. Green algae

G. Red algae

H. Stramenopiles

MULTIPLE CHOICE: *Choose the one best answer.*

1. Mixotrophs
   a. have both unicellular and multicellular parts of their life cycles.
   b. have both bacterial and archaeal features as a result of endosymbiosis.
   c. can be both autotrophic and phototrophic.
   d. can be both heterotrophic and autotrophic.
   e. can be both parasitic and free-living.

2. According to the theory of secondary endosymbiosis,
   a. multicellularity evolved when primitive cells incorporated prokaryotic cells that then took on specialized functions.
   b. the symbiotic associations found in coral result from the incorporation of dinoflagellates into coral animals.
   c. cells that had obtained their plastids through endosymbiosis were engulfed and themselves became plastids in heterotrophic eukaryotic cells.
   d. the infoldings and specializations of the plasma membrane led to the evolution of the endomembrane system.

   e. mitochondria originated first from a heterotrophic prokaryotic endosymbiont; plastids originated from a secondary endosymbiosis of a photosynthetic prokaryote.

3. The diplomonads and parabasalids are unique among eukaryotes in that
   a. they lack double membranes around their nuclei.
   b. they lack ribosomes.
   c. they lack plastids, and all other protist lineages have at least some members that are autotrophic.
   d. they have modified mitochondria and obtain energy anaerobically.
   e. they were an early branch from bacteria rather than from archaea.

4. Which of the following is evidence that the transition to multicellularity did not require the origin of large numbers of novel genes?
   a. Genomic studies indicate that multicellular *Volvox* has few novel genes compared with unicellular *Chlamydomonas*.
   b. The multicellular groups animals and fungi are unikonts. Some studies based on whether the genes for DHFR and TS are separate or fused indicate the unikonts were the first group of eukaryotes to diverge.
   c. The cadherin proteins of animals, which are involved in animal cell adherence to each other, contain only one novel domain not found in the cadherin proteins of unicellular choanoflagellates.
   d. Multicellularity evolved once and may have involved two stages: first colonies arose as dividing cells remained attached by their shared cell walls; then differentiated cells arose as some cells became specialized for reproduction.
   e. Both **a** and **c** provide evidence.

5. Which of the following is *not* true of seaweeds?
   a. They are found in the red, green, and brown algal groups.
   b. They have true roots that anchor them tightly to withstand the turbulence of waves.
   c. They may be a source of food and commercial products.
   d. Most are classified as charophytes, the closest living relatives of land plants.
   e. They are multicellular photoautotrophs.

6. The Chlorophyta, or green algae, are believed to share a common ancestor with plants because
   a. they are the only multicellular algal protists.
   b. they do not have flagellated gametes.
   c. they are the only protists with plastids that evolved from cyanobacteria.
   d. their chloroplasts are similar in structure and pigment composition to those of plants.
   e. several green algal groups are terrestrial.

7. Which protists are in the same eukaryotic super-group as animals?
   a. dinoflagellates
   b. forams
   c. parabasalids
   d. stramenopiles
   e. amoebozoans

8. Which supergroup includes the land plants?
   a. Unikonta
   b. Archaeplastida
   c. Excavata
   d. "SAR" clade
   e. viridiplantae

9. The "SAR" clade includes all of the following *except*
   a. red algae.
   b. brown algae.
   c. ciliates.
   d. dinoflagellates.
   e. forams.

10. How do diatoms affect global CO_2 levels?
    a. As surface seawater warms, the photosynthetic output of diatoms increases, removing more CO_2.
    b. Blooms of diatoms kill fish and other consumers, increasing decomposition and raising CO_2 levels.
    c. CO_2 absorbed by diatoms is "pumped" to the ocean floor when they die, lowering CO_2 levels.
    d. Because their glass-like walls are made of silica and not carbon, diatoms do not affect CO_2 levels.
    e. As the basis of all marine food webs, diatoms perform 50% of the world's photosynthesis and lower CO_2.

Chapter 26

The Colonization of Land by Plants and Fungi

Chapter Focus

This chapter describes the evolutionary history of plants and fungi, which facilitated each other's colonization of land. Plants are distinguished by walled spores produced in sporangia, apical meristems, and an alternation of generations in which the diploid sporophyte is the more conspicuous stage in all groups except the bryophytes. Lignified vascular tissues enable vascular plants to grow tall. The reduction and protection of the gametophyte within the parent sporophyte, the protection and dispersal of embryos in seeds, and the use of pollen for the transfer of sperm are reproductive adaptations of seed plants that enhance their success on land. Fungi play an essential ecological role, both as decomposers and in their mycorrhizal association with plant roots. Plants and fungi changed the biotic interactions and chemical cycling on land.

Chapter Review

For most of Earth's history, life was confined to aquatic environments. Cyanobacteria and protists colonized land by 1.2 billion years ago. Small plants and fungi began to move onto land about 500 million years ago, and forests appeared about 385 million years ago.

26.1 Fossils show that plants colonized land more than 470 million years ago

Evidence of Algal Ancestry Plants are multicellular, photosynthetic eukaryotes, as are certain groups of algae. Their cells have chloroplasts containing chlorophylls *a* and *b* and cell walls of cellulose. These characteristics, however, are also found in several algal groups. Only the green algal charophytes and land plants share the distinctive traits of rings of cellulose-synthesizing proteins and similarities in sperm cells (in those land plants that have flagellated sperm).

Results from the analysis of nuclear and chloroplast genes confirm that charophytes, particularly *Chara* and *Coleochaete*, are the closest living relatives of land plants.

Adaptations Enabling the Move to Land The tough polymer **sporopollenin** protects charophyte zygotes during fluctuations in water levels. With the accumulation of such traits, ancient charophytes living along the edges of ponds and lakes may have given rise to the first plants to colonize land.

FOCUS QUESTION   26.1

List some of the benefits and challenges of a terrestrial habitat for the first land plants.

Plant biologists debate different dividing lines for the plant kingdom: this textbook uses the traditional kingdom Plantae containing only embryophytes (plants with embryos).

Derived Traits of Plants Three derived traits distinguish land plants: alternation of generations, walled spores produced in sporangia, and apical meristems. The tough polymer sporopollenin protects spores as they disperse on land. Spores are produced and protected within multicellular **sporangia** on the sporophyte plant. Regions of cell division at the tips of roots and shoots, called **apical meristems**, produce linear growth and provide increasing access to environmental resources throughout a plant's life.

An **alternation of generations** is found in all land plants. In this life cycle, the multicellular haploid **gametophyte** alternates with the multicellular diploid **sporophyte**. The sporophyte produces haploid **spores**, which are reproductive cells that develop directly into new organisms. Plant embryos, which are protected and nourished within the female gametophyte plant, have *placental transfer cells* that facilitate the transfer of nutrients from parental tissues. This derived trait is the basis for referring to land plants as **embryophytes**.

FOCUS QUESTION   26.2

The gametophyte produces **a.** _____ by **b.** _____.
Following fertilization, the **c.** _____ divides by **d.** _____ to develop into the **e.** _____. The sporophyte produces **f.** _____ by **g.** _____. Spores germinate and develop into the **h.** _____.

The leaves and stems of many land plants are coated by a waxy **cuticle**, which protects the plants from water loss and microbial attack. Most plants also have **stomata**, pores through which gases are exchanged and water evaporates.

Early Land Plants   Fossils of spores with chemical and structural similarities to plant spores appeared 470 million years ago, and spores embedded in plant tissue have been found that date back 450 million years. By 400 million years ago, diverse plants are found in the fossil record, with key traits such as sporopollenin-containing spores, cuticles, stomata, water transport tissues, and branched sporophytes present.

26.2   Fungi played an essential role in the colonization of land

Symbiotic associations with fungi, called *mycorrhizae*, may have helped early plants, which lacked roots, colonize land.

Fungal Nutrition   Fungi are heterotrophs that obtain their nutrients by absorption; many secrete digestive enzymes into the surrounding food and absorb the resulting small organic molecules. Their **chitin**-strengthened cell walls protect cells from excessive water uptake by osmosis.

Most fungi are composed of multicellular filaments. Some are single-celled **yeasts**. Some species can grow as both filaments and yeasts. A typical multicellular fungal body consists of a network of filamentous **hyphae**, which form a mass called a **mycelium**. This body form provides an extensive surface area for absorption of nutrients.

Symbiotic fungi may penetrate plant cell walls with specialized hyphae called **haustoria**, which can extract or exchange nutrients with the plant hosts. Mutualistic relationships with plant roots are known as **mycorrhizae**. **Ectomycorrhizal fungi** form hyphal sheaths around a root and also grow into spaces in the root cortex. **Arbuscular mycorrhizal fungi** extend their hyphae through root cell walls, pushing in the plant cell membrane to form tubes within root cells.

FOCUS QUESTION   26.3

In what way(s) do mycorrhizal fungi benefit plants? In what way(s) do plants benefit the fungi?

Sexual and Asexual Reproduction   Many fungi reproduce both sexually and asexually by producing spores. In sexual reproduction, **plasmogamy**, the fusion of two parent mycelia, is later followed by **karyogamy**, the fusion of the haploid nuclei contributed by the two parents. The diploid zygotes or other structures then produce haploid spores by meiosis. Spores often germinate and grow as filamentous fungi that asexually produce haploid spores. Fungi that form visible mycelia are informally called *molds*.

The Origin of Fungi   Fungi and animals are more closely related to each other than to plants or to other eukaryotes. DNA data indicate that fungi are most closely related to **nucleariids**, a group of single-celled protists, suggesting that the ancestor of fungi was unicellular. Evidence indicates that fungi and animals evolved from different single-celled ancestors, and thus evolved multicellularity independently.

Estimates using the molecular clock suggest that the ancestors of animals and fungi diverged about 1–1.5 billion years ago. Although fungi most likely originated in water, the oldest fossils of fungi are of terrestrial species that date back 460 million years.

The Move to Land   Land may first have been colonized by cyanobacteria, algae, and fungi. Early fossils of land plants (from 405 million years ago) include evidence of mycorrhizae. Both a fungus and its plant partner must express certain genes to establish a symbiosis. Plant genes required for the formation of mycorrhizae are present in all major plant lineages, suggesting that these "*sym*" genes were present in the common ancestor of land plants.

Diversification of Fungi   Most mycologists recognize five major groups of fungi: chytrids, zygomycetes, glomeromycetes, ascomycetes, and basidiomycetes. Recent metagenomic studies, however, indicate that there may be as many as 1.5 million species and new groups of previously unknown unicellular fungi.

Chytrids have flagellated spores and may include some of the earliest fungal groups to diverge. Zygomycetes are probably a paraphyletic group that includes decomposing molds, parasites, and commensal symbionts. Glomeromycetes are arbuscular mycorrhizal fungi; about 80% of all plants form mycorrhizae with glomeromycetes. The ascomycetes, or sac fungi, are found in a wide variety of habitats. The basidiomycetes, or club fungi, include important decomposers such as mushrooms and ectomycorrhizal fungi.

FOCUS QUESTION   26.4

Describe how a fungus may produce spores both asexually and sexually. Are these spores haploid or diploid? Explain.

26.3   Early land plants radiated into a diverse set of lineages

Vascular plants have **vascular tissue**, a transport system composed of cells joined into tubes. Nonvascular plants, informally called **bryophytes**, do not have an extensive transport system.

Bryophytes: A Collection of Early Diverging Plant Lineages   Today there are three clades of small herbaceous plants: *liverworts, mosses,* and *hornworts*. Researchers think that these three clades were the first lineages to diverge from the land plant common ancestor, with fossil spores similar to those of liverworts dating back to 470 million years ago.

Bryophytes have **rhizoids**, which anchor them but do not play a major role in water and mineral absorption. The gametophyte is the dominant generation, and sperm must swim through a film of water to fertilize the egg.

Seedless Vascular Plants: The First Plants to Grow Tall   The earliest fossils of vascular plants date back 425 to 420 million years ago. Two clades of **seedless**

vascular plants are the **lycophytes** (club mosses and their relatives) and the **monilophytes** (ferns and their relatives).

In extant vascular plants, the diploid sporophyte is the larger and more complex plant. In seedless vascular plants, the sperm still swims to reach the eggs.

What types of vascular tissues are found in vascular plants? Water and minerals are conducted up from roots in **xylem**. The xylem of most vascular plants includes tube-shaped cells called **tracheids** whose cell walls are strengthened by **lignin**. **Phloem** transports sugars and other organic nutrients through cells arranged into tubes.

Unlike the rhizoids of bryophytes, **roots** both anchor vascular plants and absorb water and nutrients.

Leaves increase a plant's photosynthetic surface area. Lycophytes, the oldest lineage of extant vascular plants, have single-veined leaves known as **microphylls**. The branching vascular systems of **megaphylls**, typical of most other vascular plants, support larger leaves with greater photosynthetic capacity. Seedless vascular plants were abundant during the Carboniferous period (359 to 299 million years ago).

FOCUS QUESTION   26.5

Describe the traits of vascular plants that enable them to grow tall. What are the adaptive advantages of increased height?

26.4   Seeds and pollen grains are key adaptations for life on land

Seed plants arose about 360 million years ago. A **seed** is an embryo enclosed with a supply of nutrients within a protective coat. Seed plants include two major clades: **Gymnosperms** are called "naked seed" plants because their seeds are not enclosed in chambers. **Angiosperms** are plants in which seeds develop inside ovaries.

Terrestrial Adaptations in Seed Plants   The extremely reduced gametophytes of seed plants develop from spores within the sporangium and thus are protected from drying out and UV radiation. They are also nourished by the sporophyte plant.

The female gametophyte develops within the parent sporophyte enclosed in an **ovule**, which consists of protective **integuments** derived from sporophyte tissue, a megasporangium, and a haploid *megaspore*. The megaspore develops into an egg-producing female gametophyte.

A *microspore* develops into a male gametophyte contained in a sporopollenin-protected **pollen grain**. The transfer of pollen to the part of a plant containing ovules is called **pollination**. When pollen grains germinate, they release sperm through a pollen tube to the female gametophyte.

FOCUS QUESTION 26.6

Why is the evolution of pollen an important terrestrial adaptation?

An ovule develops into a seed, consisting of a sporophyte embryo surrounded by a food supply and enclosed within a protective coat derived from the integument of the ovule. Seeds enable plants to withstand harsh environmental conditions and provide stored food to a sporophyte embryo.

FOCUS QUESTION 26.7

Label the parts in the following generalized diagrams of an unfertilized ovule and a seed of a gymnosperm. Indicate whether structures are diploid or haploid tissues.

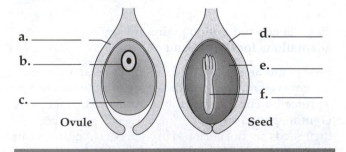

a. _____   d. _____
b. _____   e. _____
c. _____   f. _____

Ovule   Seed

Early Seed Plants and the Rise of Gymnosperms   The first seed plants to appear in the fossil record (around 360 million years ago) became extinct. The earliest gymnosperm fossils date from 305 million years ago.

During the Carboniferous period, early gymnosperms lived in ecosystems dominated by seedless vascular plants. Drier conditions during the Permian gave gymnosperms a selective advantage. Today, the cone-bearing **conifers** are found in large forests in northern latitudes.

The Origin and Diversification of Angiosperms   About 90% of all plant species—over 250,000 species—are angiosperms, or flowering plants.

The **flower**, the reproductive structure of an angiosperm, has up to four rings of modified leaves called floral organs. The outer **sepals** are usually green, whereas **petals** are brightly colored in most flowers that are pollinated by insects and birds. Pollen grains containing male gametophytes are produced in the anthers of **stamens**. **Carpels** produce female gametophytes. The carpel has a sticky stigma, which receives pollen, and a style, which leads to the **ovary**. The ovary contains ovules, which develop into seeds.

A **fruit** is a mature ovary that functions in the protection and dispersal of seeds.

FOCUS QUESTION 26.8

a. Name the four floral organs that make up a flower.

b. Name the parts of a seed.

c. What is a fruit?

Although details of their evolutionary origin remain unclear, angiosperms are thought to have originated about 140 million years ago and diversified over a period of 20 to 30 million years. Although fossils of 125-million-year-old *Archaefructus* indicate that this group was herbaceous and aquatic, all fossil seed plants that are proposed to be close relatives of angiosperms were woody.

The oldest angiosperm lineage is represented today by one woody species, *Amborella trichopoda*. The next lineage to diverge includes water lilies, with the lineage of star anise and its relatives diverging later. A lineage known as the magnoliids is more closely related to monocots and eudicots than to the basal angiosperms.

26.5 Land plants and fungi fundamentally changed chemical cycling and biotic interactions

Physical Environment and Chemical Cycling **Lichens** are symbiotic associations between a fungus and a green alga or a cyanobacterium. Lichens are important colonizers of bare rock and soil and may have helped facilitate the colonization of land by plants.

Plants also changed the physical environment by affecting the formation of soil and releasing oxygen to the atmosphere. Fungi are decomposers of organic matter, facilitating the essential recycling of chemical elements for plant growth. The removal of atmospheric CO_2 by the photosynthesis of seedless vascular plants in Carboniferous forests resulted in global cooling.

Biotic Interactions Plants, as photosynthesizers, and fungi, as decomposers, increase the availability of energy and nutrients for other organisms. Fungi function as mutualists in their mycorrhizal associations with vascular plant roots. The leaves or other parts of all plant species studied thus far also have symbiotic fungal **endophytes**, which benefit plants in various ways, such as producing toxins that deter herbivores; increasing the plant's tolerance for heat, drought, or heavy metals; or playing a role in defending against pathogens.

Fungi may be parasites on plants and animals, absorbing nutrients and, in some cases, causing serious diseases.

FOCUS QUESTION 26.9

Describe some beneficial as well as some harmful associations between fungi and other organisms. In what ways can decomposing fungi be both beneficial and harmful?

Animals influenced the evolution of plants, and vice versa. Both plant defenses against herbivory and herbivores that can overcome plant defenses would be favored by natural selection.

Pollinators can enter bilateral flowers from only one direction and thus more specifically pick up and disperse pollen to flowers of the same species. A study comparing the number of species in clades with bilaterally symmetrical flowers with the number in sister clades with radially symmetrical flowers supports the hypothesis that bilateral flower shape promotes speciation, perhaps by reducing gene flow between diverging populations.

The growing human population—with its demand for space, food, and natural resources—is pushing hundreds of species toward extinction each year. Tropical rain forests, where plant diversity is greatest, are rapidly being destroyed.

Word Roots

angio- = vessel (*angiosperm:* a flowering plant, which forms seeds inside a protective chamber called an ovary)

bryo- = moss; **-phyte** = plant (*bryophyte:* an informal name for a moss, liverwort, or hornwort; a nonvascular plant that lives on land but lacks some of the terrestrial adaptations of vascular plants)

-gamy = marriage (*plasmogamy:* in fungi, the fusion of the cytoplasm of cells from two individuals; occurs as one stage of sexual reproduction, followed later by karyogamy, the fusion of nuclei)

gymno- = naked; **-sperm** = seed (*gymnosperm:* a vascular plant that bears naked seeds—seeds not enclosed in specialized chambers)

mega- = large; **-phyll** = leaf (*megaphyll:* a leaf with a highly branched vascular system)

micro- = small (*microphyll:* in lycophytes. a small leaf with a single unbranched vein)

myco- = fungus; **rhizo-** = root (*mycorrhiza:* a mutualistic association of plant roots and fungus)

plasmo- = plasm; **-gamy** = marriage (*plasmogamy:* in fungi, the fusion of the cytoplasm of cells from two individuals; occurs as one stage of sexual reproduction, followed later by karyogamy)

rhizo- = root; **-oid** = like, form (*rhizoid:* a long tubular single cell or filament of cells that anchors bryophytes to the ground)

Structure Your Knowledge

1. The evolution of land plants involved adaptations to a terrestrial habitat. List some of these adaptations that evolved in the bryophytes, the seedless vascular plants, and the vascular plants.

2. What roles did fungi play in the colonization of land?

3. The following diagram presents a widely held view of plant phylogeny. Fill in blanks **a–e** with the informal names for the major groups, and blanks **f–k** with the names for the extant plant groups. To what events do the numbers **1, 2,** and **3** refer?

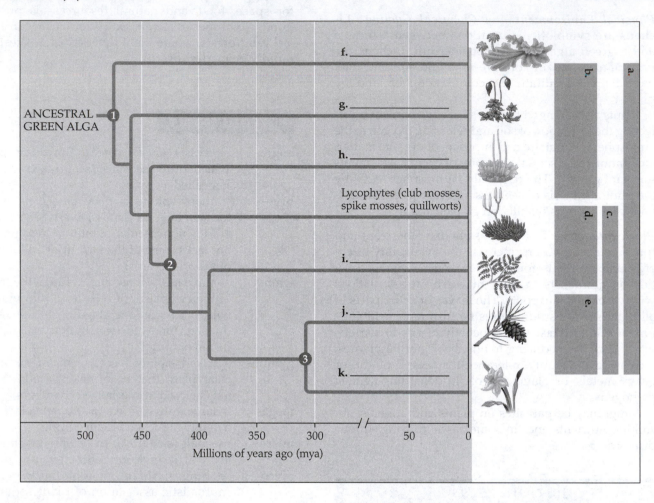

ANCESTRAL GREEN ALGA

f. _____

g. _____

h. _____

Lycophytes (club mosses, spike mosses, quillworts)

i. _____

j. _____

k. _____

500   450   400   350   300   50   0

Millions of years ago (mya)

a.

b.

c.

d.

e.

Test Your Knowledge

MULTIPLE CHOICE: *Choose the one best answer.*

1. Which of the following are adaptations for terrestrial life seen in all plants?
   a. chlorophylls *a* and *b*
   b. cell walls of cellulose and lignin
   c. sporopollenin and an embryo protected and nourished by the parent plant
   d. vascular tissue
   e. cuticle and stomata

2. Which of the following clades has been proposed by plant biologists as the clade that defines the plant kingdom?
   a. kingdom Plantae, which includes only the embryophyte clade
   b. kingdom Streptophyta, which also includes the charophytes

   c. kingdom Viridiplantae, which also includes all green algae (chlorophytes)
   d. kingdom Vasculata, which includes only the vascular plants
   e. Options **a, b,** and **c** have all been proposed as the clade that defines the plant kingdom.

3. Which of the following groups is *most* likely the closest relative of seed plants?
   a. charophytes
   b. monilophytes
   c. hornworts
   d. lycophytes
   e. mosses

4. The evolution of sporopollenin was important to the movement of plants onto land because
   a. it provided a mechanism for the dispersal of spores and pollen.

b. it provided the structural support necessary to withstand gravity.

c. it enclosed developing gametes and embryos in maternal tissue and prevented desiccation.

d. it provided a tough coating for spores so they could disperse on land.

e. it initiated the alternation of generations that is characteristic of all plants.

5. Bryophytes differ from other plant groups because
   a. their gametophyte generation is dominant.
   b. they lack gametangia.
   c. they have flagellated sperm.
   d. they are not embryophytes.
   e. all of the above are true.

6. Bryophytes were the dominant plants in the first 100 million years of plant evolution. By the Carboniferous period, seedless vascular plants formed giant forests. Why were these plants able to outcompete bryophytes?
   a. Their protected embryos were better able to withstand dry conditions, providing a selective advantage in dominating terrestrial habitats.
   b. They did not require water for their sperm to swim to fertilize the eggs, allowing them to colonize dry habitats.
   c. They were diploid, so they could grow faster and taller than haploid bryophytes.
   d. They had megaspores and microspores. This extra genetic variation enabled natural selection to favor taller growth.
   e. Their vascular tissue enabled them to grow tall, competing for light and more widely dispersing their spores.

7. Xylem and phloem are found in
   a. all plants.
   b. bryophytes, ferns, conifers, and angiosperms.
   c. only the gametophytes of vascular plants.
   d. the vascular plants, which include lycophytes, ferns, gymnosperms, and angiosperms.
   e. only the vascular plants with seeds.

8. If you could take a time machine back to the Carboniferous period, which of the following would you most likely encounter?
   a. creeping mats of low-growing bryophytes
   b. fields of tall grasses swaying in the wind
   c. swampy forests dominated by tree lycophytes, horsetails, and ferns
   d. huge forests of naked-seed trees filling the air with pollen
   e. the dominance of flowering plants

9. In which of the following groups must sperm no longer swim to reach the female gametophyte?
   a. bryophytes
   b. ferns
   c. gymnosperms
   d. angiosperms
   e. both **c** and **d**

10. A major difference between fungi and plants is that fungi
    a. have an absorptive form of nutrition.
    b. do not have cell walls.
    c. are not eukaryotic.
    d. are multinucleate but not multicellular.
    e. Both **a** and **d** are correct.

11. Fungi and animals appear to be more closely related to each other than either is to plants
    a. because neither of them are photosynthetic.
    b. based on similarities in cell structure.
    c. based on molecular analyses.
    d. because they moved onto land together.
    e. based on the homologous ultrastructure of their flagella.

12. Lichens are symbiotic associations that
    a. involve a fungus and a green alga or cyanobacterium.
    b. are found in the tissues of plants and may protect them from pathogens or herbivores.
    c. require moist environments to grow.
    d. fix nitrogen for absorption by plant roots.
    e. enabled plants to move onto land.

From the following phylogenetic tree showing the proposed relationships among fungi, choose the letter that is associated with each of the lineages named in questions 13 through 17.

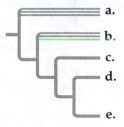

13. Ascomycetes _____
14. Basidiomycetes _____
15. Chytrids _____
16. Glomeromycetes _____
17. Zygomycetes _____

Chapter 27

The Rise of Animal Diversity

Chapter Focus

This chapter surveys the evolution of animal diversity, from an origin more than 700 million years ago, through the Cambrian explosion, the diversification in aquatic environments, to the colonization of land. This phylogeny shows the earliest appearances of selected animal groups in the fossil record. In this chapter you will learn how the clades Eumetazoa, Bilateria, Deuterostomia, Lophotrochozoa, and Ecdysozoa relate to this phylogeny.

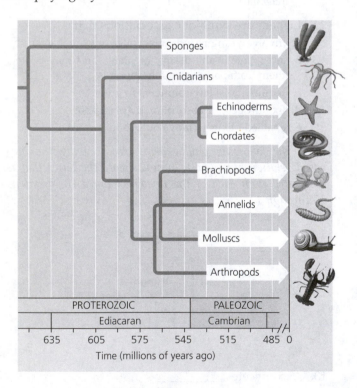

Chapter Review

Most animals are mobile, with specialized muscle and nerve cells that enable them to procure food and digest it with efficient digestive systems.

27.1 Animals originated more than 700 million years ago

Evidence indicates that animals evolved from single-celled eukaryotes that resembled choanoflagellates.

Fossil and Molecular Evidence Fossilized remains of steroids, which are today produced by a group of sponges, suggest that animals had arisen by 710 million years ago (mya). Molecular clock studies have estimated that sponges originated 700 million years ago. The earliest generally accepted fossils of animals, known as the **Ediacaran biota,** date from about 560 million years ago. Some are sponges; others may be related to cnidarians and molluscs.

Early-Diverging Animal Groups Sponges (phylum Porifera) are sessile animals mostly found in marine waters. Sponges are **filter feeders,** collecting food particles by the action of collared, flagellated **choanocytes** lining the central body cavity.

Sponges are *basal animals*, originating near the base of the animal phylogenetic tree. The resemblance between choanocytes and the cells of choanoflagellates support the molecular evidence that animals originated from a choanoflagellate-like ancestor.

Sponges lack true tissues. In addition to epidermal cells and choanocytes, their cell types include **amoebocytes.** These cells take up food, digest it, and carry nutrients to other cells.

One of the oldest lineages of *eumetazoans*, animals with true tissues, is the phylum Cnidaria. The cnidarians include hydrozoans, jellies, and sea anemones. Their simple anatomy consists of a sac with a central **gastrovascular cavity** and a single opening serving as both mouth and anus. They have simple muscles and a noncentralized nerve net.

FOCUS QUESTION 27.1

Describe how sponges and cnidarians differ in body form and feeding mechanisms.

27.2 The diversity of large animals increased dramatically during the "Cambrian explosion"

Evolutionary Change in the Cambrian Explosion
Fossils from the **Cambrian explosion** (a burst of evolutionary change about 535–525 mya) represent about half of all extant phyla and include the first animals with hard skeletons. These fossils are mostly of **bilaterians**—animals with a complete digestive tract and two-sided or bilateral symmetry. The Cambrian explosion saw the origin of large predators with novel prey-capturing adaptations and defensive adaptations in their prey.

FOCUS QUESTION   27.2

Describe the three hypotheses for the rapid diversification of animal phyla during the Cambrian period.

Dating the Origin of Bilaterians   Molecular estimates suggest that bilaterians had evolved by 670 million years ago, although the oldest fossil bilaterian is of molluscs that lived 560 million years ago. Fossils of eukaryotes with defensive structures from the Ediacaran period (635–542 mya) suggest that predatory bilaterian animals may have originated by that time, possibly between 670 and 635 mya.

27.3 Diverse animal groups radiated in aquatic environments

Animal Body Plans   An animal's set of morphological and developmental traits is often referred to as a **body plan.** Three features of animal body plans are symmetry, tissues, and body cavities.

Sponges lack body symmetry. An animal with **radial symmetry** has a round or barrel shape, and any slice through its center divides the animal into mirror images. Animals with **bilateral symmetry** have **dorsal** (top) and **ventral** (bottom) sides, distinct **anterior** (front) and **posterior** (back) ends, and left and right sides. Bilateral symmetry is associated with the concentration of sensory organs and a central nervous system in the head end, which is an adaptation facilitating unidirectional movement.

True tissues are functional groups of specialized cells separated from other tissues by membranous layers. Sponges lack true tissues. In all other animals, tissues and organs originate from *germ layers* which form in the embryo. **Ectoderm** develops into the outer body covering and, in some phyla, into the central nervous system; **endoderm** lines the digestive tract and associated organs. Cnidarians have only these two germ layers. The bilaterally symmetrical animals have a third, middle layer, the **mesoderm,** from which arise muscles and most other organs.

A fluid- or air-filled **body cavity** between the digestive tract and the outer body wall is called a *coelom.* A fluid-filled body cavity cushions internal organs, allows organs to grow and move independently of the outer body wall, and also functions as a skeleton in soft-bodied animals.

FOCUS QUESTION   27.3

Compare the body plans (symmetry, tissues, and body cavity or coelom) of sponges, cnidarians, and bilateral animals.

The Diversification of Animals   Evolutionary relationships among animals are estimated using ribosomal RNA genes, *Hox* genes, and many protein-coding genes, as well as mitochondrial genes and morphological traits.

There are five major points reflected in the current hypothesis of animal phylogeny: (1) All animals share a common ancestor and thus represent a clade called Metazoa. (2) Sponges are basal animals. (3) Eumetazoa is a clade of animals with true tissues. The **eumetazoans** include all animals except sponges and a few other groups. Basal members of the clade include the ctenophores and cnidarians, which have radial symmetry and two germ layers. (4) Most animal phyla belong to the clade Bilateria. The **bilaterians** are defined by the shared derived characters of bilateral symmetry and three germ layers. (5) Most animals are **invertebrates,** animals that lack a backbone. Only phylum Chordata includes **vertebrates.**

Bilaterian Radiation I: Diverse Invertebrates   The bilaterians diversified into three main clades: Lophotrochozoa, Ecdysozoa, and Deuterostomia. All species in clades Lophotrochozoa and Ecdysozoa are invertebrates. Indeed, 95% of known animal species are invertebrates.

Examples of animals in clade Lophotrochozoa are the ectoprocts (commonly called bryozoans), the molluscs (snails, clams, squids, and octopuses), and the annelids (segmented worms such as earthworms). Clade Ecdysozoa includes arthropods and nematodes (roundworms). The invertebrates in clade Deuterostomia are the hemichordates (such as acorn worms) and the echinoderms (sea stars, sea urchins, and sand dollars).

FOCUS QUESTION 27.4

In terms of species diversity, distribution, and vast numbers, **arthropods** are the most successful group of animals. List the three characteristics of the arthropod body plan that are most likely related to this success.

The earliest arthropods are found in the fossil record of the Cambrian explosion. The extinct *lobopods,* with identical body segments, may have been the group from which arthropods evolved. Trilobites were early arthropods with fairly uniform appendages. Studies show that onychophorans, a closely related invertebrate group, share all the arthropod *Hox* genes, including the unusual *Ubx* and *abd-A* genes. Thus, changes in the sequence or regulation of these genes (and not the origin of new *Hox* genes) may underlie the increased diversity in arthropod body segments.

Bilaterian Radiation II: Aquatic Vertebrates The deuterostome clade of bilaterians includes the **chordates.** Vertebrates are members of phylum Chordata.

Chordates have four structural trademarks, some of which may occur only during the embryonic stage: (1) a flexible rod called a **notochord** that provides skeletal support; (2) a dorsal, hollow nerve cord that develops into the brain and spinal cord; (3) **pharyngeal slits** (or **pharyngeal clefts**), which function in suspension feeding, as gills, or as parts of the head; and (4) a muscular, post-anal tail.

The ancestral chordate may have resembled a *lancelet,* the basal branch of chordates. *Tunicates,* which display the chordate traits only as larvae, are another early diverging group.

Vertebrates, which originated about 500 million years ago, are characterized by a backbone; a well-defined head with a brain, eyes, and other sense organs; and a skull. Early vertebrates include the *conodonts,* which were soft-bodied vertebrates with barbed, mineralized hooks in the mouth. Other early vertebrates also lacked jaws but had a muscular pharynx and were armored with mineralized bone. Extant jawless vertebrates include the *hagfishes* and *lampreys.*

First appearing in the fossil record about 450 million years ago, **gnathostomes,** or jawed vertebrates, became increasingly diverse. Their hinged jaws, paired fins, and tail facilitated a predatory lifestyle. Gnathostomes diverged into three lineages: chondrichthyans, ray-finned fishes, and lobe-fins (which include tetrapods).

Chondrichthyans, which have a skeleton made predominantly of cartilage, include sharks, rays, and their relatives.

Nearly all vertebrates belong to the clade of gnathostomes called Osteichthyes. Almost all living **osteichthyans** have an ossified endoskeleton and lungs or lung derivatives. Aquatic osteichthyans are mostly **ray-finned fishes** (Actinopterygii), named for the bony rays that support their fins.

The **lobe-fins** have muscular pectoral and pelvic fins supported by a series of rod-shaped bones. Only three lobe-fin lineages survive: the coelacanths, which were once thought to be extinct; the lungfishes, which have both gills and lungs; and the **tetrapods,** which are vertebrates with limbs and digits.

Complete this phylogenetic hypothesis that shows the major clades of vertebrates.

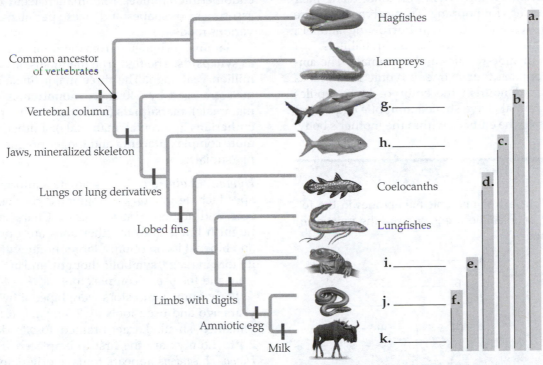

27.4　Several animal groups had features facilitating their colonization of land

Early Land Animals　Arthropods were among the first animals to colonize land, beginning about 450 million years ago, and vertebrates moved onto land 365 mya. Both the insects and the vertebrates possessed ancestral traits that facilitated the transition to terrestrial life, such as well-developed skeletal, muscular, digestive, and nervous systems.

Colonization of Land by Arthropods　Arthropod appendages have become specialized for walking, feeding, sensing, reproduction, and defense. The **cuticle** of an arthropod is a protective exoskeleton that provides attachment points for muscles. As some arthropods moved to land, the exoskeleton helped prevent desiccation and provided support. While most aquatic arthropods have gills, terrestrial species have internal surfaces for gas exchange, such as the tracheal systems of insects.

The oldest insect fossils are from about 416 million years ago. A major diversification occurred between 359–251 mya with the evolution of flight with wings that are extensions of the cuticle and the modification of mouthparts for specialized feeding on plants. A diversification of insects appears to have accompanied the radiation of flowering plants about 90 mya.

Terrestrial Vertebrates　Limbs support the weight of tetrapods on land, and feet with digits facilitate walking. A wide range of lobe-fins inhabited Devonian coastal wetlands, probably using stout fins to crawl through shallow waters and supplementing gas exchange with lungs. The fossil record documents how fins became progressively more limb-like. The first tetrapods appeared 365 million years ago. A diversity of tetrapods arose during the next 60 million years, most of which probably remained tied to the water.

Describe the 375-million-year-old "fishapod" fossil *Tiktaalik*. Is it considered to be a fish or a tetrapod?

Amphibians include salamanders, frogs, and caecilians. Salamanders, which may be aquatic or terrestrial, move with a lateral bending of the body. Frogs are more specialized for moving on land, using their powerful hind legs for hopping. Tropical caecilians are nearly blind, wormlike (legless) burrowing animals. Amphibians are most abundant in damp habitats.

Amniotes include reptiles and mammals. The **amniotic egg** contains four extraembryonic membranes that protect and nourish the embryo. Most reptiles and some mammals have shelled amniotic eggs. Most mammals retain the embryo within the mother's body.

FOCUS QUESTION 27.7

Identify the four extraembryonic membranes in the following sketch of an amniotic egg. What is the function of each membrane?

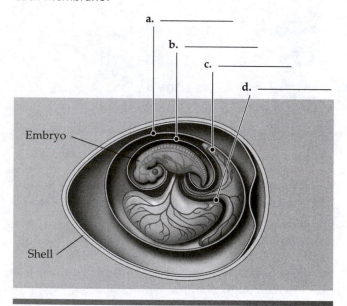

The common ancestor of living amphibians and amniotes likely lived about 350 million years ago. Over time, early amniotes expanded into drier environments. Reptiles and mammals are the two clades of amniotes today.

The **reptile** clade includes a number of extinct groups and tuataras, lizards and snakes (squamates), turtles, crocodilians, and birds. The earliest reptiles date from about 310 million years ago. Their derived characters include waterproof scales, and most have shelled amniotic eggs. Fertilization is internal.

Many reptiles are **ectothermic**—they absorb external heat rather than generating their own—but they may regulate their body temperature through behavioral adaptations. Birds are **endothermic,** warming their bodies with metabolic heat.

Mammals have mammary glands, which produce milk to nourish the young, and hair, which (along with a layer of fat under the skin) helps to insulate these endothermic animals. Mammalian teeth come in an assortment of shapes and sizes specialized for eating various foods.

Mammals originated from the amniote group known as **synapsids.** The first true mammals arose about 180 million years ago. The three major lineages of mammals emerged by 140 mya: **monotremes** (egg-laying mammals), **marsupials** (mammals with a pouch), and **eutherians** (placental mammals). Eutherians have a more complex **placenta** and longer pregnancy than do marsupials.

Human Evolution   The primates informally called apes include gibbons, orangutans, gorillas, chimpanzees, and humans (*Homo sapiens*). Characters that distinguish humans from other apes are upright stance and bipedal locomotion; a larger brain with the capacity for language, symbolic thought, and artistic expression; and the use of complex tools.

Early human ancestors were bipedal by 4.4 million years ago and used tools by 2.5 mya, but their brains remained small. Larger-brained fossils dating from 2.4 to 1.6 mya are the first to be placed in the genus *Homo. H. sapiens* appears to have originated in Africa about 200,000 years ago.

27.5   Animals have transformed ecosystems and altered the course of evolution

Ecological Effects of Animals   Evidence indicates that ocean waters were murky, had low oxygen levels, and were dominated by cyanobacteria before 600 million years ago. By 530 mya, fossil biochemical evidence indicates that the abundance of cyanobacteria had decreased, perhaps due to the evolution of filter-feeding animals. Algae became the dominant producers in the clearer waters. New feeding relationships emerged involving small animals that ate algae and larger predators that ate these smaller animals.

On land, animals transformed the simple ecosystem structure of producers and decomposers to one that included herbivores, predators, and detritivores.

Evolutionary Effects of Animals   Interacting species can exert selective pressures on one another that drive evolutionary changes in both species.

Humans have caused large environmental changes, altering the selective pressures on many species and leading to evolutionary change. Many of our actions have led to extinctions resulting from habitat loss, pollution, overharvesting, and predation or competition by introduced species.

FOCUS QUESTION 27.8

a. Give some examples of reciprocal selection pressures that have led to evolutionary changes in interacting species.

b. How might the origin of new species of animals stimulate evolutionary radiations in other groups?

Word Roots

arthro- = jointed; **-pod** = foot (*arthropod*: a segmented ecdysozoan with a hard exoskeleton and jointed appendages)

bi- = two (*bilaterian*: member of a clade of animals with bilateral symmetry and three germ layers)

choano- = a funnel; **-cyte** = cell (*choanocyte*: a flagellated feeding cell found in sponges; also called a collar cell)

ecdys- = an escape (*ecdysozoan*: member of a group of animal phyla identified as a clade by molecular evidence; many are animals that molt)

ecto- = outside; **-derm** = skin (*ectoderm*: the outermost primary germ layer in animal embryos)

endo- = within (*endoderm*: the innermost primary germ layer in animal embryos)

endo- = inner; **-therm** = heat (*endothermic*: referring to organisms that are warmed by heat generated by their own metabolism)

eu- = good (*eutherian*: Placental mammal; mammal whose young complete their embryonic development within the uterus, joined to the mother by the placenta)

gastro- = stomach; **-vascula** = a little vessel (*gastrovascular cavity*: a central cavity with a single opening in the body of certain animals (such as cnidarians) that functions in both the digestion and distribution of nutrients)

gnatho- = the jaw; **-stoma** = the mouth (*gnathostome*: member of the vertebrate clade possessing jaws)

in- = without (*invertebrate*: an animal without a backbone)

lopho- = a crest, tuft; **-trocho** = a wheel (*lophotrochozoan*: member of a group of animal phyla identified as a clade by molecular evidence; includes organisms that have lophophores or trochophore larvae)

marsupi- = a bag, pouch (*marsupial*: a mammal, such as a koala, kangaroo, or opossum, whose young complete their embryonic development inside a maternal pouch)

meso- = middle (*mesoderm*: the middle primary germ layer in a triploblastic animal embryo)

mono- = one (*monotreme*: an egg-laying mammal, such as a platypus or echidna)

noto- = the back; **-chord** = a string (*notochord*: a flexible rod that runs along the dorsal anterior-posterior axis of a chordate)

osteo- = bone; **-ichthy** = fish (*osteichthyan*: member of a vertebrate clade with jaws and mostly bony skeletons)

radia- = a spoke, ray (*radial symmetry*: symmetry in which the body is shaped like a pie or barrel and can be divided into mirror-imaged halves by any plane through its central axis)

syn- = together (*synapsid*: member of an amniote clade distinguished by a single hole on each side of the skull; includes the mammals)

tetra- = four; **-podi** = foot (*tetrapod*: a vertebrate clade whose members have limbs with digits; includes mammals, amphibians, and birds and other reptiles)

tunic- = a covering (*tunicate*: member of the clade Urochordata; sessile marine chordates that lack a backbone)

Structure Your Knowledge

1. The following simplified tree represents a current hypothesis of animal phylogeny. Fill in the blanks and then list some of the characteristics for clades **a.** through **d.** and name some representative animals in clades **e.** through **g.**

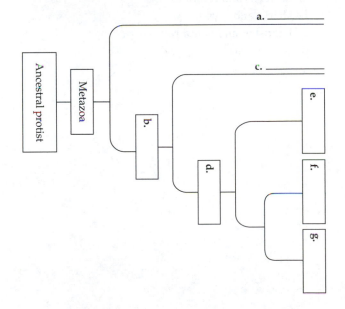

Test Your Knowledge

MULTIPLE CHOICE: *Choose the one best answer.*

1. Sponges differ from other animals in that
   a. they are completely sessile.
   b. they have radial symmetry and are suspension feeders.
   c. their simple body structure has no true tissues, and they have no symmetry.
   d. they are not multicellular.

2. Invertebrates include
   a. all animals except for those in the phylum Vertebrata.
   b. all animals without backbones.
   c. only animals that have hydrostatic skeletons.
   d. all animals without an endoskeleton.

3. Which of the following lists provides the *best* description of cnidarians?
   a. no real symmetry, two germ layers, tentacles that capture prey
   b. radial symmetry, two germ layers, gastrovascular cavity
   c. no real symmetry, without true tissues, choanocytes for trapping food particles
   d. bilateral symmetry, three germ layers, simple muscles and a nerve net

4. The basal group in clade Eumetazoa is
   a. Ectoprocta.
   b. Mollusca.
   c. Porifera.
   d. Ctenophora.

5. Which of the following groups represents the basal lineage(s) of vertebrates?
   a. lancelets and tunicates
   b. sharks and rays
   c. hagfishes and lampreys
   d. coelacanths

6. Which of the following organisms is *not* in the osteichthyan clade?
   a. bird
   b. shark
   c. cod
   d. coelacanth

7. Which of the following groups is *not* in the same lineage as the others?
   a. lizards
   b. birds
   c. dinosaurs
   d. crocodilians

8. Which of the following phenomena is *not* considered a factor in the amazing diversity of insect groups?
   a. wings and flight
   b. mouthparts specialized for feeding of plants
   c. the radiation of flowering plants, providing new food sources
   d. evolution of new *Hox* genes

9. Which of these characters was *not* an ancestral trait that facilitated the colonization of land by an animal group?
   a. exoskeleton of arthropod
   b. amniotic egg
   c. lungs
   d. vertebrate skeleton

10. Major changes in ocean conditions during the early Cambrian period were probably linked most closely to the evolution of
    a. cyanobacteria.
    b. algae.
    c. filter-feeding animals.
    d. large and motile predatory animals.

UNIT 5 Plant Form and Function

Chapter 28

Plant Structure and Growth

Chapter Focus

A rather large new vocabulary is needed to name the specialized cells and structures in a study of plant structure and growth. Focus your attention on how a plant's roots, stems, and leaves are specialized to function in absorption, support, transport, protection, and photosynthesis. The plant body is composed of dermal, vascular, and ground tissue systems. Apical meristems at the tips of roots and shoots create primary growth. The lateral meristems—vascular cambium and cork cambium—create secondary growth that adds girth to stems and roots.

Chapter Review

Angiosperms (flowering plants) are split into two major clades: *monocots* (with one cotyledon or seed leaf) and *eudicots* (with two cotyledons).

28.1 Plants have a hierarchical organization consisting of organs, tissues, and cells

A plant **organ** performs a specific function and is composed of several types of **tissues**, groups of cells with a common function.

The Three Basic Plant Organs: Roots, Stems, and Leaves As an evolutionary adaptation to the dispersed resources in a terrestrial environment, vascular plants have an underground **root system** for obtaining water and minerals from the soil and an aerial **shoot system** of stems and leaves for absorbing light and CO_2 for photosynthesis. *Photosynthates* from the shoot system nourish the roots. The shoot system depends on the water and minerals absorbed by roots.

The functions of **roots** include anchorage, absorption, and often storage of food. A *taproot system* is found commonly in tall, large plants. The one main deep **taproot** penetrates deep into the soil and gives rise to **lateral roots**. Small plants usually have a shallower *fibrous root system* in which many small roots grow from the stem. Such roots are called *adventitious*, as they grow in an unusual location. These roots then form their own lateral roots, forming mats that are good for anchoring the plant and preventing soil erosion. Most monocots have fibrous root systems.

Most absorption of water and minerals occurs through tiny **root hairs**, extensions of epidermal cells that are clustered near root tips. Most plants also form *mycorrhizal associations* with soil fungi, which increase mineral absorption. Modified roots may serve various functions, including support, storage, and oxygen absorption.

A **stem** consists of alternating **nodes**, the points at which leaves are attached, and segments between nodes, called **internodes**. Located at the tip or apex of a shoot is an **apical bud** with developing leaves, as well as compacted internodes. An **axillary bud** is found in the upper angle (axil) between a leaf and the stem. Axillary buds may develop into lateral branches.

Modifications of shoots include stolons (horizontal shoots that enable asexual reproduction), horizontally growing underground rhizomes, and food storage structures (such as tubers).

Leaves, the main photosynthetic organs of most plants, usually consist of a flattened **blade** and a **petiole**, or stalk. Monocot leaves usually have parallel major **veins** (the vascular tissue of leaves), whereas eudicot leaves have networks of branched veins. Modified leaves may function in support, protection, storage, or reproduction.

FOCUS QUESTION   28.1

Label the parts in the following diagram of a flowering plant. Is this plant a monocot or a eudicot? How can you tell?

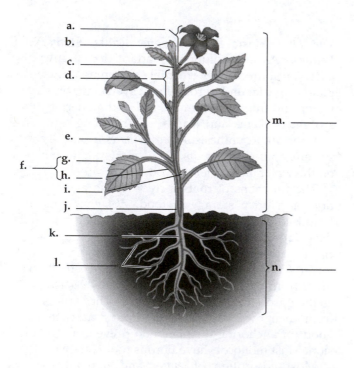

Dermal, Vascular, and Ground Tissues   Three fundamental tissue systems are continuous throughout the plant but have specific characteristics in each plant organ (root, stem, and leaf).

The **dermal tissue system** forms a protective outer layer. The **epidermis** is a single layer of tightly packed cells that covers nonwoody plants. The epidermis of leaves and most stems is covered with a **cuticle**, a waxy coating that prevents excessive water loss. Older stems and roots of woody plants are covered by a protective **periderm**. *Trichomes* are outgrowths of the shoot epidermis that may protect against herbivores and pathogens by forming a mechanical or chemical-secreting barrier.

The **vascular tissue system** provides mechanical support and consists of **xylem** and **phloem**, which function in long-distance transport of water and minerals (xylem) and of sugars (phloem). In angiosperms, the **stele**, or vascular tissue of an organ, takes the form of a solid central *vascular cylinder* in roots but is arranged in *vascular bundles* in stems and leaves.

The **ground tissue system** contains cells that function in photosynthesis, support, and storage. Ground tissue internal to vascular tissue is called **pith;** ground tissue outside the vascular tissue is called **cortex**.

Common Types of Plant Cells   Get ready for a lot of new terms while we review the five main types of plant cells. They may be easier to remember if you focus on how they are adapted for their specific functions.

Parenchyma cells carry on most of a plant's metabolic functions, such as photosynthesis and food storage. They usually lack secondary walls and have large central vacuoles. Most parenchyma cells retain the ability to divide and differentiate into other types of plant cells.

Collenchyma cells lack secondary walls but have thickened primary walls. Strands or cylinders of collenchyma cells provide flexible support for young parts of the plant and elongate along with the plant.

Sclerenchyma cells have thick secondary walls strengthened with **lignin**. The protoplasts of these specialized supporting cells often die at maturity. **Fibers** are long, tapered cells that are usually grouped in strands. **Sclereids** are shorter and irregular in shape, with very thick, lignified cell walls.

The water-conducting cells of xylem also die at functional maturity, leaving behind their secondary walls, which have interspersed thinner pits. **Tracheids**, found in all vascular plants, are long, thin, tapered cells. Water passes through pits from cell to cell. **Vessel elements**, found mainly in angiosperms, are wider, shorter, and thinner walled, with perforations in their end walls. They align to form long tubes known as **vessels**. Both tracheids and vessel elements have lignin-strengthened walls.

In the phloem of angiosperms, sugars flow through chains of cells called **sieve-tube elements**, or sieve-tube members. These cells remain alive at functional maturity but lack nuclei, ribosomes, and vacuoles. Fluid flows through pores in the **sieve plates** in the end walls between cells. The nucleus and ribosomes of an adjacent **companion cell**, which is connected to a sieve-tube element by numerous plasmodesmata, serve both cells.

FOCUS QUESTION   28.2

a. Which types of plant cells are dead at functional maturity?

b. Which types of plant cells lack nuclei at functional maturity?

28.2   Meristems generate new cells for growth and control the developmental phases and life spans of plants

Plants have tissues called **meristems**, which remain undifferentiated and can perpetually divide to form new cells.

Different Meristems Produce Primary and Secondary Growth   **Apical meristems**, located at the tips of roots and shoots and in the axillary buds of shoots, produce **primary growth**, resulting in elongation of roots and shoots. Herbaceous (nonwoody) plants usually undergo primary growth only. **Secondary growth** is an increase in diameter as new cells are produced by **lateral meristems**. **Vascular cambium** produces secondary xylem (wood) and secondary phloem; **cork cambium** produces the protective periderm. Cells that remain to divide in a meristem are called *initials* (or *stem cells*), whereas cells that are displaced from the meristem and become specialized in developing tissues are called *derivatives*.

FOCUS QUESTION   28.3

a. What is the difference between **indeterminate growth** and **determinate growth**?

b. What is the difference between primary and secondary growth?

Gene Expression and Control of Cell Differentiation   Differential gene expression within genetically identical cells leads to their differentiation into the various plant cell types. Cell-to-cell communication provides the positional information that largely determines a plant cell's fate. In the root epidermis of the model plant *Arabidopsis thaliana*, the gene *GLABRA-2* is expressed in epidermal cells that are in contact with only one underlying cortical cell, and these cells do not develop root hairs. The gene is not expressed in those cells in contact with two cortical cells, and they differentiate into root hair cells.

Meristematic Control of the Transition to Flowering and the Life Spans of Plants   At some point in the life of most angiosperms, the shoot apical meristems transition from *vegetative growth*, the production of leaves and stems, to *reproductive growth*, the production of flowers, fruits, and seeds. The timing and completeness of this switch from vegetative to reproductive growth relates to the length of a plant's life cycle—from germination to flowering to seed production to death. Flowering plants may be *annuals*, which complete their life cycle in a year or less; *biennials*, which have a life cycle spanning two years; or *perennials*, which live many years.

28.3   Primary growth lengthens roots and shoots

Herbaceous plants and the youngest parts of a woody plant consist of primary growth, which is produced by apical meristems. What does the primary growth of roots and shoots look like?

Primary Growth of Roots   The apical meristem of the root tip is protected by a **root cap**, which secretes a lubricating polysaccharide slime. The *zone of cell division* includes the apical meristem and its derivatives. In the *zone of elongation*, cells lengthen to many times their original size, which pushes the root tip through the soil. As the zone of elongation grades into the *zone of differentiation* (or maturation), cells become specialized in structure and function.

In the vascular cylinder of most eudicot roots, xylem cells radiate from the center in spokes, with phloem located between them. The vascular cylinder of a monocot may have a central core of undifferentiated parenchyma cells, located inside a ring of xylem and then a ring of phloem.

The ground tissue of roots consists of parenchyma cells in the cortex. The innermost layer of cortex is the one-cell-thick **endodermis**, which regulates the passage of materials into the vascular cylinder.

Lateral roots develop from the **pericycle**, the outer layer of cells in the vascular cylinder, and push through the cortex and epidermis.

Label the tissues in the following cross sections of a eudicot root and its vascular cylinder. Identify the functions of each of these tissues.

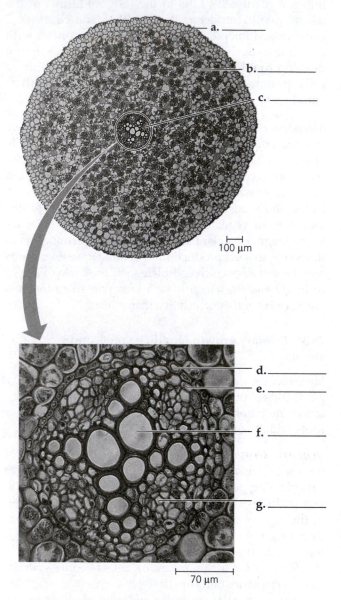

a. _____

b. _____

c. _____

100 µm

d. _____

e. _____

f. _____

g. _____

70 µm

Primary Growth of Shoots   The dome-shaped mass at the tip of the apical bud is the shoot apical meristem. **Leaf primordia** form on the sides of the apical meristem, and axillary bud meristems develop at the bases of the leaf primordia. The apical bud may exhibit **apical dominance**, inhibiting the growth of axillary buds. When axillary buds break dormancy, they develop into lateral shoots, or branches, each with its own apical bud, leaves, and axillary buds.

Elongation of the shoot occurs by cell elongation within young internodes. In grasses and some other plants, the bases of leaves and stems have *intercalary meristems,* which enable damaged leaves to regrow.

How are the tissues of a leaf organized? The cuticle-covered epidermis has tiny pores called **stomata** flanked by **guard cells**, which permit both gas exchange and evaporation of water from the leaf.

The ground tissue, or **mesophyll**, of a leaf consists of parenchyma cells containing chloroplasts. In many eudicot leaves, columnar *palisade mesophyll* is located above *spongy mesophyll,* which has loosely packed cells surrounding many air spaces.

Vascular tissue in the stem branches into the leaf and divides repeatedly, providing support and vascular tissue to the photosynthetic mesophyll. Each vein is enclosed in a *bundle sheath*, a protective ring of cells, which regulates exchange between the mesophyll and vascular tissue.

How are the tissues of a stem organized? The epidermis of the dermal tissue system covers stems. In most eudicots, vascular bundles are arranged in a ring, with the ground tissues pith inside and cortex outside the ring. Xylem is located internal to the phloem in the vascular bundles. In most monocot stems, the vascular bundles are scattered throughout the ground tissue. A layer of collenchyma just beneath the epidermis and sclerenchyma fiber cells strengthen the stem.

FOCUS QUESTION 28.5

Name the indicated structures in the following diagram of a leaf. In addition, identify the functions of these structures.

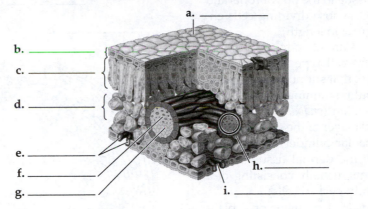

a. _____

b. _____

c. _____

d. _____

e. _____

f. _____

g. _____

h. _____

i. _____

28.4 Secondary growth increases the diameter of stems and roots in woody plants

The vascular cambium and cork cambium produce secondary growth, which occurs in all gymnosperms and in many eudicots but is rare in monocots.

The Vascular Cambium and Secondary Vascular Tissue In a woody stem, vascular cambium forms a continuous cylinder of meristematic cells that produce secondary xylem to the inside of the cylinder and secondary phloem to the outside. Some cambial initials (stem cells) produce radial lines of parenchyma cells called *vascular rays*, which function in lateral transport of water and nutrients, storage of carbohydrates, and wound repair.

Wood is the accumulation of secondary xylem cells with thick, lignified walls. Annual *growth rings* in temperate regions result from the seasonal cycle of cambium dormancy, early wood production (with larger, thinner cells in the spring), and late wood production (in the rest of the growing season). Scientists analyzing tree rings (*dendrochronology*) of Mongolian conifers from over the past five centuries have found evidence of warming temperatures during the twentieth century.

In older trees, a central column of *heartwood* consists of older xylem with resin-filled cell cavities; the *sapwood* consists of secondary xylem that still functions in transport.

Older secondary phloem is sloughed off; only that closest to the vascular cambium functions in sugar transport.

The Cork Cambium and the Production of Periderm The epidermis splits off during secondary growth and is replaced by new protective tissues produced by the cork cambium, a meristematic cylinder that first forms in the outer cortex in stems and in the outer pericycle in roots. The cork cambium produces *cork cells*, which develop *suberin*-impregnated walls and form a tightly packed protective layer. As secondary growth continually splits the outer layers of periderm, new cork cambia develop, eventually forming from parenchyma cells in the secondary phloem. **Lenticels** are spaces between cork cells through which gas exchange occurs. **Bark** refers to all tissues external to the vascular cambium: secondary phloem and periderm.

FOCUS QUESTION 28.6

Place the letters of the following tissues in the blanks so that they are in the order in which they are located in a tree trunk going from the outside in.

A. primary phloem   **E.** pith

B. secondary phloem   **F.** cork cambium

C. primary xylem   **G.** vascular cambium

D. secondary xylem   **H.** cork

___ ___ ___ ___ ___ ___ ___ ___

Word Roots

apic-   = the tip; **meristo-** = divided (*apical meristem:* embryonic plant tissue in the tips of roots and the buds of shoots, whose dividing cells enable the plant to grow in length)

coll-   = glue; **-enchyma** = an infusion (*collenchyma cell:* a flexible plant cell type that occurs in strands or cylinders that support young parts of the plant without restraining growth)

endo-   = inner; **derm-** = skin (*endodermis:* in plant roots, the innermost layer of the cortex, which surrounds the vascular cylinder)

epi-   = over (*epidermis:* the dermal tissue system of nonwoody plants, usually consisting of a single layer of tightly packed cells)

inter-   = between (*internode:* a segment of a plant stem between the points where leaves are attached)

meso-   = middle; **-phyll** = a leaf (*mesophyll:* leaf cells specialized for photosynthesis)

peri-   = around; **-cycle** = a circle (*pericycle:* the outermost layer in the vascular cylinder, from which lateral roots arise)

phloe-   = the bark of a tree (*phloem:* vascular plant tissue consisting of living cells arranged into elongated tubes that transport sugar and other organic nutrients throughout the plant)

sclero-   = hard (*sclerenchyma cell:* a rigid, supportive plant cell type usually lacking a protoplast and having thick secondary walls strengthened by lignin)

trachei-   = the windpipe (*tracheid:* a long, tapered water-conducting cell found in the xylem of nearly all vascular plants; functioning tracheids are no longer living.)

vascula-   = a little vessel (*vascular tissue system:* a transport system formed by xylem and phloem throughout a vascular plant; xylem transports water and minerals; phloem transports sugars)

xyl-   = wood (*xylem:* vascular plant tissue consisting mainly of tubular dead cells that conduct water and minerals upward from the roots to the rest of the plant)

Structure Your Knowledge

1. How does the indeterminate growth pattern of plants enable a plant to respond to the challenges and opportunities of its local environment?

2. In the following cross section of a young stem, label the indicated structures. Is this a stem of a monocot or a eudicot? How can you tell? Does this stem illustrate primary growth or secondary growth? How can you tell?

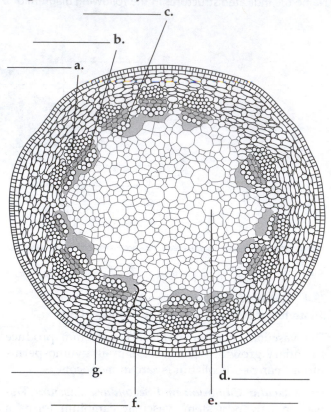

Test Your Knowledge

MULTIPLE CHOICE: *Choose the one best answer.*

1. Which of the following is an *incorrect* completion to this sentence? Monocots typically have
    a. no secondary growth.
    b. leaves with parallel veins rather than branching venation.
    c. a taproot rather than a fibrous root system.
    d. scattered vascular bundles in the stem rather than bundles in a ring.
    e. pith in the center of the vascular cylinder in the root.

2. Sieve-tube elements
    a. are responsible for lateral transport through a woody stem.
    b. control the activities of phloem cells that lack nuclei and ribosomes.
    c. have spiral thickenings that allow the cell to elongate along with a young shoot.
    d. are phloem transport cells with sieve plates in the end walls between cells.
    e. transport photosynthate from leaves through companion cells.

3. Axillary buds
    a. may exhibit apical dominance over the apical bud.
    b. form at nodes in the angle where leaves join the stem.
    c. grow out from the pericycle layer.
    d. are formed from intercalary meristems.
    e. only develop into vegetative shoots.

4. Which of the following structures provide structural support to a leaf?
    a. petioles
    b. vascular bundles
    c. suberin-impregnated cork cells
    d. epidermal cells with lignin-strengthened cell walls
    e. columnar cells of the palisade mesophyll

5. Which of the following plant structures show determinate growth?
    a. roots
    b. vegetative shoots
    c. adventitious roots
    d. eudicot leaves
    e. No parts of a plant show determinate growth.

6. Which of the following is *incorrectly* paired with the length of its life cycle?
    a. pine tree—perennial
    b. rose bush—perennial
    c. carrot—biennial
    d. marigold—annual
    e. corn—biennial

7. Which of the following is *incorrectly* paired with its function?
    a. perpendicularly oriented initials—form radial vascular rays
    b. lenticels—facilitate gas exchange in woody stems
    c. root hairs—absorb water and dissolved minerals
    d. root cap—protects the root as it pushes through soil
    e. vascular cambium—forms protective layer of cork

8. Secondary xylem and phloem are produced in a root by the
    a. pericycle.
    b. endodermis.
    c. vascular cambium.
    d. apical meristem.
    e. cork cambium.

9. As a young teenager, you carve your initials into the bark of a tree. When you return after college, where are your initials?
    a. much higher on the trunk
    b. out on the fifth or sixth branch from the main trunk
    c. at the same level as you carved them
    d. gone because secondary xylem has grown over them
    e. Both b and d could be correct.

MATCHING: *Match each of the plant tissues (1–10) with its description (A–J).*

_____ 1. cambial initials
_____ 2. collenchyma
_____ 3. endodermis
_____ 4. fibers
_____ 5. mesophyll
_____ 6. pericycle
_____ 7. periderm
_____ 8. pith
_____ 9. sclerenchyma
_____ 10. tracheids

A. layer from which lateral roots originate
B. tapered xylem cells with lignin in cell walls
C. parenchyma cells with chloroplasts in leaves
D. protective coat made of cork and cork cambium
E. bundles of long sclerenchyma cells
F. supporting cells with thickened primary walls
G. parenchyma cells inside vascular ring in eudicot stem
H. supporting cells with thickened secondary walls
I. cells that produce secondary xylem and phloem
J. cell layer in root regulating movement into vascular cylinder

Chapter 29

Resource Acquisition, Nutrition, and Transport in Vascular Plants

Chapter Focus

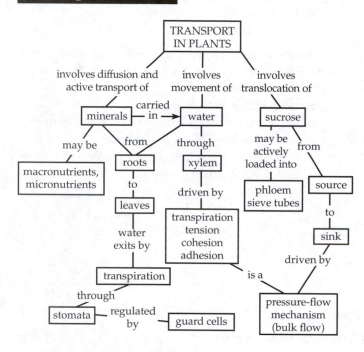

Phyllotaxy, the arrangement of leaves on the stem of a plant, may be one leaf per node (alternate, or spiral), two leaves per node (opposite), or multiple (whorled). Alternate phyllotaxy, with leaves emerging at a light-maximizing 137.5° angle, is most common in angiosperms.

The depth of the **canopy**, the layers of leaves in a plant community, affects productivity. When the canopy is too thick, the shaded, nonproductive lower leaves of a plant may be shed through *self-pruning*.

Root Architecture and Acquisition of Water and Minerals Evidence indicates that physiological mechanisms prevent a plant's roots from competing with themselves for the same limited resources. Mycorrhizae, symbiotic associations between roots and fungi, greatly increase the surface area for water and mineral absorption.

FOCUS QUESTION 29.1

a. What leaf orientation works best in low light levels? In bright sunny conditions?

b. How might root branching and physiology change in response to patches of nutrients such as nitrate?

Chapter Review

29.1 Adaptations for acquiring resources were key steps in the evolution of vascular plants

Evolutionary adaptations to life on land involved the specialization of roots to absorb water and minerals, shoots to absorb CO_2 and light, and vascular tissue to transport materials between roots and shoots. **Xylem** transports water and minerals upward from roots to shoots; **phloem** transports sugars throughout the plant to where they are stored or used.

Shoot Architecture and Light Capture The extent of branching and stem elongation represents a compromise between energy spent growing tall to avoid shading versus energy spent branching and increasing photosynthetic surface area. Increased stem thickness usually accompanies growth in height.

29.2 Different mechanisms transport substances over short or long distances

The Apoplast and Symplast: Transport Continuums What are the nonliving and living components of plants? The **apoplast** consists of plant cell walls, extracellular spaces, and interiors of dead cells (such as xylem transport cells). The protoplasts of plant cells are connected by plasmodesmata, forming a cytoplasmic continuum called the **symplast**.

Transport of water and solutes within plant tissues can occur by three routes: the *transmembrane route*, by crossing plasma membranes and cell walls as materials move from cell to cell; the *symplastic route*, moving through plasmodesmata; and the *apoplastic route*, moving along cell walls and extracellular spaces. Substances may change routes during transit.

Short-Distance Transport of Solutes Across Plasma Membranes Substances may move across the selectively permeable plasma membrane by passive or active transport, using protein carriers, ion channels, and cotransporters. Gated ion channels open or close in response to stimuli such as voltage, pressure, or chemicals.

A plant proton pump moves H^+ out of the cell, generating an energy-storing proton gradient and a membrane potential. Cotransport with H^+ can move solutes such as sucrose and ions such as NO_3^- across the plasma membrane.

Short-Distance Transport of Water Across Plasma Membranes The diffusion of free water across a membrane is called **osmosis**. **Water potential**, designated ψ *(psi)*, is a useful measurement for predicting the direction that water will move. It takes into account both solute concentration and physical pressure. Free water (not bound to solutes or surfaces) will flow from a region of higher water potential to one of lower water potential. Water potential is measured in **megapascals** (MPa). The water potential of pure water in an open container under standard conditions is assigned the value of 0 MPa.

The combined effect of solute concentration and pressure is shown by the *water potential equation*: $\psi = \psi_S + \psi_P$. This equation states that water potential is the sum of the solute potential and the pressure potential.

Solute potential (ψ_S), also called *osmotic potential*, is proportional to the molarity of a solution. Solutes lower the water potential, and a solution's ψ_S is always negative.

Pressure potential (ψ_P) measures the physical pressure on a solution and can have a positive or negative value. Negative pressure is tension. A plant **protoplast** that is expanding due to the osmotic uptake of water pushes against the cell wall, and the rigid cell wall exerts pressure against the protoplast. This positive pressure is called **turgor pressure**.

A **flaccid** plant cell ($\psi_P = 0$ MPa) bathed in a solution more concentrated than the cell will lose water by osmosis because the solution has a lower (more negative) ψ than the ψ of the cell. The cell's protoplast shrinks and pulls away from the cell wall (**plasmolysis**). A flaccid cell bathed in pure water has a ψ that is lower than the ψ of the water. Water will then enter the cell until enough turgor pressure builds up such that ψ_P and ψ_S are equal and opposite in magnitude, and $\psi = 0$ both inside and outside the cell. Net movement of water will then stop.

Plant cells are usually **turgid**; they have a greater solute concentration than their extracellular environment and turgor pressure keeps them firm. Loss of turgor causes **wilting** of a plant's leaves and stems.

Water-specific transport proteins called **aquaporins** increase the rate of water diffusion across a membrane.

FOCUS QUESTION 29.2

Practice using the water potential equation for plant cells in the following circumstances.

a. A flaccid plant cell has a water potential of –0.6 MPa. Fill in the water potential equation for this cell.

$$\psi_p = $$
$$+ \psi_s = $$
$$\overline{\psi = -0.6 \text{ MPa}}$$

b. The cell is then placed in a beaker of distilled water ($\psi = 0$ MPa). Fill in the equation for the cell after it reaches equilibrium in pure water. Explain what happens to the water potential of this cell.

$$\psi_p = $$
$$+ \psi_s = $$
$$\overline{\psi = 0 \text{ MPa}}$$

c. Explain what would happen to the same cell if it is placed in a solution that has a water potential of –0.8 MPa. Fill in the equation for the cell after it reaches equilibrium in the solution.

$$\psi_p = $$
$$+ \psi_s = $$
$$\overline{\psi = -0.8 \text{ MPa}}$$

Long-Distance Transport: The Role of Bulk Flow Long-distance transport in plants occurs by **bulk flow**, the movement of fluid driven by a pressure gradient. What physical aspects of phloem and xylem facilitate efficient bulk flow? Sieve-tube elements lack most cellular organelles, and vessel elements and tracheids are dead (and empty) at maturity.

29.3 Plant roots absorb essential elements from the soil

Water makes up 80–90% of the weight of a fresh plant. Carbon, oxygen, and hydrogen—the components of carbohydrates such as cellulose—are the most abundant elements making up the dry weight of a plant. Inorganic substances absorbed from the soil account for only 4% of the dry mass of a plant.

Macronutrients and Micronutrients **Essential elements** are those required for a plant to complete its life cycle from a seed to an adult that produces more seeds.

Hydroponic culture has been used to help identify the 17 elements that are essential in all plants.

Nine **macronutrients** are required by plants in relatively large amounts and include the six major elements of organic compounds as well as calcium, potassium, and magnesium.

Eight **micronutrients** are needed by plants in very small amounts, mainly as cofactors of enzymatic reactions.

FOCUS QUESTION   29.3

List some functions of each of the following elements, and indicate whether each is a macronutrient or a micronutrient.

a. magnesium

b. phosphorus

c. iron

Symptoms of Mineral Deficiency   Mobile nutrients move to young, growing tissues, so deficiencies of those nutrients first appear in older parts of the plant. Symptoms of a given mineral deficiency may be distinctive enough to allow diagnosis. Deficiencies of potassium, phosphorus, and especially nitrogen are most common.

FOCUS QUESTION   29.4

a. What is *chlorosis?*

b. Where in a plant would you first expect to see a deficiency of a relatively immobile element?

Soil Management   As ancient farmers learned to fertilize the soil, humans were able to establish permanent residences.

Minerals are recycled in natural ecosystems through the decomposition of **humus** (organic matter in the soil). Agriculture removes mineral nutrients; fertilization replaces them. In industrialized nations, commercially produced fertilizers (usually containing nitrogen, phosphorus, and potassium) are commonly used. Such fertilizers may rapidly leach from the soil, polluting streams and lakes. Manure, fishmeal, and compost (so-called "organic" fertilizers) contain organic material that slowly decomposes into inorganic nutrients.

The acidity of the soil affects cation exchange. It can also alter the chemical form of minerals and thus their ability to be absorbed by the plant. Managing soil pH is an important aspect of maintaining fertility and productivity.

FOCUS QUESTION   29.5

a. What is the N-P-K ratio?

b. How is soil pH related to toxic aluminum ions (Al^{3+})?

The Living, Complex Ecosystem of Soil   The texture of soil depends on particle size, which varies from coarse sand to silt to fine clay. The formation of soil begins with the weathering of rock as frozen water in crevices mechanically fractures rock and soil acids break down rocks chemically. Plant roots accelerate these processes. Topsoil is a mixture of mineral particles from rock, humus, and living organisms.

Loams, containing almost equal parts of sand, silt, and clay, are often the most fertile soils, having enough fine particles to provide a large surface area for retaining water and minerals but enough coarse particles to provide air spaces for respiring roots.

Positively charged ions, such as K^+, Ca^{2+}, and Mg^{2+}, adhere to the negatively charged surfaces of soil particles and are not easily lost by leaching. In **cation exchange**, hydrogen ions displace positively charged mineral ions from soil particles, making the ions available for absorption. Negatively charged ions, such as nitrate (NO_3^-), phosphate ($H_2PO_4^-$), and sulfate (SO_4^{2-}), tend to leach away more quickly.

Humus builds a crumbly soil that retains water, provides good aeration of roots, and supplies mineral nutrients.

The activities of the numerous soil inhabitants—bacteria, fungi, algae and other protists, insects, worms, nematodes, and plant roots—affect the physical and chemical properties of soil.

Describe the characteristics of a fertile soil.

29.4 Plant nutrition often involves relationships with other organisms

Plants form *mutualistic* relationships with many bacteria and fungi.

Soil Bacteria and Plant Nutrition The **rhizosphere** is the layer of soil surrounding plant roots. It contains high populations of soil bacteria called **rhizobacteria**, which are supported by plant secretions.

Some rhizobacteria, called *plant-growth-promoting rhizobacteria*, enhance plant growth by producing growth-stimulating chemicals or antibiotics, by absorbing toxins, or by making nutrients more available.

The **nitrogen cycle** describes the movement of nitrogen in an ecosystem, which involves soil microbes, plants, decomposers, and the atmosphere. *Ammonifying bacteria* decompose organic nitrogen in humus to ammonium (NH_4^+), and *nitrogen-fixing bacteria* convert atmospheric N_2 to NH_3. In the soil solution, ammonia forms NH_4^+. Different types of *nitrifying bacteria* perform the two-step oxidization of ammonium to nitrate (NO_3^-), the form of nitrogen most readily absorbed by roots. Most plants incorporate nitrogen into amino acids or other organic compounds in the roots and then transport these compounds or NO_3^- through the xylem to shoots. Some nitrate is lost to the atmosphere by the action of denitrifying bacteria.

Some free-living soil bacteria and bacteria of the genus *Rhizobium*, which are closely associated with the roots of legumes, are able to convert atmospheric nitrogen into ammonia through the process of **nitrogen fixation**. *Nitrogenase* is an enzyme complex that reduces N_2 by adding H^+ and electrons to form NH_3. This process is energetically expensive, and nitrogen-fixing bacteria rely on decaying material or plant roots for carbohydrates.

Fill in the types of bacteria (a–d) that participate in the nitrogen nutrition of plants. Indicate the form (e) in which nitrogen is transported in xylem to the shoot system.

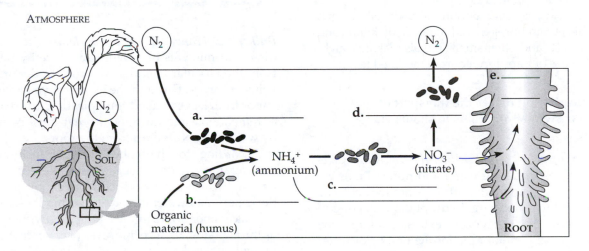

Legume roots have root swellings, called **nodules**, composed of plant cells with vesicles containing bacteria in a form called **bacteroides**. Nitrogen fixation requires an anaerobic environment. Lignified external layers of nodule cells limit gas exchange. In some nodules, the presence of oxygen-binding leghemoglobin keeps the concentration of free O_2 low and regulates the oxygen supply for the bacterial respiration required to provide ATP for nitrogen fixation. Bacteria

are supplied with carbohydrates, and nodule cells use the fixed nitrogen to make amino acids, which are then transported throughout the plant.

Fungi and Plant Nutrition Most plants have symbiotic associations between their roots and fungi called **mycorrhizae**. The fungus receives sugar from the plant while providing a large surface area for the absorption of water and minerals (especially phosphate) for the plant. The fungus secretes growth factors that stimulate root growth and branching and antibiotics that protect the plant from soil pathogens.

What are the two main types of mycorrhizae? Especially common in woody plants, **ectomycorrhizae** form a dense sheath or mantle of mycelium over the root surface. Hyphae extending from the mantle provide a huge absorptive surface area. Hyphae also grow into extracellular spaces in the root cortex, facilitating exchange between plant and fungus.

Hyphae of **arbuscular mycorrhizae**, also called endomycorrhizae, penetrate through root cell walls and form tubes that invaginate the cell membrane of cortex cells, forming dense branched structures called arbuscles that facilitate nutrient exchange. Over 85% of plant species have arbuscular mycorrhizae.

Most plants form mycorrhizae when they grow in their natural habitat. When seeds are planted in foreign soil lacking their symbiotic fungal species, the resulting plants show signs of malnutrition. The invasive plant garlic mustard has been shown to interfere with the growth of arbuscular mycorrhizae on native trees.

Epiphytes, Parasitic Plants, and Carnivorous Plants
What are some unusual nutritional adaptations of plants? **Epiphytes** are plants that grow on the surface of another plant but do not take nourishment from it. Many parasitic plants produce roots that function as haustoria that siphon nutrients from their host plant.

Living in acid bogs or other nutrient-poor soils, carnivorous plants obtain nitrogen and minerals by killing and digesting insects and other small animals that are caught in various types of traps formed from modified leaves.

29.5   Transpiration drives the transport of water and minerals from roots to shoots via the xylem

Absorption of Water and Minerals by Root Cells
Much of the absorption of water and mineral ions occurs along root tips, where root hairs are located. The soil solution flows along the hydrophilic walls of epidermal cells into the root cortex, exposing a large surface area of plasma membrane for the uptake of water and minerals. Active transport allows cells to accumulate essential minerals.

Transport of Water and Minerals into the Xylem Water and minerals that entered the symplast through a cortex or epidermal cell pass through plasmodesmata of cells of the **endodermis** and into the vascular cylinder, or stele. A ring of suberin around each endodermal cell, called the **Casparian strip**, prevents water and minerals from the apoplast from entering the stele without passing through a selectively permeable plasma membrane.

By a combination of diffusion and active transport, minerals move from endodermal cells to the apoplast and enter the nonliving xylem tracheids and vessel elements along with water.

FOCUS QUESTION   29.8

Label the following diagram of a section of a root. Letters **a** and **b** refer to transport routes of water and minerals; letters **c–i** refer to cell layers or structures.

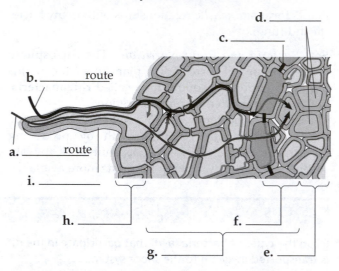

Bulk Flow Transport via the Xylem The water and dissolved minerals of **xylem sap** move by bulk flow from the vascular cylinder of a root to the branching veins of leaves. Through the evaporative loss of water vapor from leaves, called **transpiration**, the plant loses a tremendous amount of water that must be replaced by water transported up from the roots.

According to the **cohesion-tension hypothesis**, transpiration creates a negative-pressure pull on xylem sap, which is transmitted from shoots to roots by the cohesion of water molecules within the xylem.

Water vapor from saturated air spaces within a leaf exits through stomata. The thin layer of water that coats the mesophyll cells lining the air spaces begins to evaporate. The adhesion of the remaining water to the hydrophilic walls causes a curvature in the air-water interface. Because of the high surface tension of water, this curvature produces tension in the water layer. This negative pressure draws water from the cells, which is replaced by water from the xylem. Water moves along a gradient toward more negative water potentials, from xylem to neighboring cells to air spaces to the drier air outside the leaf.

The transpirational pull on xylem sap is transmitted from the leaves to the root tips by the cohesiveness of water, which results from hydrogen bonding between water molecules. Adhesion of water molecules to the hydrophilic walls of the narrow xylem cells also contributes to overcoming the downward pull of gravity.

The upward transpirational pull on the cohesive sap creates tension within the xylem, lowering the water potential so that water flows passively from the soil, across the cortex, and into the vascular cylinder.

A break in the chain of water molecules by the formation of a water vapor pocket in a xylem vessel, called cavitation, breaks the transpirational pull. The vessel cannot function in transport unless the chain of water molecules detours through adjacent vessels. *Root pressure* may allow some small plants to refill blocked vessels.

Xylem Sap Ascent by Bulk Flow: A Review The cohesion-tension mechanism results in the bulk flow of water and solutes from roots to leaves. Solar-powered tension produced by transpiration creates the pressure gradient that is responsible for long-distance transport of water and minerals.

FOCUS QUESTION 29.9

Explain the contribution of each of the following phenomena to the long-distance transport of water.

a. transpiration

b. tension

c. cohesion

d. adhesion

29.6 The rate of transpiration is regulated by stomata

For photosynthesis to occur, leaves must exchange gases through the stomata and provide a large internal surface area for CO_2 uptake, both of which increase evaporative water loss.

Stomata: Major Pathways for Water Loss A plant loses most of its water through stomata. Guard cells regulate the size of stomatal openings, which controls the rate of transpiration. Stomatal densities are both genetically and environmentally influenced.

Mechanisms of Stomatal Opening and Closing When the guard cells of most angiosperms become turgid and swell, their radially oriented microfibrils cause them to increase in length and bow outward, increasing the size of the gap between them.

What causes changes in turgor pressure in guard cells? These cells can actively accumulate potassium ions (K^+), which lowers water potential and leads to the osmotic inflow of water and an increase in turgor pressure. The flow of K^+ through specific membrane channels into the cell is coupled with the generation of a membrane potential by proton pumps that transport H^+ out of the cell. The exodus of K^+ (with water following) leads to a loss of turgor and stomatal closing.

Stimuli for Stomatal Opening and Closing The opening of stomata at dawn is related to at least three factors. First, light stimulates guard cells to accumulate K^+, triggered by the illumination of blue-light receptors that activate proton pumps. Second, the depletion of CO_2 within air spaces of the leaf as photosynthesis begins in the mesophyll stimulates stomata to open. The third factor is a daily rhythm of opening and closing that is endogenous to guard cells. Cycles that have intervals of approximately 24 hours are called **circadian rhythms**.

Environmental stress can cause stomata to close during the day. Guard cells lose turgor when water is in short supply. Also, **abscisic acid (ABA)**, a hormone produced in roots and leaves in response to a lack of water, signals guard cells to close stomata.

Effects of Transpiration on Wilting and Leaf Temperature When transpiration exceeds the water available, leaves wilt as cells lose turgor pressure. Transpiration produces evaporative cooling, maintaining a cooler temperature for critical enzymes in leaf cells.

Adaptations That Reduce Evaporative Water Loss Many **xerophytes**, plants adapted to arid climates, have leaves that are highly reduced; photosynthesis is carried out in thick, water-storing stems. Leaf cuticles may be thick, and stomata may be sheltered in depressions. A form of photosynthesis called **crassulacean acid metabolism (CAM)** is used by succulents of several plant families. These plants take in CO_2 at night, so their stomata can be closed during the day, reducing water loss.

FOCUS QUESTION 29.10

What is meant by saying that plants face a photosynthesis–transpiration compromise? Explain how a hot sunny day with a dry wind can affect this compromise in a plant.

29.7 Sugars are transported from sources to sinks via the phloem

Movement from Sugar Sources to Sugar Sinks   **Translocation**, the transport of photosynthetic products throughout the plant, occurs in the sieve-tube elements of phloem. **Phloem sap** may have a sucrose concentration as high as 30% and may also contain minerals, amino acids, and hormones.

Phloem sap flows from a **sugar source**, where it is produced by photosynthesis or the breakdown of starch, to a **sugar sink**, an organ that consumes or stores sugar. The direction of transport in any one sieve tube depends on the location of the source and sink connected by that tube.

In some species, sugar in the leaf moves through the symplast of the mesophyll cells to sieve-tube elements. In other species, sugar first moves through the symplast and then into the apoplast in the vicinity of sieve-tube elements and companion cells, which actively accumulate sugar.

Sucrose may be actively loaded into phloem using proton pumps and the cotransport of sucrose through cotransport proteins along with the returning protons. In sink tissues, the concentration gradient favors the diffusion of sugar out of sieve tubes because sugar is either used or converted into starch within sink cells.

FOCUS QUESTION   29.11

Label the components of the following diagram of the chemiosmotic mechanism involved in the active transport of sucrose into companion cells or sieve-tube elements. Which side of this diagram represents the symplast (the inside of a cell)?

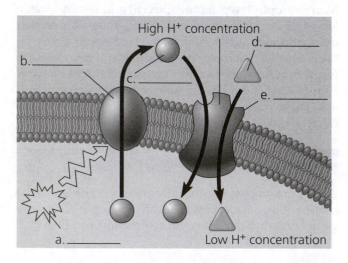

Bulk Flow by Positive Pressure: The Mechanism of Translocation in Angiosperms   The movement of phloem sap from source to sink is driven by positive pressure, called *pressure flow*. High solute concentration at the source lowers the water potential, and the resulting movement of water into the sieve tube produces positive pressure. At the sink end, the osmotic loss of water following the exodus of sucrose into the surrounding tissue results in a lower pressure.

In a phenomenon known as *self-thinning*, plants abort flowers, seeds, or fruits when they have more sinks than can be supported by sources.

■ Word Roots

apo-   = off, away; **-plast** = formed, molded (*apoplast:* everything external to the plasma membrane of a plant cell, including cell walls, intercellular spaces, and the space within dead structures such as xylem vessels and tracheids)

aqua-   = water; **-pori** = a pore, small opening (*aquaporin:* a membrane channel protein that specifically facilitates osmosis, the diffusion of water across the membrane)

arbo-   = tree; **-myco-** = a fungus; **-rhizo** = a root (*arbuscular mycorrhiza:* association of a fungus with a plant root system in which the fungus causes the invagination of the plant cells' plasma membranes)

circa-   = a circle (*circadian rhythm:* a physiological cycle of about 24 hours that persists even in the absence of external cues)

ecto-   = outside (*ectomycorrhiza:* association of a fungus with a plant root system in which the fungus surrounds the roots but does not cause invagination of the plant cells' plasma membranes)

endo-   = within, inner; **-derm** = skin (*endodermis:* in plant roots, the innermost layer of the cortex that surrounds the vascular cylinder)

macro-   = large (*macronutrient:* an essential element that an organism must obtain in relatively large amounts)

mega-   = large, great (*megapascal:* a unit of pressure equivalent to about 10 atmospheres of pressure)

micro-   = small (*micronutrient:* an essential element that an organism needs in very small amounts)

myco-   = a fungus; **-rhizo** = a root (*mycorrhiza:* a mutualistic association of plant roots and fungus)

osmo-   = pushing (*osmosis:* the diffusion of free water across a selectively permeable membrane)

sym-   = with, together (*symplast:* in plants, the continuum of cytoplasm connected by plasmodesmata between cells)

turg- = swollen (*turgor pressure*: the force directed against a plant cell wall after the influx of water and swelling of the cell due to osmosis)

xero- = dry; **-phyto** = a plant (*xerophyte*: a plant adapted to an arid climate)

Structure Your Knowledge

1. Describe the ways in which solutes may move across the plasma membrane in plants.

2. What are the differences between root nodules and mycorrhizae? How is each beneficial to plants?

3. What are the similarities between root nodules and mycorrhizae?

4. Both xylem sap and phloem sap move by bulk flow in an angiosperm. Compare and contrast the mechanisms for the movement of each.

Test Your Knowledge

MULTIPLE CHOICE: *Choose the one best answer.*

1. If a plant has a phyllotaxy of alternate leaves that emerge at a 30° angle from each previous leaf, one can surmise that
   a. the leaves are most likely very broad, indicating that the plant grows where there is sufficient water.
   b. the leaves are most likely very broad, indicating that the plant grows in low light levels.
   c. the leaves are most likely very thin, because emerging at such a close angle they would have to be thin to avoid shading each other.
   d. the leaves are most likely very thin, because the plant grows in a sunny, wet area.
   e. the canopy of the plant must be very deep with leaves spaced so close together, and lower leaves are lost through self-pruning.

2. Which of the following structures are *not* components of the symplast?
   a. sieve-tube elements
   b. xylem tracheids
   c. endodermal cells
   d. root hairs
   e. both **a** and **b**

3. Proton pumps in the plasma membranes of plant cells may
   a. generate a membrane potential that helps drive cations into the cell through specific channels.
   b. be coupled to the movement of K^+ into guard cells.

   c. drive the accumulation of sucrose in sieve-tube elements.
   d. contribute to the movement of ions through a co-transport mechanism.
   e. be involved in all of the above.

4. If a turgid plant cell placed in a solution in an open beaker becomes flaccid, which of the following statements is true?
   a. The water potential of the cell was initially higher than that of the solution.
   b. The water potential of the cell was initially equal to that of the solution.
   c. The pressure (ψ_P) of the cell was initially lower than that of the solution.
   d. The ψ_S of the cell was initially more negative than that of the solution.
   e. Turgor pressure disappeared because the cell no longer needed support.

5. Which of the following mechanisms facilitates the movement of NO_3^- into epidermal cells of the root?
   a. bulk flow of water into the root
   b. transport along with H^+ through a membrane cotransporter
   c. passage through selective channels, aided by the membrane potential created by proton pumps
   d. active transport through a nitrate pump
   e. simple diffusion across the cell membrane down its concentration gradient

6. Aquaporins are
   a. cytoplasmic connections between cortical cells.
   b. pores through the ends of sieve-tube elements through which phloem sap flows.
   c. openings in the lower epidermis of leaves through which water vapor escapes.
   d. openings into root hairs through which water enters.
   e. water-specific channels in membranes that speed the rate of osmosis.

7. The water potential of a plant cell
   a. is equal to 0 MPa when the cell is in pure water and is turgid.
   b. is equal to that of air.
   c. is equal to –0.23 MPa.
   d. becomes greater when K^+ are actively moved into the cell.
   e. becomes 0 MPa due to loss of turgor pressure in a concentrated sugar solution.

8. The most fertile type of soil is usually
   a. sand, because its large particles allow room for air spaces.
   b. loam, which has a mixture of fine and coarse particles.
   c. clay, because the fine particles provide much surface area to which minerals and water adhere.
   d. humus, which is decomposing organic material.
   e. moist and alkaline.

9. An advantage of organic fertilizers over commercially produced fertilizers is that they
   a. are more natural.
   b. release their nutrients over a longer period of time and are less likely to be lost to runoff.
   c. provide nutrients in the forms most readily absorbed by plants.
   d. are easier to mass produce and transport.
   e. are all of the above.

10. Negatively charged minerals
   a. are released from clay particles by cation exchange.
   b. are reduced by cation exchange before they can be absorbed.
   c. are converted into amino acids before they are transported through the plant.
   d. are bound when roots release acids into the soil.
   e. are leached away more easily than positively charged minerals.

11. The effects of mineral deficiencies involving fairly mobile nutrients will first be observed in
   a. older portions of the plant.
   b. new leaves and shoots.
   c. the root system.
   d. the color of the leaves.
   e. the flowers.

12. Micronutrients
   a. may be cofactors in enzymes.
   b. are required in very minute quantities.
   c. may be components of cytochromes.
   d. can be identified using hydroponic culture techniques.
   e. All of the above are true.

13. Chlorosis is
   a. yellowing leaves due to decreased chlorophyll synthesis in response to a mineral deficiency.
   b. a plant's uptake of the micronutrient chlorine, facilitated by symbiotic bacteria.
   c. the production of chlorophyll within the thylakoid membranes of a plant.
   d. a contamination of glassware in hydroponic culture.
   e. a root mold caused by wet soil conditions.

14. Nitrogenase
   a. is an enzyme complex that reduces atmospheric nitrogen to ammonia.
   b. is found in *Rhizobium* and other nitrogen-fixing bacteria.
   c. catalyzes the energy-expensive fixation of nitrogen.
   d. requires an anaerobic environment.
   e. is or does all of the above.

15. Epiphytes
   a. have haustoria for anchoring to their host plants and obtaining water and nutrients.
   b. are symbiotic relationships between specialized leaves and specific fungi.
   c. live in poor soil and capture and digest insects to obtain nitrogen.
   d. grow on other plants but do not obtain nutrients from their hosts.
   e. house nitrogen-fixing bacteria within vesicles of their root cells.

16. The nitrogen content of some agricultural soils may be improved by
   a. the synthesis of leghemoglobin by ammonifying bacteria.
   b. mycorrhizae on legumes.
   c. crop rotation with legumes that house nitrogen-fixing bacteria.
   d. cation exchange.
   e. the action of denitrifying bacteria.

17. Which of the following structures takes the form of a mycelial sheath over plant roots, from which hyphae extend and increase the surface area for absorption?
   a. root nodules
   b. haustoria
   c. arbuscular mycorrhizae
   d. ectomycorrhizae
   e. root epiphyte

18. What result would you predict if a gardener sterilizes all of her soil in an effort to reduce plant diseases?
   a. decreased plant growth due to a decrease in cation exchange
   b. increased plant growth due to lack of soil microbes draining away the photosynthetic output of the plants
   c. decreased growth due to lack of water-retaining bacteria
   d. increased growth due to the destruction of disease-causing microbes
   e. overall decreased plant growth due to destruction of mutualistic associations with microbes in the rhizosphere

19. The Casparian strip prevents water and minerals from entering the vascular cylinder through the
    a. plasmodesmata.
    b. endodermal cells.
    c. symplast.
    d. apoplast.
    e. xylem vessels.

20. Which of the following does *not* increase the surface area of a root available for absorbing water and minerals?
    a. mycorrhizae
    b. numerous branch roots
    c. root hairs
    d. cytoplasmic extensions of tracheids
    e. the large surface area of cortical cells

21. The formation of a curvature in the air-water interface along the cell walls of cells lining the air space of a leaf contributes to water transport by
    a. creating a more positive water potential than in the surrounding mesophyll cells.
    b. creating tension, thereby lowering pressure and the water potential of the leaf.
    c. raising the water potential of the surrounding saturated air.
    d. increasing the adhesion of water molecules to the cell walls.
    e. increasing the rate of transpiration from the leaf.

22. Adhesion is a result of
    a. hydrogen bonding between water molecules.
    b. the pull on the water column as water evaporates from the surfaces of mesophyll cells.
    c. tension within the xylem caused by a negative pressure.

    d. attraction of water molecules to hydrophilic walls of narrow xylem tubes.
    e. the high surface tension of water.

23. Which of the following factors does *not* stimulate the opening of stomata?
    a. daylight in a CAM plant
    b. depletion of CO_2 in the air spaces of the leaf
    c. stimulation of proton pumps that result in the movement of K^+ into the guard cells
    d. circadian rhythm of guard cell opening
    e. an increase in the turgor of guard cells

24. Your favorite houseplant is wilting. Which of the following statements describes the most likely cause and remedy for its declining condition?
    a. Water potential is too low; apply sugar water.
    b. The plant's stomata won't open; no remedy is available.
    c. The plant's cells are undergoing plasmolysis; water it.
    d. Cavitation has occurred; perform a xylem vessel bypass.
    e. Circadian rhythm has caused the stomata to open; place the plant in the dark.

25. Which of the following mechanisms explains the movement of sucrose from source to sink?
    a. evaporation of water and active transport of sucrose from the sink
    b. osmotic movement of water into the sucrose-loaded sieve-tube elements, creating a higher pressure in the source than in the sink
    c. tension created by pressure differences in the source and the sink
    d. active transport of sucrose through the sieve-tube cells driven by proton pumps
    e. the hydrolysis of starch to sucrose in mesophyll cells, which raises their water potential and drives the bulk flow of sap to the sink

Chapter 30

Reproduction and Domestication of Flowering Plants

Chapter Focus

This chapter describes the sexual and asexual reproduction of flowering plants. The flower produces spores by meiosis that grow into the haploid gametophyte stages of the life cycle: Microspores in the anther develop into male gametophytes in pollen grains, and a megaspore in the ovule produces an embryo sac. Pollination and the double fertilization of egg and polar nuclei are followed by the development of a seed with an embryo and endosperm, protected in a seed coat and housed within a fruit. Seed dormancy is broken following proper environmental cues and the imbibition of water.

Asexual reproduction allows successful plants to clone themselves. Agriculture makes extensive use of vegetative propagation with cuttings, grafts, and test-tube cloning. Plant biotechnologists are creating genetically modified (GM or transgenic) plants that have such traits as insect or disease resistance, herbicide tolerance, and improved nutritional value.

Chapter Review

Humans and their crop plants are an example of the mutually beneficial relationships common between angiosperms and animals.

30.1 Flowers, double fertilization, and fruits are unique features of the angiosperm life cycle

Plants exhibit an alternation of generations between haploid (*n*) and diploid (*2n*) plants. The diploid sporophyte produces haploid spores by meiosis. Spores develop into multicellular haploid male and female gametophytes, which produce gametes by mitosis. **Fertilization** yields zygotes that grow into new sporophyte plants. In angiosperms, the male and female gametophytes have become reduced to only a few cells.

Flower Structure and Function Flowers are reproductive shoots that usually contain four concentric whorls of floral organs—**carpels, stamens, petals, and sepals**—which attach to a part of the stem called the **receptacle**. A carpel consists of a sticky **stigma** at the top of a slender **style**, which leads to an **ovary.** The ovary encloses one or more **ovules.** A flower may have a single carpel or multiple fused carpels; either may be referred to as a **pistil.** Stamens consist of a *filament* and an **anther,** which contains microsporangia (pollen sacs). Petals are generally more brightly colored than sepals and may attract pollinators. Sepals enclose and protect the unopened floral bud.

A **complete flower** has sepals, petals, stamens, and carpels. **Incomplete flowers** lack one or more of these floral organs. Some incomplete flowers may be sterile, others may be *unisexual.* Floral variations include individual flowers or clusters of flowers called **inflorescences**, as well as diverse shapes, colors, and odors adapted to attracting different pollinators.

FOCUS QUESTION 30.1

Identify the flower parts in the following diagram.

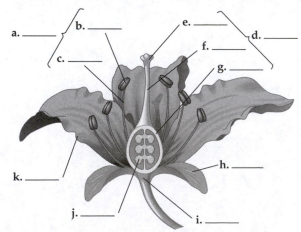

How does a flower form? A combination of environmental and internal cues triggers the transition of the shoot apical meristem to a *floral meristem.* Floral organ identity genes regulate the formation of the four whorls of floral organs. A mutation in one of these genes results in the placement of one type of floral organ where another type would normally develop. The **ABC hypothesis** of flower formation identifies three classes of organ identity genes, each of which is expressed in two adjacent whorls.

The following diagram depicting the ABC hypothesis shows the active genes in each whorl and the resulting anatomy of a wild-type flower.

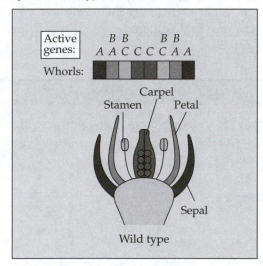

Wild type

a. Fill in the following table to indicate which organs are produced in the whorls in a normal flower. In a mutant that lacks a functional gene *A*, what gene expression pattern and resulting flower organ arrangement would be produced? (Remember that the lack of *A* activity removes the inhibition of gene *C*, and vice versa.)

Whorl	Genes Active	Organs in Normal Flower	Genes Active in Mutant A	Organs in Mutant A Flower
1	A			
2	AB			
3	BC			
4	C			

b. If you had a double-mutant plant that had no gene activity for *B* or *C*, what would the resulting flower look like?

Now that we have explored how the four floral organs form, let's see how a female gametophyte, also called an **embryo sac**, develops inside the ovary of a carpel. Two *integuments* surround each ovule except at the *micropyle*. The *megasporocyte* in the megasporangium of an ovule undergoes meiosis to form four haploid **megaspores**, only one of which survives. This megaspore divides by mitosis three times, forming the female gametophyte, which typically consists of eight nuclei contained in seven cells. At the micropylar end of the embryo sac, an egg cell is lodged between two cells called synergids, which help attract the pollen tube. Three antipodal cells are at the other end; and two nuclei, called polar nuclei, are in a large central cell.

How does the male gametophyte develop? Within each microsporangium (pollen sac) in the anther, diploid cells called *microsporocytes* undergo meiosis to form four haploid **microspores**. A microspore divides once by mitosis to produce a *tube cell* and a *generative cell*, which moves into the tube cell. The spore wall surrounding the cells thickens into the sculptured coat of the **pollen grain**. After the pollen grain lands on a receptive stigma, the tube cell begins to form the **pollen tube**.

The transfer of pollen from an anther to a stigma, called **pollination**, may be accomplished by wind, water, or animals, including bees, moths and butterflies, flies, birds, or bats.

Double Fertilization A pollen grain that lands on a receptive stigma absorbs moisture and germinates. The pollen tube grows through the style, and the generative cell divides to form two sperm. The pollen tube reaches the micropyle and releases its two sperm into the embryo sac. In **double fertilization**, one sperm fertilizes the egg to form the zygote, and the other combines with the polar nuclei to form a triploid nucleus, which will develop into a food-storing tissue called the **endosperm**.

a. Describe the male gametophyte.

b. Describe the female gametophyte.

c. What function does double fertilization serve?

Seed Development, Form, and Function Each ovule develops into a seed enclosed in the ovary, which develops into a fruit. The endosperm develops into a multicellular mass rich in nutrients that are stored for later use by the seedling. In many eudicots, the food reserves of the endosperm are transferred to the cotyledons before the seed matures.

The first mitotic division of the zygote creates a basal cell and a terminal cell. The basal cell divides to produce a thread of cells, called the *suspensor*, which anchors the embryo and transfers nutrients to it. The terminal cell divides to form a spherical proembryo, on which the cotyledons (two in eudicots, one in monocots) begin to form. The embryo elongates, and apical meristems develop at the apices of the embryonic shoot and root.

As it matures, the seed dehydrates and the embryo enters **dormancy**. The embryo and its food supply, the endosperm and/or enlarged cotyledons, are enclosed in a **seed coat** formed from the ovule integuments.

In a eudicot seed such as a bean, the embryo is an elongated embryonic axis attached to fleshy cotyledons. The axis below the cotyledonary attachment is called the **hypocotyl;** it terminates in the **radicle**, or embryonic root. The upper axis is the **epicotyl;** it leads to a shoot tip with a pair of leaves. Together, the epicotyl, young leaves, and shoot apical meristem are called the *plumule*.

The monocot seed found in grasses has a single thin cotyledon, called a *scutellum*, which absorbs nutrients from the endosperm during germination. A sheath called a **coleorhiza** covers the root, and a **coleoptile** encloses the young shoot.

Dormancy increases the chances that the seed will germinate when and where the seedling has a good chance of surviving. The specific cues for breaking dormancy vary with the environment and may include heavy rain, intense heat from fires, cold, light, or chemical breakdown of the seed coat within an animal's digestive tract. The viability of a dormant seed may vary from a few days to decades or longer.

Imbibition, the absorption of water by the dry seed, causes the seed to expand, rupture its coat, and begin a series of metabolic changes. Enzymes digest stored compounds, and nutrients are sent to growing regions.

The radicle emerges from the seed first, followed by the shoot tip. In many eudicots, a hook that forms in the hypocotyl is pushed up through the ground, pulling the delicate shoot and cotyledons behind it. Light stimulates the straightening of the hook, and the first foliage leaves begin to perform photosynthesis.

FOCUS QUESTION 30.5

How does the shoot tip break through the soil in maize and other grasses?

FOCUS QUESTION 30.4

Label the parts in the following diagrams of a bean seed and a corn seed.

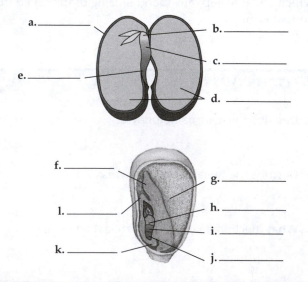

a. _____ b. _____

c. _____

e. _____ d. _____

f. _____ g. _____

l. _____ h. _____

i. _____

k. _____ j. _____

Fruit Form and Function The ovary of the flower develops into a **fruit**, which both protects and helps to disperse the seeds. Hormonal changes following fertilization cause the ovary to enlarge; its wall becomes the *pericarp*, or thickened wall of the fruit. Fruit usually does not set if a flower has not been pollinated.

A **simple fruit** is derived from a single carpel or several fused carpels; an **aggregate fruit** results from a flower with more than one separate carpel; a **multiple fruit** develops from an inflorescence whose many ovaries fuse together to become one fruit. Other floral parts (such as the receptacle) may develop into a fruit. These fruits—apples, for example—are called **accessory fruits**.

Fruits usually ripen as the seeds are completing their development.

a. List some ways in which fruits aid in seed dispersal.

b. What changes usually occur when a fleshy fruit ripens?

30.2 Flowering plants reproduce sexually, asexually, or both

Mechanisms of Asexual Reproduction Many plant species produce genetically identical copies (clones) of themselves through **asexual reproduction**. Asexual reproduction is an extension of the indeterminate growth of plants in which meristematic tissues can grow indefinitely and parenchyma cells can divide and differentiate into specialized cells. A common type of vegetative reproduction is **fragmentation**, the formation of whole plants from parts of a parent plant. In some species, the root system gives rise to many adventitious shoots that develop into a clone with separate shoot systems. Some plants, such as dandelions, can produce seeds asexually, a process called **apomixis**.

Advantages and Disadvantages of Asexual Versus Sexual Reproduction Advantages of asexual or **vegetative reproduction** include the production of clones of plants that are well suited to a certain environment and of progeny that are usually stronger than seedlings. Sexual reproduction generates variation in a population, an advantage when the environment changes. Seeds, which are almost always produced sexually, provide a means of dispersal to new locations and dormancy during harsh conditions.

a. What is an advantage of apomixis?

b. What is "selfing," and why is it a desirable trait in plants such as garden peas?

Mechanisms That Prevent Self-Fertilization Self-fertilization may be prevented by temporal or structural adaptations in flowers. In **dioecious** species, flowers are either staminate or carpellate and cannot self-fertilize because they are on separate plants.

The ability of flowers to reject their own pollen or that of closely related individuals, called **self-incompatibility**, depends on genes called *S*-genes. A plant population may have dozens of alleles of the *S*-gene. Gametophytic self-incompatibility depends on the *S*-allele of the pollen grain. RNA-hydrolyzing enzymes produced by the style may destroy the RNA of developing pollen tubes that have a matching allele. In sporophytic self-incompatibility, a signal-transduction pathway in stigma cells blocks germination of pollen with matching sporophyte S-alleles.

Further research on the mechanisms of self-incompatibility may allow plant breeders to manipulate crop species to assure hybridization.

Totipotency, Vegetative Reproduction, and Tissue Culture Many plant cells are **totipotent**, able to divide and produce a clone of the original plant. The cloning of plants by humans is called **vegetative propagation**. New plants may develop from stem cuttings when adventitious roots develop from a mass of dividing cells, called a **callus**, that forms at the cut end of the shoot. Adventitious roots can also form from a node in the shoot fragment.

Twigs or buds of one plant can be grafted onto a plant of a different variety or closely related species. The plant that provides the root system is called the **stock**, and the twig is called the **scion**. Grafting can combine the best qualities of different plants.

In test-tube cloning, whole plants can develop from pieces of tissue from the parent plant. A single plant can be cloned into thousands of plants by subdividing the undifferentiated calluses as they grow in tissue culture. Stimulated by proper hormone balances, calluses sprout shoots and roots and develop into plantlets, which can be transferred to soil to develop.

Foreign DNA may be inserted into individual plant cells, which then grow into genetically modified (GM) plants using *in vitro* culture.

How are vegetative reproduction and self-fertilization the same? How are they different?

30.3 People modify crops through breeding and genetic engineering

Almost all our crop species were first domesticated about 10,000 years ago. Natural hybridization between different species of plants is common, and humans have exploited the resulting genetic variation using artificial selection to develop and improve crops.

Plant Breeding   Breeders search for valuable traits in domesticated and wild species and hasten mutations by treating seeds or seedlings with mutagens. Wild plants with desirable traits are repeatedly crossed with the domesticated species until they are agriculturally suitable.

Plant Biotechnology and Genetic Engineering   Plant biotechnology refers both to the age-old use of plants to make products for human use and to the use of GM organisms in agriculture and industry. Genetic engineering allows plant biotechnologists to transfer genes between species, producing **transgenic** organisms.

The use of transgenic crops has increased dramatically. Cotton, maize, and potatoes have been engineered to contain genes from *Bacillus thuringiensis* that code for *Bt* toxin, reducing the need for spraying chemical insecticides. Researchers have developed other transgenic crops that are resistant to some herbicides, allowing farmers to "weed" crops without heavy tillage. Some transgenic plants are more resistant to disease or have improved nutritional quality.

Biomass from fast-growing plants, such as switchgrass and poplars, may help meet the world's energy demands. Enzymatic reactions break down plant cell walls into sugars, which are fermented and distilled to yield **biofuels**.

FOCUS QUESTION   30.9

Biofuels are considered carbon neutral, even though the burning of biofuels releases CO_2, just as the burning of fossil fuels does. Explain.

The Debate over Plant Biotechnology   GM organisms (GMOs) may present an unknown risk to human health or the environment. One concern is the transfer of allergens to a food source. Some GM crops may be safer than regular crops; for example, *Bt* maize contains lower levels of a harmful toxin that is produced by a fungus that infects insect-damaged crops.

Many ecologists are concerned about the effects of GM crops on nontarget organisms. A serious concern is the escape of herbicide- or disease-resistant genes through crop-to-weed hybridization that may create "superweeds." Various techniques are being developed to reduce the ability of transgenic crops to hybridize, such as developing male sterility so that transgenic pollen is not produced; engineering apomixis into the crop; introducing genes into chloroplast DNA so that those genes are not present in pollen; and developing flowers that self-pollinate and do not open so that pollen cannot escape.

Word Roots

a-   = without; **-pomo** = fruit (*apomixis:* the ability of some plant species to reproduce asexually through seeds without fertilization by a male gamete)

anth-   = a flower (*anther:* the terminal pollen sac of a stamen, where pollen grains containing sperm-producing male gametophytes form in an angiosperm)

carp-   = a fruit (*carpel:* the ovule-producing reproductive organ of a flower, consisting of the stigma, style, and ovary)

coleo-   = a sheath; **-rhiza** = a root (*coleorhiza:* the covering of the young root of the embryo of a grass seed)

di-   = two (*dioecious:* having male and female reproductive parts on different plants of the same species)

dorm-   = sleep (*dormancy:* a condition typified by extremely low metabolic rate and a suspension of growth and development)

endo-   = within (*endosperm:* a nutrient-rich tissue, formed by the union of a sperm with two polar nuclei during double fertilization, that provides nourishment to the developing embryo in angiosperm seeds)

epi-   = on, over (*epicotyl:* the embryonic axis above the point of attachment of the cotyledon(s) and below the first pair of miniature leaves)

hypo-   = under (*hypocotyl:* the embryonic axis below the point of attachment of the cotyledon(s) and above the radicle)

mega-   = large (*megaspore:* a spore that develops into a female gametophyte)

micro-   = small (*microspore:* a spore that develops into a male gametophyte)

stam-   = standing upright (*stamen:* the pollen-producing reproductive organ of a flower, consisting of an anther and filament)

Structure Your Knowledge

1. Draw a diagram of the major events in the life cycle of an angiosperm.
2. List the advantages and disadvantages of sexual and asexual reproduction in plants.
3. List some of the potential benefits and dangers of plant biotechnology.

Test Your Knowledge

FILL IN THE BLANKS

_____ 1. structure from which fruit typically develops
_____ 2. generation that produces spores by meiosis
_____ 3. fuel produced from organic matter
_____ 4. female gametophyte of angiosperms
_____ 5. embryonic root
_____ 6. embryonic axis above attachment point of cotyledon
_____ 7. protects a grass shoot as it breaks through the soil
_____ 8. twig or stem portion of a graft
_____ 9. spore that develops into a male gametophyte
_____ 10. mass of dividing cells at the cut end of a shoot

MULTIPLE CHOICE: *Choose the one best answer.*

1. A flower on a dioecious plant is
   a. complete.
   b. biennial.
   c. incomplete.
   d. staminate or carpellate.
   e. both **c** and **d**.

2. Which of the following structures is haploid?
   a. embryo sac
   b. anther
   c. endosperm
   d. microsporocyte
   e. both **a** and **b**

3. The terminal cell of an early plant embryo
   a. develops into the shoot apex of the embryo.
   b. forms the suspensor that anchors the embryo and transfers nutrients.
   c. develops into the endosperm when fertilized by a sperm nucleus.
   d. divides to form the proembryo.
   e. develops into the cotyledons.

4. In angiosperms, sperm are formed by
   a. meiosis in the anther.
   b. meiosis in the pollen grain.
   c. mitosis in the anther.
   d. mitosis in the pollen tube.
   e. double fertilization in the embryo sac.

5. The endosperm
   a. may have its nutrients absorbed by the cotyledons in the seeds of eudicots.
   b. is usually a triploid tissue.
   c. is digested by enzymes in monocot seeds following hydration.
   d. develops in concert with the embryo as a result of double fertilization.
   e. is or does all of the above.

6. A seed consists of
   a. an embryo, a seed coat, and a nutrient supply.
   b. an embryo sac.
   c. a gametophyte, a seed coat, and a nutrient supply.
   d. an enlarged ovary.
   e. an immature ovule.

7. Which of the following structures protects a bean shoot as it breaks through the soil?
   a. hypocotyl hook
   b. radicle
   c. coleoptile
   d. coleorhiza
   e. seed coat

8. Which of the following processes is a form of asexual or vegetative reproduction?
   a. apomixis
   b. grafting
   c. test-tube cloning
   d. fragmentation
   e. All of the above are forms of asexual reproduction.

9. Which of the following statements is *false*?
   a. Fly-pollinated flowers may smell like rotten meat.
   b. Bird-pollinated flowers are usually light-colored, strong smelling, and produce copious amounts of nectar.
   c. Insects pollinate approximately 65% of all flowering plants, and bees are the most important of these pollinators.
   d. The flowers of many temperate, wind-pollinated trees appear before the production of the leaves, which could interfere with pollen movement.
   e. Moths and bats, which are often active at night, are attracted to light-colored, fragrant flowers.

10. Self-incompatibility provides a plant
    a. a means of transferring pollen to another plant.
    b. a means of coordinating the fertilization of an egg with the development of stored nutrients.
    c. a means of destroying foreign pollen before it fertilizes the egg cell.
    d. a biochemical block to self-fertilization so that cross-fertilization is assured.
    e. a means of producing seeds without the need for fertilization.

11. Which of the following statements about fruit is *not* true?
    a. The primary functions of fruits include protection of dormant seeds and their dispersal away from the parent plant.
    b. An accessory fruit develops from the endosperm of a seed and is composed of triploid cells.
    c. Wind-dispersed fruits may have wings or "parachutes."
    d. Animals may help disperse seeds by "planting" fruits in underground caches.
    e. Most fruits develop from the ovary of a flower.

12. Which of the following techniques is being developed to reduce the likelihood that introduced genes for herbicide or insect resistance will escape to closely related weed species?
    a. engineering flowers to self-fertilize and remain closed so that pollen is not released
    b. breeding male sterility into transgenic plants so that they have no pollen to be transferred to nearby weeds
    c. engineering the gene of interest into chloroplast DNA, which is inherited from the maternal plant and is not transferred by pollen
    d. engineering crops, such as soybeans, that have no weedy relatives
    e. All of the above would reduce the risk of crop-to-weed transgene escape.

Plant Responses to Internal and External Signals

Chapter Focus

Plant hormones—auxin, cytokinins, gibberellins, brassinosteroids, abscisic acid, and ethylene—control growth, development, fruit development, and senescence as plants respond to their environments. Plant movements that occur in response to environmental stimuli include phototropism, gravitropism, and thigmotropism. The biological clock of plants controls circadian rhythms, such as the plant's stomatal opening and sleep movements. Blue-light photoreceptors are involved in phototropism and stomatal opening. Phytochromes absorb red and far-red light and are involved in seed germination and the photoperiodic control of flowering. Plants have various physiological responses to environmental stresses and pathogens.

Chapter Review

Plants are able to sense and adaptively respond to their environments, generally by altering their patterns of growth and development.

31.1   Plant hormones help coordinate growth, development, and responses to stimuli

Hormones, chemical signals transported throughout the plant body, trigger responses in target cells and tissues. Some signal molecules act only locally, and others occur at much higher concentrations than a typical hormone. Thus, many plant biologists use the term *plant growth regulators* to describe compounds that affect specific physiological processes in plants.

Very low concentrations of hormones, acting through signal transduction pathways, may control plant growth. Depending on the developmental stage of a plant and the site of action of a hormone and its relative concentration compared with other hormones, each hormone can have multiple effects on growth and development.

The Discovery of Plant Hormones   A **tropism** is a growth response of plant organs toward or away from a stimulus. The growth of a shoot toward light is called positive **phototropism**. A shoot bends toward the light when illuminated from one side because of the faster elongation of cells on the darker side than on the brighter side.

Darwin and his son observed that a grass seedling enclosed in its coleoptile would not bend toward light if its tip were removed or covered by an opaque cap. They postulated that a signal must be transmitted from the tip to the elongating region of the coleoptile. P. Boysen-Jensen demonstrated that the signal was a mobile substance, capable of being transmitted through a block of gelatin separating the tip from the rest of the coleoptile. In 1926, F. Went concluded that the chemical produced in the tip, which he called auxin, promoted growth and that it was in higher concentration on the side away from the light.

Researchers have not found a light-induced asymmetrical distribution of auxin in eudicots, but certain substances that may act as growth inhibitors have been shown to be more concentrated on the lighted sides of such stems.

A Survey of Plant Hormones   The textbook surveys six of the major classes of plant hormones that have been identified.

Auxin refers to any substance that stimulates elongation of coleoptiles. The major auxin extracted from plants is indoleacetic acid (IAA). Auxin is transported from the shoot tip apical meristem (where it is synthesized) down the shoot. This *polar transport* involves auxin transporters located only at the basal ends of cells.

FOCUS QUESTION 31.1

Fill in the following blanks to describe the *acid growth hypothesis* for the elongation of cells in response to auxin.

Auxin stimulates a plasma membrane's **a.** _____, resulting in an increased membrane potential and a **b.** _____ pH in the cell wall. This activates enzymes called **c.** _____, which break the cross-links between cellulose microfibrils. The increased membrane potential enhances ion uptake, which is followed by the **d.** _____. Thus, the two factors that cause the cell to elongate are **e.** _____ and **f.** _____.

For continued growth after this relatively fast initial elongation due to the osmotic uptake of water, the cell must produce more cytoplasm and wall material, processes that rely on changes in gene expression that are also stimulated by auxin.

Auxin is important in the *pattern formation* of a developing plant. The polar transport of auxin, produced by shoot tips, controls branching patterns. A reduced flow from a branch releases lateral buds below the branch from dormancy. Polar auxin transport also helps establish *phyllotaxy*, the arrangement of leaves. Peaks in auxin establish the sites of leaf primordium formation.

Auxin interacts with the plant hormones cytokinins and the newly discovered strigolactones in the control of apical dominance—the suppression of axillary buds.

The auxin IBA affects lateral root formation and is used commercially to enhance formation of adventitious roots at the cut base of stems or leaves. Synthetic auxins, such as 2,4-D, are used as herbicides that kill eudicot (broadleaf) weeds with a hormonal overdose. Auxin produced by developing seeds promotes fruit growth; synthetic auxins can induce fruit development in greenhouses.

Cytokinins—so named because they stimulate cytokinesis—are modified forms of adenine. Zeatin is the most common naturally occurring cytokinin. Cytokinins are produced in actively growing roots, embryos, and fruits. Acting with auxin, they stimulate cell division and affect differentiation. Cytokinins can retard aging of some plant organs.

In the 1930s Japanese scientists determined that the fungus *Gibberella* secretes a chemical that causes the hyperelongation of rice stems, or "foolish seedling disease." More than 100 different naturally occurring **gibberellins** (GAs) have now been identified.

Gibberellins, which are produced by roots and young leaves, stimulate growth by enhancing both cell elongation and division. *Bolting*, the growth of an elongated floral stalk, is caused by a surge of gibberellins.

In many plants, both auxin and gibberellins contribute to fruit set. Gibberellins are applied in the production of Thompson seedless grapes.

The release of gibberellins from the embryo signals seeds of many plants to break dormancy. GAs stimulate the production of enzymes that mobilize stored nutrients for a growing seedling.

Brassinosteroids are steroids that have effects very similar to those of auxin: They promote cell elongation and division, retard leaf abscission, and promote xylem differentiation. Identification of a brassinosteroid-deficient mutant of *Arabidopsis* helped to establish these compounds as nonauxin plant hormones.

The hormone **abscisic acid (ABA)** generally slows growth. The high concentration of ABA in maturing seeds inhibits germination and stimulates production of proteins that protect the seeds during dehydration. For dormancy to be broken in some seeds, ABA must be removed or inactivated, or the ratio of gibberellins to ABA must increase.

ABA also reduces drought stress. In a wilting plant, ABA causes stomata in the leaves to close. ABA may be produced in the roots in response to water shortage and then be transported to the leaves.

Plants produce the gas **ethylene** in response to stress and during fruit ripening and programmed cell death. Ethylene production may be induced by a high concentration of auxin.

How does a seedling pushing against an obstacle as it grows upward through the soil respond to this mechanical stress? It produces ethylene, which then initiates a growth pattern called the **triple response**. The three effects are a slowing of stem elongation, a thickening of the stem, and initiation of horizontal growth. As the effects of ethylene lessen, normal upward growth resumes. Researchers have identified *Arabidopsis* mutants that are ethylene insensitive *(ein)*, ethylene overproducing *(eto)*, or that undergo the triple response in the absence of ethylene. In these latter constitutive triple-response *(ctr)* mutants, the ethylene signal transduction pathway is permanently turned on, even though ethylene is not present.

Senescence is the programmed death of plant cells, organs, or the whole plant. On the cellular level, *apoptosis* requires the synthesis of new enzymes that break down many cellular components, which the plant may salvage. Ethylene is almost always associated with this programmed cell death.

Deciduous leaf loss protects against winter desiccation. Before leaves abscise in the autumn, many of the leaves' compounds are stored in the stem. A change in the balance between auxin and ethylene initiates changes in the abscission layer located near the base of the petiole, including the production of enzymes that hydrolyze polysaccharides in cell walls.

Ethylene initiates the breakdown of cell walls and the conversion of starches to sugars associated with fruit ripening. In an example of positive feedback, ethylene triggers ripening, and ripening triggers more ethylene production. Many commercial fruits are ripened in huge containers perfused with ethylene gas.

FOCUS QUESTION 31.2

Fill in the name of the hormone that is responsible for each of the following functions:

a. _____ stimulate cell division, growth, and germination; are anti-aging compounds

b. _____ promotes fruit ripening; initiates the triple response; is involved in apoptosis

c. _____ inhibits growth; maintains dormancy; closes stomata during water stress

d. _____ stimulates stem elongation, root branching, and fruit development; is involved in apical dominance

e. _____ promote cell elongation and division, and xylem differentiation; also retard leaf abscission

f. _____ promote stem elongation and seed germination; contribute to fruit set

---

31.2 Responses to light are critical for plant success

Photomorphogenesis The effect of light on plant morphology is called **photomorphogenesis**. The growth pattern of a sprouting potato shoot growing in darkness, called **etiolation**, facilitates the shoot breaking ground. When the shoot reaches light, stem elongation slows, leaves expand, the root system elongates, and chlorophyll production begins—all part of a process known as **de-etiolation** or greening. This transformation begins with the reception of light by a receptor protein and the triggering of signal transduction pathways.

An **action spectrum** is a graph of a physiological response across different wavelengths of light. An absorption spectrum shows the wavelengths of light a pigment absorbs. Close correlation between an action spectrum for a plant response and the absorption

spectrum of a pigment may indicate that the pigment is the photoreceptor involved in the response. **Blue-light photoreceptors** and **phytochromes**, which absorb mostly red light, are the two major classes of light receptors.

Molecular biologists have determined that plants use different types of pigments to detect blue light: *cryptochromes* are involved in inhibition of stem elongation when a seedling breaks ground, and *phototropin* is involved in phototropism and stomatal opening.

Studies of lettuce seed germination in the 1930s determined that red light (wavelength: 660 nm) increased germination the most, and that far-red light (730 nm) inhibited germination. The effects of red and far-red light are reversible, and a seed's response is determined by the last flash of light it receives.

A phytochrome is a photoreceptor that alternates between two forms; P_r absorbs red light and is converted to the P_{fr} form, and P_{fr} absorbs far-red light and is converted to the P_r form. The P_r to P_{fr} interconversion acts as a switch that controls various events in the life of the plant.

The signal that tells the plant that sunlight is present is the conversion of P_r (the form of phytochrome that the plant synthesizes) to P_{fr}. Subsequently P_{fr} triggers many plant responses to light, such as breaking seed dormancy.

FOCUS QUESTION 31.3

Explain how phytochrome can also indicate the *quality* of light available to a plant. What effect might this have on a shaded tree?

Biological Clocks and Circadian Rhythms A **circadian rhythm** is a physiological cycle with about a 24-hour frequency. These rhythms persist even when the organism is sheltered from environmental cues. Research indicates that the circadian clock is internal, although it is set (entrained) to a 24-hour period by daily environmental signals.

FOCUS QUESTION   31.4

What are free-running periods?

The molecular mechanism of the biological clock may be negative-feedback loops involving cyclical changes in the transcription of genes for transcription factors that inhibit, after a time delay, the expression of their own genes.

What is the effect of light on the biological clock? In darkness, the phytochrome ratio shifts toward P_r, in part because P_{fr} converts slowly to P_r in some plants, and in part because P_{fr} is degraded and new pigment is synthesized as P_r. When the sun rises, P_r is rapidly converted to P_{fr}, resetting the biological clock each day at dawn.

Photoperiodism and Responses to Seasons   Seasonal events in the life cycle of plants usually are cued by photoperiod, the relative length of night and day. A physiological response to photoperiod is called **photoperiodism**.

Researchers discovered that a variety of tobacco plant flowered only when day length was 14 hours or shorter. They termed it a **short-day plant**. **Long-day plants** flower when days are longer than a certain number of hours. The flowering of **day-neutral plants** is unaffected by photoperiod.

Researchers have determined that it is actually night length that controls flowering and other photoperiod responses. If the dark period is interrupted by even a few minutes of light, a short-day (long-night) plant will not flower. Photoperiodic responses thus depend on a critical night length.

Red light was found to be the most effective in interrupting a plant's perception of night length. A flash of red light breaks a dark period of sufficient length and prevents short-day plants from flowering, whereas a flash of red light during a dark period that is longer than the critical length will induce flowering in a long-day plant.

Some plants bloom after a single exposure to the required photoperiod. Others respond to photoperiod only after exposure to another environmental stimulus. **Vernalization** is the need for pretreatment with cold to induce flowering.

Leaves detect the photoperiod. The signal for flowering is called **florigen**. Recent research indicates that the gene *FLOWERING LOCUS T* (*FT*) produces the FT protein in the leaf, from where it travels to the shoot apical meristem and initiates flowering.

FOCUS QUESTION   31.5

Indicate whether a short-day plant (left column) and a long-day plant (right column) would flower or would not flower under the indicated light conditions.

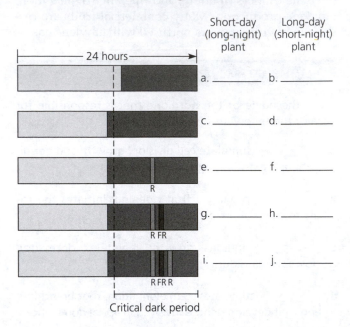

How do these results demonstrate the red/far-red photo-reversibility of phytochrome?

31.3   Plants respond to a wide variety of stimuli other than light

Gravity   Roots exhibit positive **gravitropism**, whereas shoots show negative gravitropism. According to one hypothesis, the settling of **statoliths**—plastids containing dense starch grains—in cells of the root cap triggers movement of calcium, which causes the lateral transport of auxin. Accumulating on the lower side of the growing root, auxin inhibits cell elongation, causing the root to curve downward. The settling of the protoplast and large organelles may distort the cytoskeleton and also signal gravitational direction.

Mechanical Stimuli   The stunting of growth in height and increase in girth of plants that are exposed to wind is an example of **thigmomorphogenesis**, changes in plant form in response to mechanical stimulation.

Most climbing plants have tendrils that coil rapidly around supports, exhibiting **thigmotropism** or directional growth in response to touch.

The sensitive plant *Mimosa* folds its leaves after being touched due to the rapid loss of turgor by cells in specialized motor organs called pulvini, located at the joints of the leaf. The message travels through the plant from the point of stimulation, perhaps as the result of electrical impulses called **action potentials**. These electrical messages may be used in plants as a form of internal communication.

Environmental Stresses   Plants respond to both **abiotic** environmental stresses and to **biotic** stresses such as pathogens and herbivores.

Mechanisms that reduce transpiration help a plant respond to water deficit. Guard cells lose turgor and stomata close. Abscisic acid, produced by leaves and roots in response to water deficit, acts on guard cell membranes to keep stomata closed. Wilted leaves of grasses may roll up, further reducing transpiration. These responses, however, reduce photosynthesis.

Plants adapted to wet habitats may have aerial roots that provide oxygen to their submerged roots. When roots of other plants are in waterlogged soils, oxygen deprivation may stimulate ethylene production, causing some root cells to die, which opens up air tubes within the roots.

Excess salts in the soil may lower the water potential of the soil solution, reducing water uptake by roots. Plants may respond to moderate soil salinity by producing organic solutes that lower the water potential of root cells and help prevent the accumulation of toxic salts. Special adaptations for dealing with high soil salinity have evolved in *halophytes*.

Transpiration creates evaporative cooling for a plant, but this effect may be lost on hot, dry days when stomata close. Above critical temperatures, plant cells produce **heat-shock proteins** that reduce protein denaturation.

Plants respond to cold stress by increasing the proportion of unsaturated fatty acids in membrane lipids, which maintain the fluidity of cell membranes. Subfreezing temperatures cause ice to form in cell walls, lowering the extracellular water potential and causing cells to dehydrate. Plants adapted to cold winters have adaptations to deal with freezing stress, such as changing the solute composition of the cytosol. Plants may also produce *antifreeze proteins* that reduce the formation of ice crystals.

31.4   Plants respond to attacks by herbivores and pathogens

Defenses Against Herbivores   Physical defenses, such as thorns, and chemical defenses, such as distasteful or toxic compounds, help plants cope with herbivory.

Some plants "recruit" predators of their herbivores. A combination of damage by caterpillars and a compound in caterpillar saliva stimulates leaves to release volatile compounds that attract parasitoid wasps. The wasps lay their eggs inside the caterpillar, and then the developing wasp larvae eat their host. These volatile compounds may also signal neighboring plants, making them less susceptible to herbivory.

Defenses Against Pathogens   The physical barrier of the plant's epidermis and/or periderm is the first line of defense against pathogenic viruses, bacteria, and fungi. Should pathogens enter the plant due to injuries or through openings such as stomata, the plant mounts a chemical defense.

Virulent pathogens are those to which the plant has no specific defense. Plants are generally resistant to most pathogens. These **avirulent** pathogens do not extensively harm or kill the plant.

Specific resistance is based on **gene-for-gene recognition** between a plant's resistance (R) proteins produced by the hundreds of its resistance genes and a pathogen molecule called an *effector* coded for by the pathogen's avirulence (*Avr*) gene. The recognition of pathogen effectors by a plant's R proteins triggers signal transduction pathways that produce local and systemic responses.

What are these defense responses that are activated upon detection of a pathogen? In the **hypersensitive response**, plant cells produce antimicrobial phytoalexins and *PR proteins* (pathogenesis-related proteins), which may attack the pathogen's cell wall. The hypersensitive response also stimulates strengthening of cell walls to slow the spread of the pathogen, after which the plant cells destroy themselves.

A hypersensitive response also involves the release of alarm signals, which stimulate PR protein production throughout the plant, resulting in **systemic acquired resistance**. The signaling molecule is probably *methyl-salicylic acid*, which is converted to **salicylic acid** in cells distant from the infection and activates nonspecific resistance to a diversity of pathogens for several days.

FOCUS QUESTION   31.8

a. In what ways may insects harm plants?

b. In what ways may insects benefit plants?

Word Roots

aux-   = grow, enlarge (*auxin*: a term usually referring to indoleacetic acid, a natural plant hormone that has a variety of effects, including cell elongation, root formation, secondary growth, and fruit growth)

circ-   = a circle (*circadian rhythm*: a physiological cycle of about 24 hours that persists even in the absence of external cues)

cyto-   = cell; **-kine** = moving (*cytokinin*: any of a class of related plant hormones that retard aging and act in concert with auxin to stimulate cell division, influence cell differentiation, and control apical dominance)

gibb-   = humped (*gibberellin*: any of a class of related plant hormones that stimulate growth in the stem and leaves, trigger the germination of seeds and breaking of bud dormancy, and (with auxin) stimulate fruit development)

hyper-   = excessive (*hypersensitive response*: a plant's localized defense response to a pathogen, involving the death of cells around the site of infection)

photo-   = light; **-trop** = turn, change (*phototropism*: growth of a plant shoot toward or away from light)

phyto-   = a plant; **-chromo** = color (*phytochrome*: a type of light receptor that mostly absorbs red light and regulates many plant responses, such as seed germination and shade avoidance)

stato-   = standing, placed; **-lith** = a stone (*statolith*: a specialized plastid that contains dense starch grains and may play a role in detecting gravity)

thigmo- = a touch; **morpho-** = form; **-genesis** = origin (*thigmomorphogenesis*: a response in plants to chronic mechanical stimulation, resulting from increased ethylene production; an example is thickening stems in response to strong winds)

Structure Your Knowledge

1. List some agricultural uses of plant hormones.

2. Develop a concept map to illustrate your understanding of photoperiodism and the control of flowering. Remember to include the role of the biological clock.

3. Briefly describe the steps in the hypersensitive response and in systemic acquired resistance that result from a plant's encounter with an avirulent pathogen.

Test Your Knowledge

TRUE OR FALSE: *Indicate T or F and then correct the false statements.*

_____ 1. Abscisic acid is necessary for a seed to break dormancy.

_____ 2. Cytokinins, synthesized in the root, seem to counteract apical dominance.

_____ 3. Thigmomorphogenesis is a growth response of a seedling that encounters an obstacle while pushing upward through the soil.

_____ 4. Action potentials are electrical impulses that may be used as internal communication in plants.

_____ 5. A physiological response to day or night length is called a circadian rhythm.

_____ 6. Vernalization is the pretreatment with cold before a plant flowers.

_____ 7. A virulent pathogen is one to which a plant has a specific resistance based on gene-for-gene recognition.

MULTIPLE CHOICE: *Choose the one best answer.*

1. The body form within a species of plant may vary more than that within a species of animal because
   a. growth in animals is indeterminate.
   b. plants respond adaptively to their environments by altering their patterns of growth and development.
   c. plant growth and development are governed by many hormones.
   d. plants can respond to environmental stress.
   e. all of the above are true.

2. Polar transport of auxin involves
   a. the accumulation of higher concentrations on the side of a shoot away from light.
   b. the movement of auxin from roots to shoots.
   c. the movement of auxin through transport proteins located only at the basal end of cells.
   d. the unidirectional active transport of auxin into and out of parenchyma cells.
   e. the settling of statoliths.

3. According to the acid growth hypothesis,
   a. auxin stimulates membrane proton pumps.
   b. a lowered pH outside the cell activates expansins, which break cross-links between cellulose microfibrils.
   c. the membrane potential created by the proton pumps enhances ion uptake, increasing the osmotic movement of water into cells.
   d. cells elongate when they take up water by osmosis.
   e. all of the above are involved in cell elongation.

4. Which of the following situations would most likely stimulate the development of axillary buds?
   a. a large quantity of auxin traveling down from the shoot and a small amount of cytokinin produced by the roots
   b. a reduction of auxin traveling from the shoot and thus a reduced production of strigolactone
   c. a 1:1 ratio of gibberellins to auxin
   d. the absence of cytokinins caused by the removal of the terminal bud
   e. an increase in the concentration of brassinosteroids produced by leaves

5. The growth inhibitor in seeds is usually
   a. abscisic acid.
   b. ethylene.
   c. gibberellin.
   d. a small amount of ABA combined with a larger concentration of gibberellins.
   e. a high cytokinin-to-auxin ratio.

6. A circadian rhythm
   a. is controlled by an external oscillator.
   b. is seen in the oscillation of stomatal opening and production of photosynthetic enzymes in a 24-hour period.
   c. involves an internal biological clock that is not set by daily environmental signals.
   d. provides the signal for flowering.
   e. involves all of the above.

7. A flash of far-red light during a critical-length dark period will
   a. induce flowering in a long-day plant.
   b. induce flowering in a short-day plant.
   c. not influence flowering.
   d. increase the P_{fr} level suddenly.
   e. be negated by a second flash of far-red light.

8. The conversion of P_r to P_{fr}
   a. occurs slowly at night.
   b. may be the way in which plants sense daybreak and serves to entrain the biological clock.
   c. is the molecular mechanism responsible for the biological clock.
   d. occurs when P_r absorbs far-red light.
   e. occurs in flower buds and induces flowering.

9. A plant may withstand salt stress by
   a. releasing abscisic acid, which closes stomata to prevent salt accumulation.
   b. producing organic solutes, which lower the water potential of root cells.
   c. wilting, which reduces water and salt uptake by reducing transpiration.
   d. producing solutes that reduce the toxic effect of sodium ions.
   e. releasing ethylene, which leads to apoptosis of damaged cells.

10. Which of the following hormones would be sprayed on barley seeds to speed germination in the production of malt for making beer?
    a. abscisic acid
    b. auxin
    c. cytokinin
    d. brassinosteroid
    e. gibberellin

11. Scientists have identified an *Arabidopsis eto* mutant that greatly overproduces ethylene. Which of the following statements accurately describes the growth of such a mutant?

   a. It undergoes the triple response only when it encounters an obstacle, but the response is greater than that of a wild-type plant.

   b. It does not undergo the triple response, because its ethylene receptors are blocked.

   c. *Eto* mutants undergo the triple response even out of the soil.

   d. When experimentally exposed to ethylene, the triple response of such mutants is inhibited.

   e. Treating *eto* mutants with inhibitors of ethylene synthesis would not reduce their triple response when they do not encounter an obstacle.

12. Many plants will flower in response to a specific

   a. photoperiod, which seems to be measured by the length of darkness to which the leaves of the plant are exposed.

   b. effector, which is produced in leaves and travels to a floral meristem.

   c. minimal temperature that appears to signal a seasonal change.

   d. flowering hormone that is produced in the apical bud.

   e. combination of high levels of auxin and FT protein and low levels of cytokinin and ethylene.

13. Which of the following is *not* a plant defense against herbivory?

   a. production of distasteful compounds

   b. production of toxic compounds

   c. physical defenses such as thorns

   d. initiation of a hypersensitive response with production of phytoalexins and PR proteins

   e. release of volatile compounds that attract parasitoid wasps

14. Which of the following is *not* true of the hypersensitive response?

   a. It relies on recognition between a specific plant resistance protein and a pathogen effector molecule.

   b. It increases the production of PR proteins that may have antimicrobial functions.

   c. It enhances the production of ethylene, which serves as a signal molecule that is transported throughout the plant and activates systemic acquired resistance.

   d. It contains an infection by stimulating the cross-linking of cell wall molecules and the production of lignin.

   e. The plant cells involved in the defense response die, leaving lesions that indicate the site of the contained infection.

UNIT 6

6 Animal Form and Function

Chapter 32

Homeostasis and Endocrine Signaling

Chapter Focus

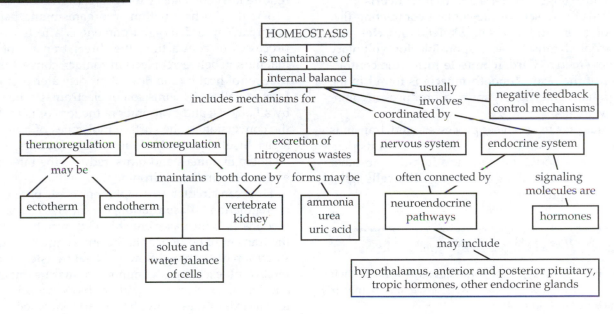

HOMEOSTASIS

is maintainance of

internal balance

includes mechanisms for

usually involves → negative feedback control mechanisms

coordinated by

thermoregulation

osmoregulation

excretion of nitrogenous wastes

nervous system

endocrine system

may be

ectotherm

endotherm

maintains

both done by

forms may be

often connected by

signaling molecules are

solute and water balance of cells

vertebrate kidney

ammonia urea uric acid

neuroendocrine pathways

hormones

may include

hypothalamus, anterior and posterior pituitary, tropic hormones, other endocrine glands

Chapter Review

Anatomy is the study of an organism's biological form; **physiology** is the study of its biological function. The diversity of animal forms and functions reflect the evolutionary adaptations favored by natural selection that increase the fit between organisms and their environments.

32.1 Feedback control maintains the internal environment in many animals

Hierarchical Organization of Animal Bodies   What are the levels of structural and functional organization?

Tissues are groups of similar cells with a common function. **Organs,** which perform specific functions, often consist of a layered arrangement of tissues. Groups of organs are integrated into **organ systems,** which carry out the main functions required for life.

Epithelial tissue lines the outer and inner surfaces of the body in sheets of tightly packed cells. The *apical* surface of an **epithelium** faces the lumen or the outside of an organ. Cells at the *basal* surface are attached to a *basal lamina,* a dense layer of extracellular matrix that separates the epithelium from underlying tissues.

Connective tissue is composed of relatively few cells suspended in an extracellular matrix, which consists of protein fibers embedded in a liquid, jellylike, or solid material. The cells within the matrix include *fibroblasts,* which secrete the protein fibers, and *macrophages,* which engulf bacteria and cellular debris.

The cells of **muscle tissue** contain contracting filaments of actin and myosin. **Skeletal muscle**—also called striated muscle—is responsible for voluntary body movements. **Cardiac muscle** forms the contractile wall of the heart. **Smooth muscle** is found in the walls of many internal organs and produces involuntary movements.

Nervous tissue receives, processes, and transmits information. A **neuron** receives impulses via its dendrites and cell body and conducts impulses to other neurons or muscles along its axons. **Glial cells (glia)** insulate and nourish neurons.

FOCUS QUESTION 32.1

Can you list several types of connective tissues and their functions? Which connective tissue has a liquid matrix and which has a solid matrix?

Regulating and Conforming   For a given environmental variable, an animal may be a **regulator,** using mechanisms to control internal fluctuations, or a **conformer,** allowing internal conditions to vary with external changes. Most animals control some of their internal conditions through regulation of the **interstitial fluid,** the fluid surrounding body cells.

Homeostasis   The maintenance of a "steady state" or relative internal balance is called **homeostasis.** Homeostatic control mechanisms involve a receptor or **sensor,** which detects a **stimulus** (a fluctuation above or below a particular **set point** for a variable), and a *control center,* which triggers a **response** that returns the variable to the set point.

Most homeostatic mechanisms operate by **negative feedback,** in which the response reduces the stimulus, and thus turns off the response. Variables may have a *normal range* within which they fluctuate instead of a single set point.

Thermoregulation: A Closer Look   Animals maintain internal temperatures within an optimal range through the process of **thermoregulation.** Animals that use metabolic heat in generating body warmth (birds and mammals) are **endothermic.** Animals that gain their heat mostly from the environment (most invertebrates and many nonavian reptiles and fishes) are **ectothermic.** Both endotherms and ectotherms may use behaviors that contribute to thermoregulation. Ectotherms require less food energy than do endotherms.

The flow of heat within an organism and between an organism and its environment is affected by the processes of conduction, the direct transfer of thermal motion between objects in contact; convection, the transfer of heat by the flow of air or water past a surface; radiation, the emission of electromagnetic waves by all objects; and evaporation, the loss of heat due to the conversion of the surface molecules of a liquid to a gas. Insulation, in the form of hair, feathers, and fat layers in mammals and birds, reduces the rate of heat exchange with the environment.

How can circulatory adaptations contribute to thermoregulation? **Vasodilation** of superficial (surface) blood vessels increases both blood flow in the skin and the transfer of body heat to the environment. *Vasoconstriction* reduces blood flow and heat transfer. In **countercurrent exchange,** common in marine mammals and birds, the close proximity of blood vessels servicing the extremities allows heat in arterial blood leaving the body core to be transferred to venous blood returning to the body core.

Acclimatization to environmental temperature can also contribute to thermoregulation. Birds and mammals often adjust the thickness of their insulating coats to meet seasonal changes. Ectotherms may adjust their cellular physiology (changes in enzymes, membranes, or production of antifreeze proteins) to acclimatize to changes in environmental (and thus body) temperature.

In mammals, a group of neurons in the **hypothalamus** functions as a thermostat that triggers heat loss or gain mechanisms in response to a body temperature outside the normal range. Mammals and birds develop a *fever* in response to certain infections, apparently due to the resetting of the hypothalamic set point.

FOCUS QUESTION 32.2

a. What cooling mechanisms are activated when the hypothalamus senses a body temperature above the set point?

b. What mechanisms are activated when body temperature falls below the set point?

32.2 Endocrine signals trigger homeostatic mechanisms in target tissues

Coordination and Control Functions of the Endocrine and Nervous Systems The **endocrine** and **nervous systems** coordinate the body's activities. Endocrine cells secrete **hormones** into the bloodstream. Body cells with matching receptors respond to these signaling molecules. Neurons deliver signals via axons directly to target cells, which include other neurons, muscle cells, endocrine cells, and exocrine cells.

FOCUS QUESTION 32.3

Compare the speed of signaling and the duration of the response in the endocrine and nervous systems. How do the functions of these two communication systems differ?

Simple Endocrine Pathways In a simple endocrine pathway, endocrine cells secrete a hormone in response to a stimulus; the hormone travels in the bloodstream to target cells, which produce a response. For example, endocrine cells of the duodenum secrete the hormone secretin in response to a low pH caused by the entry of acidic stomach contents. Target cells in the **pancreas** release bicarbonate, which raises the pH.

Neuroendocrine Pathways The hypothalamus often integrates nervous system and endocrine pathways (neuroendocrine pathways). In response to nerve signals it receives from throughout the body, the hypothalamus sends hormonal signals to the **pituitary gland,** located at its base. Hypothalamic hormones are carried through short blood vessels called portal vessels to the **anterior pituitary,** where they trigger the synthesis and release of specific hormones. Many of these hormones, called *tropic hormones*, regulate other endocrine glands.

In a hormone cascade pathway, the hypothalamus secretes a hormone that stimulates or inhibits release of a particular anterior pituitary hormone. This hormone, called a *tropic hormone or tropin*, acts on a target endocrine tissue, stimulating the release of the hormone that produces a response. For example, in response to nervous stimulation signaling a drop in thyroid hormone levels, the hypothalamus secretes thyrotropin-releasing hormone (TRH), which stimulates the anterior pituitary to secrete thyrotropin, or thyroid-stimulating hormone (TSH). TSH stimulates the thyroid gland to release thyroid hormone, a hormone that stimulates and maintains metabolic processes. Hormone cascade pathways are typically regulated by negative feedback. In this case, high levels of thyroid hormone block the secretion of TRH and TSH.

The **posterior pituitary,** which is an extension of the hypothalamus, stores and releases two hormones produced by neurosecretory cells of the hypothalamus. (Neurosecretory cells are specialized neurons that secrete hormones, often called neurohormones.) In response to nerve signals from the nipples, the hypothalamus triggers the release of the neurohormone **oxytocin,** which stimulates the mammary glands to secrete milk. **Antidiuretic hormone (ADH)** is the other posterior pituitary hormone.

Feedback Regulation in Endocrine Pathways Most endocrine pathways are regulated by negative feedback. In **positive feedback,** the response to a stimulus reinforces the stimulus, leading to a further increase in response.

FOCUS QUESTION 32.4

Explain the difference between negative feedback and positive feedback in the regulation of hormonal pathways and give an example of each.

Pathways of Water-Soluble and Lipid-Soluble Hormones   Water-soluble hormones bind to cell-surface receptors, triggering a series of changes, called *signal transduction*, that leads to a cellular response. Lipid-soluble steroid hormones move through the plasma membrane and bind to cytosolic receptors, which then move into the nucleus and influence gene expression.

Multiple Effects of Hormones   A given signal can have different effects on different target cells as a result of the types of receptors (such as the α and β epinephrine receptors) or the specific signal transduction pathways and/or effector proteins present in the cell. For example, **epinephrine,** secreted by the *adrenal glands* in response to stress, stimulates glucose release from liver cells. It also causes vasodilation of blood vessels supplying skeletal muscle cells, but vasoconstriction of blood vessels to the intestines.

Evolution of Hormone Function   The hormone *prolactin* has effects ranging from milk production and secretion in mammals, to delay of metamorphosis in amphibians, to osmoregulation in fishes. The diversity of actions in different vertebrate species suggests that prolactin is an ancient hormone whose functions diversified during the evolution of vertebrate groups.

32.3   A shared system mediates osmoregulation and excretion in many animals

Animals must regulate solute concentrations and water balance of the interstitial fluid (**osmoregulation**) as well as dispose of *nitrogenous* wastes (**excretion**).

Osmosis and Osmolarity   Osmosis is the diffusion of water across a selectively permeable membrane separating two solutions that differ in **osmolarity**—total solute concentration expressed as moles of solute per liter expressed in units of milliOsmoles (mOsm/L). *Isoosmotic* solutions are equal in osmolarity. There is a net flow of water from a *hypoosmotic* (more dilute) to a *hyperosmotic* (more concentrated) solution.

Osmoregulatory Challenges and Mechanisms   What are the two ways by which animals maintain water balance? **Osmoconformers** are isoosmotic with their surroundings and are only found among marine animals. **Osmoregulators** control their internal osmolarity, regardless of their environment.

Marine fishes are hypoosmotic to seawater. They must drink large amounts of seawater to replace the water they lose by osmosis. *Chloride cells* in the gills pump out excess Cl^- ions (and Na^+ ions then follow), and other ions are excreted in the scanty urine. Unlike a marine fish, a freshwater fish certainly doesn't "drink like a fish." In fact, it hardly drinks at all. Freshwater

fish constantly take in water by osmosis and lose salts by diffusion. They excrete large quantities of dilute urine. Salt supplies are replaced from their food or by active uptake of ions across the gills.

Water-impervious coverings help to prevent dehydration in terrestrial animals. Water is lost in urine and feces and through evaporation, and is gained by drinking and eating moist food and by metabolic production during cellular respiration.

Nitrogenous Wastes   **Ammonia,** a small and toxic molecule, is produced when proteins and nucleic acids are metabolized. Aquatic animals can excrete nitrogenous wastes as ammonia because it easily diffuses through membranes into the surrounding water.

Most terrestrial animals and many marine species excrete **urea,** a much less toxic compound than ammonia. Urea can be tolerated in a more concentrated form and excreted with less loss of water. Ammonia and carbon dioxide are combined in the vertebrate liver to produce urea.

Insects, land snails, and many reptiles, including birds, produce **uric acid,** a compound of low solubility in water that can be excreted as a semisolid with very little water loss. Its synthesis, however, is energetically expensive.

Excretory Processes   The movement of specific solutes across **transport epithelia** is involved in both osmoregulation and the disposal of metabolic wastes. Transport epithelia are usually arranged in tubular networks that provide large surface areas for exchange.

By what basic steps is the waste fluid urine produced? During **filtration,** water and small solutes are forced out of the blood or body fluids across a transport epithelium into an excretory tubule. Two mechanisms transform this **filtrate** to urine: Valuable solutes are returned from the filtrate by selective **reabsorption,** and excess ions and toxins are added to the filtrate by selective **secretion.** The osmotic movement of water into or out of the filtrate follows the direction in which solutes are pumped. The processed filtrate is released from the body by excretion.

The *protonephridia* of flatworms are branched systems of closed tubules. Water and solutes from the interstitial fluid pass into *flame bulbs* at the ends of the tubules, propelled by cilia. Urine is excreted into the external environment. The protonephridia of freshwater flatworms are primarily osmoregulatory. In parasitic flatworms, however, protonephridia function mainly in the excretion of nitrogenous wastes.

Malpighian tubules are excretory organs in insects and other terrestrial arthropods. Transport epithelia secrete solutes and wastes into the tubule, which connects with the intestines. Most of the solutes are pumped back into the hemolymph, and water follows

FOCUS QUESTION 32.5

Indicate whether the following animals are isoosmotic, hyperosmotic, or hypoosmotic to their environment. Briefly list their mechanisms of osmoregulation. Which type of nitrogenous waste product do they most likely produce?

Animal	Osmotic Relation to Environment	Osmoregulatory Mechanisms	Nitrogenous Waste Product
Marine invertebrate	a.	b.	c.
Marine bony fish	d.	e.	f.
Freshwater fish	g.	h.	i.
Bird	j.	k.	l.

by osmosis. In this water-conserving system, uric acid is eliminated as dry matter along with the feces.

The numerous excretory tubules of most vertebrates are associated with a network of capillaries and arranged into compact **kidneys.** The kidneys and the structures that carry urine out of the body constitute the vertebrate excretory system, which functions in both osmoregulation and excretion.

Exploring the Mammalian Excretory System Blood enters each of the pair of bean-shaped kidneys through a renal artery and leaves by way of a renal vein. Urine exits the kidney through a **ureter** and is temporarily stored in the **urinary bladder.** Urine exits the body through the **urethra.**

The outer **renal cortex** and inner **renal medulla** of the kidney are packed with excretory tubules and blood vessels. Urine collects in the inner **renal pelvis.** In the human kidney, 85% of the **nephrons** (the functional units of the kidney) are **cortical nephrons** located almost entirely in the renal cortex. **Juxtamedullary nephrons** have long loops of Henle that extend into the renal medulla and allow the formation of hyperosmotic urine.

A nephron consists of a long tubule, at the blind end of which is a cuplike **Bowman's capsule** enclosing a ball of capillaries called the **glomerulus.** Filtration of the blood occurs as blood pressure forces water, urea, salts, glucose, amino acids, and other small molecules through the porous capillary walls and specialized capsule cells into Bowman's capsule, forming the filtrate. The filtrate, in turn, passes through the **proximal tubule;** the **loop of Henle,** consisting of a descending limb and an ascending limb; and the **distal tubule. Collecting ducts** receive processed filtrate from many nephrons and pass the urine into the renal pelvis.

An *afferent arteriole* supplying each nephron subdivides to form the capillaries of the glomerulus, which then converge to form an *efferent arteriole.* This arteriole divides into a second capillary network—the **peritubular capillaries**—that surrounds the proximal and distal tubules. The **vasa recta** includes descending and ascending capillaries that parallel the loop of Henle.

FOCUS QUESTION 32.6

a. What substances are removed from the blood during filtration?

b. What substances remain in the capillaries?

32.4 Hormonal circuits link kidney function, water balance, and blood pressure

From Blood Filtrate to Urine: A Closer Look The numbered steps in the following discussion correspond to the figure in Focus Question 32.7.

1. Reabsorption and secretion both take place in the *proximal tubule.* Salt (NaCl) moves by facilitated diffusion and cotransport from the filtrate into the cells of the proximal tubule, which then actively transport Na^+ into the interstitial fluid; Cl^- is transported passively out to balance the positive charge; and water follows by osmosis. Salt and water diffuse from the interstitial fluid into the peritubular

capillaries. Glucose, amino acids, K⁺, and other substances are reabsorbed either actively or passively. The transport epithelium secretes drugs and poisons (processed by the liver) into the filtrate.

2. The *descending limb of the loop of Henle* has numerous **aquaporins** and is thus freely permeable to water. It is not permeable to NaCl or other small solutes. The interstitial fluid is increasingly hyperosmotic toward the inner medulla, so water exits by osmosis, leaving behind a filtrate with a high solute concentration.

3. The *ascending limb of the loop of Henle* is not permeable to water but is permeable to salt, which

diffuses out of the thin lower segment of the loop, increasing the osmolarity of the medulla. NaCl is actively transported out of the thick upper portion of the ascending limb. As salt leaves, the filtrate becomes less concentrated.

4. The *distal tubule* regulates K⁺ secretion and NaCl reabsorption, and it helps to regulate pH by secreting H⁺ and reabsorbing HCO_3^-.

5. The pair of human kidneys processes about 180 L of filtrate each day to produce about 1.5 L of urine. As the filtrate moves along the *collecting duct*, final processing produces the urine.

FOCUS QUESTION   32.7

In the following diagram of a nephron and collecting duct, label the parts on the indicated lines. Review steps 1 to 5 as described previously. Label the arrows to indicate the movement of NaCl, water, nutrients, K⁺, HCO_3^-, H⁺, NH_3, and urea into or out of the tubule.

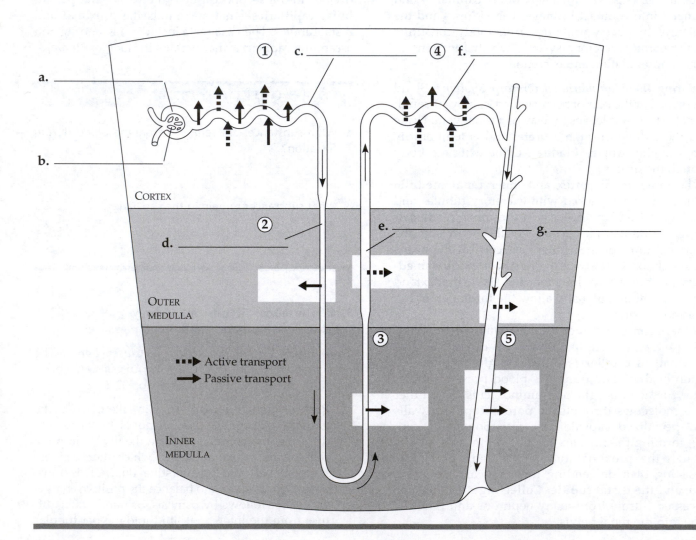

Concentrating Urine in the Mammalian Kidney Now let's see how the human kidney can produce urine up to four times as concentrated as blood.

Filtrate leaving Bowman's capsule has an osmolarity of about 300 mOsm/L, the same as blood. Both water and salt are reabsorbed in the proximal tubule; thus, the volume decreases but the osmolarity remains the same. During the filtrate's movement down the descending limb of the loop of Henle, water exits by osmosis, and the filtrate becomes more concentrated. Salt, which is now in high concentration in the filtrate, diffuses out as the filtrate moves up the salt-permeable but water-impermeable ascending limb—helping to create the osmolarity gradient.

This steep osmotic gradient is produced through a type of countercurrent system involving the loop of Henle and surrounding capillaries. Through a **countercurrent multiplier system,** the thick segment of the ascending limb expends energy to actively transport NaCl from the filtrate and generate the osmotic gradient in the medulla.

The filtrate is hypoosmotic to body fluids after salt is pumped out of the ascending limb. The filtrate makes one final pass through the medulla, this time in the collecting duct, which is permeable to water but not to salt. Water flows out by osmosis, and as the filtrate becomes more concentrated, urea diffuses into the interstitial fluid, adding to the high osmolarity of the inner medulla. The resulting concentrated urine may achieve an osmolarity as high as 1,200 mOsm/L, isoosmotic to the interstitial fluid of the inner medulla.

Adaptations of the Vertebrate Kidney to Diverse Environments Modifications of nephron structure and function correspond to osmoregulatory requirements in various habitats. Mammals in dry habitats, needing to conserve water and excrete hyperosmotic urine, have exceptionally long loops of Henle.

Birds also have juxtamedullary nephrons, but their shorter loops of Henle prevent them from producing urine as hyperosmotic as that of mammals. The production of uric acid is their main means of water conservation.

As illustrated by the widely varying osmolarity of urine produced by vampire bats as they feed and then process their blood meals, mammalian kidneys can adjust both volume and osmolarity of urine in response to water balance and urea production.

FOCUS QUESTION 32.8

a. What type of nephron enables the production of hyperosmotic urine?

b. Which animal groups have this type of nephron?

Homeostatic Regulation of the Kidney What is the role of the mammalian kidney in the regulation of blood osmolarity, blood pressure, and blood volume? Antidiuretic hormone (ADH), also called *vasopressin,* is produced in the hypothalamus and stored in the pituitary gland. When blood osmolarity rises above a set point of 300 mOsm/L, osmoreceptor cells in the hypothalamus trigger the release of ADH. This hormone binds to receptors on cells in the collecting duct, leading to an increase in aquaporin proteins in the plasma membranes. These water channels increase water permeability and thus water reabsorption from the urine. As blood osmolarity declines, negative feedback decreases the release of ADH.

The **renin-angiotensin-aldosterone system (RAAS)** is a homeostatic feedback circuit that maintains adequate blood pressure and blood volume. The **juxtaglomerular apparatus (JGA),** located around the afferent arteriole leading to the glomerulus, responds to a drop in blood pressure or volume by releasing renin. This enzyme converts the plasma protein angiotensinogen to **angiotensin II.** Angiotensin II raises blood pressure (by constricting arterioles) and stimulates the adrenal glands to release **aldosterone.** This hormone stimulates Na^+ and water reabsorption in the distal tubules and collecting ducts. Some drugs that treat hypertension target the enzyme ACE, which functions in the activation of angiotensin II.

Both ADH and the RAAS increase water absorption, but in response to different problems. ADH counters dehydration by responding to an increase in blood osmolarity. The RAAS responds to a loss of both fluid and salts by sensing a drop in blood volume or pressure.

FOCUS QUESTION   32.9

The following concept map may look a little overwhelming, but working through it should help organize your understanding of the homeostatic regulation of the kidney.

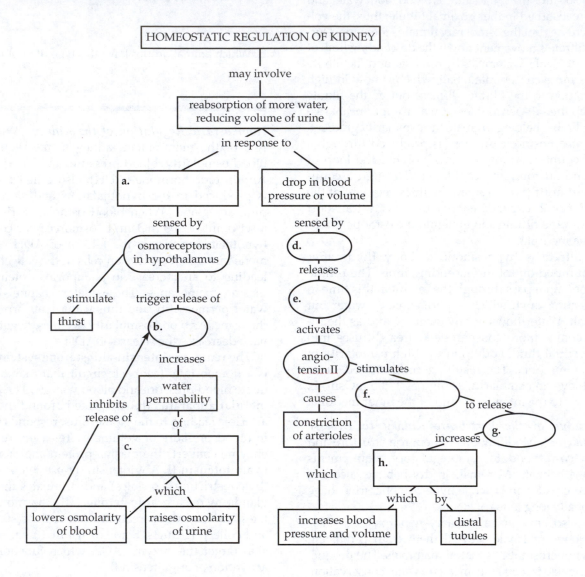

Word Roots

anti- = against; **-diure** = urinate (*antidiuretic hormone:* a hormone that promotes water retention by the kidneys. Produced in the hypothalamus and released from the posterior pituitary)

counter- = opposite (*countercurrent exchange:* the exchange of a substance or heat between two fluids flowing in opposite directions)

-dilat = expanded (*vasodilation:* an increase in the diameter of blood vessels triggered by relaxation of smooth muscles in the vessel walls)

ecto- = outside; **-therm** = heat (*ectothermic:* referring to organisms for which external sources provide most of the heat for temperature regulation)

endo- = inner (*endothermic:* referring to organisms that are warmed by heat generated by their own metabolism. This heat usually maintains a relatively stable body temperature higher than that of the external environment)

epi- = above, over (*epinephrine:* a hormone secreted by the adrenal medulla that mediates "fight-or-flight" responses to short-term stresses; also called adrenaline)

glomer- = a ball (*glomerulus:* a ball of capillaries surrounded by Bowman's capsule in the nephron and serving as the site of filtration in the vertebrate kidney)

homeo- = same; **-stasis** = standing, posture (*homeostasis:* the steady-state physiological condition of the body)

inter- = between (*interstitial fluid:* the fluid filling the spaces between cells in most animals)

juxta- = near to (*juxtaglomerular apparatus:* a specialized tissue in nephrons that releases the enzyme renin in response to a drop in blood pressure or volume)

osmo- = pushing; **-regula** = regular (*osmoregulation:* regulation of solute concentrations and water balance by a cell or organism)

oxy- = sharp, acid (*oxytocin:* a hormone produced by the hypothalamus and released from the posterior pituitary that induces contractions of the uterine muscles during labor and causes the mammary glands to eject milk during nursing)

peri- = around (*peritubular capillaries:* the network of tiny blood vessels surrounding the proximal and distal tubules in the kidney)

reni- = a kidney; **-angio** = a vessel; **-tens** = stretched (*renin-angiotensin-aldosterone system:* a hormone cascade pathway that helps regulate blood pressure and volume)

vasa- = a vessel; **-recta** = straight (*vasa recta:* the capillary system in the kidney that serves the loop of Henle)

Structure Your Knowledge

1. Organs are composed of layers of several different tissues. Which of the four animal tissues would you expect to find in all organs? In what types of organs would you predict the remaining tissues occur?

2. Explain the following statement: Endotherms can tolerate wider external temperature fluctuations; ectotherms may be able to tolerate wider internal temperature fluctuations.

3. Explain how secretin, which is released in response to a low pH in the intestines and stimulates the release of bicarbonate from the pancreas, illustrates a simple hormone pathway.

4. What is a hormone cascade pathway? Describe how thyroid regulation illustrates such a pathway.

5. After referring to the diagram in Focus Question 32.7, describe the two-solute model that enables the production of hyperosmotic urine.

Test Your Knowledge

MULTIPLE CHOICE: *Choose the one best answer.*

1. Which of the following statements is *not* true of connective tissue?
   a. It consists of a few cells surrounded by fibers in a matrix.
   b. It includes such diverse tissues as bone, cartilage, adipose, and loose connective tissue.
   c. It connects and supports other tissues.
   d. It forms the internal and external lining of many organs.
   e. It can have a matrix that is a liquid, a gel, or a solid.

2. The interstitial fluid of vertebrates
   a. is the internal environment within cells.
   b. surrounds cells and provides for the exchange of nutrients and wastes.
   c. makes up blood plasma.
   d. is not necessary in flat, thin vertebrates.
   e. is less abundant in ectotherms than in endotherms.

3. Which of the following would be *least* likely to be true of an animal that is a regulator?
   a. It can live in a variable climate because of its homeostatic mechanisms.
   b. It may have a larger geographic range than a conformer.
   c. It may acclimatize to winter by increasing the thickness of its insulating coat.
   d. It has lower energy requirements than a conformer of similar size.
   e. It may use behavioral as well as physiological mechanisms for responding to changing conditions.

4. Negative feedback loops are
   a. mechanisms that most commonly maintain homeostasis.
   b. activated only when a physiological variable rises above a set point.
   c. analogous to a radiator that heats a room.
   d. involved in maintaining contractions during childbirth.
   e. found in endotherms but not in ectotherms.

5. Which of the following mechanisms dissipates heat?
   a. countercurrent exchange between vessels that service the extremities
   b. trapping air by raising fur or feathers
   c. vasoconstriction of surface vessels
   d. vasodilation of surface vessels
   e. both a and c

6. Which of the following statements is *not* accurate?
   a. Not all hormones are secreted by endocrine glands.
   b. Hormones are transported through the circulatory system to their destinations.
   c. Target cells have specific hormone receptors.
   d. Hormones are essential to homeostasis.
   e. Because steroid hormones influence gene expression, they produce a faster response in target cells.

7. The finding that MSH (melanocyte-stimulating hormone) causes frog skin cells to darken when added to the interstitial fluid but has no effect on color when injected into the cells supports which of the following ideas?
   a. MSH is a tropic hormone.
   b. MSH is a neurohormone.
   c. MSH is a steroid hormone.
   d. MSH binds to cell surface receptors.
   e. Both c and d are reasonable conclusions.

8. Which one of the following hormones is *incorrectly* paired with its site of production?
   a. releasing hormones—hypothalamus
   b. growth hormone—anterior pituitary
   c. estrogens and progestins—ovary
   d. TSH—thyroid
   e. epinephrine—adrenal gland

9. A tropic hormone is a hormone
   a. that is produced by the hypothalamus but stored and released from the anterior pituitary.
   b. that stimulates other endocrine glands to secrete hormones.
   c. that acts by negative feedback to regulate its own levels in the body.
   d. from the hypothalamus that regulates hormone secretion of the posterior pituitary.
   e. that is released in response to nervous stimulation.

10. The anterior pituitary
    a. stores and releases oxytocin and ADH produced by the hypothalamus.
    b. receives releasing and inhibiting hormones from the hypothalamus through portal vessels connecting capillary beds.
    c. produces releasing and inhibiting hormones.
    d. is responsible for nervous and hormonal stimulation of the adrenal glands.
    e. produces only tropic hormones.

11. Transport epithelia are responsible for
    a. pumping water across a membrane.
    b. forming an impermeable boundary at an interface with the environment.
    c. the movement of solutes in osmoregulation or excretion.
    d. transporting urine in the ureter and urethra.
    e. the passive transport of H^+ and HCO_3^- for the regulation of pH.

12. A freshwater fish would be expected to
    a. pump salt out through salt glands in the gills.
    b. produce copious quantities of dilute urine.
    c. excrete urea across the epithelium of the gills.
    d. have scales that reduce water loss.
    e. do all of the above.

13. The production of uric acid is advantageous to birds because uric acid
    a. has low toxicity, requires little water to excrete, and can be safely stored as an embryo develops in an egg.
    b. is very soluble in water and takes very little energy to produce.
    c. contributes to the production of the egg shell and can thus serve two purposes.
    d. takes less energy to produce than urea.
    e. Both a and d are correct.

14. Which of the following correctly traces the pathway of urine flow in vertebrates?
    a. collecting tubule $\longrightarrow$ ureter $\longrightarrow$ bladder $\longrightarrow$ urethra
    b. renal vein $\longrightarrow$ ureter $\longrightarrow$ bladder $\longrightarrow$ urethra
    c. nephron $\longrightarrow$ urethra $\longrightarrow$ bladder $\longrightarrow$ ureter
    d. cortex $\longrightarrow$ medulla $\longrightarrow$ bladder $\longrightarrow$ ureter
    e. renal pelvis $\longrightarrow$ medulla $\longrightarrow$ bladder $\longrightarrow$ urethra

15. Which of the following statements is *incorrect*?
    a. Long loops of Henle are associated with steep osmotic gradients and the production of hyperosmotic urine.
    b. Ammonia is a toxic nitrogenous waste molecule that passively diffuses out of the bodies of aquatic invertebrates.
    c. Uric acid is the nitrogenous waste that requires the least amount of water to excrete.
    d. In the mammalian kidney, some urea diffuses out of the collecting duct and contributes to the osmotic gradient within the medulla.
    e. Ammonia is produced by a mammalian fetus, removed through the placenta, transported to the mother's liver, and then converted to urea.

16. The mechanism for the filtration of blood within the nephron involves
    a. the active transport of Na^+ and glucose, followed by osmosis.
    b. both active and passive secretion of ions and toxins into the tubule.
    c. high hydrostatic pressure of blood forcing water and small molecules out of the capillary.
    d. the high osmolarity of the medulla that was created by active and passive transport of salt from the tubule, and passive diffusion of urea from the collecting duct.
    e. a lower osmotic pressure in Bowman's capsule compared to that in the glomerulus.

17. The process of reabsorption in the formation of urine ensures that
    a. excess H^+ is removed from the blood.
    b. drugs and toxins are removed from the blood.
    c. urine is always hyperosmotic to the interstitial fluid.
    d. glucose, salts, and water are returned to the blood.
    e. pH is maintained by balancing H^+ and HCO_3^-.

18. Which of the following sections of the mammalian nephron is *incorrectly* paired with its function?
    a. Bowman's capsule and glomerulus—filtration of blood
    b. proximal tubule—secretion of toxins and H^+ into the filtrate and transport of glucose and amino acids out of tubule
    c. descending limb of loop of Henle—diffusion of urea out of the filtrate
    d. ascending limb of loop of Henle—diffusion and pumping of NaCl out of the filtrate
    e. distal tubule—regulation of pH and K^+

19. Which of the following animals exhibits the most flexible osmoregulatory and excretory systems?
    a. a marine fish
    b. a migratory salmon returning to its home stream from the ocean
    c. a parasitic flatworm
    d. a vampire bat
    e. both **b** and **d**

20. ADH and the RAAS both increase water reabsorption, but they respond to different osmoregulatory problems. Which *two* of the following statements are true?
    1. ADH will be released in response to a drop in blood pressure or volume.
    2. ADH is released when osmoregulatory cells in the hypothalamus detect an increase in blood osmolarity.
    3. The RAAS will decrease blood osmolarity due to the cooperative actions of renin, angiotensin II, and aldosterone.
    4. The RAAS is a response to a rise in blood pressure or volume.
    5. The RAAS is most likely to respond following an accident or a severe case of diarrhea.
    a. 1 and 4
    b. 1 and 5
    c. 2 and 3
    d. 2 and 4
    e. 2 and 5

21. Which of the following is an effective mechanism for a drug used to treat hypertension (high blood pressure)?
    a. increase the production of ADH
    b. increase the production of renin
    c. produce vasoconstriction of renal arteries
    d. inhibit an enzyme involved in producing angiotensin II
    e. stimulate the insertion of aquaporins in cells lining the collecting ducts

22. The production of urine that is hyperosmotic to blood and interstitial fluids requires
    a. juxtamedullary nephrons.
    b. an osmotic gradient in the medulla.
    c. the inhibition of ADH release.
    d. **a** and **b** only.
    e. **a**, **b**, and **c**.

Chapter 33

Animal Nutrition

Chapter Focus

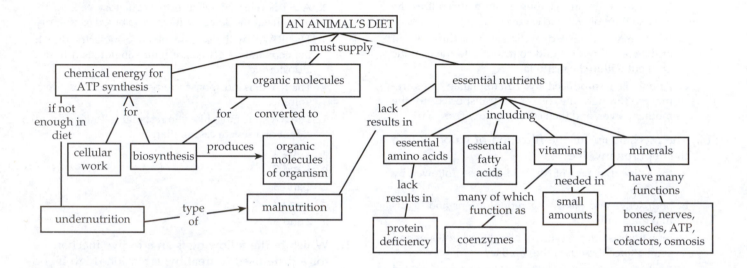

Chapter Review

Nutrition is the process by which organisms take in and use food. **Herbivores** eat plants or algae; **carnivores** eat animals; and **omnivores** consume both plants and animals. Most animals are opportunistic and eat foods from different categories when they are available.

33.1 An animal's diet must supply chemical energy, organic molecules, and essential nutrients

An animal's diet must provide fuel for ATP production and organic carbon and nitrogen molecules for biosynthesis.

Essential Nutrients An animal's diet must also provide **essential nutrients,** which are molecules that an animal requires but cannot synthesize itself. The **essential fatty acids,** such as linoleic acid, contain one or

more double bonds. Such fatty acids are readily supplied in the diet from plant sources, and deficiencies are rare.

Eight of the 20 amino acids that are needed to make proteins are **essential amino acids** required in the diet of most animals (including adult humans). Meat, eggs, and cheese contain "complete" proteins with all essential amino acids in their proper proportions. Most plant proteins are "incomplete," deficient in one or more essential amino acids.

FOCUS QUESTION   33.1

How can vegetarians obtain all their essential amino acids?

Vitamins are organic molecules required in very small amounts in the diet. Thirteen vitamins essential to humans have been identified. Water-soluble vitamins include the B complex, most of which function as coenzymes, and vitamin C, which is required for the production of connective tissue. The fat-soluble vitamins include A, incorporated into visual pigments, and D, which aids in calcium absorption and bone formation.

Minerals are inorganic nutrients that are usually needed in small amounts. Some minerals are built into the structure of enzymes and other proteins. Iodine is needed by vertebrates to make the metabolism-regulating thyroid hormones. Sodium, potassium, and chloride are important in nerve and muscle function and osmotic balance.

Dietary Deficiencies   *Malnutrition* results from a failure to obtain one or more essential nutrients or sufficient chemical energy. *Undernutrition* refers to a diet deficient in calories. With severe deficiency, the body breaks down its own proteins for energy, eventually causing irreversible damage or death.

Assessing Nutritional Needs   Research on human dietary requirements involves many challenges. Techniques include the study of genetic defects that affect nutrient uptake or use, and *epidemiology*, the study of health and disease among various populations (in this case, those with different dietary practices).

FOCUS QUESTION 33.2

a. Which two minerals do vertebrates require in relatively large amounts? Why?

b. How does undernutrition differ from other types of malnutrition?

c. What is the most common type of malnutrition among humans?

33.2   The main stages of food processing are ingestion, digestion, absorption, and elimination

Ingestion is the act of consuming food. Many aquatic animals are *filter feeders or suspension feeders*, sifting or capturing small organisms or food particles from the water. *Substrate feeders* live in or on their food, eating their way through it. *Fluid feeders* suck fluids from a living plant or animal host. Most animals are *bulk feeders*, which eat relatively large pieces of food.

Digestion splits food into small molecules. Mechanical digestion often precedes chemical digestion, which is the *enzymatic hydrolysis* of bonds between components of large molecules.

In the last two stages of food processing, **absorption** and **elimination,** small molecules are taken into the cells of an animal, and the undigested remainder of the food leaves the digestive system.

FOCUS QUESTION 33.3

Why must large molecules in food be digested into monomers or smaller components?

Digestive Compartments   Breaking down food in specialized compartments protects animals from digesting themselves. In intracellular digestion, food vacuoles formed by phagocytosis or pinocytosis fuse with enzyme-containing lysosomes. Chemical digestion then occurs safely within a membrane-enclosed compartment. Most animals break down their food, at least initially, by extracellular digestion within a separate compartment that connects to the external environment.

A **gastrovascular cavity** functions in both digestion and the transport of nutrients throughout the body. In the small cnidarian hydra, digestive enzymes secreted into the gastrovascular cavity begin food breakdown. Gastrodermal cells then engulf food particles for hydrolysis within food vacuoles. Undigested materials are expelled through the mouth.

Most animals have a *complete digestive tract* or **alimentary canal** with two openings, a mouth and an anus. Food ingested through the mouth and pharynx passes through an esophagus that may lead to a crop, stomach, or gizzard—organs specialized for storing or grinding food. In the intestine, digestive enzymes hydrolyze large molecules, and nutrients are absorbed across the intestinal lining. Undigested material exits through the anus.

What are the advantages of an alimentary canal compared to a gastrovascular cavity?

33.3 Organs specialized for sequential stages of food processing form the mammalian digestive system

The mammalian digestive system consists of the alimentary canal and accessory glands that secrete digestive juices through ducts into the canal. Rhythmic waves of muscular contraction called **peristalsis** push food along the tract. Muscular ringlike valves called **sphincters** regulate the passage of material between some segments.

The Oral Cavity, Pharynx, and Esophagus Physical and chemical digestion begins in the mouth, or **oral cavity,** where teeth grind food to expose a greater surface area to enzyme action. Saliva, released from **salivary glands,** has several components: **mucus,** which contains slippery glycoproteins that protect the mouth lining from abrasion and lubricate food for swallowing; buffers to neutralize acidity; antimicrobial agents; and **amylase,** an enzyme that begins the hydrolysis of starch and glycogen. The tongue tastes and manipulates food, and pushes the food ball, or **bolus,** into the pharynx for swallowing.

The **pharynx,** or throat region, is the intersection leading to both the esophagus and the trachea. During swallowing, the *larynx,* the upper part of the respiratory tract, moves up so that a cartilaginous flap covers the vocal cords and the opening between them. Food moves down through the **esophagus** to the stomach, squeezed along by a wave of peristalsis.

Digestion in the Stomach In the expandable **stomach,** food is combined with **gastric juice,** producing a mixture called **chyme.**

Parietal cells within gastric glands in the stomach secrete hydrochloric acid (HCl), which breaks down food tissues, kills bacteria, and denatures proteins. A **protease** called **pepsin** is secreted by *chief cells* in an inactive form called pepsinogen. It is activated by HCl and by pepsin itself, activating more of this protein-digesting enzyme.

Mucus helps protect the rapidly dividing epithelium from digestion. Gastric ulcers, damaged areas of the stomach lining, have been linked to infection by the bacterium *Helicobacter pylori.*

Smooth muscles churn the contents of the stomach. Except when a bolus arrives, the opening from the esophagus is closed to prevent backflow of the acidic chyme. A sphincter regulates passage of chyme into the intestine.

Two types of macromolecules have been partially digested by the time chyme moves into the intestine. What are these molecules, where does this digestion take place, and what enzymes are involved?

a.

b.

Digestion in the Small Intestine Digestive juices are mixed with chyme in the **duodenum,** the first section of the **small intestine.** The **pancreas** produces digestive enzymes and a bicarbonate-rich alkaline solution that offsets the acidity of the chyme. The **liver** produces **bile,** which is stored in the **gallbladder** until needed. Bile salts emulsify fats for easier digestion. Bile also contains pigments that are by-products of the breakdown of red blood cells in the liver. The lining of the duodenum secretes some enzymes; others are bound to the surfaces of epithelial cells.

Most digestion occurs in the duodenum. The *jejunum* and *ileum* function in nutrient and water absorption.

For each of the following molecules, describe their enzymatic digestion in the small intestine.

a. polysaccharides

b. polypeptides

c. nucleic acids

d. fats

Absorption in the Small Intestine Large folds of the small intestine lining are covered with fingerlike projections called **villi.** Each epithelial cell of a villus has microscopic extensions called **microvilli.** This huge surface area for absorption is called the *brush border.* Nutrients may be actively or passively transported across the membranes.

Absorbed nutrients leave the epithelial cells and enter capillaries in the core of each villus. The nutrient-laden blood from the small intestine is carried directly to the liver by the **hepatic portal vein.** The liver regulates the nutrient content of the blood and detoxifies foreign molecules.

Monoglycerides and fatty acids absorbed by epithelial cells are recombined to form fats. These fats are coated with proteins, phospholipids, and cholesterol to make tiny globules called **chylomicrons,** which move into a lymph vessel called a **lacteal** in the center of each villus. They are then transported by the lymphatic system to large veins.

Absorption in the Large Intestine The small intestine leads into the **large intestine** at a T-shaped junction. One arm of this junction is a pouch called the **cecum,** which in humans has a fingerlike extension called the **appendix.** The other arm of the junction, the **colon,** finishes the reabsorption of the large quantity of water secreted with digestive enzymes into the digestive tract.

Escherichia coli and other mostly harmless bacteria live on unabsorbed organic matter in the large intestine. Some of these bacteria produce vitamins. The **feces** contain cellulose, other undigested materials, and a large proportion of intestinal bacteria. Feces are stored in the **rectum,** the final section of the large intestine.

33.4 Evolutionary adaptations of vertebrate digestive systems correlate with diet

Dental Adaptations Dentition, the type and arrangement of teeth, correlates with diet.

Mutualistic Adaptations Many herbivorous mammals have fermentation chambers filled with symbiotic bacteria and protists. These microorganisms, often housed in the cecum, digest cellulose to simple sugars and produce a variety of essential nutrients. Rabbits and rodents ingest their feces (*coprophagy*) to capture the nutrients that their mutualistic bacteria had released in the large intestine.

Ruminants have an elaborate system involving several stomach chambers, regurgitation and rechewing of the cud, and digestion of their mutualistic microorganisms, which maximizes the nutrient yield of their grass diet.

Stomach and Intestinal Adaptations Vegetation is more difficult to digest than meat. The longer alimentary canal of herbivores facilitates the digestion of plant material.

FOCUS QUESTION 33.7

Compare and contrast the dentition and alimentary canals of carnivores and herbivores.

33.5 Feedback circuits regulate digestion, energy allocation, and appetite

Regulation of Digestion The *enteric division* of the nervous system regulates peristalsis and release of digestive juices. What role does the endocrine system play in digestion? Hormones, such as gastrin, released by the stomach, and secretin and cholecystokinin (CCK), released by the duodenum, stimulate the release of digestive secretions from the stomach, pancreas, and gallbladder.

Energy Allocation An animal's **bioenergetics** involves the overall flow and transformation of energy—the input of energy in the form of food, its use in the body, and its return to the environment in waste products and as heat. Fuel molecules are used to generate ATP for cellular work and for biosynthesis, which is needed for growth, energy storage, and reproduction.

The total energy an animal uses in a unit of time is its **metabolic rate.** Metabolic rate can be determined by measuring heat loss, the rate of oxygen consumption, the rate of CO_2 production, or the energy content of food and the energy lost in waste products.

The minimal metabolic rate for a nongrowing endotherm that is at rest, fasting, and nonstressed is called the *basal metabolic rate* (BMR). The *standard metabolic rate* (SMR) is the metabolic rate of a resting, fasting, nonstressed ectotherm determined at a specific temperature.

A human adult male's BMR is about 1,600 to 1,800 kcal/day; a female's is about 1,300 to 1,500 kcal/day. The SMR of a comparably sized ectotherm would be much, much less. Maximal metabolic rates occur during intense activity. Most terrestrial animals have an average daily rate of energy consumption that is 2 to 4 times BMR or SMR.

When an animal consumes more calories than are needed to meet its energy requirements, the excess can be stored in the liver and muscles as glycogen, or in adipose tissue as fat (when glycogen stores are full).

Glucose homeostasis is maintained by two hormones: insulin, which promotes glycogen synthesis in the liver and thus lowers blood glucose, and glucagon, which stimulates the liver to break down glycogen and thus raises blood glucose. Scattered within the exocrine tissue of the pancreas are clusters of endocrine cells called pancreatic islets. Within each islet are *alpha cells*, which secrete glucagon, and *beta cells*, which secrete insulin.

In **diabetes mellitus**, the absence of insulin in the bloodstream or the loss of response to insulin in target tissues reduces glucose uptake by cells. The body must use fats for fuel, and acids from the breakdown of fats may lower blood pH. Glucose accumulates in the blood and is excreted in the urine, with an accompanying loss of water.

Type 1 diabetes, also known as insulin-dependent diabetes, is an autoimmune disorder in which pancreatic beta cells are destroyed. This type of diabetes is treated by regular injections of insulin. More common is *type 2 diabetes*, or non-insulin-dependent diabetes, characterized by reduced responsiveness of target cells to insulin. Lack of exercise and excess weight increase the risk of type 2 diabetes.

Regulation of Appetite and Consumption *Overnourishment*, consuming more calories than the body needs, can lead to obesity. This excessive accumulation of fat is associated with a number of health problems.

Appetite-regulating hormones act on the brain's "satiety center," which generates impulses that make one feel hungry or satiated. These hormones include *ghrelin*, which is secreted by the stomach and triggers hunger, and three hormones that suppress appetite: insulin from the pancreas, *PYY* from the small intestine, and *leptin* produced by adipose tissue.

Word Roots

chylo- = juice; **micro-** = small (*chylomicron*: a lipid transport globule composed of fats mixed with cholesterol and coated with proteins)

chymo- = juice (*chyme*: the mixture of partially digested food and digestive juices formed in the stomach)

gastro- = stomach; **-vascula** = a little vessel (*gastrovascular cavity*: a central cavity with a single opening that functions in both digestion and nutrient distribution)

herb- = grass; **-vora** = eat (*herbivore*: an animal that mainly eats plants or algae)

micro- = small; **-villi** = shaggy hair (*microvilli*: the many fine, fingerlike projections of the epithelial cells in the lumen of the small intestine that increase its surface area)

omni- = all (*omnivore*: an animal that regularly eats animals as well as plants or algae)

peri- = around; **-stalsis** = a constriction (*peristalsis*: alternating waves of contraction and relaxation in the smooth muscle lining the alimentary canal that push food along the canal)

FOCUS QUESTION   33.8

Complete the following concept map on the regulation of blood glucose levels by hormones of the pancreas.

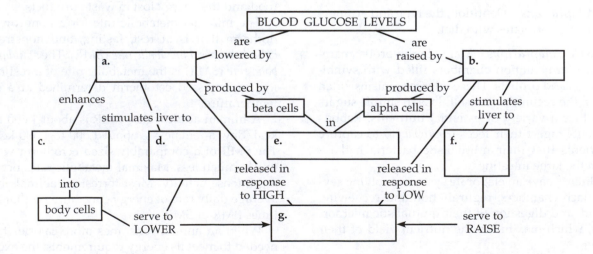

Structure Your Knowledge

1. Label the indicated structures in the following diagram of the human digestive system. Review the functions of these structures.

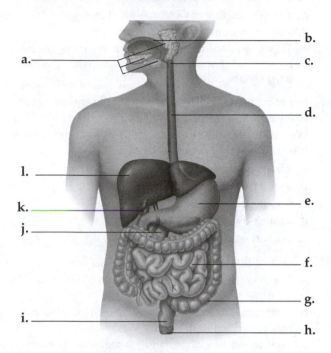

Test Your Knowledge

MULTIPLE CHOICE: *Choose the one best answer.*

1. Why do many vegetarians combine different protein sources in their diet or eat some animal products such as eggs or milk products?
   a. to make sure they obtain sufficient calories
   b. to provide sufficient vitamins
   c. to make sure they ingest all the essential fatty acids
   d. to make their diet more interesting
   e. to provide all the essential amino acids

2. Which of the following is a *true* statement about vitamins?
   a. They may be produced by intestinal microorganisms.
   b. Requirements for them are the same from one species to the next.
   c. They are stored in large quantities in the liver.
   d. They are inorganic nutrients, needed in small amounts, that often function as cofactors.
   e. Because they are fat soluble, they may be provided in insufficient quantities in a fat-free diet.

3. An amino acid that is not an essential amino acid would be one that an animal
   a. can obtain from its own proteins.
   b. has the needed enzymes for synthesizing.
   c. gets in sufficient quantities in its diet.
   d. obtains from its microbial symbionts.
   e. does not use in protein synthesis.

4. Which of the following substances is mismatched with its function?
   a. most B vitamins—coenzymes
   b. vitamin A—incorporated into visual pigment
   c. vitamin D—aids in calcium absorption
   d. iron—component of thyroid hormones
   e. phosphorus—bone formation, nucleotide synthesis

5. An organism that exclusively uses extracellular digestion
   a. is a sponge.
   b. is a hydra with a gastrovascular cavity.
   c. needs no circulatory system.
   d. would be a filter feeder.
   e. has a specialized body compartment containing digestive enzymes.

6. Substrate feeders such as earthworms
   a. eat mostly mineral substrates.
   b. filter small organisms from water.
   c. eat autotrophs.
   d. move and eat through their food.
   e. are bulk feeders.

7. In animals, a distinct advantage of extracellular digestion over intracellular digestion is that
   a. polymers are hydrolyzed to monomers by digestive enzymes.
   b. a greater surface area is available for the absorption of digested nutrients.
   c. larger pieces of food can be ingested and then digested.
   d. all four types of macromolecules can be digested instead of just carbohydrates.
   e. the products of extracellular digestion can be absorbed into all body cells, without the need for a transport system.

8. After a meal of greasy french fried potatoes, which enzymes would you expect to be most active?
   a. salivary and pancreatic amylase, disaccharidases, lipase
   b. lipase, lactase, maltase
   c. pepsin, trypsin, chymotrypsin, dipeptidases
   d. gastric juice, bile, bicarbonate
   e. sucrase, lipase, bile

9. Why does salivary amylase not hydrolyze starch in the duodenum?
   a. Starch is completely hydrolyzed into maltose in the oral cavity.
   b. The acid pH of the stomach denatures salivary amylase, and pepsin begins hydrolyzing it.
   c. Salivary amylase is produced by salivary glands and never leaves the oral cavity.
   d. Pancreatic amylase is a more effective enzyme in the pH of the duodenum.
   e. Salivary amylase can hydrolyze glycogen but not starch.

10. Villi are
   a. folds in the stomach that allow the stomach to expand.
   b. extensions of the lymphatic system that pick up digested fats for transport to the circulatory system.
   c. fingerlike projections of the small intestine lining that increase the surface area for absorption.
   d. microscopic extensions of epithelial cells lining the small intestine that provide more surface area for digestion.
   e. extensions of the intestinal capillaries that join to form the hepatic portal vein.

11. The hepatic portal vein
   a. supplies the capillaries of the intestines.
   b. carries absorbed nutrients to the liver for processing.
   c. carries blood from the liver to the heart.
   d. drains the lacteals of the villi.
   e. supplies oxygenated blood to the liver.

12. The most likely action of antidiarrhea medicine is to
   a. speed up peristalsis in the small intestine.
   b. speed up peristalsis in the large intestine.
   c. kill *E. coli* in the intestine.
   d. increase water reabsorption in the large intestine.
   e. increase salt secretion into the feces.

13. Which of the following is *not* a common component of feces?
   a. intestinal bacteria
   b. cellulose
   c. saturated fats
   d. bile pigments
   e. undigested materials

14. Ruminants
   a. eat exposed salts and minerals from the soil in order to obtain phosphorus.
   b. house microorganisms that digest cellulose.
   c. eat their feces to obtain nutrients digested from cellulose by microorganisms.
   d. house symbiotic bacteria and protists in a large cecum.
   e. get all of their nutrition from digested plant material.

15. Which of the following animals would have the largest percentage of its energy budget available for growth?
   a. a deer mouse from a temperate forest
   b. a python from tropical India
   c. a human adult from a temperate climate
   d. a penguin from Antarctica
   e. They would all have the same percentage, because most of their energy is used to maintain basal or standard metabolic rate.

16. For most terrestrial animals, the average daily rate of energy consumption is 2 to 4 times BMR or SMR. Which of the following animals has an unusually low average daily metabolic rate?
   a. a human in a developed country
   b. a wild turkey
   c. a domesticated turkey
   d. a forest-dwelling salamander
   e. a desert snake

17. Which of the following statements is *false*?
   a. The average human has enough stored fat to supply calories for several weeks.
   b. An increase in leptin levels leads to an increase in appetite and weight gain.
   c. Conversion of glucose and glycogen takes place in the liver.
   d. After glycogen stores are filled, excessive calories are stored as fat, regardless of their original food source.
   e. Carbohydrates and fats are preferentially used as fuel before proteins are used.

18. You and two friends are trying a cereal diet to lose weight. You each eat 30 g of cereal, six times a day. You choose an organic cereal with nuts, grains, and no added sugar that has 3 g of fat, 20 g of carbohydrates (4 g of which are nondigestible insoluble fiber), and 7 g of protein. Friend 1 chooses a high-fiber cereal with 1 g of fat, 27 g of carbohydrates (9 g of insoluble fiber), and 2 g of protein. Friend 2 chooses a sugary cereal with 0 g of fat, 28 g of carbohydrates (3 g of insoluble fiber), and 2 g of protein. Predict the order, from fastest to slowest, in which the three of you will lose weight.
   a. you (because you ate the fewest carbs), then Friend 1 (because 9 g of the carbohydrates eaten are not digestible), then Friend 2
   b. Friend 1 (who ate 9 g of insoluble fiber), then you, then your sugary Friend 2
   c. Friend 1, then Friend 2, then you (because each of your fat grams contains approximately two times the energy content of carbohydrate or protein)
   d. Friend 2 (sugar burns the fastest), then Friend 1, then you
   e. you (because proteins are not used as fuel, those 7 g of protein do not count), then Friend 2, then Friend 1

MATCHING: *Match the description with the correct enzyme, hormone, or other substance.*

_____ 1. enzyme that hydrolyzes peptide bonds in the stomach

_____ 2. hormone that stimulates secretion of gastric juice

_____ 3. hormone that stimulates release of bile and pancreatic enzymes

_____ 4. lipid transport globule

_____ 5. hormone secreted by adipose cells that suppresses appetite

_____ 6. substances that emulsify fats

_____ 7. enzyme that hydrolyzes fats

_____ 8. hormone that is secreted by the stomach and triggers hunger

A. carboxypeptidase
B. bile salts
C. cholecystokinin (CCK)
D. chylomicron
E. gastrin
F. ghrelin
G. insulin
H. leptin
I. lipase
J. pepsin

Chapter 34

Circulation and Gas Exchange

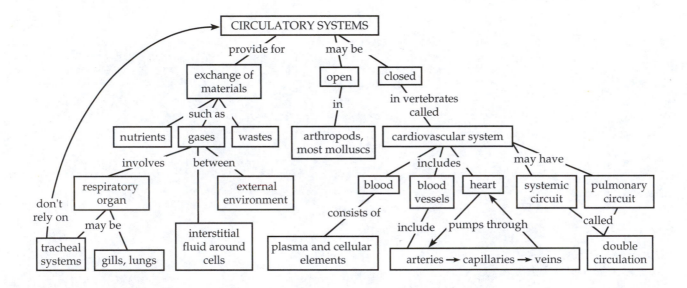

34.1 Circulatory systems link exchange surfaces with cells throughout the body

Every cell must obtain O_2 and nutrients and dispose of CO_2 and metabolic wastes. Many of these substances can enter or leave a cell by **diffusion,** moving down their concentration gradient across the cell membrane. Some animals have body shapes and sizes that allow most body cells to be in contact with the environment for exchange of materials. Because diffusion is too slow a process across distances more than a few millimeters, however, most animals have body tissues that exchange materials with the environment and a circulatory system to service body cells.

Gastrovascular Cavities The central gastrovascular cavity inside the two-cell-thick body wall of cnidarians serves for both the digestion and transport of materials. The internal fluid exchanges directly with the aqueous environment through the single opening. Flatworms also have gastrovascular cavities that branch throughout their thin, flat bodies.

Open and Closed Circulatory Systems The circulatory systems of more complex animals consist of a pumping **heart,** a circulatory fluid, and transport vessels.

Arthropods and some molluscs have an **open circulatory system.** The *interstitial fluid*, called **hemolymph,** is pumped into sinuses (spaces) between organs, where it bathes the internal tissues and provides for chemical exchange.

Cephalopods, annelids, and vertebrates have **closed circulatory systems,** in which **blood** remains in vessels and materials are exchanged between the blood and interstitial fluid bathing the cells.

---

a. How is hemolymph different from blood?

b. Compare the advantages of open and closed circulatory systems.

Organization of Vertebrate Circulatory Systems In the **cardiovascular system** of vertebrates, **arteries** carrying blood away from the heart branch into tiny *arterioles* within organs. Arterioles then divide into microscopic **capillaries,** which infiltrate tissues in networks called **capillary beds.** Capillaries converge to form *venules*, which meet to form the **veins** that return blood to the heart.

The vertebrate heart has one or two **atria,** which receive blood, and one or more **ventricles,** which pump blood out of the heart.

The ventricle of the two-chambered heart of a fish pumps blood to the gills, from which the oxygen-rich blood flows through a vessel to capillary beds in the other organs. Veins return the oxygen-poor blood to the atrium. Passage through two capillary beds slows the flow of blood, but body movements help to maintain circulation.

In the circulatory systems of amphibians, reptiles, and mammals, one side of the heart pumps blood to the **gas exchange circuit,** either a *pulmonary circuit* (lungs) or *pulmocutaneous circuit* (lungs and skin), while the other side pumps the returning oxygen-rich blood to capillary beds in the body tissues (**systemic circuit**). This double pumping assures a strong flow of oxygen-rich blood to the brain, muscles, and body organs.

In the three-chambered heart of amphibians, the single ventricle pumps blood into both the pulmocutaneous circuit, which leads to the lungs and skin and then back to the left atrium, and the systemic circuit, which carries blood to the rest of the body and back to the right atrium. A ridge in the ventricle diverts most of the blood from the left atrium (returning from the lungs and skin) into the systemic circuit. When underwater, blood flow to the lungs can be shut off.

The three-chambered heart of lizards, snakes, and turtles has a partially divided ventricle that helps to separate blood flow through the pulmonary and systematic circuits. In crocodilians, the ventricles are completely divided, although connections between arteries of the pulmonary and systemic circuits allow the lungs to be bypassed when the animal is underwater.

Delivery of oxygen for cellular respiration is most efficient in birds and mammals, which, as endotherms, have high oxygen and energy demands. The left side of the large and powerful four-chambered heart receives and pumps oxygen-rich blood only, whereas the right side receives and pumps oxygen-poor blood.

---

a. Explain the difference between a **single circulation** and a **double circulation.**

b. Describe the difference between the circulatory system of amphibians and that of mammals.

34.2 Coordinated cycles of heart contraction drive double circulation in mammals

Mammalian Circulation Review the circulation of blood through a mammalian cardiovascular system by completing Focus Question 34.3.

FOCUS QUESTION   34.3

In the following diagram of a mammalian circulatory system, label the indicated parts, color the vessels that carry oxygen-rich blood red, and then trace the flow of blood by numbering the circles from 1–11. Start numbering by assigning 1 to the right ventricle.

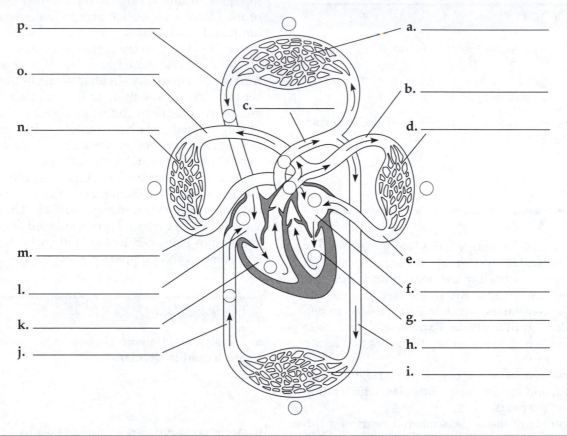

p. _____

o. _____

n. _____

c. _____

m. _____

l. _____

k. _____

j. _____

a. _____

b. _____

d. _____

e. _____

f. _____

g. _____

h. _____

i. _____

The Mammalian Heart: A Closer Look   The human heart, located just beneath the sternum, is composed mostly of cardiac muscle. The atria have relatively thin muscular walls, whereas the ventricles have thicker walls. The heart's rhythmic sequence of pumping and relaxing is called the **cardiac cycle,** and it consists of **systole,** during which the cardiac muscle contracts and the chambers pump blood, and **diastole,** when the heart chambers are relaxed and filling with blood.

How much blood is pumped by each ventricle per minute? Cardiac output depends on both the heart rate and the *stroke volume* (the quantity of blood pumped by each contraction).

Atrioventricular (AV) valves between each atrium and ventricle are snapped shut when blood is forced against them as the ventricles contract. The **semilunar valves** where the aorta and pulmonary artery exit the heart are forced open by ventricular contraction and close when the ventricles relax.

The heart sounds are caused by the recoil of blood against the closed heart valves. A **heart murmur** is the detectable hissing sound of blood leaking back through a defective valve.

Maintaining the Heart's Rhythmic Beat   The rhythm of heart contractions is coordinated by the **sinoatrial (SA) node,** or *pacemaker*, a region of specialized autorhythmic muscle cells located in the wall of the right atrium. The SA node initiates an electrical impulse that spreads through the cardiac muscle cells, and the two atria contract. After a 0.1-second delay, the **atrioventricular (AV) node,** located between the left and right atria, relays the impulse through specialized fibers to the ventricles. The electrical currents produced during the heart cycle can be detected by electrodes placed on the skin and recorded in an **electrocardiogram** (**ECG** or **EKG**).

The SA node is controlled by sympathetic and parasympathetic nerves (which speed or slow the heart rate) and is influenced by hormones and body temperature.

FOCUS QUESTION 34.4

a. Name the valve between an atrium and ventricle. _____

b. Name the valve between a ventricle and the aorta or pulmonary artery. _____

c. Which node initiates atrial contraction? _____

d. Which node initiates ventricular contraction?

_____

34.3 Patterns of blood pressure and flow reflect the structure and arrangement of blood vessels

Blood Vessel Structure and Function The wall of an artery or vein consists of three layers: an outer connective tissue layer with elastic fibers that allow the vessel to stretch and recoil; a middle layer of smooth muscle and more elastic fibers; and an inner lining of **endothelium,** a single layer of flattened epithelial cells and a basal lamina. The outer two layers are thicker in arteries than they are in veins. One-way valves in the larger veins assure that blood cannot backflow away from the heart. Capillaries have only an endothelium.

Blood Flow Velocity The enormous number of capillaries creates a large total cross-sectional area, resulting in a very slow flow of blood. Blood flow speeds up within venules and veins due to the decrease in total cross-sectional area.

Blood Pressure Blood pressure, the force that drives blood from the heart to the capillary beds, is highest during systole (*systolic pressure*). Resistance caused by the narrow openings of the arterioles impeding the exit of blood from arteries causes the swelling of the arteries during systole, which can be felt as the **pulse.** The recoiling of the stretched elastic arteries during diastole creates the *diastolic pressure*, which maintains a continuous flow of blood into arterioles and capillaries.

Arterial blood pressure may be measured with a sphygmomanometer. The first number is the systolic pressure (when blood first spurts through the artery that was closed off by the cuff), and the second number is the diastolic pressure.

Vasoconstriction, a reduction in vessel diameter due to contraction of smooth muscles in arteriole walls, increases blood pressure in the upstream arteries. **Vasodilation** of arterioles as smooth muscles relax increases blood flow into the arterioles and thus lowers blood pressure. Nervous and hormonal signals control the production of the gas nitric oxide (NO) and the peptide endothelin, which induce vasodilation and vasoconstriction, respectively.

Blood pressure is very low by the time blood exits the capillary beds. Contraction of the smooth muscle walls of venules and veins and the contraction of skeletal muscles between which veins are embedded force blood through veins. One-way valves keep blood moving toward the heart.

FOCUS QUESTION 34.5

Complete the following concept map to help you organize your understanding of blood pressure.

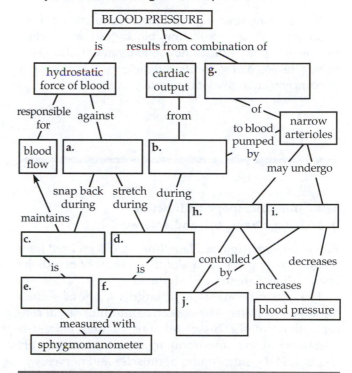

Capillary Function Only about 5–10% of the body's capillaries have blood flowing through them at any one time. The distribution of blood to capillary beds varies with need and is regulated by vasocontraction or vasodilation of arterioles and opening or closing of rings of smooth muscle at the entrance to capillary beds.

The exchange of substances between blood and interstitial fluid may occur by endocytosis and exocytosis across endothelial cells, by passive diffusion of small molecules, and by movement through microscopic pores in the capillary wall. Water, sugars, salts, and urea move through these openings, whereas blood cells and proteins are too large to fit through. Blood pressure forces fluid out of capillaries by bulk flow. The blood's *osmotic pressure* due to the proteins that remain within the capillaries tends to counter this fluid movement out of the capillaries.

Fluid Return by the Lymphatic System   The fluid lost from capillaries and any proteins that may have leaked out are returned to the blood through the **lymphatic system.** Fluid diffuses into tiny lymph vessels intermingled with blood capillaries. This fluid, now called **lymph,** moves through lymph vessels containing one-way valves, mainly as a result of the movement of skeletal muscles, and is returned to the circulatory system via large veins in the neck. In **lymph nodes,** the lymph is filtered, and white blood cells attack viruses and bacteria.

FOCUS QUESTION   34.6

Edema, swelling resulting from the accumulation of fluid in body tissues, can result from blockages that interfere with lymph flow. Explain how edema might also result from a severe dietary protein deficiency that decreases the concentration of blood plasma proteins.

34.4   Blood components function in exchange, transport, and defense

Blood Composition and Function   Cells and cell fragments make up about 45% of the volume of blood; the rest is a liquid matrix called **plasma.**

The plasma consists of a large variety of solutes dissolved in water. The collective and individual concentrations of the dissolved ions of inorganic salts (electrolytes) are important to osmotic balance, pH levels, and the functioning of muscles and nerves.

Plasma proteins function in the osmotic balance of blood and as buffers, antibodies, escorts for lipids, and clotting factors. Blood plasma from which clotting factors have been removed is called serum. Nutrients, metabolic wastes, gases, and hormones are transported in the plasma.

Red and white blood cells and platelets are produced from multipotent **stem cells** in red bone marrow. **Platelets** are involved in the clotting process.

Erythrocytes (red blood cells) transport O_2. The red blood cells of mammals lack nuclei. Their small size and biconcave shape create a large surface area of plasma membrane across which O_2 can diffuse. Erythrocytes are packed with **hemoglobin,** an iron-containing protein that binds O_2.

In **sickle-cell disease,** abnormal hemoglobin molecules cause erythrocytes to distort into a sickle shape, resulting in blocked vessels and other serious effects. Ruptured sickled erythrocytes can lead to anemia.

Erythrocytes circulate for about 4 months before being replaced. The production of red blood cells is controlled by a negative feedback mechanism involving the hormone *erythropoietin (EPO),* which is produced by the kidney in response to low O_2 supply in tissues.

There are five major types of **leukocytes,** or white blood cells, all of which fight infections. Some white blood cells are phagocytes that engulf bacteria and cellular debris; lymphocytes give rise to cells that produce the immune response. Leukocytes are also found in interstitial fluid and lymph nodes.

Platelets are pinched-off fragments of specialized bone marrow cells. The clotting process usually begins when platelets, clumped together along a damaged endothelium, release clotting factors. By a series of steps, the enzyme thrombin converts the plasma protein fibrinogen to its active form, fibrin. The threads of fibrin form a patch.

Anticlotting factors normally prevent the clotting of blood in the absence of injury. A **thrombus** is a clot that occurs within a blood vessel and blocks the flow of blood.

FOCUS QUESTION 34.7

Complete the following concept map of the components of vertebrate blood and their functions.

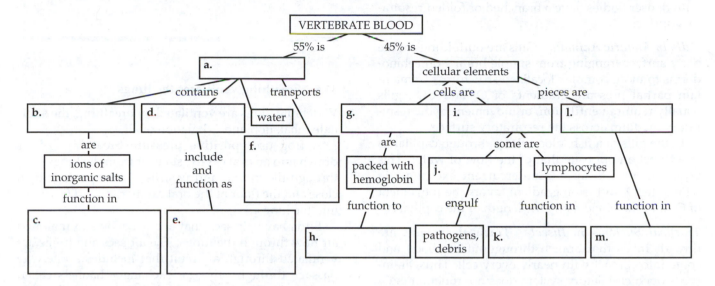

Cardiovascular Disease Cholesterol metabolism as well as *inflammation* are factors in the development of cardiovascular diseases, disorders of the heart and blood vessels. Cholesterol is carried to body cells for inclusion in cell membranes in particles called **low-density lipoproteins (LDLs). High-density lipoproteins (HDLs)** return excess cholesterol to the liver. A high LDL/HDL ratio is a risk factor for heart disease.

In the chronic cardiovascular disease known as **atherosclerosis,** fatty deposits called plaques narrow the arteries. Injury to or infection of the lining of an artery can lead to inflammation, which attracts lipid-accumulating leukocytes. The plaque grows, becomes covered with a fibrous cap of smooth muscle cells, and may obstruct the artery or rupture.

A **heart attack,** or *myocardial infarction,* is the death of cardiac muscle due to blockage of coronary arteries; a **stroke,** the death of nervous tissue in the brain, results from blockage or rupture of arteries in the head.

Statins are drugs used to lower LDL levels in individuals at risk of cardiovascular disease. High levels of C-reactive protein (CRP), indicative of acute inflammation, also indicate disease risk.

Hypertension, or high blood pressure, is thought to damage the endothelium and initiate plaque formation, promoting atherosclerosis and increasing the risk of heart attack and stroke. This condition can be easily diagnosed and controlled by diet, exercise, and drugs.

FOCUS QUESTION 34.8

What lifestyle choices have been correlated with increased risk of cardiovascular disease?

34.5 Gas exchange occurs across specialized respiratory surfaces

Gas exchange, or respiration, is the uptake of O_2 from and the discharge of CO_2 to the environment.

Partial Pressure Gradients in Gas Exchange The concentration of a gas in air or dissolved in water is that gas's **partial pressure.** At sea level, the partial pressure of O_2, which makes up 21% of the atmosphere, is 160 mm Hg (0.21 × 760 mm, which is atmospheric pressure at sea level). A gas diffuses from a region of higher partial pressure to a region of lower partial pressure.

Respiratory Media How do air and water differ as respiratory media? Air has plentiful O_2. Water has much less O_2 and greater density and viscosity. Thus, gas exchange in water is more energetically demanding and often involves adaptations for increased efficiency.

Respiratory Surfaces The respiratory surface, where gas exchange with the respiratory medium occurs, must be moist, thin, and have a large enough area to supply the animal's metabolic needs.

In animals with simple body forms, gases can easily diffuse in and out of body cells. In earthworms and some amphibians, the skin, supplied with a network of capillaries, serves as a respiratory organ. Most animals with denser bodies have a branched or folded respiratory surface.

Gills in Aquatic Animals   Gills are outfoldings of the body surface, ranging from simple bumps on echinoderms to more complex localized structures. To maintain partial pressure gradients of O_2 and CO_2, gills usually require **ventilation,** or movement of the respiratory medium across the respiratory surface.

In the gills of a fish, blood flows through capillaries in a direction opposite that of the flow of water. This arrangement sets up a **countercurrent exchange,** in which the O_2 diffusion gradient favors the movement of O_2 into the blood along the length of the capillary.

Tracheal Systems in Insects   **Tracheal systems** are tiny air tubes that branch throughout the body and come into contact with nearly every cell. Thus, an insect's open circulatory system does not function in O_2 and CO_2 transport. In large insects, rhythmic muscle contractions ventilate tracheal systems.

Lungs   **Lungs** are invaginated respiratory surfaces restricted to one location. A circulatory system transports gases between the lungs and the body.

The lungs of mammals are located in the thoracic cavity. Air entering through the nostrils is filtered, warmed, humidified, and smelled in the nasal cavity. Air passes through the pharynx and enters the respiratory tract. The **larynx** moves up and tips a flap of cartilage over the opening to the **trachea,** or windpipe, when food is swallowed. The larynx functions to produce sound in most mammals as exhaled air vibrates a pair of vocal folds (called vocal cords in humans). The trachea branches into two **bronchi,** which then branch repeatedly into *bronchioles* within each lung. Ciliated, mucus-coated epithelium lines the major branches of the respiratory tree.

Where does gas exchange actually take place? **Alveoli** are air sacs encased in a web of capillaries at the tips of the tiniest bronchioles. The thin, moist epithelium of each alveolus is coated with **surfactants,** which decrease surface tension and help keep these tiny sacs from sticking shut.

FOCUS QUESTION   34.9

What causes respiratory distress syndrome (RDS) in infants?

34.6   Breathing ventilates the lungs

Vertebrate lungs are ventilated by **breathing,** the alternate inhalation and exhalation of air.

A frog uses **positive pressure breathing.** Air is drawn into an oral cavity; stale air from the lungs exits through the mouth and nostrils; these openings then close and the floor of the oral cavity rises and forces air into the lungs.

Birds have air sacs that act as bellows to maintain air flow through the lungs. The air sacs and lungs are ventilated through a circuit that includes a one-way passage through tiny, gas-exchange channels called *parabronchi.* Two cycles of inhalation and exhalation are required to pass air through the entire system, in which fresh air and stale air do not mix.

How a Mammal Breathes   Mammals ventilate their lungs by **negative pressure breathing.** Contraction of the rib muscles and the **diaphragm** expands the chest cavity and increases the volume of the lungs. The reduction in air pressure in this increased volume draws air into the lungs. Relaxation of the rib muscles and diaphragm compresses the lungs, increasing the pressure and forcing air out.

Tidal volume is the volume of air inhaled and exhaled by an animal during normal breathing. The maximum volume during forced breathing is called **vital capacity.** The **residual volume** is the air that remains in the alveoli and lungs after forceful exhaling.

Control of Breathing in Humans   Breathing is under automatic regulation by neural circuits that form a breathing control center in the medulla oblongata. A negative-feedback mechanism involving sensors in the lungs and control circuits in the medulla prevent overexpansion of the lungs. Nerve impulses from the medulla instruct the rib muscles and diaphragm to

FOCUS QUESTION 34.10

For the following animals, indicate the respiratory medium, respiratory surface, and means of ventilation.

Animal	Respiratory Medium	Respiratory Surface	Means of Ventilation
Fish	a.		
Grasshopper	b.		
Frog	c.		
Human	d.		

contract. When sensors in the medulla detect a drop in the pH of the cerebrospinal fluid, which is caused by an increase in blood CO_2 concentration, its control circuits increase the depth and rate of breathing. Although breathing centers respond primarily to CO_2 levels, they are alerted by oxygen sensors in the aorta and carotid arteries that react to severe deficiencies of O_2.

34.7   Adaptations for gas exchange include pigments that bind and transport gases

Coordination of Circulation and Gas Exchange Blood entering the capillaries of the lungs has a lower P_{O_2} and a higher P_{CO_2} than does the air in the alveoli, so O_2 diffuses into the capillaries and CO_2 diffuses out. In the tissue capillaries, pressure differences favor the diffusion of O_2 out of the blood into the interstitial fluid and of CO_2 into the blood.

Respiratory Pigments   In most animals, O_2 is carried in the blood or hemolymph by **respiratory pigments,** which consist of a metal bound to a protein. Copper is the oxygen-binding component in the blue respiratory pigment hemocyanin, common in arthropods and many molluscs.

What is the respiratory pigment used by almost all vertebrates? Hemoglobin, contained in erythrocytes, is composed of four subunits, each of which has a cofactor called a heme group with iron at its center. The binding of O_2 to the iron atom of one subunit induces a change in shape of the other subunits, and their affinity for O_2 increases. Likewise, the unloading of the first O_2 lowers the other subunits' affinity for O_2.

The dissociation curve for hemoglobin shows the relative amounts of O_2 bound to hemoglobin under varying O_2 partial pressures. In the steep part of this S-shaped curve, a slight change in partial pressure will cause hemoglobin to load or unload a substantial amount of O_2. Because of the **Bohr shift** in response

to the lower pH of active tissues, hemoglobin releases more O_2 in such tissues.

Carbon Dioxide Transport   Most carbon dioxide is transported in blood as bicarbonate ions. CO_2 enters red blood cells, where it first combines with H_2O to form carbonic acid (catalyzed by carbonic anhydrase) and then dissociates into H^+ and HCO_3^-. Bicarbonate moves into the plasma for transport. The H^+ binds to hemoglobin and other proteins, which minimizes changes in pH during the transport of CO_2. In the lungs, the diffusion of CO_2 out of the blood shifts the equilibrium in favor of the conversion of HCO_3^- back to CO_2, which is unloaded from the blood.

FOCUS QUESTION 34.11

The following graph shows dissociation curves for hemoglobin at two different pH values. Explain the significance of these two curves.

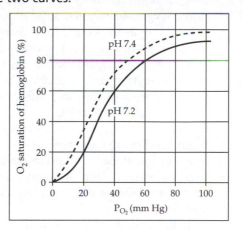

Respiratory Adaptations of Diving Mammals   Special adaptations have enabled some air-breathing animals to make sustained underwater dives. Weddell seals store twice the amount of O_2 per kilogram of body weight as do humans, mostly by having a larger volume of blood and a higher concentration of **myoglobin,** an oxygen-storing muscle protein. Oxygen-conserving adaptations during a dive include decreasing the heart rate, rerouting blood to the brain and essential organs, and restricting blood supply to the muscles, which use fermentation to produce ATP during long dives.

Word Roots

alveol-   = a cavity (*alveolus:* one of the dead-end air sacs where gas exchange occurs in a mammalian lung)

atrio-   = a vestibule; **-ventriculo** = ventricle (*atrioventricular node:* a region of specialized heart muscle tissue between the left and right atria where electrical impulses are delayed before spreading to both ventricles and causing them to contract)

cardi-   = heart; **-vascula** = a little vessel (*cardiovascular system:* the closed circulatory system characteristic of vertebrates)

counter-   = opposite (*countercurrent exchange:* exchange of a substance or heat between two fluids flowing in opposite directions)

endo-   = inner (*endothelium:* the flattened single layer of cells lining the lumen of blood vessels)

erythro-   = red; **-cyte** = cell (*erythrocyte:* a blood cell that contains hemoglobin, which transports oxygen; also called a red blood cell)

hemo-   = blood; **-lymph** = clear fluid (*hemolymph:* in invertebrates with an open circulatory system, the body fluid that bathes tissues)

leuko-   = white; **-cyte** = cell (*leukocyte:* a blood cell that functions in fighting infections; also called a white blood cell)

myo-   = muscle (*myoglobin:* an oxygen-storing, pigmented protein in muscle cells)

pulmo-   = a lung; **-cutane** = skin (*pulmocutaneous circuit:* a branch of the circulatory system in many amphibians that supplies the lungs and skin)

semi-   = half; **-luna** = moon (*semilunar valve:* a valve located at each exit of the heart, where the aorta leaves the left ventricle and the pulmonary artery leaves the right ventricle)

thrombo-   = a clot (*thrombus:* a fibrin-containing clot that forms in a blood vessel and blocks the flow of blood)

Structure Your Knowledge

1. Identify the labeled structures in the following diagram of a human heart. Draw arrows to trace the flow of blood.

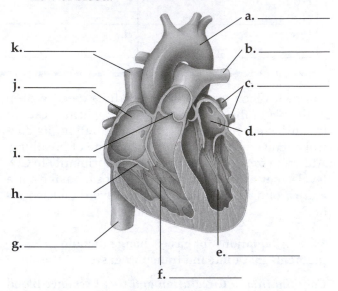

2. Trace the path of a molecule of O_2 from its inhalation into the nasal cavity to its delivery to the kidney.

Test Your Knowledge

MULTIPLE CHOICE: *Choose the one best answer.*

1. A gastrovascular cavity
    a. is found in cnidarians and annelids.
    b. pumps fluid through short vessels from which the fluid spreads throughout the body.
    c. functions in both digestion and distribution of nutrients.
    d. has a single opening for ingestion and elimination, but a separate opening for gas exchange.
    e. involves all of the above.

2. Which of the following statements does *not* describe a similarity between open and closed circulatory systems?
    a. Some sort of pumping device helps to move blood through the body.
    b. Some of the circulation of blood results from body movements.

c. The blood and interstitial fluid are distinguishable from each other.

d. All tissues come into close contact with the circulating body fluid so that the exchange of nutrients and wastes can take place.

e. All of these apply to both open and closed circulatory systems.

3. In a system with double circulation,

a. blood is usually pumped at two separate locations as it circulates through the body.

b. there is a countercurrent exchange in the gills.

c. hemolymph circulates both through a pumping blood vessel and through body sinuses.

d. blood is pumped to the gas-exchange organ, and returning blood is then pumped to the systemic circuit.

e. there are always two ventricles and often two atria.

4. Which of the following statements is the *best* explanation for the presence of four-chambered hearts in both birds and mammals?

a. They shared a common ancestor that had a four-chambered heart.

b. They are the only vertebrates with double circulation, which requires four heart chambers.

c. They are both endotherms, and the evolution of efficient circulatory systems supported the high metabolic rate of endotherms.

d. This is an example of convergent evolution, because animals that obtain their O_2 from air require both a pulmonary circuit and a systemic circuit.

e. The more inefficient single atrium of amphibians and most reptiles could not supply the higher O_2 needs of these endotherms.

5. During diastole,

a. the atria fill with blood.

b. blood flows passively into the ventricles.

c. the elastic recoil of the arteries maintains hydrostatic pressure on the blood.

d. semilunar valves are closed, but atrioventricular valves are open.

e. all of the above are occurring.

6. An atrioventricular valve prevents the backflow or leakage of blood from

a. the right ventricle into the right atrium.

b. the left atrium into the left ventricle.

c. the aorta into the left ventricle.

d. the pulmonary vein into the right atrium.

e. the right atrium into the vena cava.

7. As a general rule, blood leaving the right ventricle of a mammal's heart will pass through how many capillary beds before it returns to the right ventricle?

a. one

b. two

c. three

d. one or two, depending on the circuit it takes

e. at least two, but almost always three

8. The nurse tells you that your blood pressure is 112/70. The "70" refers to

a. your heart rate.

b. the velocity of blood during diastole.

c. the systolic pressure from ventricular contraction.

d. the diastolic pressure from the recoil of the arteries.

e. the venous pressure caused by the compression of the blood pressure cuff.

9. Blood flows more slowly in arterioles than in arteries because arterioles

a. have narrow openings that impede the exit of blood from arteries.

b. collectively have a larger cross-sectional area than do arteries.

c. must provide an opportunity for exchange with the interstitial fluid.

d. have rings of smooth muscle that restrict blood flow to capillary beds.

e. are narrower than arteries.

Use the following to answer questions 10 through 13.

1. vena cava
2. left ventricle
3. pulmonary vein
4. right atrium
5. aorta
6. pulmonary capillaries

10. Which of the following sequences represents the flow of blood through the human body? (Obviously, many structures are not included.)

a. 1-5-3-6-2-4

b. 1-2-6-3-4-5

c. 1-4-3-6-2-5

d. 1-2-3-6-4-5

e. 1-4-6-3-2-5

11. In which vessel is blood pressure the highest?

a. 1   d. 5

b. 3   e. 6

c. 4

12. In which vessel is the velocity of blood flow the lowest?

a. 1   d. 5

b. 3   e. 6

c. 4

13. Of these structures, which has the thickest muscle layer?

a. 1   d. 4

b. 2   e. 5

c. 3

14. Which of the following processes is *not* a factor in the exchange of substances in capillary beds?
    a. endocytosis and exocytosis
    b. passive diffusion
    c. blood pressure at the arterial end of a capillary
    d. bulk flow through pores in the capillary wall
    e. active transport by white blood cells

15. Which of the following processes would tend to reduce the blood supply to a capillary bed?
    a. release of EPO (erythropoietin)
    b. relaxation of smooth muscle rings at the entrance to the capillary bed
    c. vasodilation
    d. lowering of blood pH
    e. release of endothelin

16. Which of the following changes would you expect following an increase in the concentration of blood proteins?
    a. the swelling of body tissues (edema)
    b. the swelling of lymph glands
    c. an increase in the outward movement of fluid at the arterial end of capillaries
    d. a decrease in lymph flow through the lymphatic system
    e. an increase in the delivery of nutrients to body tissues

17. Fibrinogen is
    a. a blood protein that escorts lipids through the circulatory system.
    b. a cell fragment involved in the blood-clotting mechanism.
    c. a blood protein that is converted to fibrin to form a blood clot.
    d. a leukocyte involved in trapping bacteria.
    e. a lymph protein that regulates osmotic balance in the tissues.

18. A thrombus
    a. may form at a site of plaque formation in an artery.
    b. is a typical complication of hemophilia.
    c. may cause hypertension.
    d. narrows the diameter of arteries due to the buildup of plaque.
    e. is often associated with high levels of HDLs in the blood.

19. Which of the following functions is *not* associated with plasma proteins in a mammal?
    a. pH buffer
    b. osmotic balance
    c. clotting agent
    d. oxygen transport
    e. antibody

20. In countercurrent exchange,
    a. the flow of fluids in opposite directions maintains a favorable diffusion gradient along the length of an exchange surface.
    b. O_2 is exchanged for CO_2.
    c. double circulation keeps oxygenated and deoxygenated blood separate.
    d. O_2 moves from a region of high partial pressure to one of low partial pressure, but CO_2 moves in the opposite direction.
    e. the capillaries of the lung pick up more O_2 than do tissue capillaries.

21. The tracheae of insects
    a. are stiffened with chitinous rings and lead into the lungs.
    b. are filled by positive pressure breathing.
    c. ramify along capillaries of the circulatory system for gas exchange.
    d. are highly branched and come into contact with almost every cell to effect gas exchange.
    e. provide an extensive, fluid-filled system of tubes for gas exchange.

22. What do the alveoli of mammalian lungs, the gill filaments of fish, and the tracheoles of insects all have in common?
    a. They use a circulatory system to transport absorbed O_2.
    b. Their respiratory surfaces are invaginated.
    c. They function using countercurrent exchange.
    d. They have a large, thin surface area for gas exchange.
    e. They have all of the above in common.

23. The cilia in the trachea and bronchi
    a. move air into and out of the lungs.
    b. increase the surface area for gas exchange.
    c. vibrate when air rushes past them to produce sounds.
    d. filter the air that rushes through them.
    e. sweep mucus containing trapped particles up and out of the respiratory tract.

24. When you hold your breath, which of the following leads to the urge to breathe?
    a. falling O_2
    b. falling CO_2
    c. rising CO_2
    d. falling blood pH
    e. both **c** and **d**

25. Which of the following statements is the *best* explanation for why birds can fly over the Himalayas but most humans require bottled oxygen to climb these mountains?
    a. Birds are much smaller and require less O_2.
    b. Birds use positive pressure breathing, whereas humans use negative pressure breathing.

   c. With a one-way flow of air and efficient ventilation, the lungs of birds extract more O_2 from the air they breathe.

   d. The circulatory system of birds is much more efficient in delivering O_2 to body tissues.

   e. The vital capacity of bird lungs, relative to their size, is much greater than that of humans.

26. Gas exchange between maternal blood and fetal blood takes place in the placenta. Fetal hemoglobin and adult hemoglobin differ slightly in composition, and their dissociation curves are different such that O_2 can be transferred from maternal blood to fetal blood. If curve *c* in the following graph represents the dissociation curve of adult hemoglobin, which curve would represent the dissociation curve of fetal hemoglobin? (*Hint:* Must fetal hemoglobin have a higher or lower affinity for O_2 than maternal hemoglobin has?)

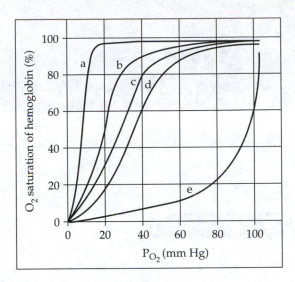

Chapter 35

The Immune System

Chapter Focus

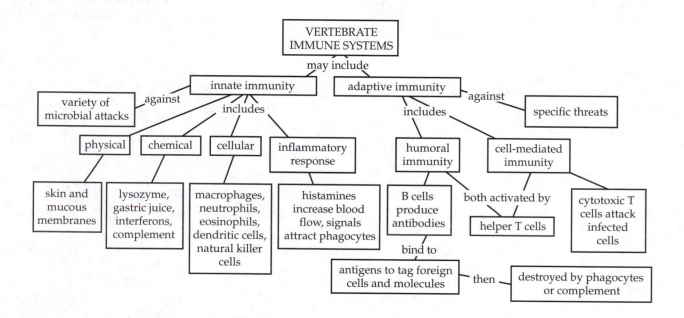

Chapter Review

The **immune system** defends against **pathogens**—disease-causing agents such as bacteria, viruses, and fungi. **Innate immunity** includes external physical barriers and internal defenses that are active immediately upon exposure to a pathogen and are the same whether or not the pathogen was previously encountered. The immune system relies on *molecular recognition* to identify internal invaders that are nonself. In innate immunity, immune cells have a small group of receptor proteins that recognize a broad range of pathogens and can trigger a defensive response. **Adaptive immunity,** also called acquired immunity, is a line of defense in vertebrates in which immune cells react specifically to pathogens. A vast array of adaptive immune receptors enables recognition and response to specific pathogens, and the response is enhanced upon reexposure to a pathogen.

35.1 In innate immunity, recognition and response rely on traits common to groups of pathogens

Innate Immunity of Invertebrates Insect defenses begin with their protective exoskeleton. The enzyme **lysozyme,** which attacks microbial cell walls, protects the digestive system. *Hemocytes* are circulating cells that destroy bacteria by **phagocytosis** or release chemicals that kill pathogens. Antimicrobial peptides disrupt the plasma membranes of fungi and bacteria. Characteristic macromolecules found on bacteria or fungi are bound by recognition proteins secreted by immune cells, triggering an innate immune response.

Innate Immunity of Vertebrates In mammals, skin and the mucous membranes lining the digestive, respiratory, urinary, and reproductive tracts are barrier defenses. The ciliated epithelial lining of the respiratory tract sweeps *mucus* containing trapped microbes upward and out. Lysozyme present in tears, saliva, and mucus kills bacteria. Skin secretions maintain a low pH, which discourages the colonization of many bacteria. The acidity of gastric juice kills most microorganisms that reach the stomach.

Cellular innate defenses involve specific **Toll-like receptors (TLRs)** that recognize molecules that are common to a set of pathogens, such as double-stranded RNA from certain viruses or lipopolysaccharide on the surface of many bacteria. As in invertebrates, binding to such a receptor triggers phagocytosis.

Neutrophils are phagocytic cells that circulate in the blood. **Macrophages** are larger phagocytic cells that migrate through the body or remain in various organs, such as the spleen or lymph nodes, where they filter pathogens from blood or lymph.

The primary role of *dendritic cells,* located in tissues in contact with the environment, is to stimulate adaptive immunity. *Eosinophils* attack multicellular parasitic invaders with destructive enzymes. **Natural killer cells** recognize abnormal surface proteins on some virus-infected or cancerous cells and release chemicals that trigger target cell death.

Various antimicrobial peptides and proteins are released in response to pathogen recognition. Virus-infected cells produce **interferons,** which stimulate neighboring cells to produce substances that inhibit viral reproduction in those cells. Some white blood cells produce an interferon that activates macrophages.

The **complement system** is a group of about 30 proteins in the blood plasma that, when activated by contact with microbes, may lyse cells, trigger inflammation, or assist in adaptive defenses.

How does the body respond to physical injury or pathogen entry? Signaling molecules released from the affected area trigger an **inflammatory response,** characterized by redness, swelling, heat, and pain. Damaged **mast cells** in connective tissue release **histamine,** a chemical that triggers dilation and leakiness of blood vessels; activated macrophages and neutrophils release **cytokines,** signaling molecules that promote blood flow to the damaged area. Vasodilation and signaling proteins result in the congregation of phagocytes and antimicrobial compounds. Pus is an accumulation of fluid containing white blood cells, dead microbes, and cell debris.

A systemic (whole body) response to an infection may include an increase in the number of circulating white blood cells and a fever. Fevers, triggered by substances released by activated macrophages, may stimulate phagocytosis and speed tissue repair. *Septic shock* is a dangerous condition resulting from an overwhelming systemic inflammatory response.

Evasion of Innate Immunity by Pathogens Certain bacteria may evade the innate immune system by means of an outer capsule that interferes with recognition and phagocytosis or by resisting breakdown and growing within the host's cells.

FOCUS QUESTION   35.1

Complete the following table that summarizes the functions of the cells and chemicals of the innate defense mechanisms of mammals.

Cells or Compounds	Functions
Neutrophils	**a.**
Macrophages	**b.**
Eosinophils	**c.**
Dendritic cells	**d.**
Natural killer cells	**e.**
Mast cells	**f.**
Histamine	**g.**
Lysozyme	**h.**
Interferons	**i.**
Complement system	**j.**
Cytokines	**k.**

35.2 In adaptive immunity, receptors provide pathogen-specific recognition

Lymphocytes are the key cells of adaptive immunity. **T cells** migrate to the **thymus** to mature, whereas **B cells** remain in the bone marrow and mature there.

B cells and T cells become activated upon binding with an **antigen,** often a bacterial or viral protein. B cells and T cells have thousands of identical membrane-bound **antigen receptors** that enable them to recognize a specific antigen. The small region of an antigen to which an antigen receptor binds is called an **epitope,** or *antigenic determinant.*

Antigen Recognition by B Cells and Antibodies   Each Y-shaped antigen receptor on a B cell consists of four polypeptide chains: two identical **light chains** and two identical **heavy chains,** linked together by disulfide bridges. Both heavy and light chains have *variable (V) regions* at the ends of the two arms of the Y, which form two identical antigen-binding sites. The *constant (C) regions* of the molecule vary little among receptors on different B cells. The constant region of the heavy chains includes an anchoring transmembrane region and a cytoplasmic tail.

Activated B cells give rise to cells that secrete **antibodies,** or **immunoglobulins (Ig),** which are soluble antigen receptors that defend against pathogens. The particular amino acid sequences of the variable regions of an antibody or B cell receptor account for the highly specific binding with a particular epitope.

Antigen Recognition by T Cells   A T cell antigen receptor, also anchored by a transmembrane region, consists of one α *chain* and one β *chain,* linked by a disulfide bridge. The variable regions at the tip of the molecule form a single antigen-binding site. T cells recognize small fragments of antigens displayed on the surface of cells. The displaying protein is called a **major histocompatibility complex (MHC) molecule.**

Host cells that are infected by or take in a pathogen cleave the antigen into small peptides. In a process called **antigen presentation,** an MHC molecule binds with an *antigen fragment* within the cell and then displays that fragment on the cell's surface, where a T cell receptor can recognize the antigen–MHC molecule complex.

FOCUS QUESTION   35.2

Describe the differences between the antigens that B cell antigen receptors recognize and the antigens that T cell antigen receptors recognize.

B Cell and T Cell Development   Four major characteristics of adaptive immunity are the immense diversity of antigen receptors, the self-tolerance of the system, the huge increase in B cells and T cells specific for an antigen following activation, and immunological memory, an enhanced response to a previously encountered foreign molecule.

Each person may have more than 1 million different B cell antigen receptors and 10 million different T cell antigen receptors. The genes coding for this remarkable diversity have numerous coding segments that are randomly and permanently rearranged. For example, in the Ig (immunoglobulin) light-chain gene, there are 40 variable (V) gene segments and five joining (J) gene segments. The J segments are followed by a single C segment, which codes for the constant region. Early in B cell development, one V gene segment is randomly linked to one J segment by an enzyme complex called *recombinase,* producing one of 200 possible light chains. These light chains combine with heavy chains, which have an even greater number of possible recombinations. Transcription, splicing, and then translation of such permanently rearranged genes produces receptors in different B cells (and T cells) that provide an enormous diversity of antigen-binding sites.

In the bone marrow and thymus, respectively, the antigen receptors of maturing B and T cells are tested for self-reactivity. Lymphocytes with receptors specific for the body's own molecules are either inactivated or destroyed by programmed cell death. This critical self-tolerance means that normally the body contains no mature lymphocytes that react against self components.

When an antigen interacts with antigen receptors on B cells or T cells that are specific for epitopes of that antigen, those particular lymphocytes are activated to divide repeatedly into a clone of identical cells. Some cells develop into short-lived **effector cells,** which combat the antigen. Other cells develop into long-lived **memory cells.** By this process of **clonal selection,** a small number of cells is selected by their interaction with a specific antigen to produce thousands of identical cells keyed to that particular antigen.

The body mounts a **primary immune response** upon a first exposure to an antigen. About 10 to 17 days are required for selected lymphocytes to proliferate and yield the maximum response produced by effector T cells and the antibody-producing effector B cells, called **plasma cells.** Should the body reencounter the same antigen, the resulting **secondary immune response** is more rapid, effective, and prolonged. The long-lived T memory cells and B memory cells are responsible for this immunological memory. The secondary immune response provides long-term protection against a previously encountered pathogen.

FOCUS QUESTION   35.3

Answer the following questions concerning the four major characteristics of adaptive immunity.

a. How is the great diversity of B and T cells produced?

b. What prevents B cells and T cells from reacting against the body's own molecules?

c. Describe clonal selection.

d. What is immunological memory?

35.3   Adaptive immunity defends against infection of body fluids and body cells

The **humoral immune response** involves the production of antibodies that help destroy toxins and pathogens in the blood and lymph. The **cell-mediated** **immune response** involves specialized T cells that identify and destroy infected body cells.

Helper T Cells: A Response to Nearly All Antigens   Both humoral and cell-mediated immune responses are triggered by **helper T cells.** Helper T cells recognize specific antigens displayed by **antigen-presenting cells,** either a dendritic cell, macrophage, or B cell.

Most body cells have only class I MHC molecules, but dendritic cells, macrophages, and B cells have both class I and class II MHC molecules. These antigen-presenting cells engulf and fragment microbes, then display the antigen fragments in class II MHC molecules. A helper T cell antigen receptor binds to the antigen fragment, and a surface protein called CD4 enhances the binding by attaching to the class II MHC molecule. Signaling between the two cells via cytokines results in the proliferation of activated helper T cells. These helper T cells secrete cytokines that activate both B cells involved in humoral immunity and cytotoxic T cells, which are part of cell-mediated immunity.

Cytotoxic T Cells: A Response to Infected Cells   Activation of **cytotoxic T cells** requires signals from nearby helper T cells and interaction with an antigen-presenting cell. How does an activated cytotoxic T cell recognize and kill an infected cell? An infected cell produces fragments of foreign proteins and displays them in its class I MHC molecules. A cytotoxic T cell receptor (with the aid of accessory protein CD8) recognizes and binds to a class I MHC-fragment-complex. The T cell then secretes proteins that destroy the infected cell.

FOCUS QUESTION   35.4

Label the components in the following diagram that shows a helper T cell being activated by interaction with an antigen-presenting cell and the helper T cell's central role in activating both humoral and cell-mediated immunity.

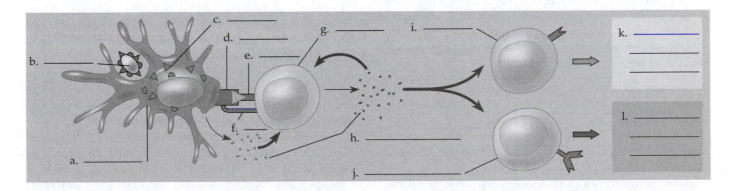

FOCUS QUESTION 35.5

a. What does an activated helper T cell release?

b. What does a cytotoxic T cell attached to an infected body cell release?

B Cells and Antibodies: A Response to Extracellular Pathogens   Upon first binding antigen, a B cell takes in and presents antigen fragments in its class II MHC molecules to a helper T cell, which had been activated by binding with a macrophage or dendritic cell displaying the same antigen. Aided by the interaction with a helper T cell and the cytokines it secretes, an activated B cell proliferates and produces memory B cells and antibody-secreting plasma cells.

Antibodies can act by *neutralization,* binding to toxins or blocking proteins on a virus and preventing these from entering a host cell. The binding of multiple antibodies may link bacterial cells, viruses, or other foreign molecules into aggregates that facilitate phagocytosis. Antigen-antibody complexes on microbes activate the complement system, which can lead to the destruction of the foreign cell.

There are five types of antibodies based on their heavy-chain C region. One is the membrane-bound B cell antigen receptor; the others are soluble antibodies.

Summary of the Humoral and Cell-Mediated Immune Responses   Adaptive immunity defends against pathogens in the blood and lymph with antibodies secreted by B cells (humoral immune response) and against intracellular pathogens and cancer cells with cytotoxic T cells (cell-mediated immune response). Both responses depend on antigen-presenting cells activating helper T cells, which secrete cytokines that activate cytotoxic T cells and B cells.

Active and Passive Immunization   **Active immunity** results when the body responds to an infection and develops immunological memory specific to that pathogen. In **passive immunity,** temporary immunity is provided by antibodies supplied through the placenta to a fetus or through milk to a nursing infant.

Active immunity can also develop from **immunization,** also called vaccination. A *vaccine* may be an inactivated toxin, a killed or weakened microbe, a portion of a microbe, or even genes for microbial proteins. In artificial passive immunization, antibodies from an immune animal are injected into a nonimmune animal to provide temporary protection.

Antibodies as Tools   Antibodies that an animal produces are *polyclonal* because they are formed by several different B cell clones—each specific for a different epitope of an antigen. A technique for making **monoclonal antibodies** can supply quantities of identical antibodies. Such antibodies are used in medical diagnosis and treatment.

Immune Rejection   The immune response to the chemical markers that determine ABO blood groups must be considered in blood transfusions. Transplanted tissues and organs are rejected because the foreign MHC molecules are antigenic and trigger immune responses. The use of closely related donors, as well as drugs that suppress immune responses, helps to minimize rejection.

Disruptions in Immune System Function   Allergies are hypersensitivities to certain antigens called **allergens.** Binding of allergens to antibodies attached to mast cells triggers the release of histamines, which can cause sneezing, a runny nose, and difficulty in breathing due to smooth muscle contractions. Antihistamines are drugs that reduce symptoms by blocking receptors for histamine. *Anaphylactic shock* is a severe allergic response in which the rapid release of inflammatory chemicals causes abrupt dilation of peripheral blood vessels and a drop in blood pressure, as well as constriction of bronchioles.

Sometimes the immune system turns against self, leading to **autoimmune diseases,** such as lupus, rheumatoid arthritis, type 1 diabetes, and multiple sclerosis. These diseases result from a loss of self-tolerance and are influenced by gender, genetics, and environment.

Mechanisms that avoid the immune system have evolved in pathogens. Some pathogens evade the body's defenses through *antigenic variation,* a change in their surface epitopes so that the immune system cannot develop an immunological memory of that pathogen. Examples include the parasite that causes sleeping sickness and the influenza virus.

Some viruses enter an inactive state called *latency,* in which viral DNA remains in a host cell's chromosome. Should conditions change, the virus can replicate and spread to neighboring cells or to new hosts.

The infectious agent responsible for AIDS (acquired immune deficiency syndrome) is **HIV (human immunodeficiency virus).** HIV infects helper T cells. It escapes the immune system through both antigenic variation (when it mutates rapidly during replication) and latency (when it integrates into host cell DNA). As

HIV infection kills helper T cells, both humoral and cell-mediated immune responses are impaired. Individuals with AIDS are highly susceptible to opportunistic infections and cancers that take advantage of a suppressed immune system.

Although new drug combinations are unable to cure HIV infection, they are slowing the progression to AIDS. HIV is transmitted by the transfer of body fluids, such as blood, semen, and breast milk, containing viral particles or infected cells.

FOCUS QUESTION 35.6

a. Why is AIDS such a deadly disease?

b. Why has it proved so difficult to prevent and cure this disease?

Cancer and Immunity   The incidence of certain cancers increases when the immune system is impaired. A healthy immune system may be able to recognize and defend against the 15–20% of human cancers that involve viruses. Vaccines are now available against hepatitis B virus, which is linked to liver cancer, and human papillomavirus (HPV), which is associated with cervical cancer.

Word Roots

anti-   = against; **-gen** = produce (*antigen:* a substance that elicits an immune response by binding to receptors of B cells or T cells, or to antibodies)

cyto-   = cell (*cytokine:* any of a group of small proteins secreted by a number of cell types, including macrophages and helper T cells, that regulate the function of other cells)

epi-   = over; **-topo** = place (*epitope:* a small, accessible region of an antigen to which an antigen receptor or antibody binds)

immuno-   = safe, free; **-glob** = globe, sphere (*immunoglobulin:* a protein secreted by plasma cells that binds to a particular antigen; also called an antibody)

macro-   = large; **-phage** = eat (*macrophage:* a phagocytic cell present in many tissues that functions in innate immunity by destroying microbes, and in adaptive immunity as an antigen-presenting cell)

neutro-   = neutral; **-phil** = loving (*neutrophil:* the most abundant type of white blood cell; neutrophils are phagocytic and tend to self-destruct as they destroy foreign invaders)

Structure Your Knowledge

This chapter contains a wealth of information that is probably fairly new to you. If you take a little time and pull out the key players of the immune system and organize them first into very basic concept clusters and then develop more interrelated concept maps, you will find that this information is both understandable and fascinating.

1. Create a concept map outlining adaptive immunity; include the cells involved in the humoral and cell-mediated responses and their functions.

2. Describe the structure of an antibody molecule, and relate this structure to its function.

3. Although innate and adaptive immunity were presented separately, these defense mechanisms interact in several ways. Describe a few of the chemical and cellular components they share.

Test Your Knowledge

MULTIPLE CHOICE: *Choose the one best answer.*

1. Which of the following statements *best* describes an insect's immune system?
   a. Insects rely on the barrier defense of an exoskeleton.
   b. Lysozyme attacks bacterial cell walls, protecting an insect's digestive system.
   c. Hemocytes can carry out phagocytosis of bacteria and foreign substances.
   d. Insects produce antimicrobial peptides in response to the binding of recognition proteins to macromolecules from a broad class of pathogens.
   e. All of the above are part of an insect's innate immunity.

2. Which of the following defense mechanisms is *incorrectly* paired with its function?
   a. gastric juice—kills bacteria in the stomach
   b. fever—may stimulate phagocytosis
   c. histamine—causes blood vessels to dilate
   d. cytokines—attract phagocytes in the inflammatory response
   e. lysozyme—attacks the cell wall of viruses

3. Which of the following cells would release interferon?

   a. a macrophage that has become an antigen-presenting cell

   b. an injured endothelial cell of a blood vessel

   c. a cell infected by a virus

   d. a mast cell that has bound an antigen

   e. a helper T cell bound to an antigen-presenting cell

4. Which of the following statements is *not* true of Toll-like receptors?

   a. They are located on phagocytic white blood cells.

   b. They recognize molecules specific to individual pathogens.

   c. They resemble Toll, an activator of innate immunity in insects.

   d. TLR proteins may be located on the plasma membrane or inside vesicles of immune cells.

   e. They trigger innate immune responses.

5. Which of the following statements correctly describes the main difference between innate immunity and adaptive immunity?

   a. Innate immunity responds only to free pathogens in a localized area; adaptive immunity responds only to pathogens that have entered body cells.

   b. Innate immunity involves only leukocytes, whereas adaptive immunity involves only lymphocytes.

   c. Innate immunity relies on phagocytes to destroy pathogens, whereas adaptive immunity does not involve phagocytes.

   d. Innate immunity recognizes molecules common to a set of pathogens, whereas adaptive immunity reacts to specific microbes on the basis of their unique antigens.

   e. Complement proteins participate in adaptive immunity but not in innate immunity.

6. Which of the following destroys a target cell by phagocytosis?

   a. a neutrophil

   b. a cytotoxic T cell

   c. a natural killer cell

   d. complement proteins

   e. a helper T cell

7. Antibodies are

   a. proteins or polysaccharides usually found on the surface of invading bacteria.

   b. proteins embedded in T cell membranes.

   c. proteins circulating in the blood that may tag foreign cells for complement destruction.

   d. proteins that consist of two light and two heavy polypeptide chains.

   e. Both **c** and **d** are correct.

8. What accounts for the huge diversity of antigens to which B cells can respond?

   a. The antibody genes have millions of alleles.

   b. The recombination within light and heavy chain genes during development results in millions of possible antigen receptors.

   c. The antigen-binding sites at the arms of the molecule can assume a huge diversity of shapes in response to the specific antigen encountered.

   d. B cells have thousands of copies of antibodies bound to their plasma membranes.

   e. B cells can be antigen-presenting cells when they take in antigens and display fragments in their class II MHC molecules.

9. A secondary immune response is more rapid and effective than a primary immune response because

   a. histamines cause rapid vasodilation.

   b. the second response is an active immunity, whereas the primary one was a passive immunity.

   c. memory cells respond to the pathogen and rapidly proliferate into effector cells.

   d. chemical signals cause the rapid accumulation of phagocytic cells.

   e. helper T cells are available to activate other white blood cells.

10. Clonal selection is responsible for the

    a. proliferation of effector cells and memory cells specific for an encountered antigen.

    b. recognition of class I MHC molecules by cytotoxic T cells.

    c. rearrangement of antibody genes for the light and heavy chains.

    d. formation of cell cultures in the commercial production of monoclonal antibodies.

    e. transformation of a clone of helper T cells into cytotoxic T cells keyed to a specific antigen.

11. Major histocompatibility complex molecules

    a. are involved in the ability to distinguish self from nonself.

    b. are a collection of cell surface proteins.

    c. may trigger T cell responses after transplant operations.

    d. present antigen fragments on infected cells.

    e. are or do all of the above.

12. In neutralization,

    a. antibodies coat microorganisms and help phagocytes bind to and engulf the foreign cell.

    b. a set of complement proteins lyses a hole in a foreign cell's membrane.

    c. antibodies coat proteins on the surface of a virus, preventing infection of a host cell.

    d. a flood of histamines is released and may result in anaphylactic shock.

    e. *V* gene segments and *J* gene segments are joined by recombinase.

13. What do some antibodies, T cell receptors, and MHC molecules have in common?
    a. They are found exclusively in cells of the immune system.
    b. They are all part of the complement system.
    c. They are antigen-presenting molecules.
    d. They are or can be membrane-bound proteins.
    e. They are involved in the cell-mediated immune response.

14. Which of the following *best* describes the relationship between antibodies and complement?
    a. They are both coded for by genes that have hundreds of alleles.
    b. They are both involved in innate defenses.
    c. They are both produced by plasma cells.
    d. Antibodies bound to antigens on a pathogen's membrane may activate complement proteins to form a pore in the membrane, causing the cell to swell and lyse.
    e. Complement proteins tag foreign cells for destruction; antibodies destroy cells.

15. In an adaptive immune response, a dendritic cell
    a. activates complement proteins.
    b. binds to an accessory protein on cytotoxic T cells to activate their production of toxic compounds.
    c. releases cytokines to activate B cells to produce clones of plasma cells.
    d. activates humoral and cell-mediated immunity by releasing interferons after engulfing a virus.
    e. presents peptide antigens of an engulfed pathogen in its class II MHC molecules to helper T cells, and releases cytokines.

16. Which of the lists in choices **a–e** places the following steps in the helper T cell activation of cell-mediated and humoral immunity in the correct order?
    1. Macrophage and helper T cell secrete cytokines.
    2. Macrophage engulfs pathogen and presents antigen in class II MHC.
    3. Plasma cells secrete antibodies, and cytotoxic T cells attack cells with class I MHC molecule-antigen complex.
    4. T cell receptor recognizes class II MHC molecule-antigen complex.
    5. Activated B cells produce plasma cells and memory cells, and activated T cells produce cytotoxic T cells and memory cells.
    a. 1, 3, 5, 2, 4
    b. 5, 1, 2, 4, 3
    c. 2, 4, 1, 5, 3

    d. 5, 2, 4, 1, 3
    e. 2, 1, 4, 5, 3

17. Which of the following statements about humoral immunity is *correct*?
    a. It is a form of passive immunity produced by vaccination.
    b. It defends against free pathogens with effector mechanisms such as neutralization, stimulation of phagocytosis, or complement activation.
    c. It protects against pathogens that have invaded body cells as well as against cancer cells.
    d. It is mounted by lymphocytes that have matured in the thymus.
    e. It requires recognition of class I MHC molecule-antigen complexes to activate its effector mechanism.

18. In which of the following circumstances would a B cell display antigens to a T cell?
    a. A B cell takes in a few antigen molecules and displays them in class II MHC molecules to activated helper T cells.
    b. A B cell engulfs bacteria and displays bacterial peptide antigens in class II MHC molecules to helper T cells.
    c. After being infected by a virus, a B cell displays viral peptides it has synthesized to cytotoxic T cells in its class I MHC molecules.
    d. A B cell binds free antigens and displays them to helper T cells while they are still attached to B cell antigen receptors.
    e. Both **a** and **c** are possible ways that a B cell could display antigens to a T cell.

19. Which of the following conditions is *not* considered a disease or malfunction of the immune system?
    a. MHC-induced transplant rejection
    b. type I diabetes
    c. lupus, multiple sclerosis
    d. AIDS
    e. allergic anaphylactic shock

20. From which of the following conditions would an AIDS patient be *least* likely to suffer?
    a. Kaposi's sarcoma or other cancers
    b. tuberculosis
    c. pneumonia
    d. rheumatoid arthritis
    e. yeast infections of mucous membranes

Chapter 36

Reproduction and Development

Chapter Focus

This chapter covers the patterns and mechanisms of animal reproduction. In sexual reproduction, fertilization may occur externally or internally. Development of the zygote may take place externally in a moist environment; in a protective, resistant egg; or internally within the female.

The human reproductive system is described, including its organs, glands, hormones, gamete formation, and sexual response. The chapter also covers basic patterns of early embryonic development and human pregnancy, birth, contraception, and reproductive technologies.

Chapter Review

36.1 Both asexual and sexual reproduction occur in the animal kingdom

In **sexual reproduction,** two meiotically formed haploid gametes fuse to form a diploid **zygote,** which develops into an offspring. A female gamete is a relatively large, nonmotile **egg;** male gametes are small, motile **sperm.** In **asexual reproduction,** a single individual produces offspring, with no fusion of egg and sperm.

Mechanisms of Asexual Reproduction Many invertebrates can reproduce asexually by *budding,* in which a new individual grows out from the parent's body; by *fission,* in which a parent is separated into two individuals of equal size; or by *fragmentation,* in which the body is broken into several pieces, each of which develops into a complete animal. *Regeneration,* the regrowth of body parts, is necessary for a fragment to develop into a new organism. In **parthenogenesis,** eggs develop without being fertilized.

Sexual Reproduction: An Evolutionary Enigma Asexual reproduction potentially produces many more offspring, yet most eukaryotic species reproduce sexually.

The shuffling of genes in sexual reproduction may allow more rapid adaptation to changing conditions.

Reproductive Cycles A combination of environmental and hormonal cues controls the timing of reproductive cycles. Cycles may be linked to favorable environmental conditions or available energy supplies.

Climate change is causing a mismatch between some animal migrations, which are cued by seasonal changes in daylight, and the new plant growth, cued by temperature, that provides nutrition for successful rearing of offspring.

Variation in Patterns of Sexual Reproduction In **hermaphroditism,** found in some sessile, burrowing, and parasitic animals, each individual has functioning male and female reproductive systems. Mating between two individuals results in fertilization of both. In some fishes, individuals reverse their sex during their lifetime.

FOCUS QUESTION 36.1

a. What adaptive advantages may asexual reproduction provide?

b. What adaptive advantages may sexual reproduction provide?

c. If a hermaphrodite self-fertilizes, is that asexual or sexual reproduction? Explain.

External and Internal Fertilization Fertilization is the union of sperm and egg. In **external fertilization,** eggs and sperm are shed, and fertilization occurs in the environment. External fertilization occurs almost exclusively in moist habitats, where the gametes and developing zygote are not in danger of desiccation.

Internal fertilization involves the placement of sperm in or near the female reproductive tract so that the egg and sperm unite internally. Behavioral cooperation, as well as copulatory organs and sperm receptacles, are required for internal fertilization.

Pheromones are small, volatile, or water-soluble chemicals that may function as mate attractants.

FOCUS QUESTION 36.2

List two mechanisms that may help to ensure that gamete release is synchronized when fertilization is external.

a.

b.

Ensuring the Survival of Offspring Species with internal fertilization usually produce fewer gametes, but a larger fraction of zygotes survive.

The shells of the eggs of birds and other reptiles protect the embryo from water loss or injury. Embryos of marsupial mammals complete development in the mother's pouch, attached to a mammary gland. Embryos of eutherian (placental) mammals develop in the uterus, nourished by the mother's blood through the placenta. Parental care of young is widespread among vertebrates and even among many invertebrates.

36.2 Reproductive organs produce and transport gametes

Precursor cells for eggs or sperm are often established early in embryogenesis and later *amplified* in number.

Variation in Reproductive Systems The simplest reproductive systems lack **gonads** to produce gametes. In polychaete annelids, eggs or sperm develop from cells lining the coelom.

Most insects have separate sexes and complex reproductive systems. Females may have *spermathecae*, or sperm-storing sacs, from which sperm may be released when conditions favor successful reproduction.

With the exception of mammals, many vertebrates have a common opening—a **cloaca**—for the digestive, excretory, and reproductive systems. Most mammals have a separate opening for the digestive tract. The reproductive and excretory openings are separate in most female mammals.

Human Male Reproductive Anatomy The external male reproductive organs include the scrotum and penis. Sperm are produced in the highly coiled **seminiferous tubules** of the **testes.** Sperm production requires a cooler temperature than the core body temperature of most mammals, so the **scrotum** suspends the testes below the abdominal cavity.

Sperm pass from a testis into the coiled **epididymis,** in which they mature and gain motility. During **ejaculation,** sperm are propelled through the **vas deferens,** into a short **ejaculatory duct,** and out through the **urethra,** which runs through the penis.

What are the three sets of accessory glands that add secretions to the sperm to form **semen**? The **seminal vesicles** contribute an alkaline fluid containing mucus, fructose (an energy source for the sperm), a coagulating enzyme, ascorbic acid, and prostaglandins. The **prostate gland** produces a secretion that contains anticoagulant enzymes and citrate. Benign enlargement of the prostate and prostate cancer are common medical problems in older men. The *bulbourethral glands* produce a mucus that neutralizes urine remaining in the urethra.

The **penis** is composed of spongy tissue that engorges with blood during sexual arousal, producing an erection that facilitates insertion of the penis into the vagina. Some mammals have a baculum, a bone that helps stiffen the penis. A fold of skin called the **prepuce,** or foreskin, covers the head, or **glans,** of the penis.

Female Reproductive Anatomy The external reproductive structures in human females include the clitoris and two sets of labia. The ovaries contain many **follicles,** each of which contains an **oocyte** surrounded by cells that nourish and protect the developing egg during its development.

The egg cell is expelled into the abdominal cavity and swept by cilia into the **oviduct,** or fallopian tube, through which it travels to the **uterus.** The uterine lining, the **endometrium,** is richly supplied with blood vessels. The neck of the uterus, called the **cervix,** opens into the **vagina,** the elastic-walled birth canal and repository for sperm during copulation.

The external genitals are collectively called the **vulva.** Two pairs of skin folds, the thick outer **labia majora** and the thin inner **labia minora,** enclose the separate openings of the vagina and urethra. The vaginal opening is initially covered by a membrane called the **hymen.** The **clitoris** is composed of erectile tissue supporting a glans, or head, covered by the prepuce.

A **mammary gland,** located within a breast, is composed of fatty tissue and small milk-secreting sacs that drain into ducts that open at the nipple. The low level of estradiol in males prevents the mammary glands from enlarging.

FOCUS QUESTION 36.3

Label the indicated structures in the following diagrams of the human male and female reproductive systems.

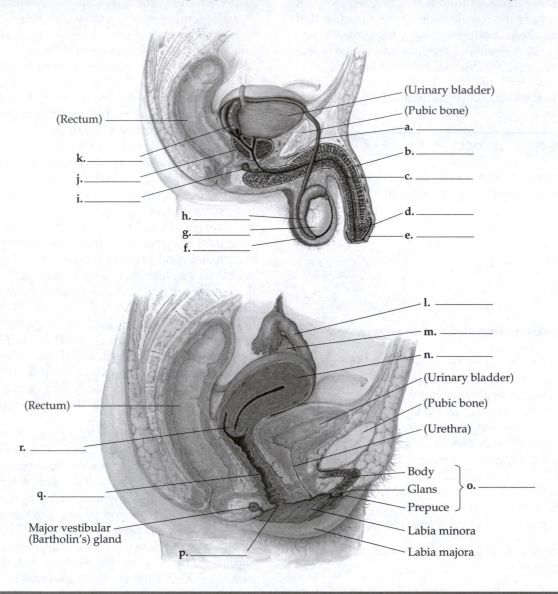

(Rectum) _____

k. _____
j. _____
i. _____

h. _____
g. _____
f. _____

(Urinary bladder)
(Pubic bone)
a. _____
b. _____
c. _____
d. _____
e. _____

l. _____
m. _____
n. _____
(Urinary bladder)
(Pubic bone)
(Urethra)

(Rectum) _____

r. _____

q. _____

Major vestibular
(Bartholin's) gland _____

p. _____

Body
Glans   } o. _____
Prepuce

Labia minora
Labia majora

Gametogenesis The production of gametes, or **gametogenesis,** differs in females and males.

Spermatogenesis occurs continuously in the seminiferous tubules of the testes. *Primordial* germ cells in male embryos differentiate into stem cells. These cells divide mitotically to form **spermatogonia,** which divide to form spermatocytes. Spermatocytes undergo meiosis to produce four spermatids. The haploid nucleus of a sperm is contained in a head, tipped with an **acrosome,** which contains enzymes that help the sperm penetrate the egg. Mitochondria provide ATP for movement of the flagellum, or tail.

Oogenesis begins in the female embryo. **Oogonia,** produced from primordial germ cells, divide and differentiate into **primary oocytes,** which are arrested in prophase I of meiosis before birth. Beginning at puberty, follicle-stimulating hormone (FSH) periodically stimulates a follicle to resume growth. Follicle cells produce the sex hormone estradiol. The primary oocyte within a follicle finishes meiosis I and develops into a **secondary oocyte,** which is arrested in metaphase II. In humans, meiosis is completed if a sperm penetrates the oocyte. Meiotic cytokinesis is unequal, producing one large egg and up to three small haploid polar bodies that disintegrate.

Following ovulation, the follicle forms a mass called the **corpus luteum**, which secretes progesterone and estradiol. If fertilization does not occur, the corpus luteum degenerates.

List three important ways in which oogenesis differs from spermatogenesis.

a.

b.

c.

36.3   The interplay of tropic and sex hormones regulates reproduction in mammals

How do hormones regulate reproduction? *Gonadotropin-releasing hormone* (GnRH) from the hypothalamus directs the anterior pituitary to produce **follicle-stimulating hormone (FSH)** and **luteinizing hormone (LH).** Both of these are **tropic hormones** (*gonadotropins*) that regulate sex hormone production and gametogenesis in the gonads of both males and females. The steroid sex hormones produced by the gonads include *androgens* (primarily **testosterone**) in males and *estrogens* (chiefly **estradiol**) and **progesterone** in females.

Androgens are responsible for the embryonic development of male primary sex characteristics. At puberty, androgens direct formation of secondary sex characteristics, such as voice deepening, pubic and facial hair, and muscle growth. In females, estrogens stimulate breast and pubic hair development at puberty. Sex hormones influence sexual behaviors in both sexes.

Hormonal Control of the Male Reproductive System   FSH acts on **Sertoli cells,** which nourish developing sperm. LH acts on **Leydig cells,** which secrete testosterone and other androgens, promoting spermatogenesis. Testosterone inhibits production of GnRH, FSH, and LH; *inhibin,* produced by Sertoli cells, reduces

FSH secretion. These feedback mechanisms maintain proper levels of androgen production.

Fill in the blanks in the following description of the control of male reproductive hormones.

a. _____, produced by the hypothalamus, regulates the release of gonadotropic hormones from the anterior pituitary.

b. _____ stimulates Sertoli cells of the seminiferous tubules, which nourish sperm.

c. _____ stimulates the production of

d. _____ by Leydig cells.

e. The production of GnRH, LH, and FSH is regulated by _____.

Hormonal Control of Female Reproductive Cycles   In females, the endometrium thickens to prepare for the implantation of the embryo and then is shed in a process called **menstruation** if fertilization does not occur. The human female reproductive cycle involves the integration of the **uterine cycle (menstrual cycle),** which involves these cyclic changes in the uterus, and the **ovarian cycle,** which involves follicle growth and ovulation.

The ovarian cycle begins with the release from the hypothalamus of GnRH, which stimulates the anterior pituitary to secrete small amounts of FSH and LH. In this **follicular phase,** FSH stimulates follicular growth, and the cells of the follicle secrete estradiol. The low level of estradiol inhibits the release of FSH and LH.

When the secretion of estradiol by the growing follicle rises sharply, the hypothalamus is stimulated to increase GnRH output, which results in a rise in LH and FSH release. The increase in LH induces maturation of the follicle, which enlarges and forms a bulge near the ovary surface. Ovulation occurs about a day after the LH surge, with the rupture of the follicle and adjacent ovary wall.

In the *luteal phase,* LH stimulates the transformation of the ruptured follicle into the corpus luteum, which secretes estradiol and progesterone. The rising level of these hormones exerts negative feedback on the hypothalamus and pituitary, inhibiting secretion of LH and FSH and thereby preventing another egg from maturing. Without LH to maintain it, the corpus luteum disintegrates, which reduces the levels of estradiol and progesterone. This drop releases the inhibition of the hypothalamus and pituitary, and FSH and LH secretion begins again, stimulating growth of new follicles and the start of the next ovarian cycle.

The uterine (menstrual) cycle is controlled by hormones secreted by the ovaries. Estradiol secreted by the growing follicles causes the endometrium to begin to thicken in the *proliferative phase*, which correlates with the follicular phase of the ovarian cycle. After ovulation, estradiol and progesterone stimulate continued development of the endometrium and of glands that secrete a nutrient fluid. This *secretory phase* of the uterine cycle correlates with the luteal phase of the ovarian cycle.

The rapid drop in ovarian hormones caused by the disintegration of the corpus luteum reduces blood supply to the endometrium and begins its disintegration, leading to the *menstrual flow phase* of the uterine cycle. By convention, the uterine and ovarian cycles begin with the first day of menstruation.

Menopause, the cessation of ovulation and menstruation, occurs in older women as the ovaries become less responsive to FSH and LH.

Humans and some other primates have menstrual cycles. Other mammals have **estrous cycles,** during which the endometrium thickens but is reabsorbed (not shed) if fertilization does not occur. Females are receptive to sexual activity only during estrus, or "heat," the period surrounding ovulation.

Human Sexual Response   The human sexual response cycle includes two types of physiological reactions: *vasocongestion*, increased blood flow to a tissue, and *myotonia*, increased muscle tension. The excitement phase involves vasocongestion and vaginal lubrication, preparing the vagina and penis for *coitus*, or sexual intercourse. The plateau phase continues vasocongestion, and breathing rate and heart rate increase. In both sexes, *orgasm* is characterized by rhythmic, involuntary contractions of reproductive structures. In males, emission moves semen into the urethra, and ejaculation occurs when the urethra contracts and semen is expelled. In the resolution phase, vasocongested organs return to normal size, and muscles relax.

36.4   Fertilization, cleavage, and gastrulation initiate embryonic development

A common sequence of developmental stages is shared across many animal species. These stages are described for sea urchins, an example of a model organism, chosen because it is easy to study in the lab.

Fertilization   Successful fertilization requires that (1) sperm penetrate any protective layers around the egg, (2) recognition molecules on the sperm bind with matching egg receptors, and (3) changes at the egg surface prevent **polyspermy.**

When a sperm comes in contact with the jelly coat of a sea urchin egg, the acrosome at the tip of the sperm discharges hydrolytic enzymes. This acrosomal reaction allows the acrosomal process to elongate through the jelly coat. Fertilization within the same species is assured when proteins on the tip of the acrosomal process bind to specific receptor proteins extending from the egg's plasma membrane.

The depolarization of the egg membrane in response to fusion of sperm and egg prevents other sperm from fusing with the egg, providing a fast block to polyspermy.

Cortical granules then secrete their contents, which clip off sperm-binding receptors and cause a fertilization envelope to form, which functions as a slow block to polyspermy. The sperm nucleus then enters the egg cytoplasm.

A rise in Ca^{2+} concentration in the fertilized egg increases cellular respiration and protein synthesis, a process known as egg activation.

FOCUS QUESTION 36.6

In the following diagram of the human female reproductive cycle, label the lines indicating the levels of the gonadotropic and the ovarian hormones in the blood and the phases of the ovarian and uterine cycles.

(a) Pituitary gonadotropins in blood

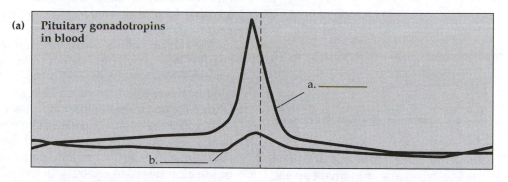

a. _____

b. _____

(b) Ovarian cycle

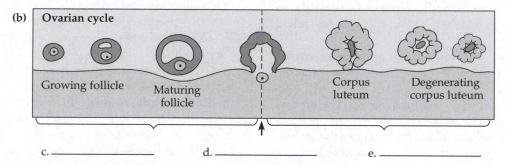

Growing follicle   Maturing follicle   Corpus luteum   Degenerating corpus luteum

c. _____   d. _____   e. _____

(c) Ovarian hormones in blood

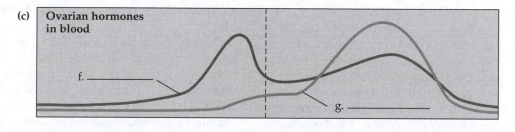

f. _____

g. _____

(d) Uterine (menstrual) cycle

Endometrium

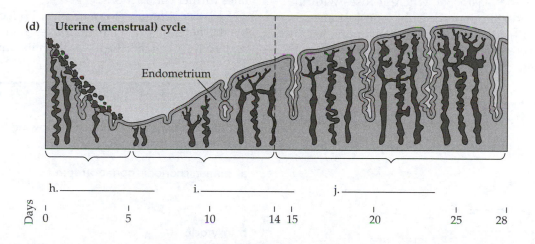

h. _____   i. _____   j. _____

Days   0   5   10   14 15   20   25   28

FOCUS QUESTION 36.7

a. What assures that only a sperm of the correct species fertilizes a sea urchin egg?

b. What assures that only one sperm will fertilize an egg?

Cleavage and Gastrulation   The rapid cell divisions of **cleavage** parcel the egg cytoplasm into smaller cells. These first divisions usually produce a **blastula**—a cluster of cells surrounding a fluid-filled cavity called the **blastocoel.**

The next embryonic stages result in **morphogenesis,** the development of body shape. During **gastrulation,** the movement of cells into the blastula produces a two- or three-layered embryo called a **gastrula.** The embryonic *germ layers* produced by gastrulation are the outer **ectoderm** and the inner **endoderm** that lines the embryonic digestive tract. Bilateral animals also have a middle germ layer called **mesoderm.** Adult tissues and organs derive from one or more of these cell layers.

In a sea urchin, gastrulation occurs as cells from the vegetal pole detach and move into the blastocoel as migratory *mesenchyme cells.* The remaining cells at that end flatten and then buckle inward. This *invagination* deepens to form a narrow pouch called the *archenteron,* which will become the digestive tube. The opening to the archenteron, the *blastopore,* develops into the anus. Extensions of mesenchyme cells pull the tip of the archenteron across the blastocoel, where it fuses with the ectoderm to form the mouth. Unlike protostomes, in which the mouth develops from the blastopore, sea urchins (echinoderms) and chordates are deuterostomes— the mouth develops from a second opening.

FOCUS QUESTION 36.8

Label the parts in the following diagram of gastrulation in a sea urchin.

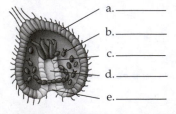

a. _____
b. _____
c. _____
d. _____
e. _____

Human Conception, Embryonic Development, and Birth   Fertilization, or **conception,** occurs in the oviduct. The cortical reaction produces a slow block to polyspermy. Cleavage produces a ball of cells by the time the embryo reaches the uterus, and the blastocyst, a hollow sphere of cells, implants in the endometrium about 7 days after conception.

The carrying of one or more embryos in the uterus is called *pregnancy* or **gestation.** In humans, pregnancy averages 266 days (38 weeks).

Human gestation can be divided into three *trimesters.* During the first trimester, the embryo secretes *human chorionic gonadotropin (hCG),* which maintains the corpus luteum's secretion of progesterone and estrogens. Splitting of the embryo in the first month can produce *monozygotic* or identical twins. In contrast, independent fertilization of two eggs can lead to *dizygotic* or fraternal twins.

Many pregnancies end in spontaneous abortion, or *miscarriage,* due to chromosomal or developmental abnormalities.

After implantation, the outer layer of the blastocyst, the **trophoblast,** grows and mingles with the endometrium, helping to form the **placenta,** a disk-shaped organ that is the site of waste removal and gas and nutrient exchange between maternal blood and embryonic capillaries. Development proceeds to **organogenesis,** and by the eighth week the embryo has all the rudimentary structures of the adult and is called a **fetus.**

During the second and third trimester, the fetus grows rapidly and is quite active.

Labor, a series of strong contractions of the uterus, leads to birth. Hormones (oxytocin and estradiol) and local regulators (prostaglandins) regulate labor. Through a positive feedback system, uterine contractions stimulate the release of oxytocin, which stimulates further contractions.

Lactation is unique to mammals. Prolactin secretion by the anterior pituitary stimulates milk production, and oxytocin triggers the release of milk during nursing.

FOCUS QUESTION 36.9

List the functions of each of the following hormones and the structures that secrete them:

a. human chorionic gonadotropin

b. oxytocin

Contraception **Contraception,** the deliberate prevention of pregnancy, can be accomplished by several methods: preventing release of egg or sperm, preventing fertilization, or preventing implantation. Complete abstinence from sexual intercourse is the most effective means of birth control. The *rhythm method,* or *natural family planning,* is based on refraining from intercourse during the period in which conception is most likely: a number of days before and after ovulation. *Coitus interruptus* is an unreliable method of preventing pregnancy.

Barrier methods of contraception include *condoms* and *diaphragms,* which, when used in conjunction with spermicidal foam or jelly, present a physical and chemical barrier to fertilization. Condoms are the only form of birth control that offers protection against sexually transmitted diseases.

IUDs (intrauterine devices) interfere with fertilization and implantation. The release of gametes may be prevented by chemical contraception, as in *birth control pills,* and by sterilization (vasectomy in males and tubal ligation in females). Most birth control pills are combinations of synthetic estrogen and progestin, which act by negative feedback to prevent the release of GnRH by the hypothalamus and thus of FSH and LH by the pituitary, resulting in a cessation of ovulation and follicle development. High doses of combination birth control pills can be taken as "morning-after" pills following unprotected intercourse. Progestin-only injections or minipills alter the cervical mucus such that it blocks sperm entry to the uterus. Cardiovascular problems are a potential effect of birth control pills.

FOCUS QUESTION 36.10

List the three general types of birth control methods, and cite examples of each. Indicate which examples are most likely to prevent pregnancy, and which are least likely to do so.

a.

b.

c.

Infertility and In Vitro Fertilization Some cases of infertility—the inability to conceive offspring—are the result of inflammatory disorders caused by *sexually transmitted diseases (STDs).* Hormone therapy may increase gamete production, and surgeries may correct blocked ducts. *In vitro* **fertilization (IVF)** involves the removal of eggs from a woman, fertilization within a culture dish, and implantation of the developing embryo in the uterus. If sperm are defective or low in number, a sperm nucleus may be directly injected into an oocyte.

Word Roots

a- = not, without (*asexual reproduction:* the generation of offspring from a single parent, without the fusion of gametes; offspring are usually genetically identical to the parent)

acro- = tip; **-soma** = body (*acrosome:* a vesicle in the tip of a sperm containing hydrolytic enzymes that help the sperm reach the egg)

coit- = a coming together (*coitus:* the insertion of a penis into a vagina; also called sexual intercourse)

contra- = against (*contraception:* the deliberate prevention of pregnancy)

ecto- = outside; **-derm** = skin (*ectoderm:* the outermost of the three primary germ layers in animal embryos; gives rise to outer covering and, in some phyla, the nervous system, inner ear, and lens of the eye)

endo- = within (*endoderm:* the innermost of the three primary germ layers in animal embryos; lines the archenteron and gives rise to the liver, pancreas, lungs, and lining of the digestive tract)

epi- = above, over (*epididymis:* a coiled tubule adjacent to the testes where sperm are stored)

fertil- = fruitful (*fertilization:* the union of haploid gametes to produce a diploid zygote)

gastro- = stomach, belly (*gastrulation:* a series of cell and tissue movements in which the blastula-stage embryo folds inward, producing a three-layered embryo, the gastrula)

labi- = lip; **major-** = larger (*labia majora:* a pair of thick, fatty ridges that encloses and protects the rest of the vulva)

menstru- = month (*menstruation:* the shedding of portions of the endometrium during a uterine (menstrual) cycle)

meso- = middle (*mesoderm:* the middle primary germ layer in an animal embryo)

minor- = smaller (*labia minora:* a pair of slender skin folds that surrounds the openings of the urethra and vagina)

oo- = egg; **-genesis** = producing (*oogenesis:* the process in the ovary that results in the production of female gametes)

partheno- = a virgin (*parthenogenesis*: asexual reproduction in which females produce offspring from unfertilized eggs)

poly- = many (*polyspermy*: fusion of an egg by more than one sperm)

tropho- = nourish (*trophoblast*: the outer epithelium of a mammalian blastocyst; forms the fetal part of the placenta)

vasa- = a vessel (*vas deferens*: the tube in the male reproductive system in which sperm travel from the epididymis to the urethra)

Structure Your Knowledge

1. Use a simple flow diagram to trace the path of a human sperm from the point of production to the point of fertilization; then briefly describe the functions of both the structures the sperm passes through and the associated glands.

2. Answer the following questions concerning the reproductive cycle of the human female.
   a. What does GnRH do?
   b. What does FSH stimulate?
   c. What causes the spike in LH level?
   d. What does this LH surge induce?
   e. What does LH maintain during the luteal phase?
   f. What inhibits the secretion of LH and FSH?
   g. What allows LH and FSH secretion to begin again?

3. Describe how birth control pills work.

Test Your Knowledge

FILL IN THE BLANKS

1. _____ type of asexual reproduction in which a new individual grows while attached to the parent's body

2. _____ small, volatile chemicals that may act as mate attractants

3. _____ development of egg without fertilization

4. _____ individual with functioning male and female reproductive systems

5. _____ common opening of digestive, excretory, and reproductive systems in nonmammalian vertebrates

6. _____ reproductive cycle in which thickened endometrium is reabsorbed

7. _____ assisted reproductive technology in which oocytes are fertilized in a culture dish and then implanted in the uterus

8. _____ common duct for urine and semen in mammalian males

9. _____ filling of a tissue with blood due to increased blood flow

10. _____ period when ovulation and menstruation cease in human females

MULTIPLE CHOICE: *Choose the one best answer.*

1. An advantage of sexual reproduction is that
   a. it is easier than asexual reproduction.
   b. offspring have a better start when produced from a union of egg and sperm than when simply budded off a parent.
   c. it produces diploid offspring, whereas asexually produced offspring are haploid.
   d. it produces variation in the offspring, and new genetic combinations may be better adapted to a changing environment.
   e. it requires both male and female members of a species to be present in a population and have behavioral interactions.

2. Which of the following animals is *least* likely to be hermaphroditic?
   a. an earthworm
   b. a barnacle (sessile crustacean)
   c. a tapeworm
   d. a grasshopper
   e. a liver fluke

3. External fertilization is *least* likely to be associated with
   a. pheromones.
   b. spermatheca.
   c. large numbers of gametes and zygotes.
   d. behavioral interaction.
   e. moist environments.

4. Which of the following structures is *incorrectly* paired with its function?
   a. seminiferous tubule—adds fluid containing mucus, fructose, and prostaglandins to semen
   b. scrotum—encases testes and suspends them below the abdominal cavity
   c. epididymis—tubules in which sperm mature
   d. prostate gland—adds fluid to semen
   e. vas deferens—transports sperm from epididymis to ejaculatory duct

5. In which location does fertilization usually take place in a human female?
   a. ovary
   b. oviduct
   c. uterus
   d. cervix
   e. vagina

6. Which of the following statements *correctly* describes how the production of human sperm and eggs differs?
    a. The meiotic production of gametes occurs before a female is born, but does not begin until puberty in males.
    b. Each meiotic division produces four sperm but only two eggs.
    c. Meiosis occurs in the testes of males but in the oviducts of females.
    d. Primary oocytes stop dividing by mitosis before birth, whereas male primary spermatocytes continue to divide throughout life.
    e. Meiosis is an uninterrupted process in males, whereas in females it resumes when a follicle matures and is only completed when a sperm penetrates the egg.

7. The function of the corpus luteum is to
    a. nourish and protect the egg cell.
    b. produce prolactin, which stimulates the milk sacs of the mammary gland.
    c. produce progesterone and estradiol during the luteal phase.
    d. eject the egg and then disintegrate following ovulation.
    e. maintain pregnancy by producing human chorionic gonadotropin.

8. The secretory phase of the uterine cycle
    a. begins with falling levels of estradiol and progesterone.
    b. corresponds with the luteal phase of the ovarian cycle.
    c. involves the initial proliferation of the endometrium.
    d. corresponds with the follicular phase of the ovarian cycle.
    e. occurs when the endometrium begins to disintegrate and menstrual flow occurs.

Choose from the following hormones to answer questions 9 through 13.

    a. estradiol
    b. progesterone
    c. LH (luteinizing hormone)
    d. FSH (follicle-stimulating hormone)
    e. hCG (human chorionic gonadotropin)

9. Which hormone stimulates ovulation and the development of the corpus luteum?
10. Which hormone is produced by the developing follicle and initiates thickening of the endometrium?
11. Which hormone is produced by the embryo during the first trimester and is necessary for maintaining a pregnancy?
12. Which hormone stimulates Leydig cells to make testosterone, which in turn stimulates sperm production?

13. Which hormone is produced by the corpus luteum and later by the placenta and is responsible for maintaining a pregnancy?

14. Which of the following hormones is *incorrectly* paired with its function?
    a. androgens—responsible for primary and secondary male sex characteristics
    b. oxytocin—stimulates uterine contractions during labor
    c. estradiol—responsible for secondary female sex characteristics
    d. FSH—acts on Sertoli cells that nourish sperm, promoting spermatogenesis
    e. prolactin—stimulates breast development at puberty

15. Examples of birth control methods that prevent the production or release of gametes are
    a. hormonal contraception and sterilization.
    b. birth control pills and IUDs.
    c. condoms and diaphragms.
    d. abstinence and hormonal contraception.
    e. the progestin minipill and spermicidal foams.

16. Certain maternal diseases, drugs, alcohol, and radiation are most dangerous to embryonic development
    a. during the first 2 weeks, when the embryo has not yet implanted and spontaneous abortion may occur.
    b. during the first 3 months, when organogenesis is occurring.
    c. during the first and second trimesters, when the embryonic liver is not yet filtering toxins.
    d. during the second trimester, when the corpus luteum no longer secretes progesterone.
    e. during the third trimester, when the most rapid growth is occurring.

17. In sea urchins, the slow block to polyspermy
    a. is essential to prevent sperm from other species from fertilizing the egg.
    b. is directly produced by the depolarization of the egg cell membrane.
    c. is a result of the formation of the fertilization envelope.
    d. is caused by the expulsion of hydrolytic enzymes from the acrosome.
    e. involves all of the above.

18. The blastocoel
    a. develops into the archenteron or embryonic gut.
    b. is a fluid-filled cavity in the blastula.
    c. opens to the exterior through a blastopore.
    d. is lined with mesoderm.
    e. Both **b** and **c** are correct.

Chapter 37

Neurons, Synapses, and Signaling

Chapter Focus

This chapter describes the transmission of nervous impulses along and between neurons. Ion channels establish the membrane potential of a neuron, and the rapid flow of ions through voltage-gated channels creates the rapid depolarization and repolarization of an action potential. The release of neurotransmitters at synapses converts electrical signals to chemical signals that pass information to receiving cells.

Chapter Review

Neurons transmit long-distance electrical signals throughout the body and usually communicate between cells using short-distance chemical signals. The higher-order processing of nervous signals may involve clusters of neurons called **ganglia** or more structured groups of neurons organized into a **brain.**

37.1 Neuron structure and organization reflect function in information transfer

Neuron Structure and Function A neuron consists of a **cell body,** which contains the nucleus and organelles, as well as numerous extensions. The highly branched, short **dendrites** (along with the cell body) *receive* signals from other neurons; the single, longer **axon** *transmits* signals to other cells. Signals that travel down an axon originate from a region of the cell body called the *axon hillock.*

Information is transmitted to another cell at a junction called a **synapse.** At the branching ends of axons are *synaptic terminals,* which usually release **neurotransmitters** that relay signals to another neuron, a muscle, or a gland cell. The transmitting cell is called the *presynaptic cell;* the receiving cell is the *postsynaptic cell.*

The very numerous supporting cells called **glial cells,** or **glia,** give structural integrity and physiological support to neurons.

Introduction to Information Processing The three stages of information processing are sensory input,

integration, and motor output. Sensors detect external stimuli or internal conditions. **Sensory neurons** transmit this information to the brain or ganglia, where **interneurons** integrate it and send output that triggers muscle or gland activity. **Motor neurons** carry impulses to muscle cells.

A **central nervous system (CNS),** found in many animals, is involved in integration and includes the brain and a longitudinal nerve cord. The **peripheral nervous system (PNS)** consists of neurons, often bundled into **nerves,** that carry information to and from the CNS.

FOCUS QUESTION   37.1

Label the indicated structures on the following diagram of a neuron. Indicate the direction of impulse transmission. What happens at part e?

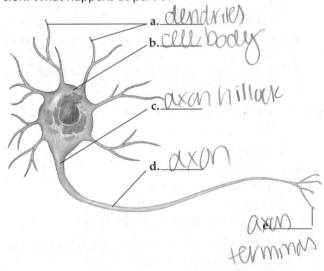

a. dendrites
b. cell body
c. axon hillock
d. axon
e. axon terminals

37.2 Ion pumps and ion channels establish the resting potential of a neuron

Due to the separation of charges, a voltage, or **membrane potential,** exists across a cell's plasma membrane. In a typical neuron at rest, the membrane potential—called the **resting potential**—is between –60 and –80 mV.

Formation of the Resting Potential In most neurons, the concentration of K⁺ is higher inside the cell than outside, and the Na⁺ concentration is greater outside than inside. *Sodium-potassium pumps* maintain these gradients, moving three Na⁺ ions out of the cell for every two K⁺ ions it pumps in.

   Ion channels that are *selectively permeable* allow specific ions to diffuse across the membrane. The diffusion of K⁺ out of a cell through the many open potassium channels results in a buildup of negative charge within the neuron, creating the membrane potential.

Modeling the Resting Potential Ions diffuse through channels down their concentration gradients until their chemical gradient is balanced by the electrical gradient across the membrane. The membrane voltage at this equilibrium, or **equilibrium potential (E_{ion}),** is determined for a single ion with a 1+ charge at 37°C by the Nernst equation: $E_{ion} = 62$ mV (log ([ion]$_{outside}$/[ion]$_{inside}$). The equilibrium potential for K⁺ (E_K) across a membrane is –90 mV. Using the concentration gradient for Na⁺, E_{Na} is +62 mV (the inside of the membrane is more positive than outside).

   Neurons at rest have more open K⁺ channels than open Na⁺ channels, and the resting potential is closer to E_K than to E_{Na}.

FOCUS QUESTION   37.2

a. What is the principal cation inside the cell? Outside the cell?

b. Which side of the membrane has a negative charge?

c. What change in the permeability of the cell's membrane to K⁺ and/or Na⁺ would cause the membrane potential to shift from –70 mV to –80 mV?

37.3   Action potentials are the signals conducted by axons

You have learned that the ungated potassium and sodium ion channels create a neuron's resting potential. But neurons also have **gated ion channels,** which open

or close in response to stimuli and alter the membrane potential. Electrophysiologists measure membrane potential by placing microelectrodes connected to a voltage recorder inside and outside a cell.

Hyperpolarization and Depolarization A stimulus that opens gated potassium channels will result in **hyperpolarization,** as K⁺ flows out and the membrane potential moves toward E_K (–90 mV). When gated sodium channels open and Na⁺ flows in, a **depolarization** occurs as the inside of the cell becomes less negative and shifts toward E_{Na} (+62 mV).

Graded Potentials and Action Potentials A shift in membrane potential can result in a **graded potential,** whose magnitude is proportional to the strength of the stimulus. But what is an **action potential?** This massive change in membrane voltage has a constant magnitude and spreads to adjacent regions of the membrane as the signal conducted along an axon. **Voltage-gated ion channels** open or close in response to a change in membrane potential. When a depolarization opens voltage-gated sodium channels, Na⁺ inflow increases depolarization and causes more sodium channels to open. Once this positive feedback results in depolarization of a neuron to a certain membrane voltage called the **threshold,** an action potential is triggered. The action potential is an *all-or-none* event, always creating the same positive voltage spike once the threshold is reached.

Generation of Action Potentials: A Closer Look Both Na⁺ and K⁺ voltage-gated channels are involved in an action potential. Sodium channels open rapidly in response to depolarization but become inactivated when a loop of the channel protein blocks the opening. Potassium channels open slowly in response to depolarization, but remain open throughout the action potential.

   As a stimulus depolarizes the membrane, Na⁺ channels open; the influx of Na⁺ causes further depolarization, which opens more channels. If the depolarization reaches the threshold, an action potential is triggered. The continuing positive-feedback cycle of Na⁺ influx during the *rising phase* brings the membrane potential close to E_{Na}. The inactivation of voltage-gated Na⁺ channels and the opening of voltage-gated K⁺ channels rapidly bring the membrane potential back toward E_K during the *falling phase.*

   During the *undershoot,* the membrane's permeability to K⁺ is higher than at rest, and the continued outflow of K⁺ temporarily hyperpolarizes the membrane. During the **refractory period,** which occurs while the Na⁺ channels remain inactivated, the neuron cannot respond to another stimulus.

   Action potentials last only about 1–2 milliseconds. The frequency of action potentials increases with the intensity of the stimulus.

FOCUS QUESTION   37.3

The following diagram shows the changes in voltage-gated channels during an action potential. Label the channels and inactivation loop, the ions, and the components of the graph. Name and describe the five phases of the action potential. Place numbers on the graph to show where each phase is occurring.

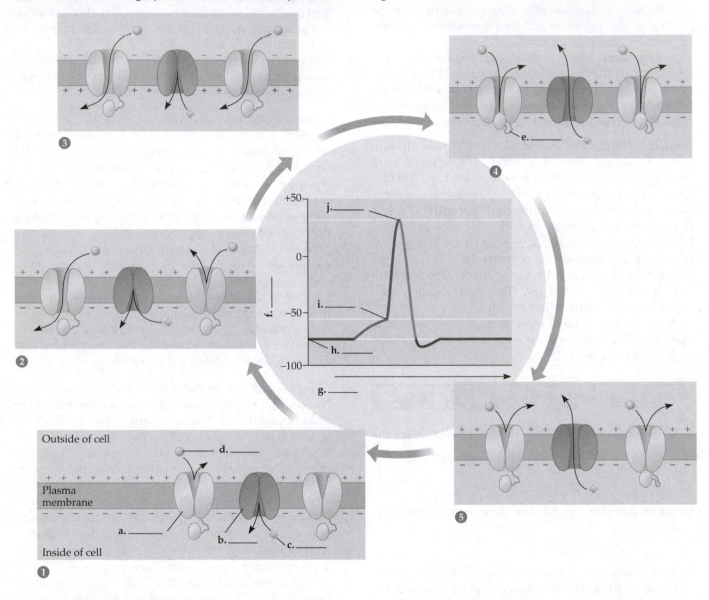

Conduction of Action Potentials   Na⁺ influx in the rising phase depolarizes adjacent sections of the membrane, bringing them to threshold. Local depolarizations and action potentials across the membrane result in the propagation of serial action potentials along the length of the neuron. Because of the brief refractory period, the action potential is propagated in one direction only.

How fast do action potentials travel? Resistance to current flow is inversely proportional to the cross-sectional area of the conducting "wire." The greater the axon diameter, the faster action potentials are propagated. Some invertebrates, such as squid, have giant axons that conduct impulses very rapidly.

In vertebrates, **oligodendrocytes** (in the CNS) and **Schwann cells** (in the PNS) insulate axons in multiple layers of membranes called a **myelin sheath.** Voltage-gated ion channels are concentrated in the **nodes of Ranvier,** small gaps between successive Schwann cells.

Action potentials can be generated only at these nodes, and a nerve impulse "jumps" from node to node, resulting in a faster mode of transmission known as **saltatory conduction.**

FOCUS QUESTION   37.4

Return to the diagram of a neuron in Focus Question 37.1 and draw in a myelin sheath. Label a Schwann cell and node of Ranvier.

37.4   Neurons communicate with other cells at synapses

At *electrical synapses,* an electrical current flows directly from cell to cell via gap junctions. Electrical synapses are found in the giant axons of squid and lobsters and in many neurons of the vertebrate brain.

Most synapses are *chemical synapses,* which involve the release of neurotransmitters. A synaptic terminal contains *synaptic vesicles,* in which neurotransmitter is stored. The depolarization of the presynaptic membrane opens voltage-gated calcium channels in the membrane. The influx of Ca^{2+} causes the synaptic vesicles to fuse with the presynaptic membrane and release neurotransmitter into the *synaptic cleft.*

FOCUS QUESTION   37.5

Identify the components of this chemical synapse following the depolarization of the synaptic terminal.

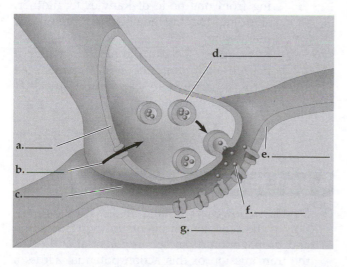

Generation of Postsynaptic Potentials   Binding of neurotransmitter to a receptor on a **ligand-gated ion channel** (often called an *ionotropic receptor*) in the postsynaptic cell membrane allows ions to cross the membrane, creating a graded *postsynaptic potential.* If both Na^+ and K^+ are able to diffuse through the channel, the net inflow of positive charge depolarizes the membrane, creating an **excitatory postsynaptic potential (EPSP)** that brings the membrane potential closer to threshold. If binding of neurotransmitter opens K^+ or Cl^- channels, the membrane hyperpolarizes, producing an **inhibitory postsynaptic potential (IPSP).**

The effect of neurotransmitters is brief because the molecules diffuse away, are taken up and repackaged into synaptic vesicles, or are broken down by enzymes.

Summation of Postsynaptic Potentials   Postsynaptic potentials are graded potentials. Their magnitude depends on the number of neurotransmitter molecules that bind to receptors and the distance a synapse is from the axon hillock. The membrane potential of the axon hillock at any given time is determined by the sum of all EPSPs and IPSPs.

FOCUS QUESTION   37.6

a. _____ **summation** occurs with repeated release of neurotransmitter from a synaptic terminal before the postsynaptic potential returns to its resting potential.

b. _____ **summation** occurs when several different presynaptic terminals, usually from different neurons, release neurotransmitter simultaneously.

Modulated Signaling at Synapses   Some neurotransmitters bind to *metabotropic receptors,* which trigger a signal transduction pathway in the postsynaptic cell. This modulated synaptic transmission begins more slowly but lasts longer. Binding of neurotransmitter to a metabotropic receptor often leads to the production of cAMP as a second messenger and to the amplified phosphorylation of channel proteins, opening or closing many ion channels.

Neurotransmitters   A neurotransmitter may have many different ionotropic and metabotropic receptors and may produce very different effects in postsynaptic cells. The more than 100 identified neurotransmitters can be divided into five groups: acetylcholine, amino acids, biogenic amines, neuropeptides, and gases.

Acetylcholine is a common neurotransmitter in invertebrates and vertebrates. In *neuromuscular junctions*, acetylcholine released from a motor axon produces an EPSP in a muscle cell. The enzyme acetylcholinesterase hydrolyzes the neurotransmitter. The same ionotropic receptor binds nicotine, producing its physiological and psychological stimulant effect.

The amino acids **glutamate** and **gamma-aminobutyric acid (GABA)** function as neurotransmitters in the CNS. Glutamate, the most common CNS neurotransmitter, is excitatory and plays a key role in long-term memory formation. GABA is the most common inhibitory transmitter in the brain. Glycine is inhibitory in parts of the CNS outside the brain.

Biogenic amines are neurotransmitters derived from amino acids. **Norepinephrine,** derived from tyrosine, is an excitatory neurotransmitter in the autonomic nervous system. It also acts as a hormone, as does the similar biogenic amine *epinephrine.* **Dopamine** is also derived from tyrosine. **Serotonin** (synthesized from tryptophan) and dopamine affect sleep, mood, attention, and learning. Imbalances of these neurotransmitters have been associated with several nervous system disorders.

Neuropeptides are short chains of amino acids that function via metabotropic receptors. *Substance P* is an excitatory neurotransmitter that functions in pain perception. **Endorphins** are produced in the brain during physical or emotional stress, and their effects include pain reduction, euphoria, and other physiological effects. Opiates bind to endorphin receptors in the brain.

Some neurons use nitric oxide (NO) and carbon monoxide as local regulators or neurotransmitters. The release of NO triggers relaxation of smooth muscle cells and vessel dilation. CO produced by neurons in the brain affects the release of hypothalamic hormones.

FOCUS QUESTION   37.7

Acetylcholine stimulates skeletal muscle contraction but inhibits or slows cardiac muscle contraction. How can this neurotransmitter have such opposite effects?

Word Roots

bio- = life; **-genic** = producing (*biogenic amine:* a neurotransmitter derived from an amino acid)

de- = down, out (*depolarization:* a change in a cell's membrane potential such that the inside of the membrane is made less negative relative to the outside)

dendro- = tree (*dendrite:* one of the usually numerous, short, highly branched extensions of a neuron that receive signals from other neurons)

endo- = within (*endorphin:* any of several neurotransmitters produced in the brain that inhibit pain perception)

glia = glue (*glia:* cells of the nervous system that support, regulate, and augment the functions of neurons)

hyper- = over, above, excessive (*hyperpolarization:* a change in a cell's membrane potential such that the inside of the membrane becomes more negative relative to the outside)

inter- = between (*interneurons:* an association neuron; a nerve cell within the central nervous system that forms synapses with sensory and/or motor neurons and integrates sensory input and motor output)

neuro- = nerve; **trans-** = across (*neurotransmitter:* a molecule that is released from the synaptic terminal of a neuron at a chemical synapse, diffuses across the synaptic cleft, and binds to the postsynaptic cell, triggering a response)

oligo- = few, small (*oligodendrocyte:* a type of glial cell that forms insulating myelin sheaths around the axons of neurons in the central nervous system)

salta- = leap (*saltatory conduction:* rapid transmission of a nerve impulse along an axon resulting from the action potential jumping from one node of Ranvier to another, skipping the myelin-sheathed regions of membrane)

syn- = together (*synapse:* the junction where a neuron communicates with another cell across a narrow gap via a neurotransmitter or an electrical coupling)

Structure Your Knowledge

1. Develop a flowchart or diagram or write a description of the sequence of events in the (1) creation and propagation of an action potential and (2) in the transmission of this action potential across a chemical synapse.

Test Your Knowledge

MULTIPLE CHOICE: *Choose the one best answer.*

1. Which of the following nervous system components is *incorrectly* paired with its function?
   a. axon hillock—region of neuron where an action potential originates
   b. Schwann cells—produce a myelin sheath around axons in the PNS
   c. synapse—space between presynaptic and postsynaptic cells into which neurotransmitter is released
   d. synaptic terminal—receptor that is part of an ion channel that is keyed to a specific neurotransmitter
   e. dendrite—receives signals from other neurons

2. Interneurons
   a. may connect sensory and motor neurons.
   b. are more common in the PNS than the CNS.
   c. are involved in the integration of sensory information.
   d. typically have more axons than dendrites.
   e. Both **a** and **c** are correct.

3. Which of the following statements describes the role of the sodium-potassium pump in establishing a cell's membrane potential?
   a. By pumping 3 Na^+ out of a cell for every 2 K^+ pumped in, it establishes a membrane potential of about –70 mV across the membrane.
   b. It is responsible for maintaining the gradients of higher K^+ concentration inside the cell and higher Na^+ concentration outside the cell.
   c. Phosphate groups attached to the sodium-potassium pump produce a high-energy membrane.
   d. It is a voltage-gated channel that opens when a cell's membrane potential deviates from its resting value, returning the ion concentrations to their proper concentration gradients.
   e. The equilibrium potential of potassium and sodium establish a cell's membrane potential; the sodium-potassium pump is only involved in reestablishing the resting potential following the refractory phase of an action potential.

4. Which of the following statements concerning the resting potential of a typical neuron is *not* true?
   a. The inside of the cell is more negative than is the outside.
   b. The concentration of sodium is higher outside the cell and that of potassium is higher inside the cell.
   c. The resting potential is about –70 mV and can be measured by using microelectrodes placed inside and outside the cell.
   d. The resting potential is formed by the sequential opening of sodium and then potassium voltage-gated channels.
   e. The diffusion of K^+ through numerous open potassium channels (coupled with the very few open sodium channels) is the main source of the separation of charges across the membrane.

5. The threshold of a membrane
   a. is exactly midway between E_K and E_{Na}.
   b. will, once it is reached, open voltage-gated channels and permit the rapid outflow of sodium ions.
   c. is the depolarization that is needed to generate an action potential.
   d. is a graded potential that is proportional to the strength of a stimulus.
   e. is an all-or-none event.

6. How is an increase in the strength of a stimulus communicated by a neuron?
   a. The spike of the action potential reaches a higher voltage.
   b. The frequency of action potentials generated along the neuron increases.
   c. The length of an action potential (the duration of the rising phase) increases.
   d. The action potential travels along the neuron faster.
   e. Both **a** and **c** are correct.

7. After the rapid depolarization of an action potential, the fall in the membrane potential occurs due to the
   a. closing of voltage-gated sodium channels.
   b. closing of potassium and sodium channels.
   c. refractory period, during which the membrane is hyperpolarized.
   d. delay in the action of the sodium-potassium pump.
   e. opening of voltage-gated potassium channels and the closing of sodium inactivation loops.

8. Nodes of Ranvier are
   a. gaps where Schwann cells abut and action potentials are generated.
   b. neurotransmitter-containing vesicles located in the synaptic terminals.
   c. the parts of neurons where action potentials are initiated.
   d. clusters of receptor proteins located on the postsynaptic membrane.
   e. ganglia adjacent to the spinal cord.

9. Movement of an action potential in only one direction along a neuron is a function of
   a. saltatory conduction.
   b. the pathway from dendrite to axon.
   c. the refractory period, when sodium channels are still inactivated.
   d. the localized depolarization of the surrounding membrane.
   e. the reaching of threshold, which creates an all-or-none firing.

10. Signal transmission is faster in myelinated axons because
    a. these axons are thinner and thus present less resistance to voltage flow.
    b. these axons use electrical synapses rather than chemical synapses.
    c. the action potential can jump from node to node along the insulating myelin sheath.
    d. these axons are thicker and thus present less resistance to voltage flow.
    e. these axons have higher depolarization values than do unmyelinated axons.

11. Which of the following statements concerning chemical synapses is *not* true?
    a. Synaptic terminals at the ends of branching axons contain synaptic vesicles, which enclose the neurotransmitter.
    b. The influx of sodium when an action potential reaches the presynaptic membrane causes synaptic vesicles to release their neurotransmitter into the cleft.
    c. The binding of neurotransmitter to receptors on the postsynaptic membrane usually opens (or sometimes closes) ligand-gated ion channels.
    d. An excitatory postsynaptic potential forms when sodium channels open and the membrane potential moves closer to threshold.
    e. Neurotransmitter is often rapidly degraded in the synaptic cleft.

12. An inhibitory postsynaptic potential occurs when
    a. sodium flows into the postsynaptic cell.
    b. enzymes do not break down the neurotransmitter in the synaptic cleft.
    c. binding of neurotransmitter opens potassium channels, resulting in the membrane becoming hyperpolarized.
    d. a neurotransmitter binds to a metabotropic receptor.
    e. norepinephrine is the neurotransmitter.

13. Which of the following substances does *not* function as a neurotransmitter or a local regulator released by neurons?
    a. an amino acid such as glutamate
    b. a neuropeptide such as an endorphin
    c. a biogenic amine such as dopamine
    d. a steroid
    e. NO

14. The main effect of the neurotransmitter GABA in the CNS is to
    a. increase pain.
    b. create inhibitory postsynaptic potentials.
    c. create excitatory postsynaptic potentials.
    d. induce sleep.
    e. decrease pain and induce euphoria.

15. If the binding of a neurotransmitter to its receptor opens Cl^- channels, what would be the effect on the postsynaptic cell? (Cl^- is in higher concentration outside the cell.)
    a. It would hyperpolarize, producing an IPSP.
    b. It would reach threshold, and an action potential would be generated.
    c. It would depolarize and form an EPSP but probably not generate an action potential.
    d. It would initiate a signal transduction pathway as part of modulated signaling.
    e. Its membrane potential would not change because neither sodium nor potassium channels opened.

Chapter 38

Nervous and Sensory Systems

Chapter Focus

This chapter describes the organization of the vertebrate nervous system. The branches of the peripheral nervous system include the sensory afferent neurons and the efferent motor system and autonomic nervous system. The central nervous system includes the spinal cord, which mediates reflexes and transmits information, and the regionally specialized brain, which regulates homeostatic functions, integrates sensory information, and controls voluntary movement and cognitive functions. Sensory receptors include nociceptors, thermoreceptors, mechanoreceptors (touch, hearing, and equilibrium), electromagnetic receptors (vision), and chemoreceptors (taste and smell).

Chapter Review

38.1 Nervous systems consist of circuits of neurons and supporting cells

Mechanisms to sense and respond to changes in the environment evolved in prokaryotes. Modification of this ability allowed for the development of communication between the cells of multicellular organisms.

Cnidarians have a simple diffuse **nerve net** of interconnected, individual neurons. In the nervous system of more complex animals, axons of multiple neurons are bundled into **nerves**.

Cephalization, the concentration of sensory neurons and interneurons at the anterior end of the body (or head), evolved in bilaterally symmetrical animals. Flatworms have the simplest **central nervous system (CNS)** consisting of a small brain and two longitudinal nerve cords. Arthropods have a more complicated brain and a ventral nerve cord with segmentally arranged clusters of neurons called *ganglia*. The vertebrate CNS has a brain and a dorsal spinal cord; nerves and ganglia make up the **peripheral nervous system (PNS)**.

Glia   Glia in the CNS include ciliated ependymal cells, which line the ventricles; microglia, which

protect against pathogens; and **astrocytes**, which facilitate information transfer at synapses and increase blood flow to active neurons. During development, astrocytes induce the formation of the *blood–brain barrier*, which restricts the passage of most substances from the blood into the brain. *Radial glia* form tracks that guide the embryonic growth of neurons.

In 1998, researchers discovered neural stem cells in the adult human brain. Mouse neural stem cells are astrocytes, which produce new cells that play a role in learning and memory.

FOCUS QUESTION   38.1

What is the difference between oligodendrocytes and Schwann cells?

Organization of the Vertebrate Nervous System   The spinal cord carries information to and from the brain and controls **reflexes**, which are automatic responses to stimuli. Neuron cell bodies and glia make up the **gray matter** of the brain and spinal cord. **White matter** is named for the white color of the myelin sheaths of bundles of axons.

Spaces in the brain called *ventricles* are continuous with the narrow *central canal* of the spinal cord, and all are filled with *cerebrospinal fluid*. This fluid, formed by filtration of the blood, carries out circulatory functions.

The Peripheral Nervous System   Sensory information is carried to the CNS along *afferent* PNS neurons. *Efferent* neurons of the PNS carry instructions to muscles, glands, and endocrine cells. Most nerves contain both afferent and efferent neurons.

Functionally, the PNS has two efferent components. The **motor system** carries signals to skeletal muscles, whose movement is largely under conscious control. The **autonomic nervous system** maintains involuntary control over smooth and cardiac muscles.

The autonomic nervous system is separated into three divisions. The **sympathetic division** accelerates heart rate and metabolic rate, generating energy and arousing an organism for action. The **parasympathetic division** carries signals that enhance self-maintenance activities that conserve energy, such as digestion and slowing the heart rate. The **enteric division** includes complex networks of neurons involved in the control of the digestive tract, pancreas, and gallbladder.

FOCUS QUESTION   38.2

Complete the following concept map to help you review the organization of the peripheral nervous system.

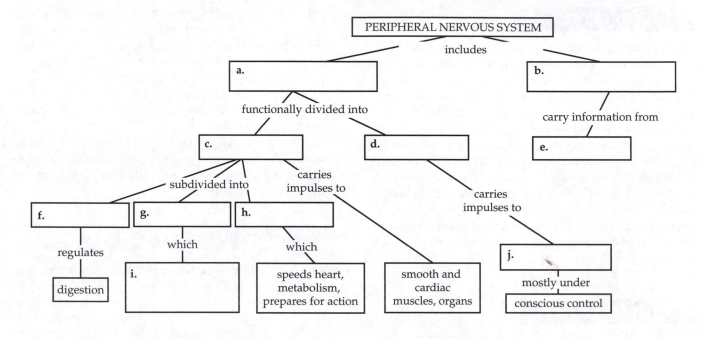

38.2   The vertebrate brain is regionally specialized

Exploring the Organization of the Human Brain   As a human embryo develops, the **forebrain, midbrain,** and **hindbrain** become evident as three anterior bulges of the neural tube. The midbrain and part of the hindbrain give rise to the **brainstem,** which joins the base of the brain to the spinal cord. The rest of the hindbrain gives rise to the **cerebellum.** The forebrain develops into the diencephalon and the telencephalon, which becomes the **cerebrum.** During the second and third months of development, the outer portion of the cerebrum, called the cortex, rapidly expands over much of the brain.

The cerebrum is divided into right and left **cerebral hemispheres.** The **cerebral cortex** is involved in perception, voluntary movement, and learning. The right side of the cortex receives information from and controls the movement of the left side of the body, and vice versa. Communication between the two hemispheres travels through the **corpus callosum,** a thick band of axons. The *basal nuclei,* embedded in the inner white matter, are important in planning and learning movements.

The diencephalon gives rise to the thalamus, hypothalamus, and epithalamus. The **thalamus** is the major sorting center for sensory information going to the cerebrum. The **hypothalamus** is the major brain region for homeostatic regulation. It produces the posterior pituitary hormones and the releasing hormones that control the anterior pituitary. The *epithalamus* includes the pineal gland, which produces melatonin, and one of several clusters of capillaries that produce cerebrospinal fluid.

The cerebellum is involved in learning and remembering motor skills, as well as in coordination and error checking during motor and perceptual functions. The cerebellum integrates information from the auditory and visual systems with sensory input from the joints and muscles as well as motor commands from the cerebrum to provide coordination of movements and balance.

The brainstem consists of the midbrain, the **pons,** and the **medulla oblongata,** or *medulla.* The midbrain receives and integrates sensory information, such as that involved in hearing and visual reflexes, and relays it to specific regions of the forebrain. The pons and medulla transfer information between the PNS and the rest of the brain. They help coordinate large-scale movements, and most axons of motor neurons cross in the medulla, so that the right side of the brain controls much of the movement of the left side of the body, and vice versa.

The medulla contains control centers for such homeostatic functions as breathing, swallowing, heart and blood vessel actions, and digestion. The pons functions with the medulla in some of these activities.

Arousal and Sleep   Arousal is a state in which an individual is aware of the external world. While asleep, an individual does not consciously perceive external stimuli, although EEGs record active brain waves. Sleep and arousal are partly controlled by a diffuse network of neurons in the midbrain and pons.

Biological Clock Regulation   Most organisms appear to have daily cycles of biological activity. Studies show that an internal timekeeper, called the **biological clock**, is important for maintaining these circadian rhythms. In mammals, the **suprachiasmatic nucleus (SCN)**, located in the hypothalamus, functions as the biological clock. Sensory neurons provide visual information to the SCN to keep the clock synchronized with natural cycles of light and dark.

Emotions   The amygdala, hippocampus, and sections of the thalamus border the brainstem and are called the *limbic system*. This system interacts with sensory and other areas of the cerebrum in generating emotions such as laughing and crying, as well as emotional feelings associated with the survival-related functions of the brainstem, such as aggression, feeding, and sexuality. The amygdala functions in creating emotional memories.

The Brain's Reward System and Drug Addiction   The reward system motivates activities such as eating, drinking, and sexual activity. It involves neurons of the *ventral tegmental area (VTA)*, located near the base of the brain, that release dopamine at synapses in regions of the cerebrum, including the nucleus accumbens.

Drug addiction is the compulsive consumption of drugs that increase the activity of the brain's reward system. Addictive drugs such as cocaine, heroin, nicotine, amphetamines, and alcohol affect the reward system in different ways—by stimulating dopamine-releasing VTA neurons, blocking the removal of dopamine from synapses, or decreasing the activity of inhibitory neurons in the pathway. Long-lasting changes in the reward circuitry result in increased craving for the drug.

Functional Imaging of the Brain   Functional magnetic resonance imaging (fMRI) is used to identify active parts of the brain in the study of emotion, brain injury, and cognition.

FOCUS QUESTION   38.3

Identify the structures (a–g) in the following illustration of the human brain. Then match the functions (1–7) to these structures.

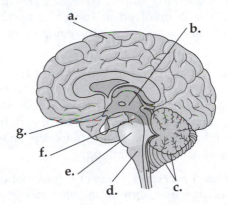

Structure   Function

a. _____   _____

b. _____   _____

c. _____   _____

d. _____   _____

e. _____   _____

f. _____   _____

g. _____   _____

Functions

1. Coordinates balance and movement

2. Aids the medulla in some functions; conducts information between the brain and spinal cord

3. Sorts and relays information to the cerebrum

4. Regulates breathing, heart rate, and digestion

5. Integrates sensory and motor information, learning, emotion, memory, and perception

6. Produces hormones; functions in homeostatic regulation

7. Sends sensory information to the forebrain; is involved in hearing and visual reflexes

38.3 The cerebral cortex controls voluntary movement and cognitive functions

Each side of the cerebral cortex is divided into four lobes—frontal, temporal, occipital, and parietal—each with specialized functions.

Language and Speech Studies of brain injuries and PET studies of brain activity show that several areas in different lobes are involved in understanding and generating language. *Broca's area* in the left frontal lobe is involved in speaking words; *Wernicke's area* in the left temporal lobe is involved in speech comprehension. Other areas are activated when seeing or generating words.

Lateralization of Cortical Function Lateralization results in the two hemispheres differing in function. The left hemisphere usually becomes specialized for language, math, and logic, whereas the right hemisphere is involved in pattern recognition, spatial perception, and nonverbal thinking.

Information Processing Sensory information from sensory organs and individual *somatosensory* receptors, which provide touch, pressure, temperature, and positional information, is directed to primary sensory areas in different lobes of the cortex. Association areas near the primary sensory areas integrate ("associate") the different aspects of sensory inputs. The cerebral cortex generates motor commands to skeletal muscles.

Frontal Lobe Function Injury to the frontal lobe often results in flawed decision making and diminished emotional responses.

Evolution of Cognition in Vertebrates The *neocortex,* the outer six layers of neurons running along the cerebrum, is highly convoluted, with greatly increased surface area in primates and cetaceans. These convolutions were once thought to be required for *cognition,* the perception and reasoning associated with knowledge. The now-documented cognitive ability of birds is linked to the *pallium,* the outer portion of the avian brain, which contains clusters of nuclei. It is thought that the ancestral nuclear organization of the pallium was transformed into a layered arrangement early in mammalian evolution.

Neural Plasticity The nervous system can be remodeled in response to its activity, a capacity called **neural plasticity**. The developmental disorder *autism* may involve a disruption of activity-dependent synapse remodeling.

Memory and Learning Human memory consists of **short-term memory**, information held briefly, and **long-term memory**, the retention of this information for later recall. Transferring information from short-term memory, which involves links in the hippocampus, to long-term memory requires the making of more permanent connections within the cerebral cortex. Long-term memories become integrated with existing knowledge.

FOCUS QUESTION   38.4

As an example of neural plasticity, describe two mechanisms by which synaptic connections can change depending on activity level at a synapse.

38.4 Sensory receptors transduce stimulus energy and transmit signals to the central nervous system

Sensory Reception and Transduction Sensory **reception** is the detection of energy from a particular stimulus by sensory cells, which are usually modified neurons or cells that regulate neurons. The term **sensory receptor** can refer to an organ, cell, or structure within a cell that detects specific stimuli from the external or internal environment.

Sensory **transduction** is the conversion of a physical or chemical stimulus into a **receptor potential**, a graded change in the membrane potential of a receptor cell that is proportional to the strength of the stimulus. Receptor potentials result from the opening or closing of ion channels in response to a stimulus.

Transmission Transmission of sensory information to the CNS occurs as action potentials. The magnitude of a receptor potential correlates with either the frequency of action potentials generated (when the sensory receptor is a neuron) or the quantity of neurotransmitter released (when a non-neuronal sensory receptor synapses with a sensory neuron). Many sensory neurons generate action potentials spontaneously at a low rate, and reception of a stimulus changes the frequency of action potentials.

Perception Impulses arriving from sensory neurons are routed to different parts of the brain that interpret them, producing **perceptions** of various stimuli.

Amplification and Adaptation The strengthening of the energy of a stimulus, either by accessory structures of sense organs or as part of signal transduction pathways, is called **amplification**. Ongoing stimulation may lead to **sensory adaptation** in which continuous stimulation results in a decline in sensitivity of the receptor cell.

FOCUS QUESTION 38.5

List and describe the four functions common to all sensory pathways.

a.

b.

c.

d.

Types of Sensory Receptors

Types of Sensory Receptors Sensory receptors can be grouped into five categories.

Mechanoreceptors respond to the mechanical energy of pressure, touch, stretch, motion, and sound. Bending or stretching of external (or internal) cell structures linked to ion channels increases their permeability to ions, leading to depolarization or hyperpolarization.

Electromagnetic receptors respond to various forms of electromagnetic energy, such as visible light, electricity, and magnetic fields.

In humans, **thermoreceptors** in the skin and hypothalamus respond to heat or cold and send information to the body's thermostat. Researchers found that receptor proteins for capsaicin, a component of hot peppers, also respond to hot temperatures.

Pain receptors (nociceptors) in humans are naked dendrites. These receptors respond to excess heat, pressure, or chemicals. Prostaglandins released from injured tissues increase pain by increasing nociceptor sensitivity.

Chemoreceptors include both general receptors that monitor total solute concentration and specific receptors that respond only to a single kind of important molecule. For terrestrial animals, the sense of **gustation** (taste) detects chemicals called **tastants** present in a solution, whereas **olfaction** (smell) detects airborne **odorants**. Both rely on chemoreceptors. With about 3% of human genes coding for odorant receptors, humans can distinguish thousands of different odors.

Taste buds contain receptor cells that are specific for five types of tastants—sweet, sour, salty, bitter, and umami.

38.5 The mechanoreceptors responsible for hearing and equilibrium detect moving fluid or settling particles

Sensing of Gravity and Sound in Invertebrates Most invertebrates have **statocysts**, which often consist of a layer of ciliated receptor cells lining a chamber containing *statoliths,* dense granules or grains of sand. The stimulation of mechanoreceptors beneath the statoliths provides positional information to the animal.

Most insects sense sounds with body hairs that vibrate in response to sound waves of specific frequencies. A tympanic membrane stretched over an internal air-filled chamber provides a vibration-sensing organ in some insect species.

Hearing and Equilibrium in Mammals In mammals and most terrestrial vertebrates, the sensory organs for hearing and balance are in the ear. In hearing, the ear transduces pressure waves in the air, which are detected by **hair cells**, into nerve impulses that the brain perceives as sound.

The ear consists of three regions: The external pinna and the auditory canal make up the **outer ear**. The **tympanic membrane** (eardrum), which separates the outer ear from the **middle ear**, transmits sound pressure waves to three small bones—the malleus (hammer), incus (anvil), and stapes (stirrup)—which conduct the waves to the inner ear by way of a membrane called the **oval window**. The **Eustachian tube**, connecting the pharynx and the middle ear, equalizes pressure within the middle ear. The **inner ear** contains fluid-filled channels. The **semicircular canals** function in equilibrium, and the **cochlea** functions in hearing.

The coiled cochlea has two large, connected, fluid-filled canals—an upper vestibular canal and a lower tympanic canal—separated by the smaller cochlear duct. The **organ of Corti**, located on the basilar membrane that forms the floor of the cochlear duct, contains mechanoreceptors called hair cells. Their hairs extend into the cochlear duct, and some attach to the tectorial membrane, which overhangs the organ of Corti.

Pressure waves in air—transmitted by the tympanic membrane, the three bones of the middle ear, and the oval window—produce pressure waves in the perilymph, which travel from the vestibular canal through the tympanic canal and dissipate when they strike the **round window**. These pressure waves vibrate the basilar membrane, bending the hairs of its receptor cells against the tectorial membrane, generating receptor potentials.

Volume is a result of the amplitude, or height, of the sound wave; a stronger wave bends the hairs more and results in more action potentials. *Pitch* is related to the frequency of sound waves. Different regions of the basilar membrane vibrate in response to different frequencies, and neurons associated with the vibrating region transmit action potentials to those auditory regions of the cortex where a particular pitch is perceived.

Within the inner ear are two chambers, the **utricle** and **saccule**, and three semicircular canals responsible for detection of body position and movement. Hair cells in the utricle and saccule project into a gelatinous material containing calcium carbonate particles called otoliths. These heavy particles exert a pull on the hairs in the direction of gravity; different head angles stimulate different hair cells, thereby altering the information about head angle or linear acceleration to the brain. Clusters of hair cells in the swellings at the base of the three semicircular canals project into a gelatinous mass called the cupula and sense angular motion in any direction by detecting the pressure of the moving perilymph against the cupula.

FOCUS QUESTION 38.6

Label the parts in the following diagram of the human ear.

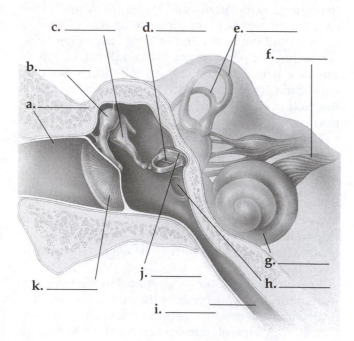

c. _____   d. _____   e. _____

b. _____   f. _____

a. _____

                    j. _____   g. _____

k. _____   h. _____

          i. _____

1. Which of these structures are filled with air?

2. Which of these structures are filled with fluid?

3. Which structure houses the organ of Corti?

4. Which structure is an organ of equilibrium that detects angular movements?

38.6   The diverse visual receptors of animals depend on light-absorbing pigments

Evolution of Visual Perception   All animal light detectors involve **photoreceptors** containing light-absorbing pigment molecules. All photoreceptors share common developmental genes, indicating that photoreceptors were present in the earliest bilaterian animals.

Most invertebrates have light-detecting organs. The ocelli or eyespots of planarians, which detect light intensity and direction, enable the animal to navigate a path away from a light source.

The **compound eye** of insects, crustaceans, and some polychaete worms contains up to thousands of light detectors called **ommatidia**, each with its own lens. Compound eyes are adept at detecting movement.

Some jellies, polychaetes, spiders, and many molluscs have a **single-lens eye**, in which light is focused through a movable single lens onto a layer of photoreceptors. The **iris** changes the diameter of the **pupil** to adjust the amount of light entering the eye.

The Vertebrate Visual System   The human eye consists of a tough, outer connective tissue layer called the sclera, and a thin, pigmented inner layer called the choroid. At the front of the eye, the sclera becomes the transparent *cornea*, and the choroid forms the colored *iris*, which regulates the amount of light entering through the pupil. The **retina**, the innermost layer of the eyeball, contains neurons and photoreceptors. The optic nerve attaches to the eye at the optic disk, forming a "blind spot" on the retina.

The transparent **lens** focuses an image onto the retina. The *aqueous humor* fills the anterior eye cavity; jellylike *vitreous humor* fills the posterior cavity.

Rods and **cones** are the photoreceptors in the retina. Several rods or cones relay information to each *bipolar cell*, several of which synapse with *ganglion cells*, whose axons form the optic nerve. *Horizontal* and *amacrine cells* in the retina help to integrate visual information.

Rods are more light sensitive and enable night vision, whereas cones distinguish colors.

FOCUS QUESTION 38.7

Label the parts of the vertebrate eye in the following illustration.

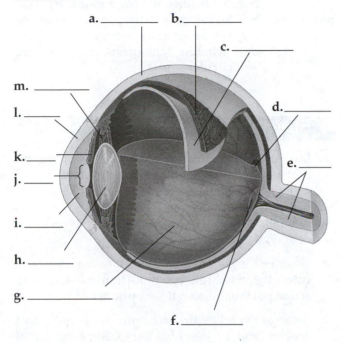

a. _____ b. _____

c. _____

m. _____

l. _____

d. _____

k. _____

j. _____

e. _____

i. _____

h. _____

g. _____

f. _____

Both rods and cones have *visual pigments* embedded in a stack of disks in the outer segment of each cell. The light-absorbing molecule **retinal** is bound to a membrane protein called an **opsin**, forming **rhodopsin**, the visual pigment in rod cells. When retinal absorbs light, it changes from the *cis* isomer to the *trans* isomer, activating rhodopsin.

Light-activated rhodopsin activates a membrane-bound G protein called transducin, which then activates an enzyme that detaches cGMP bound to Na^+ channels in the plasma membrane. Sodium channels close, hyperpolarizing the cell. Except in bright light when rhodopsin remains active, enzymes convert retinal to its original shape, and the signal transduction pathway shuts off.

Visual processing begins in the retina, where rods and cones synapse with bipolar cells. In the dark, rods and cones are depolarized and continually release glutamate into synapses with bipolar cells. When rods and cones hyperpolarize after absorbing light, glutamate release ceases. In response, the bipolar cells reverse their polarization, either becoming hyperpolarized or depolarized, depending on the receptor.

Signals from rods and cones may follow a pathway directly from photoreceptors to bipolar cells to ganglion cells. Other pathways involve horizontal cells, which carry signals from one receptor to others and to

several bipolar cells. Amacrine cells may connect one bipolar cell to several ganglion cells.

The left and right optic nerves meet at the *optic chiasma* at the base of the cerebral cortex. Information from the left visual field of both eyes travels to the right side of the brain, whereas what is sensed in the right field of view goes to the left side.

At least 30% of the cerebral cortex, including dozens of integrating centers, may be devoted to creating the three-dimensional perception of what we see. Most fishes, amphibians, and reptiles have good color vision. The relative proportion of rods and cones correlates with the activity pattern of a mammal. In humans, red, green, and blue cones, each with its own type of opsin that binds with retinal to form a *photopsin*, are named for the color of light they optimally absorb. As the genes for the red and green pigments are on the X chromosome, color blindness is more common in males.

When you focus on close objects, the lens becomes rounder. In the human eye, rods are most concentrated toward the edge of the retina, whereas the center of the visual field, the **fovea**, is filled with cones.

FOCUS QUESTION 38.8

The *receptive field* of a ganglion cell includes the rods and cones that supply information to it.

a. Would a large or a small receptive field produce the sharpest image?

b. Where on the retina are the smallest receptive fields found?

Word Roots

astro- = a star; **-cyte** = cell (*astrocyte:* a glial cell with diverse functions, including providing structural support for neurons, regulating the interstitial environment, facilitating synaptic transmission, and assisting in regulating the blood supply to the brain)

auto- = self (*autonomic nervous system:* an efferent branch of the vertebrate peripheral nervous system that regulates the internal environment; consists of the sympathetic, parasympathetic, and enteric divisions)

chemo- = chemical (*chemoreceptor:* a sensory receptor that responds to a chemical stimulus)

coch- = a snail (*cochlea:* the complex, coiled organ of hearing that contains the organ of Corti)

electro- = electricity (*electromagnetic receptor:* a receptor of electromagnetic energy, such as visible light, electricity, or magnetism)

fovea- = a pit (*fovea:* the place on the retina at the eye's center of focus, where cones are highly concentrated)

gusta- = taste (*gustation:* the sense of taste)

hypo- = below (*hypothalamus:* the ventral part of the vertebrate forebrain; functions in maintaining homeostasis, especially in coordinating the endocrine and nervous systems)

mechano- = an instrument (*mechanoreceptor:* a sensory receptor that detects physical deformation in the body's environment associated with pressure, touch, stretch, motion, and sound)

noci- = harm (*nociceptor:* a sensory receptor that responds to noxious or painful stimuli; also called a pain receptor)

olfact- = smell (*olfaction:* the sense of smell)

omma- = the eye (*ommatidium:* one of the facets of the compound eye of arthropods and some polychaete worms)

para- = near (*parasympathetic division:* one of three divisions of the autonomic nervous system; generally enhances body activities that gain and conserve energy)

photo- = light (*photoreceptor:* an electromagnetic receptor that detects the radiation known as visible light)

rhodo- = red (*rhodopsin:* a visual pigment consisting of retinal and opsin)

sacc- = a sack (*saccule:* in the vertebrate ear, a chamber in the vestibule behind the oval window that participates in the sense of balance)

semi- = half (*semicircular canals:* a three-part chamber of the inner ear that functions in maintaining equilibrium)

stato- = standing; **-cyst** = sac (*statocyst:* a type of mechanoreceptor that functions in equilibrium in invertebrates by use of statoliths, which stimulate hair cells in relation to gravity)

supra- = above, over (*suprachiasmatic nucleus (SCN):* a group of neurons in the hypothalamus of mammals that functions as a biological clock)

thermo- = heat (*thermoreceptor:* a receptor stimulated by either heat or cold)

tympan- = a drum (*tympanic membrane:* another name for the eardrum, the membrane between the outer and the middle ear)

utric- = a leather bag (*utricle:* in the vertebrate ear, a chamber in the vestibule behind the oval window that opens into the three semicircular canals)

Structure Your Knowledge

1. List the location and functions of the following important brain nuclei or functional systems.
   a. suprachiasmatic nucleus
   b. basal nuclei
   c. limbic system
   d. amygdala and hippocampus

2. Describe the path of a light stimulus from where it enters the vertebrate eye to its transmission as an action potential through the optic nerve.

3. Arrange the following structures in an order that enables you to describe the passage of sound waves from the external environment to the perception of sound in the brain.

round window	basilar membrane
oval window	cerebral cortex
pinna	vestibular and tympanic canals
auditory nerve	tectorial membrane
cochlea	tympanic membrane
auditory canal	malleus, incus, stapes
cochlear duct	organ of Corti

Test Your Knowledge

MULTIPLE CHOICE: *Choose the one best answer.*

1. Which of the following statements concerning the autonomic nervous system is *not* true?
   a. It is a subdivision of both the central and peripheral nervous systems.
   b. It consists of the sympathetic, parasympathetic, and enteric divisions.
   c. Two of its divisions have generally antagonistic effects.
   d. It controls smooth and cardiac muscles.
   e. Control is generally involuntary.

2. If you needed to obtain nuclei from the cell bodies of neurons for an experiment, you would want to make preparations of
   a. the corpus callosum.
   b. motor nerves.
   c. the cerebral cortex.
   d. ganglia near the spinal cord.
   e. either **c** or **d**.

3. The amygdala and hippocampus
   a. control biorhythms and are found in the thalamus.
   b. are the parts of the arousal system found in the midbrain.
   c. are located in the parietal and frontal lobes, respectively, and are involved in language and speech.
   d. are nuclei in the midbrain involved in hearing and vision.
   e. are found in the inner cortex and are involved in memory storage.

4. Which of the following structures is *incorrectly* paired with its function?
   a. pons—conducts information between the spinal cord and the brain
   b. cerebellum—contains the tracts by which motor neurons cross from one side of the brain to the other side of the body
   c. thalamus—sorts and relays incoming impulses to the cerebrum
   d. corpus callosum—band of axons connecting the left and right hemispheres
   e. hypothalamus—homeostatic regulation, pleasure centers

5. The white matter of the spinal cord is composed of
   a. the ventricles.
   b. myelinated sheaths of axons.
   c. cerebrospinal fluid.
   d. motor and interneuron cell bodies.
   e. sympathetic ganglia.

6. Lateralization of the cerebrum refers to the fact that
   a. visceral functions are controlled by the hindbrain, whereas thinking and language are centered in association areas of the cortex.
   b. the right side of the brain controls the left side of the body, and vice versa.
   c. somatosensory information is received in one region of the cerebral cortex, whereas another region controls motor output to the body.
   d. the left hemisphere usually controls verbal and analytic ability, whereas the right hemisphere is specialized for nonverbal thinking and pattern recognition.
   e. memory formation involves circuits that include the limbic system, hippocampus, and association areas.

7. Which of the following nervous system components does *not* contain efferent neurons?
   a. a spinal nerve
   b. the motor system
   c. the enteric division
   d. sensory neurons
   e. All of the above contain efferent neurons.

8. The pallium of birds is involved in _____ and is homologous to the _____ of mammals.
   a. circadian rhythms, suprachiasmatic nuclei
   b. flight, cerebellum
   c. cognition, cerebral cortex
   d. parenting behavior, the reward system
   e. vision, optic lobes

9. Sensory perception is
   a. the reception of a stimulus by a sensory cell.
   b. the conversion of stimulus energy into a receptor potential by a sensory receptor or sensory neuron.
   c. the interpretation of a stimulus according to the region of the brain that receives the nerve impulse.
   d. the amplification and transmission of sensory data to the CNS.
   e. the emotional response to a stimulus.

10. Which of the following sensory receptors are *not* present in human skin?
    a. thermoreceptors
    b. mechanoreceptors
    c. pain receptors
    d. electromagnetic receptors
    e. neither **b** nor **d**

11. The function of the three bones of the middle ear is to
    a. respond to different frequencies of sound waves.
    b. transmit pressure waves to the oval window.
    c. equalize pressure within the inner ear by communicating through the Eustachian tube.
    d. sense rotation in three different planes (*x*, *y*, and *z* axes).
    e. support the tympanic membrane (eardrum).

12. The volume of a sound is determined by
    a. the area in the brain that receives the signal.
    b. the section of the basilar membrane that vibrates.
    c. the number of hair cells that are stimulated.
    d. the frequency of the sound wave.
    e. the degree to which hair cells are bent and the resulting increase in action potentials in the sensory neuron.

13. Rotation of the head of a mammal is sensed by
    a. the organ of Corti located in the utricle and saccule.
    b. changes in the action potentials of hair cells in response to bending against the tectorial membrane.
    c. the bending of hair cells embedded in a cupula in a semicircular canal in response to the movement of internal fluid.
    d. hair cells in response to movement of fluid in the coiled cochlea.
    e. horizontal and amacrine cells in the retina.

14. What do statocysts in invertebrates and the cochlea of your ear all have in common?
    a. They are used to sense sound or pressure waves.
    b. They are organs of equilibrium.
    c. They use hair cells as mechanoreceptors.
    d. They use a second-messenger pathway of signal transduction.
    e. They use granules to stimulate their receptor cells.

15. The compound eye of insects and crustaceans
    a. is composed of thousands of prisms that focus light onto a few photoreceptor cells.
    b. cannot sense colors.
    c. contains many ommatidia, each of which contains a lens that focuses light from a tiny portion of the field of view.
    d. may detect infrared and ultraviolet radiation.
    e. All of the above are true.

16. Which of the following statements is *not* true?
    a. The iris regulates the amount of light entering the pupil.
    b. The choroid is the thin, pigmented inner layer of the eye that contains the photoreceptor cells.
    c. Rounding of the lens enables focus on nearby objects.
    d. Thin aqueous humor fills the anterior eye cavity.
    e. The sclera is the tough outer connective tissue layer of the eye.

17. Which of the following structures is *incorrectly* paired with its function?
    a. cones—respond to different light wavelengths, producing color vision
    b. horizontal cells—integrate information between receptor cells and bipolar cells
    c. bipolar cells—relay impulses between rods or cones and ganglion cells
    d. ganglion cells—synapse with the optic nerve at the optic disk or blind spot
    e. fovea—center of the visual field, containing only cones in humans

18. The absorption of light by rhodopsin in a rod cell leads to
    a. an increase in the release of neurotransmitter.
    b. the opening of sodium channels caused by the binding of cGMP.
    c. a hyperpolarized membrane and a decrease in the release of neurotransmitter.
    d. the bleaching of rhodopsin, which increases the sensitivity of rod cells to light.
    e. the closing of sodium channels and the release of neurotransmitter.

19. A receptive field in the retina involves a *single*
    a. bipolar cell.
    b. photoreceptor (rod or cone) cell.
    c. amacrine cell.
    d. horizontal cell.
    e. ganglion cell.

20. Which of the following statements regarding the senses of gustation and olfaction is *not* true?
    a. There is no distinction between these senses in aquatic animals.
    b. There are as many as 1,000 more genes for odorant receptors than for tastant receptors.
    c. Olfactory sensory cells are neurons; gustatory receptor cells are modified epithelial cells that synapse with sensory neurons.
    d. Each taste cell has a single type of receptor; each olfactory cell can have receptors that respond to several odorant molecules.
    e. Taste receptors of insects are located in sensory hairs on the feet and mouthparts; olfactory hairs are located on the antennae.

Chapter 39

Motor Mechanisms and Behavior

Chapter Focus

This chapter begins with muscle contraction and skeletal systems, which together provide for animal movement. It then introduces the complex and fascinating subject of animal behavior. Behaviors can range from simple fixed-action patterns in response to specific stimuli to problem-solving behaviors in novel situations. Behaviors result from interactions among environmental stimuli, experience, and individual genetic makeup. Selection for survival and reproductive success can explain most behaviors.

Chapter Review

Animal behavior, which is essential for survival and reproduction, is based on physiological processes and acted on by natural selection. An individual **behavior** is carried out by muscles, which are under the control of the nervous system. We begin this chapter by looking at how muscles function and then consider the control, development, and evolution of animal behavior.

39.1   The physical interaction of protein filaments is required for muscle function

Muscle cell contraction involves the interaction between **thin filaments,** which consist of two coiled strands of actin proteins, and **thick filaments,** which are arrays of myosin molecules.

Vertebrate Skeletal Muscle   Vertebrate **skeletal muscle** consists of a bundle of multinucleated muscle cells, called fibers, running parallel to the length of the muscle. Each fiber contains a bundle of **myofibrils,** each containing thin and thick filaments.

The regular arrangement of filaments produces repeating light and dark bands; thus, skeletal muscle is also called *striated muscle.* The repeating sections are called **sarcomeres,** which are the contractile units of skeletal muscle. Thin filaments attach to the Z lines

at the borders of a sarcomere and project toward the center. Thick filaments lie in the center of a sarcomere, anchored in the M line.

According to the **sliding-filament model**, the thick and thin filaments do not change length during muscle contraction but simply slide past each other, increasing their area of overlap.

The mechanism for the sliding of filaments is the hydrolyzing of ATP by the globular head of a myosin molecule, which then changes to a high-energy form that binds to actin and forms a cross-bridge. The myosin head then bends back to its low-energy configuration and pulls the attached thin filament toward the center of the sarcomere. When a new molecule of ATP binds to the myosin head, it breaks its bond to actin, and the cycle repeats with the myosin head attaching to an actin molecule farther along on the thin filament.

Energy for muscle contraction is stored in creatine phosphate, which regenerates ATP, and in glycogen, which is broken down to glucose. Glucose can generate ATP by glycolysis or aerobic respiration.

FOCUS QUESTION   39.1

Identify the components in the following diagram of a section of a skeletal muscle fiber. Explain how and why this diagram will look different when this muscle fiber is fully contracted.

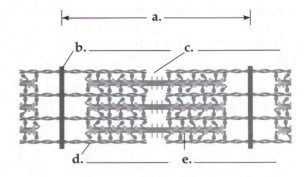

When a muscle fiber is at rest, the myosin-binding sites of the actin molecules are blocked by the regulatory protein **tropomyosin**, which is associated with another set of regulatory proteins, the **troponin complex**. When troponin binds to Ca^{2+}, the regulatory proteins change shape; myosin-binding sites on the thin filament are then exposed, and muscle contraction can occur.

Calcium ions are actively transported into the **sarcoplasmic reticulum (SR)** of a muscle cell. When an action potential of a motor neuron causes the release of acetylcholine into the synapse with a muscle fiber, the fiber depolarizes and an action potential spreads into the **transverse (T) tubules,** infoldings of the muscle cell's plasma membrane. The action potential opens Ca^{2+} channels in the sarcoplasmic reticulum, releasing Ca^{2+} into the cytosol. Calcium binding with troponin exposes myosin-binding sites, and contraction begins. The fiber relaxes when motor neuron input stops, the sarcoplasmic reticulum pumps Ca^{2+} back out of the cytosol, and the tropomyosin–troponin complex again blocks the binding sites.

The response of a single muscle fiber to an action potential of a motor neuron is a brief twitch. Graded contraction of whole muscles may result from involving additional fibers. A single motor neuron and all the muscle fibers it innervates make up a **motor unit**. The strength of a muscle contraction depends on the size and number of motor units involved. Muscle force (tension) can be increased by the activation of additional motor units, called *recruitment*.

Repeated, rapid-fire action potentials produce summation and a state of smooth and sustained contraction called **tetanus**.

What are the different types of skeletal muscle fibers? Oxidative fibers, which rely mostly on aerobic respiration, have many mitochondria, a good blood supply, and **myoglobin,** a pigment that extracts oxygen from the blood and stores it. Glycolytic fibers primarily use glycolysis to produce ATP. **Fast-twitch fibers** produce rapid and powerful contractions. **Slow-twitch fibers,** common in muscles that must sustain long contractions, have less sarcoplasmic reticulum and pump calcium more slowly; thus, calcium remains in the cytosol longer and twitches last longer.

FOCUS QUESTION   39.2

a. Identify the three distinct types of skeletal muscle fibers.

b. Which type of fiber has few mitochondria, has little myoglobin, and produces brief, powerful contractions?

c. Which type of fiber would you expect to be present in the highest proportion in postural muscles?

Other Types of Muscle   Vertebrate **cardiac muscle** is found in the heart. Ion channels in the plasma membrane of a cardiac muscle cell produce rhythmic depolarizations that generate action potentials, which last longer than those of skeletal muscle fibers. Cardiac muscle cells are connected by **intercalated disks** through which action potentials spread to all the cells of the heart.

The scattered arrangement of actin and myosin filaments in **smooth muscle** accounts for its nonstriated appearance. Smooth muscle fibers lack T tubules, a troponin complex, and a well-developed sarcoplasmic reticulum. Calcium ions enter the cell in response to an action potential and bind to calmodulin, which activates an enzyme that phosphorylates myosin. Contractions are slower and may be triggered by the autonomic nervous system or by the muscle cells themselves.

Invertebrates have muscle cells similar to the skeletal and smooth muscle cells of vertebrates. The flight muscles of insects are capable of independent and rapid contraction.

39.2   Skeletal systems transform muscle contraction into locomotion

Skeletons function in support, protection, and movement. Because muscles can only contract, antagonistic muscle pairs are required to move body parts in opposite directions.

Types of Skeleton Systems   Fluid under pressure in a closed body compartment creates a **hydrostatic skeleton**. As muscles change the shape of the fluid-filled compartment, the animal elongates or moves. Earthworms and other annelids move by **peristalsis,** using rhythmic waves of contractions of circular and longitudinal muscles.

Typical of molluscs and arthropods, **exoskeletons** are hard coverings deposited on the surface of animals. The cuticle of an arthropod contains fibrils of the polysaccharide **chitin** embedded in a protein matrix.

The supporting elements of an **endoskeleton** are embedded in the soft tissues of the animal. The endoskeletons of chordates are composed of cartilage and/or bone. The human skeleton consists of a skull, vertebral column, rib cage, the pectoral and pelvic girdles, and the limb bones. Joints, where bones are connected by ligaments, allow for flexibility in body movement. Tendons attach muscles to bones.

FOCUS QUESTION 39.3

List an advantage and a disadvantage of each of the following types of skeletons, and indicate in which animal groups each type is found.

a. hydrostatic skeleton

b. exoskeleton

c. endoskeleton

Body proportions of small and large animals vary, partly as a result of the physical relationship between the diameter of a support (strength increases with the square of its diameter) and the strain produced by increasing weight (weight increases with the cube of its dimensions). Posture is an important structural factor in supporting body weight.

Types of Locomotion **Locomotion,** the active movement from place to place, requires energy to overcome the forces of friction and gravity.

A flying animal uses its wings to provide enough lift to overcome the force of gravity. Wings are shaped as airfoils and attach to lightweight, fusiform bodies.

Walking, running, or hopping on land requires the animal to support itself and move against gravity; however, air offers little resistance to movement. Powerful leg muscles and strong skeletal support are important for supporting an animal and propelling it forward. A running or hopping animal may store some energy in its tendons to facilitate the next step or hop. Maintaining balance is another requirement

for upright movement. A crawling animal must overcome the friction produced by its contact with the ground.

Overcoming gravity is less of a problem for swimming animals due to the buoyancy of water. Most fast swimmers have a fusiform shape that reduces friction or drag caused by the density of water. Animals may swim by using their legs as oars, using water to jet-propel themselves, or by undulating their bodies and tails from side to side or up and down.

39.3 Discrete sensory inputs can stimulate both simple and complex behaviors

Niko Tinbergen's set of questions to guide behavioral studies emphasizes the importance of both *proximate causation*, the immediate cause of a behavior in terms of the stimuli that trigger it, the mechanisms that produce it, and also how experience modifies a response; and *ultimate causation*, which concerns the benefit to survival and reproduction and the evolutionary basis of the behavior. **Behavioral ecology** is the study of the ecological and evolutionary basis of animal behavior.

Fixed Action Patterns A **fixed action pattern** is a sequence of unlearned behaviors that, once begun, is usually carried through to completion. It is triggered by a simple external stimulus called a **sign stimulus**.

Migration Many animals use environmental cues to guide regular, long-distance changes in location called **migrations**. Some migratory animals alter their orientation to the sun during the day with the aid of their *circadian clock*. Some animals navigate using Earth's magnetic field.

Behavioral Rhythms Many behaviors follow a circadian rhythm mediated by the circadian clock. *Circannual rhythms* are linked to the yearly cycle of seasons and are influenced by changing lengths of daylight and darkness.

Animal Signals and Communication **Communication** between animals involves the transmission and reception of special stimuli called **signals**. Courtship often involves a *stimulus-response chain*, in which each response is the stimulus for the next behavior. Modes of communication may be visual, auditory, chemical, or tactile, depending on the lifestyle and sensory specializations of a species.

Pheromones are chemical signals commonly used by mammals and insects in reproductive behavior to attract mates and to trigger specific courtship behaviors. Pheromones can also function as alarm signals.

FOCUS QUESTION   39.4

Why is most communication among mammals olfactory and auditory, whereas communication among birds is visual and auditory?

39.4   Learning establishes specific links between experience and behavior

Behavior that is performed virtually the same by all individuals in a population is developmentally fixed and called **innate behavior**.

Experience and Behavior   In a **cross-fostering study,** young of one species are raised by adults of another species. Such studies examine how an animal's social and physical surroundings influence behavior. Male California mice provide extensive parental care and are always highly aggressive toward other mice. However, when cross-fostered in nests of white-footed mice, California mice showed reduced aggression and reduced parental behavior. Early experiences that alter parental behavior extend this environmental influence to the next generation.

Human **twin studies** involving identical twins raised apart or in the same household have been used to study various behavioral disorders in humans.

Learning   Learning is the modification of behavior as a result of specific experiences.

Imprinting is characterized by a limited **sensitive period** during which learning may occur and often involves a behavioral response to an individual that is usually long lasting. The ability to respond is innate; the environment provides the *imprinting stimulus.* Imprinting on parents may be required for future *pair-bonding* with a mate of the proper species.

The establishment of a memory of the environment's spatial structure is called **spatial learning**. Animals may learn and use a particular set of landmarks, or location indicators, to find their way within their area.

More complicated than a set of learned landmarks, **cognitive maps** are neural representations of the spatial relationships of objects in an animal's surroundings. Evidence for cognitive maps comes from research with nutcrackers, birds that are able to retrieve hidden food stores.

In **associative learning,** animals learn to associate one stimulus with another. An animal's nervous system enables the association of behaviors with those stimuli that are most likely to occur in the animal's natural habitat.

Cognition refers to an animal's ability to learn using awareness, reasoning, recollection, and judgment. Research with bees indicates that they are capable of categorizing environmental objects as "same" or "different."

Problem solving, the ability to achieve an end in the face of obstacles, is most often observed in mammals, especially in primates and dolphins. Such behavior has also been documented in some bird species.

Many animals use the behavior of others as information in problem solving. **Social learning,** learning through the observation of others, forms the basis of **culture**—a system of information transfer that involves teaching and/or social learning and influences behavior in a population.

FOCUS QUESTION   39.5

Indicate the type of learning illustrated by the following examples:

a. Ewes will adopt and nurse a lamb shortly after they give birth to their own lamb but will butt and reject a foreign lamb introduced a day or two later.

b. A dog whose early "accidents" were cleaned up with paper towels accompanied with harsh discipline hides any time a paper towel is used in the household.

c. In experiments, honey bees have the ability to distinguish between patterns that are the same or different.

d. Tinberger showed that female digger wasps use landmarks to locate their hive.

e. Young chimpanzees learn to crack oil palm nuts with two stones by observing others.

39.5   Selection for individual survival and reproductive success can explain most behaviors

Foraging is behavior involved with searching for, recognizing, obtaining, and consuming food.

Evolution of Foraging Behavior   Natural selection is expected to refine behaviors that improve foraging. Laboratory studies of *Drosophila* have documented an evolutionary change in foraging path length in populations of high or low density. The frequency of the for^R (rover) allele increased in high-density populations, whereas the frequency of the for^S (sitter) allele increased in low-density populations.

Mating Behavior and Mate Choice In some species, no strong pair-bonds form between mates. Longer-lasting relationships may be **monogamous** or **polygamous**. Polygamous relationships may be *polygynous* (one male and many females) or *polyandrous*. Monogamous species are less likely to exhibit *sexual dimorphism*.

The needs of offspring are an important factor in the evolution of reproductive patterns. If young require more food than one parent can supply, a male may increase his reproductive fitness by helping to care for offspring rather than going off in search of more mates. With mammals, the female often provides all the food, and males are often polygynous.

Certainty of paternity also influences mating behavior and parental care. With internal fertilization, the acts of mating and egg laying or birth are separated, and paternity is less certain than when eggs are fertilized externally.

FOCUS QUESTION 39.6

Exclusive male parental care is observed much more frequently in species with external fertilization. Explain how such behavior could evolve.

Sexual selection may be *intrasexual*, involving competition among members of one sex for mates, or *intersexual*, in which mates are chosen by one sex on the basis of particular characteristics.

Female choice in stalk-eyed fruit flies has been a selection factor in the evolution of long eyestalks, which correlate with male health.

Agonistic behavior involves an often-ritualized contest to determine which competitor gains access to a resource, such as a mate.

39.6 Inclusive fitness can account for the evolution of behavior, including altruism

Genetic Basis of Behavior Male prairie voles form pair-bonds, help care for the young, and are aggressive to intruders. Male meadow voles do not show these behaviors. The neurotransmitter vasopressin is released during mating. The gene that codes for vasopressin receptors is highly expressed in the brains of prairie voles but not in those of meadow voles. Male meadow voles with inserted prairie vole receptor genes developed more vasopressin receptors in their brains and showed mating behaviors similar to those of male prairie voles.

Genetic Variation and the Evolution of Behavior Variations in behavior between populations within a species may correlate with variations in the environment.

Most laboratory-born garter snakes from coastal areas (where banana slugs are an abundant source of prey) ate slugs when they were offered, whereas few laboratory-born snakes from inland populations ate the slugs. The researcher proposed that snakes in coastal areas with the ability to recognize slugs by chemoreception had higher fitness, leading to the evolution of this difference in prey selection behavior.

FOCUS QUESTION 39.7

Why do experiments examining behavioral variations among natural populations often raise and test animals in the laboratory?

Altruism Many behaviors are selfish, benefiting one individual's reproductive success at the expense of others. Selflessness, or **altruism,** is behavior that reduces an animal's individual fitness while increasing the fitness of other individuals.

Inclusive Fitness Natural selection favors traits that increase reproductive success, thereby propagating the genes for those traits. W. Hamilton explained altruistic behavior in terms of **inclusive fitness,** the ability of an individual to pass on its genes by producing its own offspring and by helping close relatives produce their offspring.

A quantitative measure, called **Hamilton's rule,** predicts that natural selection favors altruistic acts among related individuals if $rB > C$, where B and C are the benefit to the recipient and the cost to the altruist, respectively, as measured by the change in the average number of offspring produced as a result of the altruistic act, and r is the **coefficient of relatedness,** the fraction of genes that are shared by the two individuals. **Kin selection** is the term for the natural selection of altruistic behavior that enhances the reproductive success of related individuals.

FOCUS QUESTION   39.8

a. According to kin selection, would an individual be more likely to exhibit altruistic behavior toward a parent, a sibling, or a first (full) cousin?

b. Explain your answer in terms of the coefficient of relatedness and Hamilton's rule.

Word Roots

endo- = within (*endoskeleton:* a hard skeleton buried within the soft tissues of an animal)

exo- = outside (*exoskeleton:* a hard encasement on the surface of an animal, such as the shell of a mollusc or the cuticle of an arthropod, that provides protection and points of attachment for muscles)

hydro- = water (*hydrostatic skeleton:* a skeletal system composed of fluid held under pressure in a closed body compartment; the main skeleton of most cnidarians, flatworms, nematodes, and annelids)

inter- = between; **-cala** = insert (*intercalated disk:* a specialized junction between cardiac muscle cells that provides direct electrical coupling between the cells)

mono- = one; **-gamy** = reproduction (*monogamous:* a type of relationship in which one male mates with just one female)

myo- = muscle; **-fibro** = fiber (*myofibril:* a longitudinal bundle in a muscle cell (fiber) that contains thin filaments of actin and regulatory proteins and thick filaments of myosin)

peri- = around; **-stalsis** = a constriction (*peristalsis:* a type of movement on land produced by rhythmic waves of muscle contraction passing from front to back, as in many annelids)

poly- = many (*polygamous:* a type of relationship in which an individual of one sex mates with several individuals of the other sex)

sarco- = flesh; **-mere** = a part (*sarcomere:* the fundamental, repeating unit of striated muscle, delimited by the Z lines)

tetan- = rigid, tense (*tetanus:* the maximal, sustained contraction of a skeletal muscle, caused by a very high frequency of action potentials elicited by continual stimulation)

tropo- = turn, change (*tropomyosin:* the regulatory protein that blocks the myosin-binding sites on actin molecules)

Structure Your Knowledge

1. Trace the sequence of events in muscle contraction from an action potential in a motor neuron to the relaxation of the muscle.

2. How does the nature-versus-nurture controversy apply to the study of behavior?

3. How does the concept of evolutionary fitness apply to all aspects of behavior?

Test Your Knowledge

MULTIPLE CHOICE: *Choose the one best answer.*

1. When striated muscle fibers contract,
   a. the Z lines are pulled closer together.
   b. the sarcomere expands.
   c. the thin filaments become shorter.
   d. the thick filaments become longer.
   e. **a** and **c** occur.

2. Tetanus
   a. is muscle fatigue resulting from a lack of ATP and loss of ion gradients.
   b. is the all-or-none contraction of a single muscle fiber.
   c. is the result of stimulation of additional motor neurons.
   d. is the result of a volley of action potentials, whose summation produces an increased and sustained contraction.
   e. is the rigidity of muscles following death.

3. The role of ATP in muscle contraction is
   a. to form cross-bridges between thick filaments and thin filaments.
   b. to release the myosin head from actin when it binds to myosin, and to provide energy when hydrolyzed to form myosin's high-energy form.
   c. to remove the tropomyosin–troponin complex from blocking the binding sites on actin.
   d. to bend the cross-bridge and pull the thick filaments toward the center of the sarcomere.
   e. to replace the supply of creatine phosphate required for the movement of myosin past actin.

4. How does calcium affect muscle contraction?
   a. It is released from the T tubules in response to an action potential and initiates contraction.
   b. The binding of acetylcholine opens calcium channels in the plasma membrane, creating an action potential that travels down the T tubules.
   c. It binds to tropomyosin and helps to stabilize cross-bridge formation.
   d. Its binding to troponin causes tropomyosin to move away from the myosin-binding sites on the actin filament.
   e. Its release from the sarcoplasmic reticulum changes the membrane potential of the muscle cell so that contraction can occur.

5. A motor unit is
   a. a sarcomere that extends from one Z line to the next Z line.
   b. a motor neuron and the muscle fibers it innervates.
   c. the myofibrils of a fiber and the transverse tubules that connect them.
   d. the muscle fibers that make up a muscle.
   e. the antagonistic set of muscles that flex and extend a body part.

6. Which of the following characteristics does *not* pertain to cardiac muscle?
   a. scattered arrangement of actin and myosin filaments
   b. intercalated disks that spread action potentials between cells
   c. action potentials that last a long time
   d. ability to generate action potentials without nervous input
   e. striations

7. Smooth muscle contracts and relaxes relatively slowly because
   a. the only ATP available is supplied by fermentation.
   b. its contraction is stimulated by hormones, not motor neurons.
   c. it does not have a well-developed sarcoplasmic reticulum, and Ca^{2+} enters the cell through the plasma membrane during an action potential.
   d. it is not striated.
   e. it is composed exclusively of slow-twitch muscle fibers.

8. Hydrostatic skeletons are used for movement by all of the following animals *except*
   a. cnidarians.
   b. arthropods.
   c. nematodes.
   d. annelids.
   e. flatworms.

9. When you pull up your lower arm and "make a muscle" in your biceps, you are
   a. contracting a flexor.
   b. relaxing a flexor.
   c. contracting an extensor.
   d. contracting a tendon.
   e. producing a simple muscle twitch.

10. Behavioral ecology is the
    a. study of the behavior of animals, focusing on stimulus and response.
    b. application of human emotions and thoughts to other animals.
    c. study of animal cognition.
    d. study of animal behavior from an ecological and evolutionary perspective.
    e. behavioral study of ecology.

11. Proximate causes
    a. explain the evolutionary significance of a behavior.
    b. are immediate causes of behavior such as environmental stimuli.
    c. are environmental, whereas ultimate causes are genetic.
    d. are endogenous, although they may be set by exogenous cues.
    e. show that nature is more important than nurture.

12. Which of the following descriptions is an example of a fixed-action pattern?
    a. a crane in a captive-breeding program imprinting on its human caregiver
    b. a male stickleback chasing a red-bellied object from its territory
    c. a songbird migrating to winter feeding grounds using the position of the sun.
    d. the claw-waving behavior of a male fiddler crab
    e. a digger wasp returning to its nest with the aid of landmarks

13. A sensitive period
    a. is the time right after birth.
    b. usually follows the reception of a sign stimulus.
    c. is a limited time in which imprinting can occur.
    d. is the period during which birds can learn to fly.
    e. is the time during which mate selection occurs.

14. In associative learning,
    a. an animal improves its performance of a fixed-action pattern.
    b. an animal learns as a result of associating a benefit or harm with an action.
    c. an animal learns a behavior by watching others.
    d. an animal forms a cognitive map of its surroundings.
    e. an animal imprints on a particular individual, which is usually a parent providing care.

15. Which of the following types of intraspecies communication signals is best suited to a fruit fly's stimulus response chain during mating behavior?
    a. an auditory signal
    b. a visual signal
    c. a chemical signal
    d. a tactile signal
    e. all of the above

16. Which of the following examples of behavior provides evidence of animal cognition?
    a. a chimpanzee stacking up boxes to reach a banana overhead
    b. ravens pulling up string to obtain a food item attached to it
    c. trained honeybees that can discriminate between the concepts of "same" and "different" by matching colors or patterns
    d. a biology student using various resources to study for an exam
    e. All of the above show evidence of information processing and animal cognition.

17. In a species in which females provide all the needed food and protection for the young,
    a. males are likely to be promiscuous.
    b. mating systems are likely to be monogamous.
    c. mating systems are likely to be polyandrous.
    d. females will bond in social groups.
    e. females will have higher fitness than males.

18. A crow that aids its parents in raising its siblings is increasing its
    a. survival success.
    b. altruistic behavior.
    c. inclusive fitness.
    d. coefficient of relatedness.
    e. certainty of paternity.

19. According to the concept of kin selection,
    a. an animal would be more likely to aid a stranger if the "kindness" could be reciprocated.
    b. an animal would aid its parent before it would help its sibling.
    c. animals are more likely to choose close relatives as mates.
    d. examples of altruism usually involve close relatives and increase an animal's inclusive fitness.
    e. evolution is the ultimate cause of animal behavior.

20. The cross-fostering of California mice in white-footed mice nests provides evidence for
    a. the genetic control of aggression and parenting behavior.
    b. the relationship between the distribution of vasopressin receptors and parenting behavior.
    c. an imprinting period during which behaviors related to aggression and parenting are set.
    d. the influence of the early social environment on the expression of aggressive and parental behaviors.
    e. the cognitive ability of mice to change their behavior in different environments.

UNIT 7

Ecology

Chapter 40

Population Ecology and the Distribution of Organisms

Chapter Focus

This chapter explores the scope of ecology, global climate patterns, the major terrestrial and aquatic biomes, the factors that help determine the distribution of species, and the growth and regulation of populations.

Chapter Review

Ecology is the study of the interactions of organisms with each other and with their environment. Rigorous experimental designs are now commonly used to investigate complex ecological questions. These

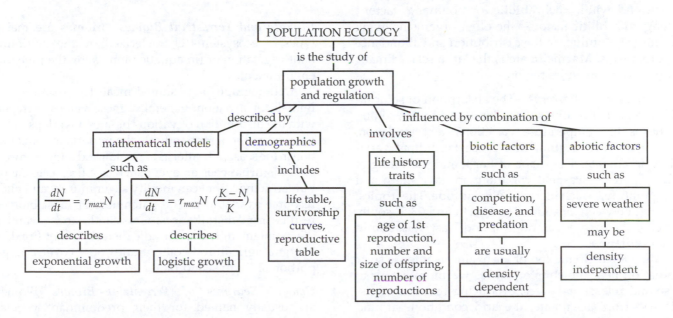

305

questions may focus on different levels of the biological hierarchy, as outlined in the following text.

The **biosphere** includes all of Earth's ecosystems and landscapes. **Global ecology** is the study of the effect of regional energy and material exchanges on the distribution and functioning of organisms across the biosphere.

A **landscape** or seascape consists of several connected ecosystems; **landscape ecology** is the study of the flow of energy, materials, and organisms across ecosystems.

An **ecosystem** includes all the organisms in an area and the abiotic factors with which they interact; **ecosystem ecology** addresses such topics as the flow of energy and chemical cycling.

A **community** includes the populations of different species in an area; **community ecology** looks at such interactions as predation and competition and how those interactions between species affect community structure.

Population ecology is concerned with the factors that affect the size of **populations,** which are groups of individuals of the same species occupying a particular area.

Organismal ecology, which may include the disciplines of behavioral, physiological, and evolutionary ecology, considers the responses and adaptations of an organism to its environment.

40.1 Earth's climate influences the structure and distribution of terrestrial biomes

The **climate,** or prevailing weather conditions of an area, is influenced by temperature, precipitation, sunlight, and wind. Such **abiotic**, or nonliving, factors, along with **biotic** factors—the other organisms living in the area—influence the distribution and abundance of organisms. **Macroclimate** is the large-scale climatic pattern over an entire region.

Global Climate Patterns   The absorption of solar radiation heats the atmosphere, land, and water, setting patterns for temperature variations, air circulation, and water evaporation that cause latitudinal variations in climate. The shape of Earth and the tilting of its axis create seasonal variations in day length and temperature, which increase with latitude. The **tropics,** located between latitudes 23.5° north and 23.5° south, receive the greatest amount of and least variation in solar radiation.

The global circulation of air begins as intense solar radiation near the equator causes warm, moist air to rise and release its water, producing the characteristic wet tropical climate, the arid conditions around 30° north and south as dry air descends, the fairly wet

though cool climate at about 60° latitude as air rises again, and the cold and rainless climates of the polar regions. Air flowing along Earth's surface produces global wind patterns, such as the cooling trade winds blowing from east to west in the tropics and the prevailing westerlies in temperate zones.

Regional Effects on Climate   The changing angle of the sun over the year produces wet and dry seasons around 20° north and 20° south latitude. Seasonal wind pattern changes affect ocean currents, sometimes causing upwellings of cold, nutrient-rich water.

Coastal areas are generally wetter than inland areas, and large bodies of water moderate the climate. For example, water warmed at the equator flows as the Gulf Stream toward the North Atlantic and warms the coast of western Europe.

FOCUS QUESTION   40.1

Describe how mountains affect local climate with respect to the following three factors:

a. sunlight

b. temperature

c. rainfall

Climate and Terrestrial Biomes   **Biomes** are major types of ecosystems characterized by the predominant vegetation or (in aquatic biomes) by the physical environment.

A **climograph** plots annual mean temperature and rainfall for a region; generally, these values correlate with the distribution of various biomes. Overlaps of biomes on a climograph indicate the importance of variation in the seasonal patterns of rainfall and temperatures.

A **disturbance** is an event such as a fire, a storm, or human activity that can modify a community and alter resource availability. The patchiness of most biomes results from disturbances. Grasslands and savannas are maintained by the periodic disturbance of fire. Urban and agricultural communities now cover a large portion of Earth's land mass.

General Features of Terrestrial Biomes   Biomes are usually named for their predominant vegetation and major climatic features. Each biome also has characteristic microorganisms, fungi, and animals. The area of transition between biomes is called an **ecotone**.

Terrestrial biomes have vertical stratification, such as the layers in a forest from upper **canopy**, low-tree layer, shrub understory, herbaceous plant ground layer, forest floor (litter), to root layer. Vertical stratification of vegetation provides diverse habitats for animals.

Exploring Terrestrial Biomes What are the main characteristics of the eight terrestrial biomes?

Tropical forests occur in equatorial and subequatorial regions. Variations in rainfall result in **tropical dry forests,** where rainfall is seasonal, and **tropical rain forests,** where rainfall is more constant and abundant. Temperatures are uniformly high. The tropical rain forest has pronounced vertical stratification. Animal diversity is higher than in any other terrestrial biome. Agriculture and development are destroying many tropical forests.

Savannas are equatorial and subequatorial grasslands with scattered trees and long dry seasons. Fires are common, and the dominant grasses and forbs (small nonwoody plants) are fire-adapted and drought-tolerant. Large grazing mammals and their predators are common, although insects are the dominant herbivores. Cattle ranching and overhunting have reduced the large-mammal populations of savannas.

Characterized by low and variable precipitation, **deserts** occur around 30° north and south latitude and in the interior of continents. Deserts may be hot or cold, and temperature varies seasonally and daily. Plants may use C_4 and CAM photosynthesis and have reduced leaf surface area, feature water storage adaptations, and produce protective spines and toxins. Desert animals have physiological and behavioral adaptations to dry conditions. Irrigated agriculture and urbanization are now common in deserts, reducing natural biodiversity.

Chaparral, common along coastlines in midlatitude regions, has cool, rainy winters and hot, dry summers. The dominant vegetation—evergreen shrubs and small trees—is maintained by and adapted to periodic fires. Browsing animals and small mammals and birds are common. Urbanization and agriculture have reduced areas of chaparral.

Temperate grasslands are maintained by fire, seasonal drought, and grazing by large mammals. Winters are generally cold and dry; summers are hot and wet. Soils are deep and fertile, and many grasslands have been converted to farmland.

Characterized by broad-leaved deciduous trees, **temperate broadleaf forests** grow in midlatitude regions in the Northern Hemisphere, with smaller areas on other continents. Winters are cold and summers hot and humid. During the cold winters, many mammals hibernate and birds migrate. Humans have heavily logged these forests, clearing land for agriculture and development.

The largest terrestrial biome, the **northern coniferous forest,** or *taiga,* is characterized by harsh winters with heavy snowfall. Summers may be hot. Birds and large mammals are common animals. Old-growth stands of coniferous trees are rapidly being logged.

Tundra, covering large areas of the Arctic, is characterized by cold winters, cool summers, and dwarfed or herbaceous vegetation. A layer of frozen soil called permafrost restricts root growth. Migratory large mammals and birds are common. The *alpine tundra,* found at all latitudes on very high mountains, has similar plant communities. Mineral and oil extraction are becoming common in areas of arctic tundra.

FOCUS QUESTION 40.2

Temperature and precipitation are two of the key factors that influence the vegetation found in a biome. On the following climograph, label the North American biomes (arctic and alpine tundra, northern coniferous forest, desert, temperate grassland, temperate broadleaf forest, and tropical forest) represented by each plotted area of temperature and precipitation.

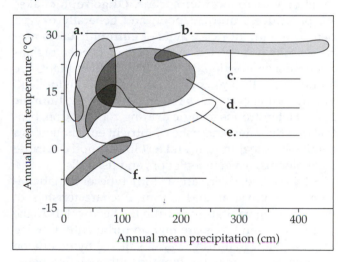

40.2 Aquatic biomes are diverse and dynamic systems that cover most of Earth

One of the chemical differences in aquatic biomes is salt concentration—less than 0.1% for freshwater biomes versus an average of 3% for marine biomes. Three-fourths of Earth is covered by oceans, which influence global rainfall, climate, and wind patterns. Marine algae and photosynthetic bacteria produce a large portion of the world's O_2 and consume enormous amounts of CO_2. Freshwater biomes are influenced by the surrounding terrestrial biome.

Zonation in Aquatic Biomes   Many aquatic biomes are layered both vertically and horizontally. The **photic zone** receives sufficient light for photosynthesis, whereas little light penetrates into the lower **aphotic zone**. The **pelagic zone** includes the waters of the photic and aphotic zone. The bottom substrate, called the **benthic zone**, is home to organisms collectively called **benthos**.

In the ocean and in most lakes, a narrow layer of abrupt temperature change called a **thermocline** separates warmer surface waters from the cold bottom layer.

Exploring Aquatic Biomes   What are the characteristics of the seven aquatic biomes?

Wetlands are areas covered with water, often enough to support plants adapted to water-saturated soil. Where a river meets the ocean, an **estuary** is formed. Nutrients from upstream make these habitats among the most productive of biomes, and they support a rich diversity of animal species. Water and soils may be periodically oxygen-poor due to high rates of decomposition. These habitats filter nutrients and chemical pollutants. Up to 90% of wetlands have been lost to draining and filling, and estuaries have been disrupted by pollution, filling, and dredging.

Lakes have photic and aphotic zones and often a seasonal or year-round thermocline. **Oligotrophic lakes** are often deep, nutrient poor, and generally oxygen rich. The nutrient-rich waters of **eutrophic lakes** may be seasonally depleted of oxygen in deeper regions due to decomposition. Rooted and floating plants are found in the **littoral zone** close to shore; phytoplankton, including cyanobacteria, are found in the **limnetic zone**. Heterotrophs include drifting zooplankton, benthic invertebrates, and fishes. Nutrient enrichment as a result of waste dumping and fertilizer runoff can cause algal blooms, oxygen depletion, and fish kills.

Streams and rivers are a third type of aquatic biome. The physical and chemical characteristics of these flowing habitats vary from the headwaters to the mouth. Oxygen levels are high in turbulently flowing water and low in murky, warm waters. Phytoplankton or rooted plants may be common, although overhanging vegetation provides most nutrient content in forest streams. Unpolluted rivers and streams are home to diverse fishes and invertebrates. Human impact on streams and rivers includes pollution and damming.

In **intertidal zones** the daily cycle of tides exposes the shoreline to variations in water, nutrients, and temperature, and to the mechanical force of wave action. Rocky intertidal communities have diverse marine algae and animals adapted to attaching firmly to the hard substrate. Sandy intertidal zones are home to burrowing worms, clams, and crustaceans. Oil pollution and construction of protective barriers have disrupted many intertidal areas.

Found in the photic zone of clear tropical waters, **coral reefs** are highly diverse and productive biomes. The structure of the reef is produced by the calcium carbonate skeletons of the coral (various cnidarians) and serves as a substrate for red and green algae. The coral animals are nourished by symbiotic unicellular algae. Overfishing, coral collecting, pollution, and global warming are destroying coral reefs.

The water of the **oceanic pelagic biome** is typically nutrient poor but oxygen rich. In temperate and high-latitude areas, seasonal mixing of waters brings nutrients to surface waters. Phytoplankton in the photic region provide half of global productivity. They are grazed on by numerous types of zooplankton. Free-swimming animals include squid, fishes, and marine mammals. Overfishing and waste dumping have damaged the Earth's oceans.

The **marine benthic zone** consists of the ocean floor. Except for shallow areas, the marine benthic zone receives no sunlight, and nutrients fall as detritus from the waters above. Inhabitants of the abyssal zone are adapted to cold and to high water pressure. Chemoautotrophic prokaryotes form the basis of a collection of organisms adapted to the hot, low-oxygen environment surrounding **deep-sea hydrothermal vents**. Overfishing has greatly reduced many benthic fish populations, and organic pollution has created oxygen-deprived areas.

FOCUS QUESTION   40.3

Different aquatic environments can be classified on the basis of light penetration, distance from shore, and open water vs. bottom. Match the following zones to their corresponding numbers on the diagram of a lake:

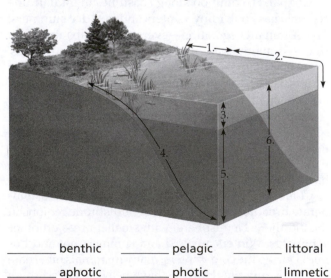

_____ benthic   _____ pelagic   _____ littoral

_____ aphotic   _____ photic   _____ limnetic

40.3 Interactions between organisms and the environment limit the distribution of species

Interactions between organisms and their environments occur within *ecological time*. The cumulative effects of these interactions are seen in the adaptation of organisms to their environment over an *evolutionary time* frame. The distribution of species reflects these ecological and evolutionary interactions with both biotic and abiotic factors. Ecologists often work through a series of logical questions to determine what factors limit the geographic distribution of a species.

Dispersal and Distribution Does dispersal limit a species distribution? **Dispersal** is the movement of individuals away from their original area. Transplants of a species to a previously inaccessible area can indicate whether dispersal limits its distribution. A successful transplant shows that the *potential* range of a species is larger than its *actual* range. Species that are introduced to new areas, either purposely or accidentally, often disrupt their new ecosystem.

Biotic Factors Is the distribution of a species limited by biotic factors such as predation, disease, parasitism, or competition? Removal experiments can test whether other species limit the distribution of a species. An Australian study showed that sea urchins limited the abundance and distribution of seaweeds.

Abiotic Factors Is a species' geographic distribution influenced by abiotic factors? Temperature is an important environmental factor because of its effects on metabolism. Most organisms function best within a narrow range of environmental temperatures.

The availability of water in different habitats can vary greatly. The distribution of organisms reflects their ability to obtain sufficient water and avoid desiccation in terrestrial habitats. Oxygen concentrations can be low in deep waters, in sediments rich in organic matter, and in flooded soils.

The ability of organisms to osmoregulate determines their distribution in freshwater, saltwater, or other high-salinity habitats.

Light energy drives almost all ecosystems. The quantity of sunlight is a limiting factor in aquatic environments and on forest floors.

Soils, which vary in their physical structure, pH, and mineral composition, affect the distribution of plants and, in turn, the distribution of animals. Substrate composition in aquatic environments influences water chemistry and the types of organisms that can inhabit those areas.

FOCUS QUESTION 40.4

List and give examples of the three factors that ecologists examine to understand the geographic distribution of a species.

a.

b.

c.

40.4 Dynamic biological processes influence population density, dispersion, and demographics

Population ecology studies the influence of the abiotic and biotic environment on population size, density, and distribution. A population is a localized group of individuals of the same species. Members of a population use the same resources and have a high probability of interacting and breeding with each other. Populations may be described by their size and geographic boundaries; ecologists define such boundaries based on the type of organism and the research question being asked.

Density and Dispersion The number of individuals per unit area or volume is a population's **density**; the pattern of spacing of those individuals is referred to as **dispersion**.

Scientists use a variety of sampling techniques to estimate population density. Indirect indicators of population density include burrows, nests, or fecal droppings.

Changes in population density reflect both additions of members through birth (including all forms of reproduction) and **immigration,** and removal of members through death (mortality) and **emigration**.

What types of dispersal patterns may be found in a population's geographic range? *Clumping* may indicate a heterogeneous environment or social interactions between individuals.

Uniform distribution may be related to competition for resources and result from interactions between individuals. **Territoriality,** the defense of a bounded physical space, can lead to uniform dispersion. *Random* spacing, indicating the absence of strong interactions among individuals or a fairly uniform habitat, is less common.

Demographics The study of the vital statistics of a population, such as birth and death rates, is called **demography**.

A **life table** presents age-specific survival data for a population. It can be constructed by following a **cohort** of organisms from birth to death and calculating the proportion of the cohort surviving in each age group.

A **survivorship curve** shows the number or proportion of members of a cohort still alive at each age. Survivorship curves are often based on a convenient beginning cohort of 1,000 individuals and typically use a logarithmic scale on the *y*-axis and a relative scale on the *x*-axis, so that species with different life spans can be compared on the same graph. There are three general types of survivorship curves. Type I has low mortality during early and middle age and a rapid increase with old age. In a Type II curve, death rate is relatively constant throughout the life span. A Type III curve has high initial mortality, with the few offspring that survive likely to reach adulthood. Many species show intermediate or more complex survivorship patterns.

FOCUS QUESTION   40.5

Identify the types of survivorship curves in the following graph. For each type of curve, describe the characteristics of representative species and then cite some examples of those species.

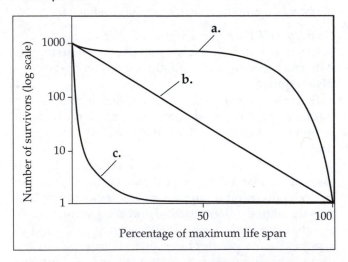

In sexually reproducing species, demographers usually follow only the reproduction of females through the generations. A **reproductive table,** or fertility schedule, gives the age-specific reproductive rates in a population.

40.5 The exponential and logistic models describe the growth of populations

Per Capita Rate of Increase A small population in a very favorable environment may exhibit unrestricted growth. Ignoring immigration and emigration, the change in population size during a specific time period is equal to the number of births minus the number of deaths.

Births and deaths can be expressed in terms of the average number per individual during a time period—that is, as a *per capita birth rate (b)* and a *per capita death rate (m, for mortality)*. These rates can be calculated from estimates of population size (N) and from data in life tables and reproductive tables. The population growth equation using per capita birth and death rates is $\Delta N/\Delta t = bN - mN$. The *per capita rate of increase (r)* is the difference between the per capita birth and death rates: $r = b - m$. **Zero population growth (ZPG)** occurs when $r = 0$, when birth and death rates are equal. The equation describing the change in the population instantaneously uses differential calculus and is written as $dN/dt = r_{inst}N$.

Exponential Growth Under ideal conditions, a population may exhibit **exponential population growth**. Expressed as $dN/dt = r_{max}N$, exponential population growth produces a J-shaped growth curve when graphed. The larger the population (N) becomes, the faster the population grows. Periods of exponential growth may occur in some populations that exploit an unfilled environment or rebound from a catastrophic event.

FOCUS QUESTION   40.6

Define r_{max}.

Carrying Capacity A population may grow exponentially for only a short time before its increased density limits the resources available to its members. The **carrying capacity (K)** is the maximum population size that a particular environment can maintain. Crowding and resource limitations may lead to decreased per capita birth rates and increased per capita death rates.

The Logistic Growth Model The per capita rate of increase decreases from its maximum at low population size to zero as carrying capacity is reached. The mathematical model of **logistic population growth** is $dN/dt = r_{max}N(K - N)/K$. The equation includes the expression $(K - N)/K$ to reflect the impact of increasing N on the per capita rate of increase as the population approaches the carrying capacity.

When N is small, $(K - N)/K$ is close to 1, and growth is approximately exponential ($r_{max}N$). As population size approaches the carrying capacity, the $(K - N)/K$ term becomes a smaller fraction, and per capita rate of increase is small. When N reaches K, the term $(K - N)/K$ is 0, and the population stops growing. The logistic model produces an S-shaped growth curve, and maximum increase in population numbers occurs when N is intermediate in size.

FOCUS QUESTION 40.7

Label the following exponential and logistic growth curves, and write the equation associated with each curve. What is K for the population shown in curve b?

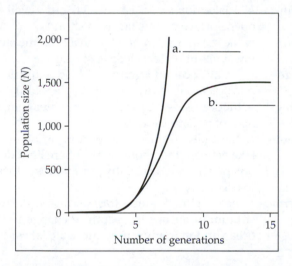

The Logistic Model and Real Populations Some laboratory populations of small animals and microorganisms show logistic growth. Natural populations often overshoot and then fluctuate around carrying capacity.

40.6 Population dynamics are influenced strongly by life history traits and population density

The **life history** of an organism involves traits that reflect tradeoffs between survival and reproduction. Life history variables include the age at first reproduction, how often an organism reproduces, and the number of offspring produced in each reproductive episode.

"Tradeoffs" and Life Histories Can organisms maximize all life history traits simultaneously? The costs of reproduction often include a reduction in survival. The production of large numbers of offspring is related to the selective pressures of high mortality rates of offspring in uncertain environments or from intense predation. Parental investments in the physical size of offspring (or seeds) and extended care increase the survival chances of offspring.

Natural selection favors different life history traits at different population densities. Populations at high density, close to their carrying capacity, may experience **K-selection** or density-dependent selection for traits such as competitive ability and efficient resource use. In environments in which population density is low, **r-selection** (or density-independent selection) would favor traits that maximize population growth, such as the production of numerous, small offspring.

FOCUS QUESTION 40.8

a. Explain why the life history of an organism cannot be to reproduce early and often, to have large numbers of offspring at a time, and to live a long life.

b. Indicate whether the following life history traits would be considered to be *r*-selected or *K*-selected:

early age at first reproduction; many small offspring produced

few, relatively large offspring produced every year

Population Change and Population Density A birth rate or a death rate that does not change as population density changes is said to be **density independent**. The death rate is **density dependent** if it rises with increasing population density; the birth rate is density dependent if it falls with increasing population density. An equilibrium density may be reached in a population as long as birth rate or death rate, or both, are density dependent.

Mechanisms of Density-Dependent Population Regulation As population density increases, density-dependent decreases in birth rate and increases in death rate operate through negative feedback to slow or stop population growth.

Competition for resources such as nutrients or food may limit reproductive output. The availability of territorial space may be the limiting resource for some animals. Increased population densities may affect the transmission rate of diseases. Predation may be a density-dependent factor when a predator feeds preferentially on a prey population that has reached a high density. The accumulation of toxic wastes may also be a limiting factor for some organisms.

Density-dependent regulation may also involve intrinsic factors. Studies of mice have shown that even when food or shelter is not limiting, aggressive interactions and hormonal changes inhibit reproduction when the population is at high density.

Population Dynamics   All populations show fluctuations in numbers. The study of such **population dynamics** looks at the biotic and abiotic factors that cause them. Fluctuations in populations of some large mammals, such as moose on Isle Royale, may be linked to several factors, such as the severity of winter and thus the availability of food and predation by wolves. Increasing pressure from parasites may also occur when population densities are high.

Emigration may increase with population density. A group of populations occupying separated suitable habitats may be linked through immigration and emigration to form a **metapopulation**. Emigrants may serve to repopulate habitats in which local populations become extinct.

FOCUS QUESTION   40.9

a. List some density-dependent factors that may limit population growth.

b. List some abiotic factors that may cause population fluctuations.

Word Roots

a-   = without; **bio-** = life (*abiotic:* nonliving; referring to physical and chemical properties of an environment)

bentho- = the depths of the sea (*benthic zone:* the bottom surface of an aquatic environment)

co-   = together (*cohort:* a group of individuals of the same age in a population)

demo-   = people; **-graphy** = writing (*demography:* the study of changes over time in the vital statistics of populations, especially birth and death rates)

estuar- = the sea (*estuary:* the area where a freshwater stream or river merges with the ocean)

eu-   = good, well; **troph-** = food, nourishment (*eutrophic lake:* a lake that has a high rate of productivity supported by a high rate of nutrient cycling)

hydro-   = water; **therm-** = heat (*deep-sea hydrothermal vent:* a dark, hot, oxygen-deficient environment associated with volcanic activity on or near the seafloor; the producers in a vent community are chemoautotrophic prokaryotes)

inter-   = between (*intertidal zone:* the shallow zone of ocean adjacent to land and between the high- and low-tide lines)

limn-   = a lake (*limnetic zone:* in a lake, the well-lit, open surface waters far from shore)

littor- = the seashore (*littoral zone:* in a lake, the shallow, well-lit waters close to shore)

oligo-   = small, scant (*oligotrophic lake:* a nutrient-poor, clear lake with few phytoplankton)

pelag-   = the sea (*pelagic zone:* the open-water component of aquatic biomes)

-photo   = light (*aphotic zone:* the part of an ocean or lake beneath the photic zone, where light does not penetrate sufficiently for photosynthesis to occur)

thermo- = heat; **-clin** = slope (*thermocline:* a narrow stratum of abrupt temperature change in the ocean and in many temperate-zone lakes)

Structure Your Knowledge

1. **a.** Define ecology.
    b. How does ecology relate to evolutionary biology?

2. **a.** What are biomes?
    b. What accounts for the similarities in life-forms found in the same type of biome in geographically separated areas?

3. Create a concept map to organize your understanding of the exponential and logistic equations—the mathematical models of population growth.

4. What is the best collection of life history traits that would maximize reproductive success?

Test Your Knowledge

MATCHING: *Match the biotic description with its biome.*

Biome

____ 1. chaparral
____ 2. desert
____ 3. savanna
____ 4. northern coniferous forest
____ 5. temperate broadleaf forest
____ 6. temperate grassland
____ 7. tropical rain forest
____ 8. tundra

Biotic Description

A. broad-leaved deciduous trees
B. lush growth, vertical layers
C. evergreen shrubs, fire-adapted vegetation
D. scattered trees, grasses, and forbs
E. tall stands of cone-bearing trees
F. low shrubby vegetation, lichens
G. grasses adapted to fire and drought
H. widely scattered shrubs, cacti, and euphorbs

MULTIPLE CHOICE: *Choose the one best answer.*

1. Which level of ecology considers the effects of predation, parasitism, and competition on species distribution?
   a. landscape
   b. community
   c. ecosystem
   d. organismal
   e. population

2. Ecologists often use mathematical models and computer simulations because
   a. ecological experiments are always too broad in scope to be performed.
   b. most ecologists are mathematicians.
   c. ecology is becoming a more descriptive science.
   d. these approaches allow ecologists to study the interactions of multiple variables and to simulate large-scale experiments.
   e. variables can be manipulated with computers but cannot be manipulated in field experiments.

3. The ample rainfall of the tropics and the arid areas around 30° north and south latitudes are caused by
   a. ocean currents that flow clockwise in the northern hemisphere and counterclockwise in the southern hemisphere.
   b. the global circulation of air initiated by intense solar radiation near the equator that produces wet and warm air.
   c. the tilting of Earth on its axis and the resulting seasonal changes in climate.
   d. the heavier rain on the windward side of mountain ranges and the drier climate on the leeward side.
   e. the location of tropical rain forests and deserts.

4. Which of the following would be most affected if Earth were not tilted on its axis relative to its orbit around the sun?
   a. the number of hours in a day
   b. the number of days in a year
   c. the distinct seasons in temperate and polar regions
   d. the amount of solar radiation that the tropics receives
   e. global air circulation and precipitation patterns

5. Why do the tropics and the windward side of mountains receive more rainfall than areas around 30° latitude or the leeward side of mountains?
   a. Rising air expands, cools, and drops its moisture.
   b. Descending air condenses and drops its moisture.
   c. The tropics and the windward side of mountains are closer to the ocean.
   d. Solar radiation is greater in the tropics and on the windward side of mountains.
   e. The rotation of Earth determines global wind patterns.

6. The permafrost of the arctic tundra
   a. prevents plants from getting established and growing.
   b. protects small animals during the long winters.
   c. anchors plant roots in the frozen soil, helping them withstand the area's high winds.
   d. prevents plant roots from penetrating deep into the soil.
   e. Both **b** and **c** are correct.

7. Many plant species have adaptations for dealing with the periodic fires typical of a
   a. savanna.
   b. chaparral.
   c. temperate grassland.
   d. temperate broadleaf forest.
   e. a, b, and c.

8. The best way to explain how two communities can have the same annual mean temperature and rainfall but very different biota and characteristics is that the communities
   a. are found at different altitudes.
   b. are composed of species that have very low dispersal rates.
   c. are found on different continents.
   d. receive different amounts of sunlight.
   e. have different seasonal temperatures and patterns of rainfall throughout the year.

9. Phytoplankton are the basis of the food chain in
   a. the headwaters of streams.
   b. wetlands.
   c. the oceanic photic zone.
   d. rocky intertidal zones.
   e. deep-sea hydrothermal vents.

10. Which of the following aquatic zones is *incorrectly* paired with its description?
    a. aphotic zone—region of lake or ocean where there is not sufficient light penetration for photosynthesis
    b. benthic zone—bottom surface
    c. limnetic zone—shallow waters close to shore in a lake
    d. intertidal zone—shallow area of the ocean adjacent to land
    e. pelagic zone—area of open water

11. In which of the following biomes would you expect to find organisms with adaptations for tolerating changes in salinity?
    a. desert
    b. wetland
    c. deep-sea hydrothermal vent
    d. estuary
    e. coral reef

12. Which of the following factors affect the distribution of a species?
    a. dispersal ability
    b. interactions with other organisms
    c. climate
    d. availability of water, oxygen, and other physical factors
    e. All of the above factors influence where species are found.

13. An experiment that removed a species from an area would enable an ecologist to test
    a. whether moving the species to a new, varied habitat could lead to its successful transplantation.
    b. whether competition or predation from that species was limiting the local distribution or abundance of other species.
    c. whether dispersal ability was limiting the range of the species.
    d. whether abiotic factors were limiting the dispersal of the species.
    e. whether the species was capable of recolonizing the area.

14. In an area with a heterogeneous distribution of suitable habitats, the dispersion pattern of a population is probably
    a. clumped.
    b. uniform.

   c. random.
   d. unpredictable.
   e. dense.

15. Which of the following statements about life tables is *not* true?
    a. They were first used by life insurance companies to estimate survival patterns.
    b. They show the age-specific death rate for a population.
    c. They are used to predict logistic growth.
    d. They can be used to construct survivorship curves.
    e. They are often constructed by following a cohort from birth to death.

16. A Type I survivorship curve is level at first, with a rapid increase in mortality in old age. This type of curve is
    a. typical of many invertebrates that produce large numbers of offspring.
    b. typical of humans and other large mammals.
    c. found most often in *r*-selected populations.
    d. almost never found in nature.
    e. typical of most species of birds.

17. The middle of the S-shaped growth curve in the logistic growth model
    a. shows that at middle densities, individuals of a population do not affect each other.
    b. is best described by the term rN.
    c. is the point where population growth begins to increase.
    d. is the period when competition for resources is the highest.
    e. is the period when the population growth rate is the highest.

18. The term $(K - N)/K$
    a. is the carrying capacity for a population.
    b. is greatest when K is very large.
    c. is zero when population size equals carrying capacity.
    d. increases in value as N approaches K.
    e. accounts for the overshoot of carrying capacity.

19. The carrying capacity for a population is estimated at 500; the population size is currently 400; and r_{max} is 0.1. What is dN/dt?
    a. 0.01
    b. 0.8
    c. 8
    d. 40
    e. 50

20. In order to maintain the largest sustainable fish harvest, fishing efforts should
    a. take only postreproductive fish.
    b. maintain the population close to its carrying capacity.

c. reduce the population to a very low number to take advantage of exponential growth.

d. maintain the population density close to ½ K.

e. be prohibited.

21. Immigration and emigration are likely to play a role in population dynamics in
    a. metapopulations.
    b. exponential growth.
    c. populations experiencing *r*-selection.
    d. logistic growth.
    e. territoriality.

22. Which of the following factors is *not* a density-dependent factor limiting a population's growth?
    a. increased specialization by a predator
    b. a limited number of available nesting sites
    c. a stress syndrome that alters hormone levels
    d. a very early fall frost
    e. intraspecific competition

In questions 23 through 25, use the following choices to indicate how these factors would be affected by the described changes.
    a. an increase
    b. a decrease
    c. no change
    d. the effect cannot be predicted

23. In a population showing exponential growth, what change would occur in dN/dt with an increase in N?

24. As a population approaches zero population growth, what change would occur in the per capita rate of increase (r)?

25. For an *r*-selected population regulated by density-independent factors, what change would occur in the population growth rate if the carrying capacity of its environment were increased?

Chapter 41

Species Interactions

Chapter Focus

Communities are composed of populations of various species that may interact through competition, predation, herbivory, symbiosis, or facilitation. The structure of a community—its species composition and relative abundance—is determined by these interactions between species. Disturbances keep most communities in a state of nonequilibrium. Species diversity relates to community size and geographic location. Identifying the hosts and vectors of pathogens helps to combat zoonotic diseases.

Chapter Review

The collection of populations of different species living close enough to interact is called a biological **community**. Ecologists study the interactions between species and the factors involved in determining a community's structure—the composition and the relative abundance of its species.

41.1 Interactions within a community may help, harm, or have no effect on the species involved

Interspecific interactions occur between the different species living in a community. The positive or negative effect of these interactions on the survival and reproduction of a population can be signified by + or – signs.

Competition   What if populations of two species use the same limited resource? Such **interspecific competition** may negatively affect both populations.

G. F. Gause's laboratory experiments showed that two species of *Paramecium* that rely on the same limited resource could not coexist in the same community. The **competitive exclusion** principle predicts that the less efficient competitor will be eliminated locally.

An organism's **ecological niche** is described as its role in an ecosystem—its use of biotic and abiotic resources. The competitive exclusion principle holds that two species with identical niches cannot coexist permanently in a community.

Resource partitioning, slight variations in niche that allow ecologically similar species to coexist, provides indirect evidence that competition was a selection factor in the evolution of a species' niche. Due to competition, a species' *realized niche* might be smaller than its *fundamental niche*. **Character displacement** of some morphological trait or in resource use enables closely related sympatric species to avoid competition. When these species are allopatric (geographically separate), their differences may be much smaller.

FOCUS QUESTION   41.1

Whenever these two spiny mouse species coexist, *Acomys cahirinus* is nocturnal, whereas *A. russatus* is active during the day. When all *A. cahirinus* were removed from a research site, diurnal *A. russatus* mice became nocturnal. How were resources partitioned between these species when they coexisted, and what do these results indicate about the niche of *A. russatus*?

Predation   **Predation** involves a predator killing and eating prey. Predator adaptations include acute senses, speed and agility, camouflage coloration, and physical structures such as claws, fangs, teeth, and stingers.

Animals can defend against predation by hiding, fleeing, or forming herds or schools. Potential prey may use camouflage in the form of **cryptic coloration**. Mechanical and chemical defenses discourage predation. Some animals can synthesize toxins or passively accumulate them from the food they eat. Bright, conspicuous, **aposematic coloration** in prey species "warns" predators that the prey possesses chemical defenses. Prey may use mimicry to exploit the warning coloration of other species.

Name the type of mimicry described in each of the following descriptions:

a. harmless species resembling a poisonous or distasteful species

b. mutual imitation by two or more unpalatable species

Herbivory In **herbivory**, an herbivore eats parts of a plant or alga. Most herbivores are small invertebrates such as insects, which may have chemical sensors that recognize nontoxic and nutritious plants. Herbivores may have teeth or digestive systems adapted for processing vegetation.

Plants may defend themselves with mechanical devices, such as thorns, or chemical compounds. Distasteful or toxic chemicals include such well-known compounds as strychnine, nicotine, and various spices.

Symbiosis A relationship between organisms of two species that live in direct contact is defined as **symbiosis**. What are the three types of symbiotic relationships?

In **parasitism**, a **parasite** obtains its nourishment from its **host**. Parasites that live within a host are called **endoparasites**; those that feed on the surface of a host are called **ectoparasites**. Parasitoid insects lay eggs on or in hosts, on which their larvae then feed.

Parasites may have complex life cycles with a number of hosts. Parasites can have a substantial effect on the density of their host population.

In **mutualism**, interactions between species benefit both participants. Mutualistic interactions may involve the coevolution of related adaptations in both species.

In **commensalism**, only one member appears to benefit from the interaction. Examples include "hitchhiking" species and species that feed on food incidentally exposed by another species.

Facilitation In **facilitation**, a species positively affects the survival and reproduction of other species without the intimate association of symbiosis. This type of interaction is common in plant ecology, in which one plant species improves the soil in ways that benefit other species.

Name and give examples of the interspecific interactions symbolized in the table.

	Interspecific Interaction	Examples
– / –	a.	
+ / +	b.	
+ / 0	c.	
+ / –	d.	
+ / –	e.	
+ / –	f.	
+ / + or 0/ +	g.	

41.2 Diversity and trophic structure characterize biological communities

Species Diversity The **species diversity** of a community is determined both by **species richness**, the number of different species present, and by **relative abundance**, the proportional abundance of the different species.

A widely used quantitative measure of diversity is the **Shannon diversity index (*H*)**, which is based on species richness and relative abundance:

$$H = -(p_A \ln p_A + p_B \ln p_B + p_C \ln p_C + \ldots)$$

where *p* is the proportion of each species (A, B, C, and so on) in the community, and ln is the natural log. Thus, Shannon diversity index (*H*) for a community is the negative sum of $p \ln p$ for all species present.

Tide pool 1 has three species of sea urchins with the following numbers: A = 8, B = 6, C = 6; tide pool 2 has four species of sea urchins with the following numbers: A = 14, B = 2, C = 2, D = 2.

a. Compute the Shannon diversity index of the sea urchin communities in the two pools.

b. Which pool has the greater species richness? Which community is the more diverse according to the Shannon diversity index?

Estimating the number and relative abundance of species requires various sampling techniques and may be difficult due to the rarity of most species in a community. The diversity and richness of bacterial communities may be determined using molecular tools such as RFLP analysis.

Diversity and Community Stability   Long-term experiments on plant diversity have shown that increased diversity correlates with increased productivity (greater **biomass**) and stability of a community. The tunicate experiment in Long Island Sound indicated that more diverse communities use more of the available resources, making it more difficult for an **invasive species** to become established outside its native range.

Trophic Structure   The feeding relationships in a community determine its **trophic structure**. A **food chain** shows the transfer of food energy from one trophic level to the next: from producers to herbivores (primary consumers) to carnivores (secondary, tertiary, or quaternary consumers). The decomposers in a community feed on all trophic levels.

A **food web** diagrams the complex trophic relationships within a community. The complicated connections of a food web arise because many consumers feed at various trophic levels. Food webs can be simplified by grouping species into functional groups (such as primary consumers) or by isolating partial food webs.

Species with a Large Impact   A **dominant species** in a community has the greatest abundance or largest biomass and is a major influence on the presence and distribution of other species. A species may become dominant due to its more competitive use of resources or its success at avoiding predation or disease. Invasive species may reach high biomass levels due to the lack of natural predators and pathogens.

A **keystone species** has a large impact on community structure as a result of its ecological role. Paine's study of a predatory sea star demonstrated its role in maintaining species richness in an intertidal community by reducing the density of mussels, a highly competitive prey species.

Ecosystem engineers or "foundation species" influence community structure by changing the physical environment.

Bottom-Up and Top-Down Controls   Arrows can be used to indicate the effect an increase in the biomass in one trophic level has on another trophic level. $V \rightarrow H$ indicates that an increase in vegetation (V) increases the number of herbivores (H); $V \leftarrow H$ indicates that an increase in herbivores decreases vegetation biomass. $V \leftrightarrow H$ means that each trophic level is affected by changes in the other.

What are the two models of community organization? According to the **bottom-up model**, $N \rightarrow V \rightarrow H \rightarrow P$, an increase in mineral nutrients (N) yields an increase in biomass at each succeeding trophic level: vegetation, herbivores, and predators (P). The **top-down model**, $N \leftarrow V \leftarrow H \leftarrow P$, assumes that predation controls community organization, with a series of +/− effects cascading down the trophic levels. According to this *trophic cascade model*, an increase in predator abundance decreases herbivore abundance, which results in increased vegetation and lowered nutrient levels.

Many freshwater communities appear to be organized along the top-down model. What actions might ecologists take if they wanted to use **biomanipulation** to control excessive algal blooms in a lake with four trophic levels (algae, zooplankton, primary predator fish, and top predator fish)?

41.3   Disturbance influences species diversity and composition

Traditionally, biological communities were viewed as existing in a state of equilibrium maintained by interspecific interactions. The ability of a community to reach and maintain this *climax community* with its relatively constant species composition is known as *stability*. The **nonequilibrium model**, in contrast, emphasizes that communities are constantly changing as a result of **disturbances**, such as fire, drought, storms, overgrazing, or human activities.

Characterizing Disturbance   According to the **intermediate disturbance hypothesis**, moderate intensity or frequency of disturbance may result in greater species diversity than either low or high levels of disturbance. Small-scale disturbances may enhance environmental patchiness, helping to maintain species diversity.

Ecological Succession   What happens in a community following a severe disturbance? The process of sequential transitions in species composition is known as **ecological succession**. If no soil was originally present, as on a new volcanic island or on the moraine left by a retreating glacier, the process is called **primary succession**. **Secondary succession** occurs when a community is disrupted by fire, logging, or farming, but the soil remains intact.

Primary succession may take hundreds or thousands of years. A series of colonizers usually begins with autotrophic and heterotrophic prokaryotes and protists and then moves through lichens, mosses, grasses, shrubs, and trees until the community reaches its prevalent form of vegetation.

Early colonizers may either *facilitate* the arrival of other species by improving the environment or *inhibit* the establishment of later species. Species may also be independent in their colonization and *tolerate* the conditions created by early species.

Ecologists have studied moraine succession during the retreat of glaciers at Glacier Bay in Alaska. The first pioneering plant species include mosses and fireweed; *Dryas,* a mat-forming shrub, dominates after about 30 years. Alder thickets are the dominant vegetation a few decades later, followed by Sitka spruce. The community becomes a spruce-hemlock forest by the third century after glacial retreat, except in flat, poorly drained areas, where sphagnum mosses invade. These mosses make the soil waterlogged and acidic, killing the trees and creating sphagnum bogs.

FOCUS QUESTION   41.6

Describe the effects of the alder stage on soil fertility during the succession following the retreat of a glacier.

Human Disturbance   Human activities, such as converting land to agriculture, logging, clearing for urban development, and ocean trawling, have altered the structure of communities all over the world. A common result is a reduction in species diversity.

41.4   Biogeographic factors affect community diversity

Latitudinal Gradients   Surveys of plant and animal species have documented much greater numbers of species in tropical habitats than in temperate and polar regions. Tropical communities are older, partly because of their longer growing season and partly because they have not had to "start over" after glaciation, as has been the case several times for many polar and temperate communities.

Solar radiation and precipitation are important climatic explanations for the latitudinal gradient in diversity. **Evapotranspiration**, the amount of water evaporated from soil and transpired by plants, is determined by solar energy, temperature, and water availability. *Potential evapotranspiration* is determined by just solar radiation and temperature. Evapotranspiration rates have been shown to correlate with species richness for both plants and animals.

FOCUS QUESTION   41.7

Why would the fact that tropical communities are "older" than temperate or polar communities contribute to greater species diversity?

Area Effects   A **species-area curve** represents the correlation between the geographic size of a community and the number of species in it. In general, the larger the area, the greater is the diversity of habitats, and the greater the species richness. Use of such curves in conservation biology can enable scientists to predict how a loss of habitat may affect diversity.

Any habitat surrounded by a significantly different habitat is considered an "island" and allows ecologists to study factors that affect species diversity. In the 1960s, R. MacArthur and E. O. Wilson developed an *island equilibrium model*, which states that the size of the island and its closeness to the mainland (or source of dispersing species) are important variables directly correlated with the number of species. When the rates of immigration and extinction become equal, species diversity reaches an equilibrium, although species composition may continue to change.

FOCUS QUESTION   41.8

a. How do the rates of immigration and extinction change as the number of species on an island increases?

b. In what physical ways would an island with a high number of species probably differ from one with a low number of species?

41.5 Pathogens alter community structure locally and globally

Effects on Community Structure Pathogens are disease-causing organisms and viruses. They can be particularly virulent when introduced into a new habitat, where host species have not yet evolved resistance or where natural controls are lacking. Pathogens have altered terrestrial ecosystems when dominant tree species are killed and are changing the community structure in coral reef communities.

Community Ecology and Zoonotic Diseases Many human diseases are caused by **zoonotic pathogens**, which are disease-causing agents transferred to humans from other animals, often by means of a **vector** such as ticks or mosquitoes. Understanding species interactions and identifying both hosts and vectors help to combat and control the spread of such diseases. Researchers have identified two inconspicuous shrew species as the primary host of the Lyme pathogen. Infected ticks then transmit Lyme disease to people. Global travel and trade contributes to the spread of pathogens.

FOCUS QUESTION 41.9

Why are ecologists trapping and testing migrating birds in Alaska?

Word Roots

crypto- = hidden, concealed (*cryptic coloration:* camouflage that makes a potential prey difficult to spot against its background)

ecto- = outer (*ectoparasite:* a parasite that feeds on the external surface of a host)

endo- = inner (*endoparasite:* a parasite that lives within a host)

herb- = grass; **-vora** = eat (*herbivory:* an interaction in which an organism eats parts of a plant or alga)

inter- = between (*interspecific competition:* competition for resources between individuals of two or more species when resources are in short supply)

mutu- = reciprocal (*mutualism:* a symbiotic relationship in which both participants benefit)

Structure Your Knowledge

1. Complete the following concept map to organize your understanding of the important factors that structure a community.

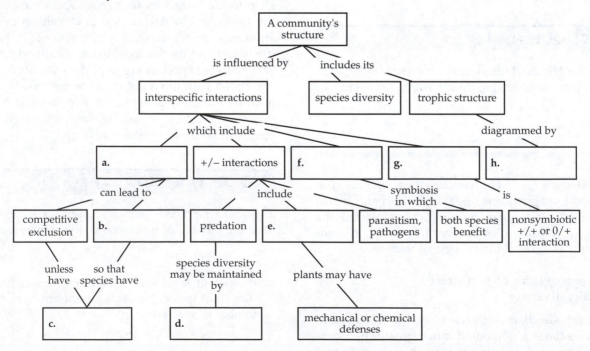

2. Community ecologists develop models or hypotheses to describe community structure and the factors that contribute to such structure. Briefly explain the following models described in this chapter.

    a. Competitive exclusion

    b. Bottom-up model

    c. Top-down (trophic cascade) model

    d. Nonequilibrium model

    e. Intermediate disturbance hypothesis

    f. Island equilibrium model

Test Your Knowledge

MULTIPLE CHOICE: *Choose the one best answer.*

1. Two allopatric species of Galápagos finches have beaks of similar size. The observation that significant differences in beak size occur when the two species occur on the same island is an example of

    a. competitive exclusion.

    b. species diversity.

    c. commensalism.

    d. character displacement.

    e. a trophic cascade.

2. Aposematic coloration is most commonly found in

    a. prey whose body morphology is cryptic.

    b. predators who are able to sequester toxic plant compounds in their bodies.

    c. prey species that have chemical defenses.

    d. palatable prey that resemble each other.

    e. plants that have toxic secondary compounds.

3. Two species, A and B, occupy adjoining environmental patches that differ in several abiotic factors. When species A is experimentally removed from a portion of its patch, species B colonizes the vacated area and thrives. When species B is experimentally removed from a portion of its patch, species A does not successfully colonize the area. From these results you might conclude that

    a. both species A and species B are limited to their range by abiotic factors.

    b. species A is limited to its range by competition, and species B is limited by abiotic factors.

    c. both species are limited to their range by competition.

    d. species A is limited to its range by abiotic factors, and species B is limited to its range because it cannot successfully compete with species A.

    e. species A is a predator of species B.

4. Through resource partitioning,

    a. two species can compete for the same prey item.

    b. slight variations in niche allow closely related species to coexist in the same habitat.

    c. two species can share identical niches in a habitat.

    d. competitive exclusion results in the success of the superior species.

    e. two species with identical niches do not share the same habitat and thus avoid competition.

5. A palatable (good-tasting) prey species may defend against predation by

    a. Müllerian mimicry.

    b. Batesian mimicry.

    c. producing secondary compounds.

    d. cryptic coloration.

    e. either **b** or **d**.

6. The species richness of a community refers to

    a. the relative numbers of individuals in each species.

    b. the number of different species found in the community.

    c. the feeding relationships or trophic structure within the community.

    d. the species diversity of that community.

    e. the community's stability or ability to persist through disturbances.

7. Which of the following organisms is *mismatched* with its trophic level?

    a. algae—producer

    b. phytoplankton—primary consumer

    c. carnivorous fish larvae—secondary consumer

    d. eagle—tertiary or quaternary consumer

    e. Both **b** and **c** are mismatched.

8. Ecologists survey the tree species in two forest plots. Plot 1 has six different species, and 95% of all trees belong to just one species. Plot 2 has five different species, each of which is represented by approximately 20% of the trees. Compared with plot 1, plot 2 has

    a. higher species richness.

    b. greater species diversity.

    c. lower relative abundance.

    d. lower species richness.

    e. Both **b** and **d** are correct.

9. When one species was removed from a tide pool, the pool's species diversity was significantly reduced. The removed species was probably

    a. a strong competitor.

    b. a potent parasite.

    c. the dominant species.

    d. a keystone species.

    e. an ecosystem engineer.

10. Invasive species often reach a large biomass because
    a. they are better competitors than native species.
    b. they are usually producers and not top predators.
    c. they often lack natural predators or pathogens.
    d. their superior ability to disperse enables them to spread to new niches.
    e. they are often protected by the humans who have introduced them.

11. According to the top-down (trophic cascade) model of community control, which trophic level would you *decrease* if you wanted to *increase* the vegetation level in a community?
    a. nutrients
    b. vegetation
    c. secondary consumers (carnivores)
    d. tertiary consumers
    e. omnivores

12. During succession, inhibition by early species
    a. may prevent the achievement of a stable community.
    b. may slow down both the rate of immigration and the rate of extinction, depending on the size of the area and the distance from the source of dispersing species.
    c. results from the frequent disturbances that often eliminate early colonizers.
    d. may slow down the successful colonization by other species.
    e. may involve changes in nitrogen levels or accelerated accumulation of humus.

13. Which of the following organisms is *mismatched* with its community role?
    a. beaver—ecosystem engineer
    b. a black rush *Juncus* in a salt marsh—facilitator
    c. a sea star in an intertidal community—keystone predator
    d. trees in a spruce-hemlock forest—dominant species
    e. alder and *Dryas* (a mat-forming shrub)—inhibitors

14. A major explanation for the decline in species richness along an equatorial-polar gradient is the correlation of high levels of solar radiation and water availability with diversity. Which of the following factors is also thought to contribute to the high species richness of tropical communities?
    a. the inverse relationship between diversity and evapotranspiration
    b. the greater age of these communities (longer growing season and fewer climatic setbacks), providing more time for speciation events
    c. the larger area of the tropics and corresponding richness predicted by the species-area curve
    d. the lack of disturbances in tropical areas
    e. the greater immigration rate and lower extinction rate found on large tropical islands

15. Which of the following descriptions *best* describes a zoonotic pathogen?
    a. a pathogen that affects insects
    b. a pathogen that requires a vector to spread from animal to animal
    c. a disease-causing agent that is transmitted to humans from other animals
    d. a pathogen that is common in zoos due to the unnatural habitat for animals in zoos
    e. an ectoparasite that is transferred from animals to humans

Chapter 42

Ecosystems and Energy

Chapter Focus

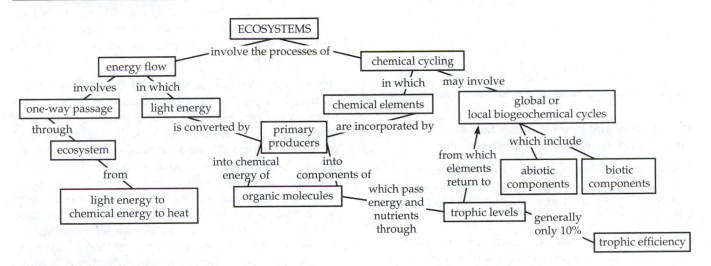

```
                        ┌─────────────┐
                        │ ECOSYSTEMS  │
                        └─────────────┘
              involve the processes of
   ┌─────────────┐                      ┌──────────────────┐
   │ energy flow │                      │ chemical cycling │
   └─────────────┘                      └──────────────────┘
   involves    in which                  in which    may involve
┌──────────────┐  ┌──────────────┐  ┌──────────────────┐  ┌─────────────────────┐
│ one-way      │  │ light energy │  │ chemical elements│  │ global or           │
│ passage      │  └──────────────┘  └──────────────────┘  │ local biogeochemical│
└──────────────┘                                          │ cycles              │
 through      is converted by    are incorporated by      └─────────────────────┘
┌───────────┐     ┌────────────┐                           which include
│ ecosystem │     │ primary    │                    from which
└───────────┘     │ producers  │                    elements  ┌────────────┐ ┌───────────┐
  from            └────────────┘                    return to │ abiotic    │ │ biotic    │
                into chemical   into                          │ components │ │ components│
┌────────────────┐ energy of    components of                 └────────────┘ └───────────┘
│ light energy to│     ┌──────────────────┐  which pass
│ chemical energy│     │ organic molecules│  energy and  ┌────────────────┐
│ to heat        │     └──────────────────┘  nutrients   │ trophic levels │
└────────────────┘                           through     └────────────────┘
                                                           generally  only 10%
                                                                 ┌────────────────────┐
                                                                 │ trophic efficiency │
                                                                 └────────────────────┘
```

Chapter Review

An **ecosystem** consists of all the organisms in a given area and the abiotic factors with which they interact. Most ecosystems are powered by energy from sunlight, which is transformed to chemical energy by autotrophs, passed to heterotrophs in the organic compounds of food, and continually dissipated in the form of heat. Chemical elements are cycled among the abiotic and biotic components of the ecosystem as autotrophs incorporate them into organic compounds and the processes of metabolism and decomposition return them to the soil, air, and water.

42.1 Physical laws govern energy flow and chemical cycling in ecosystems

Conservation of Energy According to the first law of thermodynamics, energy can neither be created nor destroyed, only transferred or transformed. Energy flows through ecosystems from its input as solar radiation to its conversion into chemical energy to its dissipation as heat. The second law of thermodynamics states that every energy exchange increases entropy. Thus, in each energy conversion, some energy is lost as heat.

Conservation of Mass The **law of conservation of mass** states that matter cannot be created or destroyed. Ecosystem ecologists measure the inputs and outputs of elements as well as chemical cycling within ecosystems.

FOCUS QUESTION 42.1

What may happen if the input of a mineral nutrient into an ecosystem is less than the output of that element from the ecosystem?

minerals will limit production

Energy, Mass, and Trophic Levels Most **primary producers,** or autotrophs, use light energy to synthesize organic compounds, which they use as fuel and building materials. Heterotrophs depend on autotrophs for their organic compounds. **Primary consumers** are herbivores; **secondary consumers** are carnivores that eat herbivores. **Tertiary consumers** eat other carnivores.

Detritivores, or **decomposers,** consume **detritus,** which includes organic wastes, fallen leaves, and dead organisms. Detritivores themselves are eaten by consumers. Fungi and prokaryotes are important detritivores in most ecosystems, converting organic materials from all trophic levels to inorganic compounds that can be recycled by autotrophs.

FOCUS QUESTION 42.2

Compare the movement of energy and chemical elements in ecosystems.

[handwritten: through trophic levels (oneway trip) abiotic & biotic factors]

42.2 Energy and other limiting factors control primary production in ecosystems

Primary production is the amount of light energy converted to chemical energy during a period of time—that is, the photosynthetic output of an ecosystem's autotrophs.

Ecosystem Energy Budgets Only a small portion of incoming solar radiation strikes photosynthetic organisms; of that, only about 1% is converted to chemical energy. Nevertheless, worldwide photosynthetic production is about 150 billion metric tons of organic material per year.

Net primary production (NPP), which represents the chemical energy available to consumers in an ecosystem, is equal to an ecosystem's **gross primary production (GPP)** minus the energy used by autotrophs in their cellular respiration (R_a): NPP = GPP $-$ R_a. NPP is usually about half of GPP.

Net primary production can be expressed as energy per unit area per unit time ($J/m^2 \cdot yr$) or as new biomass produced, expressed as dry weight of vegetation ($g/m^2 \cdot yr$). *Standing crop* is the total biomass of photosynthetic organisms in an ecosystem.

Because photosynthesizing vegetation absorbs more wavelengths of visible light, scientists can estimate primary production using data from satellites that compare the wavelengths of visible and near-infrared light reflected from Earth's surface. Primary production and the contribution to Earth's total production vary by ecosystem.

Net ecosystem production (NEP), which reflects *total biomass accumulation* in an ecosystem, deducts the energy used in respiration by all autotrophs and

heterotrophs (R_T) from gross primary production: NEP = GPP $-$ R_T. Scientists estimate NEP by measuring the net flux of CO_2 from terrestrial ecosystems, or the net flux of CO_2 or O_2 in oceans.

Primary Production in Aquatic Ecosystems The depth to which light penetrates and the availability of nutrients affect primary production in oceans and lakes. Nitrogen or phosphorus is often the **limiting nutrient** in the photic zone of the ocean. Nutrient enrichment experiments can identify, for example, whether nitrogen or phosphate pollution is causing algal "blooms," or which nutrient limits phytoplankton growth in nonproductive areas of the ocean.

In freshwater lakes, nutrient limitation also affects production. **Eutrophication,** the rapid growth of cyanobacteria and algae in response to increased nutrients, has been linked to phosphorus pollution from sewage and fertilizer runoff.

Primary Production in Terrestrial Ecosystems There is usually a positive correlation between primary production and precipitation. Another predictor of primary production—*actual evapotranspiration,* the annual amount of water evaporated from a landscape and transpired by plants—reflects both precipitation and temperature.

Nutrients affect primary production, and nitrogen and/or phosphorus are the limiting nutrients in many terrestrial ecosystems. Plant adaptations that increase uptake of nutrients include symbiotic relationships between roots and nitrogen-fixing bacteria, as well as mycorrhizal associations with fungi that provide phosphorus and other elements to plants. Plant root hairs increase absorptive surface area, and roots release substances that increase the availability of nutrients.

FOCUS QUESTION 42.3

a. List some ecosystems with high primary production.

b. List some ecosystems with low primary production.

c. The oceans have low primary production yet contribute as much global net primary production as terrestrial ecosystems. Explain.

d. Areas of the Antarctic Ocean are often more productive than tropical seas, even though they are colder and receive lower light intensity. Explain.

42.3 Energy transfer between trophic levels is typically only 10% efficient

What is the difference between primary production and secondary production? The first is the conversion of light energy to chemical energy by primary producers. **Secondary production** is the amount of energy contained in consumers' food that is converted to new biomass during a given time.

Production Efficiency Herbivores consume only a fraction of the plant material produced; they cannot digest all they eat; and much of the energy they do absorb is used for cellular respiration. Only the chemical energy stored as growth or in offspring is available as food to secondary consumers. The proportion of assimilated food energy (not including losses in feces) that is used for net secondary production (growth and reproduction) is a measure of the efficiency of energy transformation: **production efficiency** = (net secondary production × 100%)/assimilation of primary production. Production efficiencies vary from 1–3% for endothermic birds and mammals, to around 40% for insects and microorganisms.

Trophic Efficiency and Ecological Pyramids **Trophic efficiency** is the percentage of the production of one trophic level that makes it to the next level; it typically is around 10%. A *pyramid of net production* shows this multiplicative loss of energy.

A *biomass pyramid* illustrates the standing crop biomass of organisms at each trophic level. This pyramid usually narrows rapidly from producers to the top trophic level. Some aquatic ecosystems have inverted biomass pyramids in which zooplankton (consumers) outlive and outweigh the highly productive, but heavily consumed, phytoplankton. Phytoplankton have a short **turnover time,** the time required to replace a population's standing crop; turnover time is determined by dividing standing crop biomass by production. The production pyramid for such an aquatic ecosystem, however, is normal in shape.

FOCUS QUESTION 42.4

a. Why is production efficiency higher for insects than for birds and mammals?

b. Assuming a 10% trophic efficiency (transfer of energy to the next trophic level), approximately what proportion of the chemical energy produced in photosynthesis makes it to a tertiary consumer?

42.4 Biological and geochemical processes cycle nutrients and water in ecosystems

Chemical elements are passed between abiotic and biotic components of ecosystems through **biogeochemical cycles**.

Decomposition and Nutrient Cycling Rates Temperature and the availability of water influence decomposition rates; thus, nutrient cycling times vary in different ecosystems. For example, nutrients cycle rapidly in a tropical rain forest; the soil contains only about 10% of the ecosystem's nutrients. Decomposition is slower in temperate forests, and 50% of the ecosystem's organic material may be found in detritus and soil. Decomposition rates are slow where oxygen is limited, such as in peat lands and aquatic sediments.

Biogeochemical Cycles What are the two general categories of **biogeochemical cycles**? Gaseous forms of carbon, oxygen, sulfur, and nitrogen have global cycles. Heavier elements, such as phosphorus, potassium, and calcium, have a local cycle, especially in terrestrial ecosystems.

Ecologists study the movement of elements by tracking added radioactive isotopes or by following naturally occurring nonradioactive isotopes through ecosystems.

Exploring Water and Nutrient Cycling The water cycle involves solar-energy-driven evaporation from oceans and evapotranspiration from land, condensation of water vapor into clouds, and precipitation. Runoff and groundwater flow return water to the oceans.

In the carbon cycle, producers use CO_2 from the atmosphere in photosynthesis, producing organic compounds that are used by producers and consumers. Organisms release CO_2 in cellular respiration. Fossil fuel combustion is increasing atmospheric CO_2. Reservoirs of carbon include fossil fuels, dissolved carbon compounds in oceans, plant and animal biomass, CO_2 in the atmosphere, and sedimentary rocks such as limestone.

Plants require nitrogen in the form of ammonium (NH_4^+) or nitrate (NO_3^-). Animals obtain nitrogen in organic form from plants or other animals. The major reservoir is the atmosphere, and nitrogen enters ecosystems through *nitrogen fixation*, as when soil bacteria convert N_2 into compounds that can be assimilated by plants (or through lightning and volcanic activity). Some bacteria reduce nitrate and return N_2 to the atmosphere through denitrification. Fertilizers and the planting of legume crops now add more nitrogen to ecosystems than do natural processes. Large quantities of reactive nitrogen gases are released to the atmosphere by human activities.

Weathering of rock adds phosphorus to the soil in the form of $PO_4^?$, which is absorbed by plants. Organic phosphate is transferred from plants to consumers and returned to the soil through the action of decomposers or by animal excretion. Soil particles bind phosphate, keeping it available locally for recycling. Sedimentary rocks of marine origin are the largest reservoirs.

FOCUS QUESTION   42.5

Describe the biological importance of water, carbon, nitrogen, and phosphorus.

Case Study: Nutrient Cycling in the Hubbard Brook Experimental Forest   A team of scientists has looked at nutrient cycling in the Hubbard Brook Experimental Forest ecosystem since 1963. The mineral budget for each of six valleys was determined by measuring the input of key nutrients in rainfall and their outflow through the creek that drained each watershed. Most minerals were recycled within the forest ecosystem.

The effect of deforestation on nutrient cycling was measured for 3 years in a valley that was completely logged. Compared with a control area, water runoff from the deforested valley increased 30–40%. Nitrate increased in concentration in the creek 60-fold, removing this critical soil nutrient and contaminating the water.

FOCUS QUESTION   42.6

a. In which natural ecosystem do nutrients cycle the fastest? Why?

b. In which natural ecosystems do nutrients cycle slowly? Why?

c. What is the effect of loss of vegetation on nutrient cycling?

42.5   Restoration ecologists help return degraded ecosystems to a more natural state

Areas degraded by farming, mining, or environmental pollution are often abandoned. In order to speed the successional processes involved in a community's recovery, restoration ecologists attempt to identify and manipulate the factors that most limit recovery time. In some cases, the physical structure of a site must first be reconstructed.

Bioremediation   **Bioremediation** uses prokaryotes, fungi, or plants to detoxify polluted ecosystems. Some plants are able to extract metals from contaminated soils; harvesting those plants removes the metals from the ecosystem. Some prokaryotes can metabolize dangerous elements to less soluble forms.

Biological Augmentation   **Biological augmentation** uses organisms to add essential substances to degraded ecosystems. Restoration ecologists may also reintroduce animals to restored sites.

Restoration Projects Worldwide   Restoration ecologists often apply adaptive management, in which they experiment with promising management approaches and learn as they work in each unique and complex disturbed ecosystem.

FOCUS QUESTION   42.7

Give an example of bioremediation and of biological augmentation.

Word Roots

bio-   = life; **geo-** = Earth (*biogeochemical cycles*: the various chemical cycles involving both biotic and abiotic components of ecosystems)

detrit-   = wear off; **-vora** = eat (*detritivore*: a consumer that derives its energy and nutrients from nonliving organic material)

Structure Your Knowledge

1. What is primary production? What factors limit primary production in aquatic ecosystems? In terrestrial ecosystems?

2. What is secondary production? How does it relate to trophic efficiency?

3. What processes mediate the interconversion of the nutrients contained in organic materials and inorganic materials?

Test Your Knowledge

MULTIPLE CHOICE: *Choose the one best answer.*

1. Which of the following components are absolutely essential to the functioning of any ecosystem?
   a. producers and primary consumers
   b. producers and secondary consumers
   c. primary, secondary, and tertiary consumers
   d. primary consumers and detritivores
   e. producers and detritivores

2. Which of the following statements about ecosystems is true?
   a. Energy is recycled through the trophic structure.
   b. Energy is usually converted from sunlight by primary producers, passed to secondary producers in the form of organic compounds, and lost to detritivores in the form of heat.
   c. Chemicals are recycled between the biotic and abiotic sectors, whereas energy makes a one-way trip through the food web and is eventually dissipated as heat.
   d. There is a continuous process by which energy is lost as heat, and chemical elements leave the ecosystem through runoff.
   e. A food web shows that all trophic levels may feed off each other.

3. Primary production
   a. is equal to the standing crop of an ecosystem.
   b. is limited by light, nutrients, and moisture in all ecosystems.
   c. is the amount of light energy converted to chemical energy per unit time in an ecosystem.
   d. is inverted in some aquatic ecosystems.
   e. is all of the above.

4. The open ocean and tropical rain forest are the two largest contributors to Earth's net primary production because
   a. both have high rates of net primary production.
   b. both cover huge surface areas of Earth.
   c. nutrients cycle rapidly in these two ecosystems.

   d. the ocean covers a huge surface area and the tropical rain forest has a high rate of production.
   e. both **a** and **b** are correct.

5. Production in terrestrial ecosystems is affected by
   a. temperature.
   b. light intensity.
   c. availability of nutrients.
   d. availability of water.
   e. all of the above.

6. Which of the following statements concerning net ecosystem production and net primary production is *false*?
   a. NEP will always be less than NPP because it takes the respiration of heterotrophs and decomposers into account.
   b. NEP is a means of measuring the total biomass accumulation in an ecosystem over a period of time.
   c. If measurements indicate that more CO_2 enters an ecosystem than leaves, NEP would be positive.
   d. If measurements indicate that more O_2 leaves an ecosystem than enters, NEP would be negative.
   e. An increase in the rate of decomposition in an ecosystem would lower NEP, but NPP would probably remain the same or perhaps increase.

7. Secondary production
   a. is measured by the standing crop.
   b. is the rate of biomass production in consumers.
   c. is greater than primary production.
   d. is 10% less than primary production.
   e. is the gross primary production minus the energy used for respiration.

8. Which of the following statements concerning a pyramid of net production is *not* true?
   a. Only about 10% of the energy in one trophic level passes into the next level.
   b. Because of the loss of energy at each trophic level, most food chains are limited to three to five links.
   c. The pyramid of production of some aquatic ecosystems is inverted because of the large zooplankton primary consumer level.
   d. Eating grain-fed beef is an inefficient means of obtaining the energy trapped by photosynthesis.
   e. A biomass pyramid is usually the same shape as a pyramid of production.

9. In which of the following groups of organisms would you expect production efficiency to be the greatest?
   a. amphibians
   b. mammals
   c. insects and microorganisms
   d. primates
   e. birds

10. Biogeochemical cycles are global for elements
    a. that are found in the atmosphere.
    b. that are found mainly in the soil.
    c. such as carbon, nitrogen, and phosphorus.
    d. that are dissolved in water.
    e. Both a and c are correct.

11. Which of these processes is *incorrectly* paired with its description?
    a. nitrification—oxidation of ammonium in the soil to nitrite and nitrate
    b. nitrogen fixation—reduction of atmospheric nitrogen to ammonia
    c. denitrification—return of N_2 to air; occurs when denitrifying bacteria metabolize nitrate
    d. ammonification—decomposition of organic compounds to ammonium
    e. deforestation—increased supply of nitrate in the soil

12. Clear-cutting tropical forests yields agricultural land with limited productivity because
    a. it is too hot in the tropics for most food crops.
    b. the tropical forest regrows rapidly and chokes out agricultural crops.
    c. few of the ecosystem's nutrients are stored in the soil; most are in the forest trees.
    d. phosphorus, not nitrogen, is the limiting nutrient in those soils.
    e. decomposition rates are high but primary production is low in the tropics.

13. Which of the following was *not* shown by the Hubbard Brook Experimental Forest study?
    a. Most minerals recycle within a forest ecosystem.
    b. Deforestation results in a large increase in water runoff.
    c. Mineral losses from a valley were great following deforestation.
    d. Nitrate was the mineral that showed the greatest loss.
    e. The ecosystem could not recover from the effects of deforestation.

14. Which of the following actions is an example of bioaugmentation?
    a. adding fertilizer to degraded soils
    b. restoring the natural flow of river channels
    c. planting nitrogen-fixing lupines in soils disturbed by mining
    d. establishing habitat corridors to connect restored sites with undisturbed sites
    e. adding ethanol to stimulate the growth of uranium-reducing bacteria in contaminated groundwater

Chapter 43

Global Ecology and Conservation Biology

Chapter Focus

Biodiversity at the genetic, species, and ecosystem level is crucial to human welfare. This chapter explores the threats to biodiversity and several of the approaches to preserving the diversity of species on Earth. Conservation biologists focus on the population size, genetic diversity, and habitat needs of endangered species and on establishing and managing nature reserves that are often in human-dominated landscapes. Human activities are causing global change through nutrient enrichment, the release of toxins, and increasing CO_2 levels in the atmosphere. The human population growth rate has finally declined, but the global population continues to increase rapidly. Ecological research and our biophilia may help to achieve the goal of sustainable development—the long-term perpetuation of human societies *and* the ecosystems that support them.

Chapter Review

Conservation biology integrates all areas of biology in the effort to sustain ecosystem processes and biodiversity.

43.1 Human activities threaten Earth's biodiversity

Three Levels of Biodiversity   The current high rate of extinction threatens Earth's biological diversity. At the first level of diversity, loss of the genetic diversity within and between populations lessens a species' adaptive potential.

A second level of biodiversity is species diversity. An **endangered species**, according to the U.S. Endangered Species Act, is one that is "in danger of extinction throughout all or a significant portion of its range." A **threatened species** is defined as one that is likely to become endangered. There are many well-documented examples of recent extinctions and endangered species in most taxonomic groups.

The third level of biodiversity is ecosystem diversity. A loss of a species can negatively affect other species in the ecosystem. Human activities have altered many ecosystems.

Biodiversity and Human Welfare   There are both moral and practical reasons for preserving biodiversity. A loss of biodiversity is a loss of the genetic potential held in the genomes of species, such as genes for traits that could improve crops. Biodiversity is a natural resource that can provide medicines, fibers, and food.

Humans depend on Earth's ecosystems. **Ecosystem services** include such things as purification of air and water, detoxification and decomposition of wastes, nutrient cycling, flood control, pollination of crops, and soil creation and preservation. The monetary value of ecosystem services is huge.

Threats to Biodiversity   The greatest threat to biodiversity is habitat destruction, caused by agriculture, urban development, forestry, mining, pollution, and global climate change. According to the International Union for Conservation of Nature and Natural Resources (IUCN), habitat destruction is implicated in 73% of species that have been designated as extinct, endangered, vulnerable, or rare. Fragmentation of natural habitats is a common occurrence and almost always leads to species loss. Both terrestrial ecosystems and aquatic habitats have been damaged.

Introduced species, sometimes called exotic species, compete with or prey upon native species. Humans have transplanted thousands of species, intentionally and unintentionally, with huge economic costs in terms of damage and control efforts.

Overharvesting involves harvesting wild plants or animals at rates higher than the populations' abilities to reproduce. Species of large animals with low reproductive rates and species on small islands are particularly vulnerable. Overfishing, particularly using new harvesting techniques, has drastically reduced populations of many commercially important fish species. The tools of molecular genetics enable conservation biologists to determine whether tissues have come from threatened or endangered species.

Global change, the fourth major threat to biodiversity, includes alterations in climate, atmospheric chemistry, and ecological systems. The burning of fossil fuels and wood, for example, releases oxides of sulfur and nitrogen, which form sulfuric and nitric acid in the atmosphere. These acids contribute to *acid precipitation*, defined as rain, snow, sleet, or fog with a pH less than 5.2. Drifting emissions from factories create acid precipitation, which has damaged forests and fish populations in lakes across North America and Europe. New regulations and technologies have reduced sulfur dioxide emissions, and the acidity of precipitation has gradually been reduced.

FOCUS QUESTION   43.1

Give an example of how each of the following threats to biodiversity has reduced population numbers or caused extinctions.

a. habitat destruction

b. introduced species

c. overharvesting

d. global change

43.2   Population conservation focuses on population size, genetic diversity, and critical habitat

Small-Population Approach   Genetic variation is the key consideration of the small-population approach of population conservation. The inbreeding and genetic drift characteristic of a small population may draw it down an **extinction vortex,** in which the loss of genetic variation leads, by positive feedback loops, to smaller and smaller numbers until the population becomes extinct. Low genetic diversity, however, does not always doom a population.

The story of the greater prairie chicken provides an example of an averted extinction vortex. As agriculture fragmented their habitat, the number of prairie chickens in Illinois declined from millions in the 19th century to 50 in 1993. A comparison with DNA from museum specimens indicated decreased genetic variation in the threatened population. After importing birds from larger populations in other states,

researchers noted an increase in egg viability, and the Illinois population rebounded.

Computer models that integrate many factors are used to estimate the **minimum viable population (MVP),** the minimum population size necessary to sustain a population.

The **effective population size** (N_e) is based on a population's breeding potential and is determined by a formula that includes the number of individuals that breed and the sex ratio of the population: $N_e = (4N_f N_m)/(N_f + N_m)$. Alternative formulas take into account life history or genetic factors. Conservation efforts should be based on maintaining the minimum number of *reproductively active* individuals needed to prevent extinction.

FOCUS QUESTION   43.2

Is the effective population size usually larger or smaller than the actual number of individuals in the population? Explain.

M. Shaffer performed a population viability analysis as part of a long-term study of grizzly bears in Yellowstone National Park. He estimated that the minimum viable population size for threatened grizzly bear populations was 100 bears. Current estimates of the Yellowstone grizzly population is about 500. The effective population size, however, is only 25% of the total population size, or 125 bears. Genetic analyses indicate that genetic variability within the population is low.

FOCUS QUESTION   43.3

Explain the basic premise of the small-population approach. What conservation strategy is recommended for preserving small populations?

Declining-Population Approach   The emphasis of the declining-population approach is to identify populations that may be declining, determine the environmental factors that caused the decline, and then recommend corrective measures.

Several factors have driven the red-cockaded woodpecker into decline. For example, logging and agriculture have fragmented its mature pine forest habitats, and strict fire control has resulted in excessively thick undergrowth around the pine trees. Management strategies now include protection of some longleaf pine forests, controlled fires, and the excavation of breeding cavities in unoccupied habitats to encourage establishment of new breeding groups.

FOCUS QUESTION   43.4

Describe the declining-population approach to the conservation of endangered species.

Weighing Conflicting Demands   Preserving habitat for endangered species often conflicts with the economic and recreational desires of humans. However, keystone species exert more influence on community structure and ecosystem processes, and prioritizing the species to be saved on the basis of their ecological role may be crucial to the survival of whole communities.

43.3   Landscape and regional conservation help sustain biodiversity

Conservation efforts increasingly are directed at sustaining the biodiversity of whole communities and ecosystems. A goal of landscape ecology is to make biodiversity conservation a part of ecosystem management and land-use planning.

Landscape Structure and Biodiversity   Species often use more than one ecosystem or live on borders between ecosystems. Landscapes include ecosystems separated by boundaries or *edges,* which have their own sets of physical conditions and communities of organisms. The increase in edge communities due to human fragmentation of habitats may serve to reduce biodiversity as edge-adapted species become predominant. The long-term Biological Dynamics of Forest Fragments Project has shown that species adapted to forest interiors decline the most when habitat fragments are small.

Movement corridors are narrow strips or clumps of habitat that connect isolated patches. Artificial corridors are sometimes constructed when habitat patches have been separated by major human disruptions.

FOCUS QUESTION   43.5

What are some potential benefits of corridors? How might they be harmful?

Establishing Protected Areas   Currently, about 7% of Earth's land area has been set aside as reserves. Conservation biologists consider landscape dynamics in the designation and management of these protected areas. **Biodiversity hot spots**—small areas with very high concentrations of endemic, threatened, and endangered species—are good choices for nature reserves. However, global change will complicate the use of the hot-spot approach to conservation as climatic factors in those regions are expected to change.

Even though nature reserves provide islands of protected habitat, the nonequilibrium model of natural disturbances applies to them. Patch dynamics, edges, and corridor effects must be considered in the design and management of reserves.

New information on the requirements for minimum viable population sizes indicates that most national parks and reserves are much too small—the area needed to sustain a population is usually much larger than the legal boundary set in a reserve.

FOCUS QUESTION   43.6

What factors would favor the creation of larger, extensive reserves? What factors favor smaller, unconnected reserves?

Zoned reserves have protected core areas surrounded by buffer zones in which the human social and economic climate is stable and activities are regulated to promote the long-term viability of the protected zones. Costa Rica has established eight zoned reserves, but deforestation has continued in some buffer zones. The Florida Keys National Marine Sanctuary, a zoned marine reserve established in 1990, has increased marine life and generated revenue from recreational divers.

43.4   Earth is changing rapidly as a result of human actions

Nutrient Enrichment   The harvesting of crops removes nutrients that would otherwise be recycled in the soil. Once the organic and inorganic reserves of soils become depleted, crops require the addition of fertilizers. The addition of nitrogen fertilizers, increased legume cultivation, and burning of fossil fuels has more than doubled Earth's supply of fixed nitrogen.

Nitrogen in excess of the **critical load,** the amount of added nutrient that can be absorbed by plants without damaging the ecosystem, can contaminate groundwater, degrade lakes and rivers, and drain into the ocean. Elevated nitrogen levels create phytoplankton blooms. The decomposition of these phytoplankton lowers oxygen levels, creating "dead zones" in coastal waters. Eutrophication of lakes due to nutrient runoff can kill fish and harm other organisms.

Toxins in the Environment   Humans release a huge variety of toxic chemicals into the environment. Organisms absorb these toxins from food and water and may retain them within their tissues. In a process known as **biological magnification,** the concentration of such compounds increases in each successive link of the food chain. Pesticides such as DDT and industrial polychlorinated biphenyls (or PCBs) have been implicated in endocrine system problems in many animal species.

Over-the-counter and prescription drugs, as well as drugs given to farm animals, are a type of toxin increasingly found in rivers and lakes. Sex steroids such as the estrogens used for birth control may alter sexual differentiation in fish and disrupt their reproduction.

Many toxic chemicals dumped into ecosystems are nonbiodegradable; others, such as mercury, may become more harmful as they react with other environmental factors.

Greenhouse Gases and Climate Change   The concentration of CO_2 in the atmosphere has been increasing since the Industrial Revolution as a result of the combustion of fossil fuel and deforestation. Through a phenomenon known as the **greenhouse effect,** CO_2, methane, water vapor, and other gases in the atmosphere absorb infrared radiation reflected from Earth and re-reflect it back, thus warming Earth.

Scientific models predict a doubling of CO_2 levels and a temperature rise of 3°C by the end of the 21st century. The correlation between CO_2 levels and temperatures in prehistoric times supports these models.

To predict the impact of increasing temperatures, ecologists study changes in the geographic ranges of trees following the retreat of glaciers as the last ice age ended. Because many plants disperse slowly, their ranges may not shift rapidly enough for them to survive global climate change.

Controlling the levels of CO_2 emissions in increasingly industrialized societies is a huge international challenge.

FOCUS QUESTION   43.7

List some of the ways by which we may slow global warming.

43.5   The human population is no longer growing exponentially but is still increasing rapidly

The Global Human Population   After 1650, it took 200 years for the global population to double to 1 billion. In the next 80 years, it doubled to 2 billion; it doubled again to 4 billion in the next 45 years; and now, some 37 years later, it is more than 7 billion. The rate of growth has begun to slow, partly due to diseases such as AIDS and voluntary population control, although ecologists project a population of 8.1 to 10.6 billion by 2050.

The world's annual population growth rate is regionally variable; it is near equilibrium (0.1%) in industrialized nations, with the reproductive rate below the replacement level of 2.1 children per female in some countries. Less industrialized nations, where 80% of the world's population lives, contribute the most to the global growth rate of 1.1%.

Human population growth is unique in that it can be consciously controlled by voluntary contraception and family planning. In many cultures, women are receiving more education and delaying marriage and reproduction, thus slowing population growth.

Global Carrying Capacity   Estimates of Earth's carrying capacity have varied greatly and have averaged about 10–15 billion. These estimates involve different assumptions, such as the logistic equation, the amount of habitable land, or food as the limiting factor.

The concept of an **ecological footprint** takes into account multiple human needs and the land and water area required to meet those needs and absorb wastes. The amount of ecologically productive land area per person on Earth (reserving some land for conservation) is estimated to be about 1.7 hectares (ha). In the United States, a typical person, with an ecological footprint of about 10 ha, is using an unsustainable share of Earth's resources.

Ecological footprints may also be based on the amount of energy used. A typical person in the United States consumes roughly 30 times the energy consumed by an individual in central Africa. In most developed nations, 80% of the energy used comes from nonrenewable fossil fuels.

FOCUS QUESTION 43.8

What factors may determine the ultimate carrying capacity of Earth? How might the continued growth of the human population actually lower the carrying capacity of Earth?

43.6 Sustainable development can improve human lives while conserving biodiversity

Sustainable Development Economic development that meets the needs of people today without limiting the ability of future generations to meet their needs is known as **sustainable development**.

Partnerships between the government, nongovernment organizations (NGOs), and citizens have contributed to the success of conservation in Costa Rica. Living conditions in the country have improved, as evidenced by a decrease in infant mortality, an increase in life expectancy, and a high literacy rate.

The Future of the Biosphere E. O. Wilson calls our attraction to Earth's diversity of life and our affinity for natural environments *biophilia*. By coming to know and understand nature through the study of biology, we may become better able to appreciate and preserve the processes and diversity of the biosphere.

Word Roots

bio- = life (*biodiversity hot spot:* a relatively small area with numerous endemic species and a large number of endangered and threatened species)

Structure Your Knowledge

1. Describe the four major threats to biodiversity.

2. How does the loss of biodiversity threaten human welfare?

3. What do edges and movement corridors have to do with habitat fragmentation?

4. What is an ecological footprint?

Test Your Knowledge

MULTIPLE CHOICE: *Choose the one best answer.*

1. According to the Endangered Species Act, a threatened species is
   a. an exotic species that cannot successfully compete with native organisms.
   b. an endemic species that is found nowhere else in the world.
   c. a species that is found in disturbed habitats.
   d. a species that is in danger of extinction in all or a large part of its range.
   e. a species that is likely to become endangered.

2. Ecosystem services include all of the following *except*
   a. pollination of crops.
   b. production of antibiotics and drugs.
   c. purification of air and water.
   d. decomposition of wastes.
   e. reduced impact of weather extremes.

3. The most serious threat to biodiversity is
   a. competition from introduced species.
   b. environmental toxins.
   c. habitat destruction.
   d. overharvesting.
   e. disruptions of community dynamics.

4. Some grassland and conifer forest reserves have effective fire prevention programs. The most likely result of such programs is
   a. an increase in species diversity because fires are prevented.
   b. a change in community composition because fires are natural disturbances that maintain the community structure.
   c. the preservation of endangered species in the area.
   d. no change in the species composition of the community.
   e. succession to a deciduous forest.

5. According to the small-population approach, the most important remedy for preserving an endangered species is to
   a. establish a large nature reserve around its habitat.
   b. control the populations of its natural predators.
   c. determine the reason for its decline.
   d. encourage dispersal and an increase in genetic diversity.
   e. set up artificial breeding programs.

6. Which of the following characteristics is typical of biodiversity hot spots?
   a. a large number of endemic species
   b. a high rate of habitat degradation
   c. little species diversity
   d. a large land or aquatic area
   e. large populations of migratory birds

7. Which of the following phenomena may occur when a population falls below its minimum viable population size?
   a. genetic drift
   b. a further reduction in population size
   c. inbreeding
   d. a loss of genetic diversity
   e. All of the above are characteristics of an extinction vortex that the population may enter.

8. Movement corridors are
   a. strips or clumps of habitat that connect isolated habitats.
   b. the routes taken by migratory animals.
   c. connections within a landscape that includes several different ecosystems.
   d. the areas forming the boundary or edge between two ecosystems.
   e. buffer zones for human activity that promote the long-term viability of protected areas.

9. What does it mean if a population's effective population size (N_e) is the same as its actual population size?
   a. The population is not in danger of becoming extinct.
   b. The population has high genetic diversity.
   c. All the members of the population breed.
   d. The population's minimum viable population will not sustain the population.
   e. The population is being drawn into an extinction vortex.

10. The focus of the declining-population approach to conservation is to
    a. predict a species' minimum viable population size.
    b. transplant members from other populations to increase genetic diversity.
    c. perform a population viability analysis to predict the long-term viability of a population in a particular habitat.
    d. determine the cause of a species' decline and take remedial action.
    e. establish zoned reserves that ensure that human landscapes surrounding reserves support the protected habitats.

11. Given their limited resources, conservation biologists working to preserve biodiversity should assign the highest priority to
    a. the northern spotted owl.
    b. declining keystone species in a community.
    c. a commercially important species.
    d. endangered and threatened vertebrate species.
    e. all declining species.

12. The finding of harmful levels of DDT in grebes (fish-eating birds) following years of trying to eliminate bothersome gnat populations in a lakeshore town is an example of
    a. eutrophication.
    b. biological magnification.
    c. nutrient enrichment.
    d. an edge effect of a landscape ecosystem.
    e. increasing resistance to pesticides.

13. An ecological footprint is an estimate of
    a. the carrying capacity of a nation.
    b. the number of offspring an adult produces and the resulting demand on resources.
    c. the land and water area needed per person or nation to meet the current demand on resources.
    d. the size of a population in relationship to the resources it uses.
    e. how much energy is used to produce food for a vegetarian versus a meat eater.

14. Sustainable development
    a. uses nature reserves to save endangered species so that the rest of Earth's ecosystems can be used for human development.
    b. requires the establishment of zoned reserves.
    c. is economic development that meets the needs of people today without limiting the ability of people to meet their needs in the future.
    d. is only achievable if the global population is reduced.
    e. sustains the productivity of agricultural ecosystems through the use of chemical and organic fertilizers and irrigation.

15. Which of the following statements reflects a concern about the effects of global climate change on tree species?
    a. The increased ozone levels may damage leaf cells, reducing photosynthetic rates.
    b. Warmer temperatures may speed tree growth, producing trees that are too tall and spindly.
    c. The combination of acid precipitation and worsening droughts may threaten many tree species.
    d. Trees may not be able to disperse fast enough to reach new habitats that meet their climatic requirements.
    e. All of the above are concerns regarding trees and climate change.

Answer Section

CHAPTER 1: INTRODUCTION: EVOLUTION AND THE FOUNDATIONS OF BIOLOGY

FOCUS QUESTIONS

1.1. **a.** The biosphere includes all life on Earth and the environments inhabited by life—most regions of land and water, the sediments and rocks below Earth's surface, and the atmosphere up to several kilometers.
b. An ecosystem includes all the living organisms in an area and the nonliving parts of the environment with which they interact.
c. A community consists of all the organisms inhabiting a particular area. Each diverse form of life is called a *species*.
d. A population includes all the individuals of a single species in an area.
e. An organism is an individual living entity.
f. Organs, found in more complex organisms, are body parts that perform a particular function. Organ systems are groups of organs that perform a larger function.
g. Tissues are groups of cells that collectively perform a specialized function. Several tissues make up an organ.
h. Cells are the fundamental units of life.
i. Organelles are the functional components of cells.
j. Molecules are composed of two or more atoms. Molecules are the chemical units that make up living organisms.

1.2. The order of nucleotides in a DNA molecule that makes up a gene "spells" the instructions, which are transcribed into RNA and then translated into a protein with a specific function.

1.3. Chemical nutrients are recycled as photosynthesizing plants absorb water, minerals, and CO_2, producing sugar and releasing O_2; consumers eat plants and other organisms; and decomposers return nutrients to the ecosystem. Energy flows through ecosystems, entering as solar energy, being transformed to chemical energy that drives cellular work, and exiting as heat.

1.4. These organisms are characterized to a large extent by their mode of nutrition. Plants are photosynthetic, fungi absorb their nutrients from decomposing organic material, and animals ingest other organisms.

1.5. Most species tend to produce more offspring than can survive. Organisms with heritable traits best suited to the environment will tend to survive and leave more offspring. Over time, favorable adaptations will accumulate in a population. New species may arise as small populations are exposed to different environments and natural selection favors different traits.

1.6. **a.** The control groups were the camouflaged mice models placed in their native habitat. The experimental group in each habitat was the non-camouflaged mice models.
b. This experiment could not control for the number of predators in each area and thus for the total number of attacks on non-camouflaged mice. By presenting data as the predation rate or proportion of total attacks, the effect of this variable was eliminated.

1.7. **a.** A hypothesis is less broad in scope than a theory; it is a tentative explanation for a smaller set of observations. A theory can generate many testable hypotheses and is supported by a large body of evidence. Both hypotheses and theories are revised when new data fail to support their predictions.
b. Science seeks to understand natural phenomena; technology is the practical use of scientific knowledge.

SUGGESTED ANSWERS TO STRUCTURE YOUR KNOWLEDGE

1. **a.** With every increase in organizational level, the organization and interactions of component parts lead to new, emergent properties of the dynamic system. Cells are an organism's basic units of structure and function. They come in two distinct forms: prokaryotic and eukaryotic.
    b. DNA is the molecule of inheritance; it codes information for proteins and the functions of a cell and is passed on from one generation to the next.
    c. All organisms require energy. Energy flows through ecosystems from sunlight to chemical energy in producers and consumers and then escapes as heat. Chemical nutrients are recycled within an ecosystem.
    d. An organism interacts with other organisms. These interactions may be mutually beneficial, or one or both species may be harmed. Organisms also interact with the physical environment, and the environment is affected by the organisms that live there.
    e. Evolution explains the unity and diversity of life. All organisms are modified descendants of common ancestors. Darwin's theory of natural selection leading to unequal reproductive success accounts for the adaptation of populations to Earth's varying environments.

ANSWERS TO TEST YOUR KNOWLEDGE

Multiple Choice:

1. d	3. b	5. e	7. e
2. d	4. d	6. c	

CHAPTER 2: THE CHEMICAL CONTEXT OF LIFE

FOCUS QUESTIONS

2.1. calcium, phosphorus, potassium, sulfur

2.2. neutrons; 15; 15; 16; 31 daltons

2.3. absorb; loses

2.4. **a.** Hydrogen, $_1$H   **c.** Oxygen, $_8$O

b. Carbon, $_6$C   **d.** Sodium, $_{11}$Na

2.5. *Remember, you would benefit even more from creating your own map.*

   a. protons
   b. atomic number
   c. element
   d. neutrons
   e. mass number or atomic mass
   f. isotopes
   g. electrons
   h. electron shells
   i. valence shell

2.6. H = 1, O = 2, N = 3, C = 4

2.7. **a.** Nonpolar; even though N has a high electronegativity, the three pairs of electrons are shared equally between the two N atoms because each atom has an equally strong attraction for the electrons.
   b. Polar; N is more electronegative than H and pulls the shared electrons in each covalent bond closer to itself.
   c. Nonpolar; C and H have similar electronegativities and share electrons fairly equally between them.
   d. The C=O bond is polar because O is more electronegative than C; the C—H bonds are relatively nonpolar.

2.8. **a.** $CaCl_2$
   b. Ca^{2+} is the cation.

2.9.

$$\delta^- \quad O \quad H \quad \delta^+ \cdots \delta^- \quad O \quad H \quad \delta^+$$
$$H \quad \delta^+ \qquad H \quad \delta^+$$

2.10. 6, 6, 6

2.11. **a.** polar water molecules
   b. absorbed
   c. released
   d. specific heat

e. heat of vaporization
f. evaporative cooling
g. solar heat
h. rain
i. ice forms

2.12. **a.** The molecular mass of $C_3H_6O_3$ is 90 d, the combined atomic masses of its atoms. A mole of lactic acid = 90 g. A 0.5 *M* solution would require ½ mol or 45 g.
b. ½ mol is ½ Avogadro's number of molecules: 3.01×10^{23}

2.13.

$[H^+]$	$[OH]$	pH	Acidic, Basic, or Neutral?
10^{-8}	10^{-6}	8	basic
10^{-7}	10^{-7}	7	neutral
10^{-1}	10^{-13}	1	acidic

2.14. carbonic acid   bicarbonate   hydrogen ion

$$H_2CO_3 \rightleftharpoons HCO_3^- + H^+$$

H^+ donor H^+ acceptor

a. Bicarbonate acts as a base to accept excess H^+ ions when the pH starts to fall; the reaction moves to the left.
b. When the pH rises, H^+ ions are donated by carbonic acid, and the reaction shifts to the right.

2.15. **a.** $CO_2 + H_2O \rightleftharpoons H_2CO_3 \rightleftharpoons HCO_3^- + H^+$
Increasing $[CO_2]$ will drive these reactions to the right, increasing $[H^+]$.
b. A lower pH means an increasing $[H^+]$, which will drive this reaction ($HCO_3^- \rightleftharpoons CO_3^{2-} + H^+$) to the left, thus decreasing $[CO_3^{2-}]$.
c. $CO_3^{2-} + Ca^{2+} \rightleftharpoons CaCO_3$. With less CO_3^{2-} available to react with Ca^{2+}, calcification rates would be expected to decrease.

SUGGESTED ANSWERS TO STRUCTURE YOUR KNOWLEDGE

1.

Particle	Charge	Mass	Location
Proton	+1	1 dalton	nucleus
Neutron	0	1 dalton	nucleus
Electron	−1	negligible	orbitals in electron shells

2. **a.** The atoms of each element have a characteristic number of protons in their nuclei, referred to as the *atomic number*. In a neutral atom, the atomic number also indicates the number of electrons. The *mass number* is an indication of the approximate mass of an atom and is equal to the number of protons and neutrons in the nucleus.

The *atomic mass* is equal to the mass number and is measured in daltons. Protons and neutrons both have a mass of approximately 1 dalton.

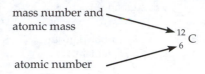

b. The *valence*, an indication of bonding capacity, is the number of electrons needed to complete an atom's valence shell.
c. The valence of an atom is most related to the chemical behavior of an atom because it is an indication of the number of bonds the atom will make, or the number of electrons the atom must share in order to reach a filled valence shell.

3. **a.** Cohesion, adhesion
b. A water column is pulled up through plant vessels.
c. Heat is absorbed or released when hydrogen bonds break or form. Water absorbs or releases a large quantity of heat for each degree of temperature change.
d. High heat of vaporization
e. Solar heat is dissipated from tropical seas.
f. Evaporative cooling
g. Evaporation of water cools surfaces of plants and animals.
h. Hydrogen bonds in ice space water molecules farther apart, making ice less dense.
i. Floating ice insulates bodies of water so they don't freeze solid.
j. Versatile solvent
k. Polar water molecules surround and dissolve ionic and polar solutes.

ANSWERS TO TEST YOUR KNOWLEDGE

Multiple Choice:

1. b	**10.** c	**19.** b	**28.** a	**37.** a
2. d	**11.** b	**20.** a	**29.** e	**38.** b
3. b	**12.** e	**21.** e	**30.** e	**39.** c
4. d	**13.** c	**22.** c	**31.** a	**40.** e
5. e	**14.** e	**23.** c	**32.** e	**41.** d
6. d	**15.** e	**24.** b	**33.** a	**42.** e
7. a	**16.** d	**25.** c	**34.** e	**43.** c
8. b	**17.** b	**26.** d	**35.** b	**44.** d
9. c	**18.** d	**27.** d	**36.** e	

CHAPTER 3: CARBON AND THE MOLECULAR DIVERSITY OF LIFE

FOCUS QUESTIONS

3.1.

a.

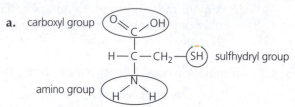

b.

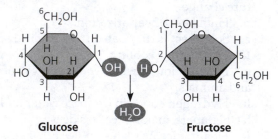

phosphate group

3.2. dehydration reactions/removal; hydrolysis/addition

3.3.
a. hydroxyl
b. carbonyl
c. aldehyde (carbonyl at end of carbon skeleton)
d. ketone
e. rings
f. *-ose*

3.4.

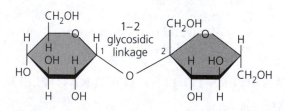

3.5.
a. monosaccharides
b. energy compounds
c. carbon skeletons, monomers

d. disaccharides
e. glycosidic linkages
f. polysaccharides
g. glycogen
h. animals
i. starch
j. cellulose
k. chitin

3.6.
a. fats, triacylglycerides
b. phospholipids
c. glycerol
d. fatty acids
e. unsaturated: has some C=C bonds
f. saturated: no C=C
g. phosphate group
h. cell membranes
i. steroids
j. animal cell membrane component (cholesterol), hormones

3.7. **a.**

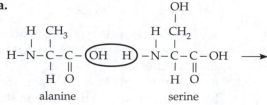

alanine   serine

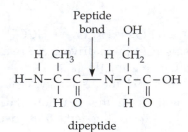

dipeptide

b. serine's R group is polar; alanine's R group is nonpolar
c. a polypeptide backbone

3.8.
a. hydrogen bond
b. hydrophobic interactions and van der Waals interactions
c. disulfide bridge
d. ionic bond

These interactions between R groups produce tertiary structure.

3.9.

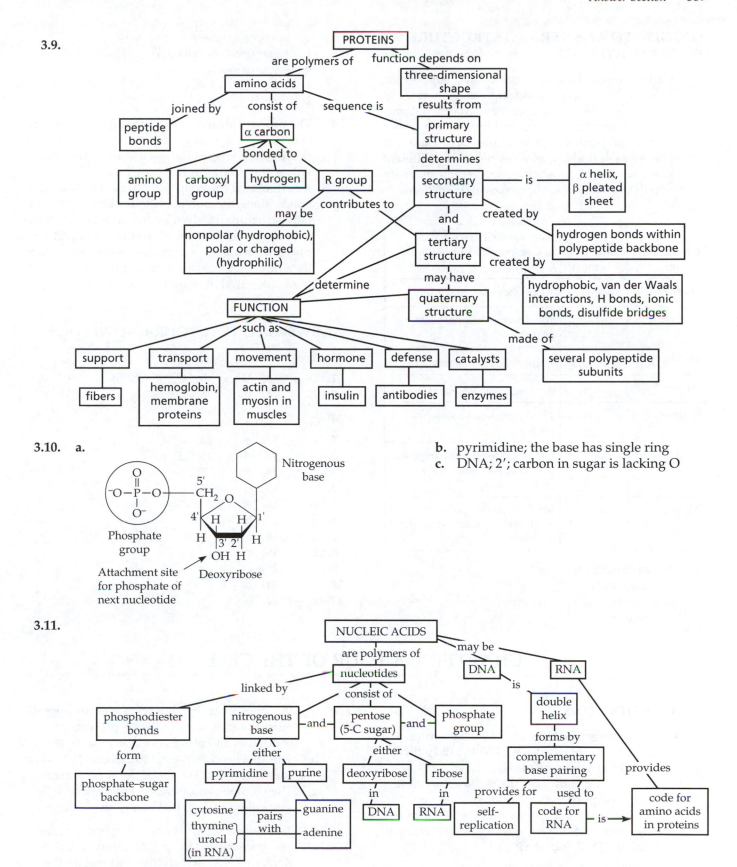

3.10. **a.**

Phosphate group

Attachment site for phosphate of next nucleotide

Deoxyribose

Nitrogenous base

b. pyrimidine; the base has single ring
c. DNA; 2′; carbon in sugar is lacking O

3.11.

SUGGESTED ANSWERS TO STRUCTURE YOUR KNOWLEDGE

1. Carbon forms four covalent bonds, either with other carbon atoms, producing chains or rings of various lengths and shapes, or with other atoms, such as characteristic chemical groups that confer specific properties on a molecule. Organisms can link a small number of monomers into different linear sequences, producing an incredible diversity of polymers with specific shapes and functions.

2.

Chemical Group	Molecular Formula	Name and Characteristics of Compounds Containing Group
Hydroxyl	–OH	Alcohols; polar group
Carbonyl	$>C=O$	Aldehyde or ketone; polar group
Carboxyl	–COOH	Carboxylic acid; release H^+
Amino	$-NH_2$	Amines; basic, accept H^+
Sulfhydryl	–SH	Thiols; cross-links stabilize proteins
Phosphate	$-OPO_3{}^{2-}$	Organic phosphates; involved in energy transfers, adds negative charge
Methyl	$-CH_3$	Methylated compounds; addition may alter expression of genes

3. a. amino acid (glycine)
   b. fatty acid
   c. nitrogenous base (purine)
   d. glycerol

c. phosphate group
f. sugar (pentose, ribose)
g. sugar (triose)

1. b, d	**3.** c, e, f	**5.** a
2. a	**4.** f, g	**6.** b

4. The primary structure of a protein is the specific, genetically coded sequence of amino acids in a polypeptide chain. The secondary structure involves the coiling (α helix) or folding (β pleated sheet) of the protein, stabilized by hydrogen bonds along the polypeptide backbone. The tertiary structure involves interactions between the side chains (R groups) of amino acids and produces a characteristic three-dimensional shape for a protein. Quaternary structure occurs in proteins composed of more than one polypeptide.

ANSWERS TO TEST YOUR KNOWLEDGE

Matching:

1. A	**3.** D	**5.** C	**7.** B	**9.** A
2. B	**4.** C	**6.** D	**8.** C	**10.** A

Multiple Choice:

1. c	**12.** a	**23.** d
2. b	**13.** d	**24.** d
3. d	**14.** a	**25.** c
4. a	**15.** d	**26.** b
5. d	**16.** e	
6. a	**17.** b	
7. e	**18.** c	
8. c	**19.** c	
9. c	**20.** c	
10. c	**21.** b	
11. d	**22.** b	

CHAPTER 4: A TOUR OF THE CELL

FOCUS QUESTIONS

4.1. a. Cytology is the study of cell structure.
     b. Cell biologists use a TEM to study the internal fine structure of cells.
     c. An SEM shows the three-dimensional surface topography of a specimen.
     d. Light microscopy enables the study of living cells and may introduce fewer artifacts than do TEM and SEM.

4.2. a. The molecular structure of the plasma membrane is a phospholipid bilayer with the hydrophobic tails clustered in the interior and the phosphate heads facing the hydrophilic outside and inside of the cell; proteins are embedded in and attached to the membrane.
     b. The area is proportional to a linear dimension squared; 10^2, or 100 times the surface area.
     c. The volume is proportional to a linear dimension cubed; 10^3, or 1,000 times the volume.

4.3. The genetic instructions for specific proteins are transcribed from DNA into messenger RNA (mRNA), which then passes into the cytoplasm to complex with ribosomes, where it is translated into the primary structure of proteins. Ribosomal subunits are synthesized in nucleoli from rRNA (transcribed from DNA) and proteins imported from the cytoplasm.

4.4. **a.** smooth ER—in different cells, it may house enzymes that synthesize lipids; metabolize carbohydrates; detoxify drugs and alcohol; store and release calcium ions in muscle cells

b. nuclear envelope—double membrane that encloses nucleus; pores regulate passage of materials

c. rough ER—attached ribosomes produce proteins that enter cisternae; produces secretory proteins and membranes

d. transport vesicle—carries products of ER and Golgi apparatus to various locations

e. Golgi apparatus—processes products of ER; makes polysaccharides; packages products in vesicles targeted to specific locations

f. plasma membrane—selective barrier that regulates passage of materials into and out of the cell

g. lysosome—houses hydrolytic enzymes that digest macromolecules

4.5.

Mitochondrion

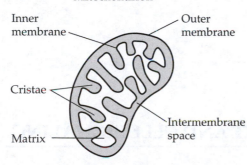

4.6. **a.** hollow tube, formed from columns of tubulin dimers; 25-nm diameter

b. cell shape and support (compression resistant), tracks for moving organelles, chromosome movement in cell division, beating of cilia and flagella

c. two twisted chains of actin subunits; 7-nm diameter

d. muscle contraction, maintain and change cell shape, amoeboid movement, cytoplasmic streaming

e. supercoiled fibrous proteins of keratin family; 8–12 nm diameter

f. reinforce cell shape (tension bearing); anchor nucleus; form nuclear lamina

4.7.

Two adjacent plant cells

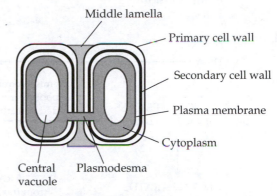

4.8. **a.** proteoglycan complex
b. collagen
c. fibronectin
d. integrin
e. microfilament

SUGGESTED ANSWERS TO STRUCTURE YOUR KNOWLEDGE

1. **a.** nucleus, chromosomes, centrioles, microtubules (move chromosomes), microfilaments (pinch apart animal cells)

b. nucleus, chromosomes, DNA → mRNA (to ribosomes) → enzymes and other proteins

c. mitochondria

d. ribosomes, rough and smooth ER, Golgi apparatus, transport vesicles

e. smooth ER (peroxisomes also detoxify substances)

f. lysosomes, food vacuoles, vesicles enclosing damaged organelles

g. peroxisomes

h. cytoskeleton: microtubules, microfilaments, intermediate filaments; extracellular matrix

i. cilia and flagella (microtubules), microfilaments (actin) in muscles and amoeboid movement

j. plasma membrane, transport vesicles

k. desmosomes, tight and gap junctions, ECM

2. **a.** structural support, middle lamella glues cells together

b. storage, waste disposal, protection (toxic compounds), growth

c. photosynthesis (production of sugars)

d. starch storage

e. cytoplasmic connections between cells

Chloroplast

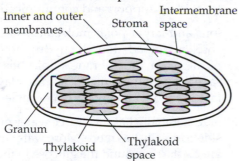

3. **a.** rough endoplasmic reticulum   **j.** mitochondrion
        b. smooth endoplasmic reticulum   **k.** lysosome
        c. chromatin   **l.** cytoskeleton
        d. nucleolus   **m.** microtubules
        e. nuclear envelope   **n.** intermediate filaments
        f. nucleus   **o.** microfilaments
        g. ribosomes   **p.** microvilli
        h. Golgi apparatus   **q.** peroxisome
        i. plasma membrane   **r.** centrosome (contains a pair of centrioles)
                                                                s. flagellum

4.

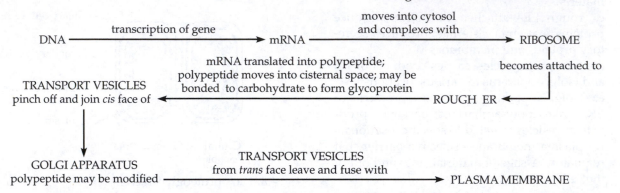

ANSWERS TO TEST YOUR KNOWLEDGE

Multiple Choice:

1. b	**5.** e	**9.** c	**13.** a	**17.** d	**21.** b
2. a	**6.** b	**10.** d	**14.** b	**18.** d	**22.** a
3. a	**7.** e	**11.** c	**15.** c	**19.** c	
4. c	**8.** b	**12.** e	**16.** e	**20.** e	

CHAPTER 5: MEMBRANE TRANSPORT AND CELL SIGNALING

FOCUS QUESTIONS

5.1. **a.** phosphate head (hydrophilic)
         b. hydrocarbon tail (hydrophobic)
         c. phospholipid bilayer
         d. hydrophobic region of integral protein
         e. hydrophilic region of integral protein
         f. peripheral protein (hydrophilic)

5.2. **a.** In hybrid human/mouse cells, membrane proteins rapidly intermix (Frye and Edidin, 1970).
         b. The cell membranes of the cold lake species would have a higher proportion of unsaturated fatty acids in their phospholipids than that of the species living in warm temperatures.

5.3. transport, enzymatic activity, signal transduction, cell–cell recognition, intercellular

joining, attachment to cytoskeleton and ECM (providing support and communication)

5.4. Ions and larger polar molecules, such as glucose, are impeded by the hydrophobic center of the plasma membrane's lipid bilayer. Passage through the center of a lipid bilayer is not fast even for small, polar water molecules.

5.5. Side A initially has fewer free water molecules; side B initially has more. More water molecules are clustered around the glucose in the 1 M solution than around the fructose and sucrose, whose combined concentration is 0.9 M. Water will move by osmosis from side B to side A.

5.6. **a.** The protist will gain water from its hypotonic environment. It has contractile vacuoles that expel excess water. Its plasma membrane is also less permeable to water.

b. Animal cells fare best in isotonic environments, in which they neither gain nor lose water. Plant cells are healthiest in hypotonic environments, in which the inward movement of water creates turgor pressure, providing mechanical support to the cell.

5.7. Although it speeds diffusion, facilitated diffusion is still passive transport because the solute is moving down its concentration gradient; the process is driven by the concentration gradient and not by energy expended by the cell.

5.8. Three Na^+ are pumped out of the cell for every two K^+ pumped in, resulting in a net movement of positive charge from the cytoplasm to the extracellular fluid.

5.9. **a.** Human cells use receptor-mediated endocytosis to take in cholesterol, which is used for the synthesis of other steroids and membranes.
b. In people with this disease, LDL receptor proteins in the plasma membrane are defective, and low-density lipoprotein particles cannot bind and be transported from the blood into the cell.

5.10. Large, hydrophobic signaling molecules cannot cross the plasma membrane. They bind to cell-surface receptors such as G protein-coupled receptors or ligand-gated ion channels. Small, hydrophobic signaling molecules, such as steroid hormones or the gas NO, can diffuse across the plasma membrane and bind with receptors within the cytoplasm or nucleus of target cells.

5.11. **a.** A protein kinase transfers a phosphate group from ATP to a protein; adding a charged phosphate group causes a shape change that usually activates the protein.
b. A protein phosphatase is an enzyme that removes a phosphate group from a protein, usually inactivating the protein. Protein phosphatases effectively shut down signaling pathways when the initial signal is no longer present.
c. A phosphorylation cascade is a series of protein kinase relay molecules that sequentially phosphorylate the next kinase in the pathway.

5.12. **a.** signaling molecule (first messenger)
b. G protein-coupled receptor
c. activated G protein (GTP bound)
d. adenylyl cyclase
e. ATP
f. cAMP (second messenger)
g. protein kinase A
h. phosphorylation cascade to cellular response

SUGGESTED ANSWERS TO STRUCTURE YOUR KNOWLEDGE

1.

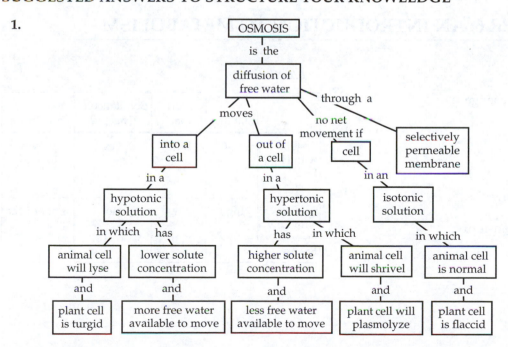

2. **a.** II represents facilitated diffusion. The solute is moving through a transport protein and down its concentration gradient. The cell does not expend energy in this transport. Polar molecules and ions may move by facilitated diffusion.

b. III represents active transport because the solute is clearly moving against its concentration gradient and the cell is expending ATP to drive this transport against the gradient.

c. In order to diffuse through the lipid bilayer, solute molecules must be hydrophobic (nonpolar) or very small polar molecules.

d. I and II. Both diffusion through the lipid bilayer and facilitated diffusion are considered passive transport because the solute moves down its concentration gradient and the cell does not expend energy in the process.

3. Cell signaling occurs through signal transduction pathways that include reception, transduction, and response. First, a signaling molecule binds to a specific receptor. The message is transduced as the receptor activates a protein that may relay the message through a sequence of activations, finally leading to the specific cellular response.

4. A signal transduction pathway often results in the activation of cellular proteins. When the signal is transduced to activate a transcription factor, however, the cellular response is a change in gene expression and the production of new proteins.

ANSWERS TO TEST YOUR KNOWLEDGE

Multiple Choice:

1. b	4. b	7. e	10. d**	13. d	16. b	19. b	22. d
2. c	5. d	8. a*	11. a	14. a	17. d	20. a	23. c
3. e	6. c	9. c	12. b	15. e	18. a	21. b	24. c

*Explanation for answer to question 8: This problem involves both osmosis and diffusion. Although the solutions are initially equal in molarity, glucose will diffuse down its concentration gradient until it reaches dynamic equilibrium with a 1.5 *M* concentration on both sides. The increasing solute concentration on side A will cause water to move into this side, and the water level will rise.

**Explanation for answer to question 10: As the solute in the solution crosses the cell membrane, it increases the concentration of solutes within the cell, reducing the hypertonicity of the solution. As the solute reaches an equal concentration inside and outside the cell, it no longer causes osmotic changes in the cell.

CHAPTER 6: AN INTRODUCTION TO METABOLISM

FOCUS QUESTIONS

6.1.
- **a.** capacity to cause change
- **b.** kinetic
- **c.** motion
- **d.** potential
- **e.** position
- **f.** conserved or constant
- **g.** created nor destroyed
- **h.** first
- **i.** transformed or transferred
- **j.** entropy
- **k.** second

6.2.

	Stability	Work Capacity	Spontaneous Change?	Equilibrium
System with High Free Energy	low	high	yes	moves toward
System with Low Free Energy	high	low	no	is at or near

6.3. A cell is provided with a steady supply of reactants, and the products of each reaction are siphoned off (as reactants for the next reaction or as waste products to be expelled).

6.4. **a.** adenine
b. ribose
c. three phosphate groups
d. A hydrolysis reaction breaks the terminal phosphate bond and releases a molecule of inorganic phosphate: ATP + H_2O → ADP + P_i + energy
e. The negatively charged phosphate groups are crowded together, and their mutual repulsion makes this area unstable. The chemical change to a more stable state of lower free energy accounts for the relatively high release of energy.

6.5. Energy coupling is the use of an exergonic process to drive an endergonic one. The ATP cycle itself illustrates this coupling in that energy-releasing processes of catabolism are coupled with the energy-consuming process of cellular work. But each side is also an example: The endergonic synthesis of ATP from ADP and P_i is coupled to the exergonic reactions of catabolism (left side), and the exergonic hydrolysis of ATP to ADP and P_i provides the immediate energy source for cellular work (right side).

6.6. **a.** free energy
b. transition state
c. E_A (free energy of activation) without enzyme
d. E_A with enzyme
e. ΔG of reaction

6.7.

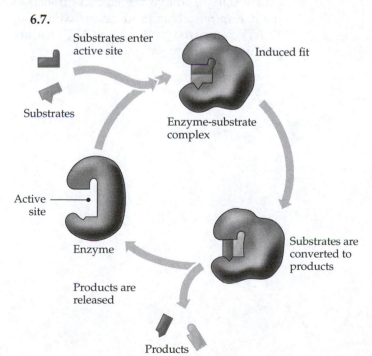

Substrates enter active site

Induced fit

Substrates

Enzyme-substrate complex

Active site

Enzyme

Substrates are converted to products

Products are released

Products

6.8. A competitive inhibitor would mimic the shape of the substrates and compete with them for the active site. A noncompetitive inhibitor would be a shape that could bind to another site on the enzyme molecule and would change the shape of the enzyme such that the active site functions less effectively.

6.9. ATP would act as an inhibitor to catabolic pathways, slowing the breakdown of fuel molecules if the supply of ATP exceeds demand. If ATP supplies drop, ADP (or its breakdown product AMP) would act as an activator of these catabolic enzymes, and more ATP would be produced.

SUGGESTED ANSWERS TO STRUCTURE YOUR KNOWLEDGE

1. Metabolism is the totality of chemical reactions that take place in living organisms. To create and maintain the structural order required for life requires an input of free energy—from sunlight for photosynthetic organisms and from energy-rich food molecules for other organisms. A cell couples catabolic, exergonic reactions ($-\Delta G$) with anabolic, endergonic reactions ($+\Delta G$), using ATP as the primary energy shuttle between the two.

2. Enzymes are essential for metabolism because they lower the activation energy of the specific reactions they catalyze and allow those reactions to occur extremely rapidly at a temperature conducive to life. By regulating the enzymes it produces, a cell can regulate which of the myriad of possible chemical reactions take place at any given time. Metabolic control also occurs through allosteric regulation and feedback inhibition. The compartmental organization of a cell facilitates a cell's metabolism.

ANSWERS TO TEST YOUR KNOWLEDGE

Fill in the Blanks:

1.	metabolism	**7.**	competitive inhibitors
2.	anabolic	**8.**	coenzymes
3.	potential	**9.**	feedback inhibition
4.	thermal energy	**10.**	phosphorylated intermediate
5.	entropy		
6.	activation energy		

Multiple Choice:

1.	c	**6.**	b	**11.**	e	**16.**	b
2.	c	**7.**	c	**12.**	d	**17.**	c
3.	b	**8.**	c	**13.**	b	**18.**	a
4.	e	**9.**	e	**14.**	d		
5.	a	**10.**	e	**15.**	c		

CHAPTER 7: CELLULAR RESPIRATION AND FERMENTATION

FOCUS QUESTIONS

7.1. $C_6H_{12}O_6$; 6 CO_2; energy (ATP + heat)

7.2.
  a. oxidized
  b. reduced
  c. donates (loses)
  d. oxidizing agent
  e. accepts (gains)

7.3.
  a. O_2
  b. glucose
  c. Some is stored in ATP and some is released as heat.

7.4.
  a. electron acceptor (or carrier or shuttle). It is a coenzyme that works with enzymes called dehydrogenases.
  b. NADH

7.5.
  a. 2 ATP
  b. 2 three-carbon sugars (glyceraldehyde-3-phosphate)
  c. 2 NAD^+
  d. 2 NADH + 2H^+
  e. 4 ATP
  f. 2 pyruvate

7.6.
  a. pyruvate
  b. CO_2
  c. NADH + H^+
  d. coenzyme A
  e. acetyl CoA
  f. oxaloacetate
  g. citrate
  h. NADH + H^+
  i. CO_2
  j. CO_2

  k. NADH + H^+
  l. GTP (may make ATP)
  m. $FADH_2$
  n. NADH + H^+

7.7.
  a. intermembrane space
  b. inner mitochondrial membrane
  c. mitochondrial matrix
  d. electron transport chain
  e. NADH
  f. NAD^+
  g. $FADH_2$
  h. 2 H^+ + ½ O_2
  i. H_2O
  j. chemiosmovsis
  k. ATP synthase
  l. ADP + $\textcircled{P}_i$
  m. ATP

7.8.
  a. -2
  b. 4
  c. citric acid cycle
  d. 26 or 28
  e. 32
  f. 2
  g. 6
  h. 2
  i. 2

7.9. Respiration yields up to 16 times more ATP than does fermentation. By oxidizing pyruvate to CO_2 and passing electrons from NADH (and $FADH_2$) through the electron transport chain, respiration can produce a maximum of 32 ATP compared to the 2 net ATP that are produced by fermentation.

SUGGESTED ANSWERS TO STRUCTURE YOUR KNOWLEDGE

1. Use Focus Questions 7.5, 7.6, and 7.7 to help you review these pathways.

2.

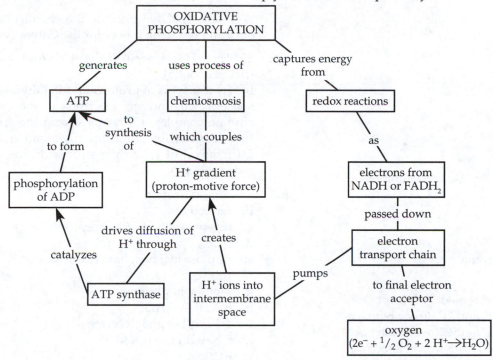

3.

Process	Brief Description	Inputs	Output
Glycolysis	Oxidation of glucose to 2 pyruvate, production of 2 ATP net	glucose 2 ATP	2 pyruvate 4 ATP (2 net) 2 NADH
Pyruvate to acetyl CoA and citric acid cycle	Oxidation of pyruvate to acetyl CoA, which combines with oxaloacetate → citrate. Citrate is cycled back as redox reactions produce NADH and $FADH_2$ and CO_2 is released. ATP is formed by substrate-level phosphorylation.	2 pyruvate 2 oxaloacetate	$6 CO_2$ 8 NADH $2 FADH_2$ 2 ATP
Oxidative phosphorylation (Electron transport and chemiosmosis)	NADH and $FADH_2$ transfer electrons to an electron transport chain. In a series of redox reactions, H^+ is pumped into intermembrane space, and electrons pass to O_2. Proton-motive force drives H^+ through ATP synthase to make ATP.	10 NADH $2 FADH_2$ $H^+ + O_2$	H_2O 28 ATP (max)
Fermentation	Anaerobic catabolism: glycolysis followed by oxidation of NADH to NAD^+ so glycolysis can continue. Pyruvate is either reduced to ethyl alcohol and CO_2 or to lactate.	See glycolysis above 2 pyruvate 2 NADH	2 ATP $2 NAD^+$ 2 ethanol and $2 CO_2$ or 2 lactate

ANSWERS TO TEST YOUR KNOWLEDGE

Multiple Choice:

1. a	**4.** e	**7.** b	**10.** a	**13.** e	**16.** d	**19.** e	**21.** b
2. d	**5.** c	**8.** b	**11.** a	**14.** d	**17.** b	**20.** c	**22.** c
3. c	**6.** d	**9.** d	**12.** d	**15.** e	**18.** c		

CHAPTER 8: PHOTOSYNTHESIS

FOCUS QUESTIONS

8.1.
  a. outer membrane
  b. granum
  c. inner membrane
  d. thylakoid space
  e. thylakoid
  f. stroma

8.2.
  a. light
  b. H_2O
  c. light reactions in thylakoid membranes
  d. O_2
  e. ATP
  f. NADPH
  g. CO_2
  h. Calvin cycle in stroma
  i. $[CH_2O]$ (sugar)

8.3. The y-axes should be labeled, "Absorption of light by chlorophyll a" and "Rate of photosynthesis." The solid line is the absorption spectrum; the dotted line is the action spectrum. Some wavelengths of light, particularly in the blue and the yellow-orange range, result in a higher rate of photosynthesis than would be indicated by the absorption of those wavelengths by chlorophyll a. These differences are partially accounted for by accessory pigments, such as chlorophyll b and carotenoids, which absorb light energy from different wavelengths and make that energy available to drive photosynthesis.

8.4. A photosystem contains light-harvesting complexes of pigment molecules (chlorophyll a, chlorophyll b, and carotenoids) bound to particular proteins, and a reaction center complex, which includes a pair of chlorophyll a molecules (P680 or P700) and a primary electron acceptor.

8.5.
  a. photosystem II
  b. photosystem I
  c. water (H_2O)
  d. oxygen ($\frac{1}{2} O_2$)
  e. P680, reaction-center chlorophyll a
  f. primary electron acceptor
  g. electron transport chain
  h. photophosphorylation by chemiosmosis
  i. ATP
  j. P700, reaction-center chlorophyll a
  k. primary electron acceptor
  l. $NADP^+$ reductase
  m. NADPH

ATP and NADPH provide the chemical energy and reducing power for the Calvin cycle.

8.6.
  a. The pH is lowest in the thylakoid space (pH of about 5).
  b. (1) transport of protons into the thylakoid space as Pq transfers electrons to the cytochrome complex; (2) protons from the splitting of water remain in the thylakoid space; (3) removal of H^+ in the stroma during the reduction of $NADP^+$.

8.7.
  a. carbon fixation
  b. reduction
  c. regeneration of CO_2 acceptor (RuBP)
  d. $3 CO_2$
  e. ribulose bisphosphate (RuBP)
  f. rubisco
  g. 3-phosphoglycerate
  h. $6 ATP \rightarrow 6 ADP$
  i. 1,3-bisphosphoglycerate
  j. $6 NADPH \rightarrow 6 NADP^+$
  k. $6 \; \textcircled{P}_i$
  l. glyceraldehyde-3-phosphate (G3P)
  m. G3P
  n. glucose and other organic compounds
  o. $3 ATP \rightarrow 3 ADP$
  9 ATP and 6 NADPH are required to synthesize one G3P.

8.8. Photorespiration may be an evolutionary relic from the time when there was little O_2 in the atmosphere and the ability of rubisco to distinguish between O_2 and CO_2 was not critical. Photorespiration appears to protect plants from damaging products of the light reactions that build up when the Calvin cycle slows due to a lack of CO_2.

8.9.
  a. The Calvin cycle takes place in the bundle-sheath cells.
  b. Carbon is initially fixed into a four-carbon compound in the mesophyll cells. When this compound is broken down in the bundle-sheath cells, CO_2 is maintained at a high enough concentration that rubisco does not accept O_2 and cause photorespiration.
  c. CO_2 is initially fixed at night when the stomata of CAM plants are open. The resulting 4-carbon acids are broken down during the day and provide CO_2 to the Calvin cycle, which takes place in the mesophyll cells—as it does in C_3 plants.

SUGGESTED ANSWERS TO STRUCTURE YOUR KNOWLEDGE

1.

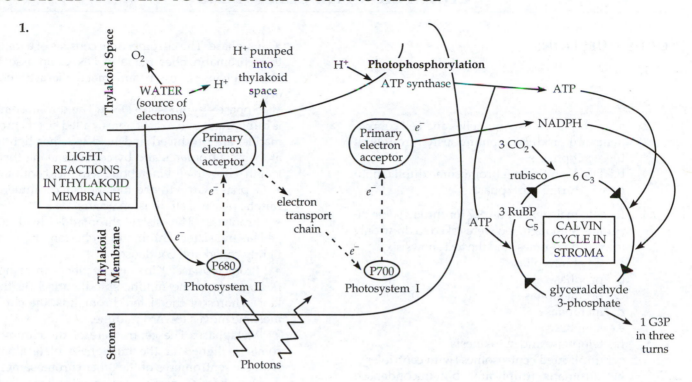

2.

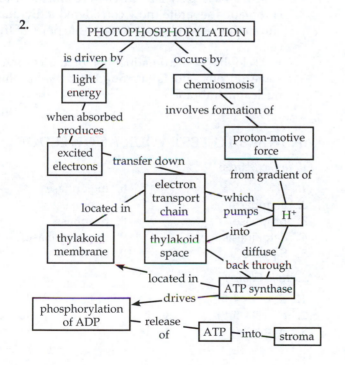

ANSWERS TO TEST YOUR KNOWLEDGE

Multiple Choice:

1. d	**6.** b	**11.** a	**16.** e	**21.** a
2. a	**7.** d	**12.** e	**17.** d	**22.** d
3. d	**8.** e	**13.** d	**18.** c	
4. a	**9.** e	**14.** d	**19.** a	
5. b	**10.** c	**15.** a	**20.** b	

CHAPTER 9: THE CELL CYCLE

FOCUS QUESTIONS

9.1. **a.** 46
b. 23
c. 92

9.2. **a.** Growth—most organelles and cell components are produced continuously throughout these subphases.
b. DNA synthesis (chromosome duplication) occurs during the S phase.

9.3. Refer to main text Figure 9.7 for chromosome diagrams (although you were asked to draw only four chromosomes, and the text shows six).
a. G_2 of interphase
b. prophase
c. prometaphase
d. metaphase
e. anaphase
f. telophase and cytokinesis
g. duplicated centrosomes (with centrioles)
h. chromatin (duplicated but uncondensed chromosomes)
i. nuclear envelope
j. nucleolus
k. early mitotic spindle
l. aster
m. nonkinetochore microtubules
n. metaphase plate
o. spindle
p. cleavage furrow
q. nuclear envelope forming

9.4. The G_1 checkpoint or "restriction point" appears to be most important. If a cell is not to divide, then it would be beneficial for it not to waste resources duplicating its chromosomes.

SUGGESTED ANSWERS TO STRUCTURE YOUR KNOWLEDGE

1. **a.** interphase
   b. telophase
   c. anaphase
   d. metaphase
2. Interphase: 90% of cell cycle; growth and DNA replication.

- G_1 phase: The chromosome consists of a long, thin chromatin fiber made of DNA and associated proteins. Growth and metabolic activities occur.
- S phase—synthesis of DNA: The chromosome is duplicated; two exact copies, called sister chromatids, are produced and held together tightly at their centromeres and by cohesins along their length. Growth and metabolic activities continue.
- G_2 phase: Growth and metabolism continue. Mitotic phase: Cell division occurs.
- Prophase: The sister chromatids, held together by sister chromatid cohesion, become tightly coiled and condensed.
- Prometaphase: Kinetochore fibers from opposite ends of the mitotic spindle attach to the kinetochores of the sister chromatids; the chromosome moves toward midline.
- Metaphase: The centromere of the chromosome is aligned at the metaphase plate along with the centromeres of the other chromosomes.
- Anaphase: Cohesins are cleaved and the sister chromatids separate (now considered individual chromosomes) and move to opposite poles of the cell.
- Telophase: Chromatin fiber of the chromosome uncoils and is surrounded by reforming nuclear membrane.

ANSWERS TO TEST YOUR KNOWLEDGE

Fill in the Blanks:

1. G_0
2. anaphase
3. prophase
4. cytokinesis
5. S phase
6. metaphase
7. telophase
8. prophase
9. prometaphase
10. G_1 phase

Multiple Choice:

1. d
2. c
3. d
4. e
5. b
6. c
7. c
8. a
9. b
10. a
11. c
12. e
13. d

CHAPTER 10: MEIOSIS AND SEXUAL LIFE CYCLES

FOCUS QUESTIONS

10.1. **a.** 14; 7
b. 28; 1
c. 56. Sister chromatids are produced when a chromosome duplicates before cell division. They are joined at the centromere and attached along their lengths. Nonsister chromatids are found on different chromosomes of a homologous pair (homologs).

10.2. **a.** meiosis
b. fertilization
c. zygote
d. mitosis
e. gametes
f. fertilization
g. zygote
h. meiosis
i. spores
j. gametes
k. fertilization
l. $2n$
m. zygote
n. meiosis

10.3. The diploid number for this animal cell ($2n$) is 6.
a. metaphase II
b. prophase I
c. anaphase I
d. metaphase I
e. anaphase II
f. telophase I, beginning of cytokinesis
Proper sequence: b. d. c. f. a. e.

10.4. **a.** 2^{23}, approximately 8.4 million
b. about 70 trillion ($2^{23} \times 2^{23}$)

c. This number of combinations does not take into account the additional variation of recombinant chromosomes produced by crossing over.

SUGGESTED ANSWERS TO STRUCTURE YOUR KNOWLEDGE

1. **a.** sister chromatids
b. centromere
c. pair of homologous chromosomes
d. nonsister chromatids

2. **a.** Chromosome duplication; sister chromatids attached at centromere and by sister chromatid cohesion along their lengths.
b. Chromosomes condense. Synapsis of homologous pairs (held by synaptonemal complex); crossing over (exchange of corresponding DNA segments) is evident at chiasmata.
c. Homologous pairs line up independently at metaphase plate (orientation of maternal and paternal homologs is random).
d. Homologous pairs of chromosomes separate and homologs move toward opposite poles; sister chromatids remain attached at centromere.
e. Haploid set of chromosomes, each consisting of two sister chromatids, aligns at metaphase plate; sister chromatids not identical due to crossing over.
f. Sister chromatids separate and move to opposite poles as individual chromosomes.

3.

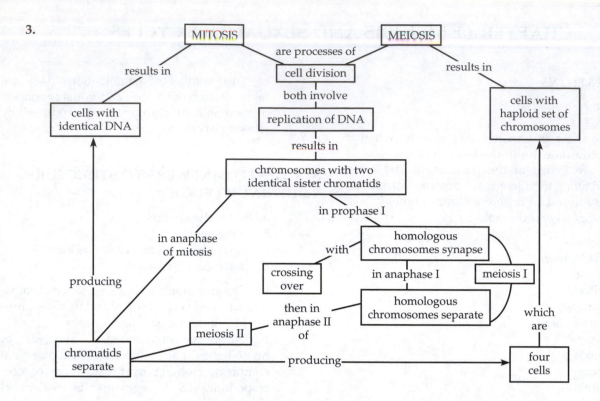

ANSWERS TO TEST YOUR KNOWLEDGE

Multiple Choice:

1. c	**4.** e	**7.** c	**10.** e	**13.** b	**16.** b	**19.** b
2. e	**5.** b	**8.** b	**11.** b	**14.** d	**17.** d	**20.** e
3. b	**6.** a	**9.** b	**12.** c	**15.** b	**18.** e	**21.** c

CHAPTER 11: MENDEL AND THE GENE IDEA

FOCUS QUESTIONS

11.1. **a.** *R*
  b. *r*
  c. F₁ Generation
  d. *Rr*
  e. F₂ Generation
  f. *R*
  g. *r*
  h. *Rr*
  i. *Rr*
  j. *rr*
  k. 3 round:1 wrinkled
  l. 1 *RR*:2*Rr*:1*rr*

11.2. **a.** all tall (*Tt*) plants, because the tall plant can only contribute a *T* allele and the dwarf plant can only contribute a *t* allele

b. 1:1 tall (*Tt*) to dwarf (*tt*), because the tall plant can contribute either a *T* or a *t* allele
You can use a Punnett square to determine the expected outcome of a testcross, but a shortcut is to use only one row or column for the recessive individual's gametes, since they produce only one type of gamete. For example, in Figure 11.7 you could use only one column for sperm because all sperm contain the recessive *p* allele.

11.3. **a.** all tall purple plants
  b. *TtPp*
  c. *TP, Tp, tP, tp*

d.

sperm

	$\frac{1}{4}\,TP$	$\frac{1}{4}\,Tp$	$\frac{1}{4}\,tP$	$\frac{1}{4}\,tp$
$\frac{1}{4}\,TP$	TTPP	TTPp	TtPP	TtPp
$\frac{1}{4}\,Tp$	TTPp	TTpp	TtPp	Ttpp
$\frac{1}{4}\,tP$	TtPP	TtPp	ttPP	ttPp
$\frac{1}{4}\,tp$	TtPp	Ttpp	ttPp	ttpp

eggs

e. 9 tall purple:3 tall white:3 dwarf purple:1 dwarf white

f. 12:4 or 3:1 tall to dwarf; 12:4 or 3:1 purple to white

11.4. **a.** Consider the outcome for each gene as a monohybrid cross. The probability that a cross of $Aa \times Aa$ will produce an $A\_$ offspring is ¾. The probability that a cross of $Bb \times bb$ will produce a $B\_$ offspring is ½. The probability that a cross of $cc \times CC$ will produce a $C\_$ offspring is 1. To have all of these events occur simultaneously, multiply their probabilities: $¾ \times ½ \times 1 = ⅜$.

b. First determine what genotypes would fill the requirement of at least two dominant traits. Offspring could be $A\_bbC\_$, $aaB\_C\_$, or $A\_B\_C\_$. The genotype A_B_cc is not possible. Can you see why?

Probability of $A\_bbC\_ = ¾ \times ½ \times 1 = ⅜$
Probability of $aaB\_C\_ = ¼ \times ½ \times 1 = ⅛$
Probability of $A\_B\_C\_ = ¾ \times ½ \times 1 = ⅜$

Probability of offspring showing at least two dominant traits is the sum of these independent probabilities, or ⅞.

c. There is only one type of offspring that can show only one dominant trait ($aabbCc$). Can you see why? Its probability is $¼ \times ½ \times 1 = ⅛$. As you may notice, this would have been an easier way to determine the answer to **b**. With problems such as these, always consider what the question is asking from as many different angles as you can.

11.5. **a.** A: $I^A I^A$ and $I^A i$
b. B: $I^B I^B$ and $I^B i$
c. AB: $I^A I^B$
d. O: ii

11.6. The ratio of offspring from this $MmBb \times MmBb$ cross would be 9:3:4, a common ratio when one gene is epistatic to another. All epistatic ratios are modified versions of 9:3:3:1.

Phenotype	Genotype	Ratio
Black	$M\_B\_$	$¾ \times ¾ = {}^{9}/_{16}$
Gray	M_bb	$¾ \times ¼ = {}^{3}/_{16}$
White	mm	$¼ \times 1 = ¼$ or ${}^{4}/_{16}$

11.7. **a.** The parental cross produced 25-cm tall F$_1$ plants; all $AaBbCc$ plants with 3 units of 5 cm added to the base height of 10 cm.

b. As a general rule, in the polygenic inheritance of a quantitative character the number of phenotypes resulting from a cross of heterozygotes equals the number of alleles involved plus one. In this case, six alleles ($AaBbCc$) + 1 = 7. So, there will be seven different phenotypes in the F$_2$ among the 64 possible combinations of the eight types of F$_1$ gametes. (See why you wouldn't want to go through a Punnett square to figure that out!) These phenotypes will range from six dominant alleles (40 cm), five dominant (35 cm), four dominant (30 cm), and so on, to all six recessive alleles (10 cm), yielding a total of seven phenotypes.

11.8. **a.** This trait is recessive. If it were dominant, then albinism would be present in every generation, and it would be impossible to have albino children with two nonalbino (and thus homozygous recessive) parents.

b. father Aa; mother Aa, because neither parent is albino and they have albino offspring (aa)

c. mate #1 AA (probably); mate #2 Aa; grandson #4 Aa

d. The genotype of son #3 could be AA or Aa. If his wife is AA, then he could be Aa (since both his parents are carriers), and the recessive allele never would be expressed in his offspring. Even if he and his wife were both carriers (heterozygotes), there would be a ${}^{243}/_{1024}$ ($¾ \times ¾ \times ¾ \times ¾ \times ¾$) or 24% chance that all five children would be normally pigmented.

11.9. **a.** ¼
b. ⅔. Of offspring with a normal phenotype, ⅔ would be predicted to be heterozygotes and, thus, carriers of the recessive allele.

11.10. Both sets of prospective grandparents must have been carriers. The prospective parents do not have the disorder, so they are not homozygous recessive. Thus, each has a $2/3$ chance of being a heterozygote carrier. The probability that both parents are carriers is $2/3 \times 2/3 = 4/9$; the chance that two heterozygotes will have a recessive homozygous child is $1/4$. The overall chance that a child will inherit the disease is $4/9 \times 1/4 = 1/9$. The fact that the first two children are unaffected does not establish the genotype of the parents. Thus, the third child would also have a $1/9$ chance of inheriting the disorder. Should the third child have the disease, however, this would establish that both parents are carriers, and the chance that a subsequent child would have the disease is now estimated at $1/4$.

SUGGESTED ANSWERS TO STRUCTURE YOUR KNOWLEDGE

1. Mendel's law of segregation occurs in anaphase I, when alleles segregate as homologs move to opposite poles of the cell. The two cells formed from this division have one-half the number of chromosomes and one copy of each gene (although sister chromatids still have to separate in anaphase II). Mendel's law of independent assortment relates to the lining up of homologous chromosome pairs at the equatorial plate in a random fashion during metaphase I. Genes on different chromosomes will assort independently into gametes.

2. Aa = 2 different gametes $AaBb$ = 4 gametes
   $AaBbCc$ = 8 gametes $AABbCc$ = 4 gametes
   A general formula is 2^n, where n is the number of gene loci that are heterozygous. Note that in the last example, $AABbCc$, all gametes will contain a dominant A allele.

3. Always look to the F_1 heterozygote resulting from a cross of true-breeding parents. If alleles show complete dominance/recessiveness, then the heterozygote will have a phenotype identical to one parent. In incomplete dominance, the F_1 phenotype will be intermediate between that of the parents. If codominant, the phenotype of both alleles will be exhibited in the heterozygote.

ANSWERS TO GENETICS PROBLEMS

1. White alleles are dominant to yellow alleles. If yellow were dominant, then you should be able to get white squash from a cross of two yellow heterozygotes.

2. **a.** $1/4$ [$1/2$ (to get AA) $\times$ $1/2$ (bb)]
   b. $1/8$ [$1/4$ (aa) $\times$ $1/2$ (BB)]

c. $1/2$ [1 (Aa) $\times$ $1/2$ (Bb) $\times$ 1 (Cc)]
d. $1/32$ [$1/4$ (aa) $\times$ $1/4$ (bb) $\times$ $1/2$ (cc)]

3. Since flower color shows incomplete dominance, use symbols such as C^R and C^W for those alleles. There will be six phenotypic classes in the F_2 instead of the normal four classes found in a 9:3:3:1 ratio. You could find the answer with a Punnett square, but multiplying the probabilities of the monohybrid crosses is more efficient.

Tall red	$T\_C^RC^R$	$3/4 \times 1/4 = 3/16$
Tall pink	$T\_C^RC^W$	$3/4 \times 1/2 = 3/8$ or $6/16$
Tall white	$T\_C^WC^W$	$3/4 \times 1/4 = 3/16$
Dwarf red	$tt\ C^RC^R$	$1/4 \times 1/4 = 1/16$
Dwarf pink	$tt\ C^RC^W$	$1/4 \times 1/2 = 1/8$ or $2/16$
Dwarf white	$tt\ C^WC^W$	$1/4 \times 1/4 = 1/16$

4. Determine the possible genotypes of the mother and child. Then find the blood groups for the father that could not have resulted in a child with the indicated blood group.
   a. no groups exonerated
   b. A or O
   c. A or O
   d. AB only
   e. B or O

5. The parents are $CcBb$ and $Ccbb$. Right away you know that all cc offspring will die and no BB black offspring are possible because one parent is bb. Only four phenotypic classes are possible. Determine the proportion of each type by applying the law of multiplication.

Lethal ($cc\_\_$)	$1/4$ of all offspring die
Normal brown ($CCBb$)	$1/4 \times 1/2 = 1/8$
Normal white ($CCbb$)	$1/4 \times 1/2 = 1/8$
Deformed brown ($CcBb$)	$1/2 \times 1/2 = 1/4$
Deformed white ($Ccbb$)	$1/2 \times 1/2 = 1/4$
Ratio of viable offspring:	1:1:2:2

6. Father's genotype must be Pp since polydactyly is dominant and he has had one normal child. Mother's genotype is pp. The chance of the next child having normal digits is $1/2$ or 50% because the mother can only donate a p allele and there is a 50% chance that the father will donate a p allele.

7. The genotypes of the puppies were $3/8\ B\_S\_$, $3/8\ B\_ss$, $1/8\ bbS\_$, and $1/8\ bbss$. Because recessive traits show up in the offspring, both parents had to have had at least one recessive allele for both genes. Black:chestnut occurs in a 6:2 or 3:1 ratio, indicating a heterozygous cross. Solid:spotted occurs in a 4:4 or 1:1 ratio, indicating a cross between a heterozygote and a homozygous recessive. Parental genotypes were $BbSs \times Bbss$.

8. **a.** First figure out possible genotypes: $\_\_E\_$ = golden (any B combination with at least one E),

B_ee = black, *bbee* = brown. All you know at the start is that both parents are _ _*E*_. Since you see non-golden offspring, you know that both parents had to be heterozygous for *Ee*, and at least one was heterozygous for *Bb* (to get black and brown offspring). Since black and brown are in a 1:1 ratio (like a testcross ratio), then one hamster was *Bb* and the other was *bb*. You need to consider the results from the second cross to know which hamster was *Bb*.

b. For the second cross, you now know that the black hamster 3 is *B?ee* and the golden hamster 2 is *?bEe*. A ratio of approximately 3:1 black to brown looks like the results of a monohybrid cross, so both parents must be *Bb*. So hamster 3 is *Bbee*, and hamster 2 must be *BbEe*. That means that hamster 1 must be *bbEe*.

9.

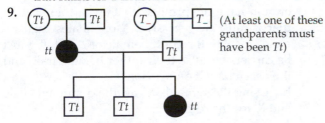

(At least one of these grandparents must have been *Tt*)

10. Since the parents were true-breeding for two characters, the F₁s would be dihybrids. Since all F₁s had red, terminal flowers, those two traits

must be dominant, and their genotypes could be represented as *RrTt*. One would predict an F₂ phenotypic ratio of 9 red, terminal:3 red, axial:3 white, terminal:1 white, axial. If 100 offspring were counted, one would expect approximately 19 (³⁄₁₆ × 100) plants with red, axial flowers.

ANSWERS TO TEST YOUR KNOWLEDGE

Multiple Choice:

1. c	**5.** a	**9.** c	**13.** b
2. b	**6.** c	**10.** d***	**14.** b
3. c	**7.** d*	**11.** d	**15.** a
4. c	**8.** d**	**12.** e	**16.** c

*There are three different ways to get this outcome: HHT, HTH, THH. Each outcome has a probability of ⅛.

**Remember that only homozygotes are true-breeding. There are four possible homozygous genotypes: AABB, aabb, AAbb, and aaBB, with a probability of ¹⁄₁₆ for each one.

***The 34-cm plant would be quadruply homozygous dominant, and the F₁ would be quadruply heterozygous. The number of phenotypic classes in the F₂ would equal the number of alleles plus 1.

CHAPTER 12: THE CHROMOSOMAL BASIS OF INHERITANCE

FOCUS QUESTIONS

12.1.

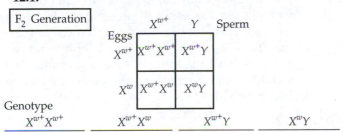

Genotype			
$X^{w+}X^{w+}$	$X^{w+}X^{w}$	$X^{w+}Y$	$X^{w}Y$

Phenotype

red-eyed female	red-eyed female	red-eyed male	white-eyed male

12.2. The gene is X-linked, so a good notation is X^N, X^n, and Y so that you will remember that the Y does not carry the gene. Capital *N* indicates normal sight. Genotypes are:
1. X^NY
2. X^NX^n
3. X^NX^N or X^NX^n (probably X^NX^N since four sons are X^NY)

4. X^NX^n
5. X^NY
6. X^nY
7. X^NX^n

12.3. **a.** The phenotypes of offspring that are parental types are tall, purple-flowered and dwarf, white-flowered.
b. The phenotypes of offspring that are recombinants are tall, white-flowered and dwarf, purple-flowered.

12.4. If linked genes have their loci close together on the same chromosome, they travel together during meiosis and *more* parental offspring are produced. Recombinants are the result of crossing over between nonsister chromatids of homologous chromosomes.

12.5. Solving a linkage problem is often a matter of trial and error. Sometimes it helps to lay out the loci with the greatest distance between them and fit the other genes between or on either side by adding or subtracting map unit distances.

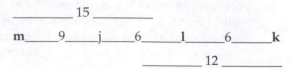

12.6. **a.** An organism with a trisomy ($2n + 1$) has an extra copy of one chromosome, usually caused by a nondisjunction during meiosis. A triploid organism ($3n$) has an extra set of chromosomes, possibly caused by a total nondisjunction in gamete formation.
b. A trisomy would probably disrupt the genetic balance more than having a complete extra set of chromosomes.

12.7. translocation (O-P-Q), deletion (D-E-F), and inversion (J-I-H).

12.8. Aneuploidies of sex chromosomes appear to upset genetic balance less, perhaps because relatively few genes are located on the Y chromosome and extra X chromosomes are inactivated as Barr bodies.

SUGGESTED ANSWERS TO STRUCTURE YOUR KNOWLEDGE

1. Genes that are not linked assort independently, and the ratio of offspring from a testcross with a dihybrid heterozygote should be 1:1:1:1 (*AaBb* × *aabb* gives *AaBb*, *Aabb*, *aaBb*, *aabb* offspring and a recombination frequency of 50%—50% of offspring are recombinants). Genes that are linked and do not cross over should produce a 1:1 ratio in such a testcross (*AaBb* and *aabb* because a heterozygote derived from a parental cross of *AABB* × *aabb* produces only *AB* and *ab* gametes). If crossovers between distant genes almost always occur, the heterozygote will produce equal quantities of *AB*, *Ab*, *aB*, and *ab* gametes, and the genotype ratio of offspring will be 1:1:1:1, the same as it is for unlinked genes. Because each 1% recombination frequency is equal to 1 map unit, this measurement ceases to be meaningful at relative distances of 50 or more map units. However, crosses with intermediate genes on the chromosome could establish both that the genes are on the same chromosome and that they are a certain map unit distance apart.

2. A cross between a mutant female fly and a normal male should produce all normal females (who get a wild-type allele from their father) and all mutant male flies (who get the mutant allele on the X chromosome from their mother). $X^m X^m \times X^{m+} Y$ produces $X^{m+} X^m$ and $X^m Y$ offspring (normal females and mutant males).

3. The serious phenotypic effects that are associated with these chromosomal alterations indicate that normal development and functioning are dependent on genetic balance. Most genes appear to be vital to an organism's existence, and extra copies of genes upset genetic balance. Inversions and translocations, which do not disrupt the balance of genes, can alter phenotype because of the effect of neighboring genes on gene functioning.

ANSWERS TO GENETICS PROBLEMS

1. **a.** The trait is recessive and probably X-linked. Two sets of unaffected parents in the second generation have offspring with the trait, indicating that it must be recessive. More males than females display the trait. Females #3 and #5 must be carriers because their father has the trait, and they each pass it on to a son.
   b. Using the symbols X^T for the dominant allele and X^t for the recessive:
   1. $X^t Y$
   2. $X^T X^t$
   3. $X^T X^t$
   4. $X^T Y$
   5. $X^T X^t$
   6. $X^T X^t$ or $X^T X^T$
   7. $X^T X^t$
   c. Female #6 has a brother who has the trait and her mother's father had the trait, so her mother must be a carrier of the trait, meaning there is a ½ probability that #6 is a carrier. If she is mated to a phenotypically normal male, none of her daughters will show the trait (0 probability). Her sons have a ½ chance of having the trait if she is a carrier. There is a ½ × ½ = ¼ probability that her sons will have the trait. There is a ½ chance that a child would be male, so the probability of an affected child is ½ × ¼, or ⅛.

2. c, a, b, d

3. The genes appear to be linked because the parental types appear most frequently in the offspring. Recombinant offspring represent 10 out of 40 total offspring for a recombination frequency of 25%, indicating that the genes are 25 map units apart.

4. One of the mother's X chromosomes carries the recessive lethal allele. One-half of male fetuses would be expected to inherit that chromosome and spontaneously abort. Assuming an equal sex ratio at conception, the ratio of girl to boy children would be 2:1, or six girls and three boys.

5. **a.** For female chicks to be black, they must have received a recessive allele from the male parent. If all female chicks are black, the male parent must have been Z^bZ^b. If the male parent was homozygous recessive, then all male offspring will receive a recessive allele, and the female parent would have to be Z^BW to produce all barred males.

b. For female chicks to be both black and barred, the male parent must have been Z^BZ^b. If the female parent were Z^BW, only barred male chicks would be produced. To get an equal number of black and barred male chicks, the female parent must have been Z^bW.

ANSWERS TO TEST YOUR KNOWLEDGE

Multiple Choice:

1.	e	**5.**	d	**9.**	c	**13.**	d
2.	a	**6.**	d	**10.**	c	**14.**	e
3.	b	**7.**	e	**11.**	a	**15.**	d
4.	d	**8.**	b	**12.**	b		

CHAPTER 13: THE MOLECULAR BASIS OF INHERITANCE

FOCUS QUESTIONS

13.1. **a.** They grew phage with radioactive sulfur, which tagged phage proteins.

b. They grew phage with radioactive phosphorus, which labeled the phage DNA.

c. Radioactivity was found in the liquid, indicating that the phage protein did not enter the bacterial cells.

d. In the samples with the labeled DNA, most of the radioactivity was found in the bacterial cell pellet.

e. They concluded that viral DNA is injected into the bacterial cells and serves as the hereditary material for viruses.

13.2. **a.** sugar-phosphate backbone
b. 3′ end of chain
c. hydrogen bonds
d. cytosine (pyrimidine)
e. guanine (purine)
f. adenine (purine)
g. thymine (pyrimidine)
h. 5′ end of chain
i. nucleotide
j. deoxyribose
k. phosphate group
l. 3.4 nm
m. 0.34 nm
n. 2 nm

13.3.

DNA molecules — Density bands

First generation (grown on light ^{14}N medium)

Second generation (grown on light ^{14}N medium)

13.4. The phosphate end of each strand is the 5′ end, and the hydroxyl group extending from the 3′ carbon of the sugar marks the 3′ end. See Figure 13.7 in the text.

13.5. **a.** single-strand binding protein
b. DNA pol III
c. leading strand
d. 5′ end of parental strand
e. 3′ end
f. helicase
g. RNA primer
h. primase
i. DNA pol III
j. Okazaki fragment
k. DNA pol I (replacing primer)
l. lagging strand
m. DNA ligase

13.6. nucleosomes (10-nm fiber of nucleosomes and linker DNA); 30-nm fiber; looped domains (300-nm fiber); coiling and folding of looped domains into highly condensed metaphase chromosome.

13.7. The third sequence, because it has the same sequence running in opposite directions. The enzyme would probably cut between G and A, producing AATT and TTAA sticky ends.

13.8.
  a. bacterial chromosome
  b. plasmid
  c. gene of interest
  d. chromosome source of "foreign" DNA
  e. recombinant plasmid
  f. recombinant bacterium
  g. clone of recombinant bacteria
  1. Gene of interest is inserted into plasmid.
  2. Recombinant plasmid put into bacterial cell.
  3. Recombinant bacteria grown in culture to form a clone of cells containing cloned gene. Copies of the cloned gene can be used in basic research or genetic engineering; protein product of the gene may be harvested from bacteria and used in various applications.

13.9.
  a. Primers may be synthesized to include a restriction site that matches a restriction site on the cloning vector, facilitating the production of recombinant plasmids. By amplifying the gene prior to cloning, the later task of identifying clones carrying the desired gene is simplified.
  b. There is a limit to the number of accurate copies that can be made due to the accumulation of relatively rare copying errors. Large quantities of a gene are better prepared by DNA cloning in cells.

SUGGESTED ANSWERS TO STRUCTURE YOUR KNOWLEDGE

1. Watson and Crick used Franklin's X-ray diffraction photo to deduce that DNA was a helix, 2 nm wide, with nitrogenous bases stacked 0.34 nm apart, and making a full turn every 3.4 nm. Franklin had concluded that the sugar-phosphate backbones were on the outside of the helix with the bases extending inside. Using molecular models of wire, Watson and Crick experimented with various arrangements and finally paired a purine base with a pyrimidine base, which produced the proper diameter. Specificity of base pairing (A with T and C with G) is ensured by hydrogen bonds.

2. Replication bubbles form where proteins recognize specific base sequences and open up the two strands. *Helicase*, an enzyme that works at the replication fork, unwinds the helix and separates the strands. *Single-strand binding proteins* support the separated strands while replication takes place. *Topoisomerase* eases the twisting ahead of the replication fork by breaking, untwisting, and rejoining DNA strands. *Primase* synthesizes a primer of 5–10 RNA bases to start the new strand. After a proper base pairs up on the exposed template, *DNA polymerase III* joins the nucleotide to the 3′ end of the new strand. On the lagging strand, short Okazaki fragments are formed by primase and DNA pol III (again moving 5′ → 3′). *DNA polymerase I* replaces the primer with DNA. *Ligase* joins the 3′ end of one fragment to the 5′ end of its neighbor. Proofreading enzymes check for mispaired bases, whereas nucleases, DNA polymerase, ligase, and other enzymes repair damage or mismatches.

3. Complementary single strands of DNA will base-pair, enabling plasmids and foreign DNA cut by the same endonuclease to match up and be joined by ligase, as well as the RNA primers used in PCR to bind with flanking ends of a region of DNA that is to be copied. Complementary base pairing also allows the sequencing of DNA as monitors determine the complementary nucleotides that sequentially bind in the replication of a new DNA strand.

ANSWERS TO TEST YOUR KNOWLEDGE

Multiple Choice:

1.	c	7.	a	13.	a	19.	a	25.	a
2.	a	8.	c	14.	d	20.	d	26.	c
3.	b	9.	a	15.	b	21.	c	27.	d
4.	d	10.	e	16.	c	22.	e	28.	d
5.	b	11.	d	17.	e	23.	d		
6.	b	12.	e	18.	c	24.	b		

CHAPTER 14: GENE EXPRESSION: FROM GENE TO PROTEIN

FOCUS QUESTIONS

14.1. **a.** RNA differs from DNA in three ways: The sugar component of RNA nucleotides is ribose, rather than deoxyribose; uracil (U) replaces thymine as one of its nitrogenous bases; and RNA is usually single stranded.

b. DNA $\xrightarrow{\text{transcription}}$ RNA $\xrightarrow{\text{translation}}$ protein

14.2. Met Pro Asp Phe Lys stop

14.3. **a.** Initiation: Transcription factors bind to promoter and facilitate the binding of RNA polymerase II, forming a transcription initiation complex; RNA polymerase II separates DNA strands and RNA synthesis begins at the start point.

b. Elongation: RNA polymerase II moves along the DNA strand, connecting RNA nucleotides that have paired to the DNA template to the 3' end of the growing RNA strand.

c. Termination: After polymerase transcribes past a polyadenylation signal sequence, the pre-mRNA is cut and released.

14.4. A 5' cap consisting of a modified guanine nucleotide is added to the 5' UTR. A poly-A tail consisting of up to 250 adenine nucleotides is attached to the 3' UTR. Spliceosomes have cut out the introns and spliced the exons together.

14.5.

DNA Triplet 3'→5'	mRNA Codon 5'→3'	Anticodon 3'→5'	Amino Acid
TAC	AUG	UAC	methionine
GGA	CCU	GGA	proline
TTC	AAG	UUC	lysine
ATC	UAG	AUC	stop

14.6. **1.** Codon recognition: An elongation factor (not shown) helps an aminoacyl tRNA into the A site where its anticodon base-pairs to the mRNA codon; hydrolysis of GTP increases accuracy and efficiency.

2. Peptide bond formation: Ribosome catalyzes peptide bond formation between new amino acid and polypeptide held in the P site.

3. Translocation: The empty tRNA in the P site is moved to the E site and released; the tRNA now holding the polypeptide is moved from the A to the P site, taking the mRNA with it; GTP is required.

4. Termination: A release factor binds to stop codon in the A site and promotes hydrolysis of bond between polypeptide and the tRNA in the P site. Free polypeptide is released and leaves through the exit tunnel. Ribosomal subunits and other assembly components separate. GTP is required.

a. amino end of growing polypeptide
b. aminoacyl tRNA
c. large subunit
d. A site
e. small subunit
f. 5' end of mRNA
g. peptide bond formation
h. E site
i. release factor
j. stop codon
k. P site
l. free polypeptide

14.7. A ribosome that is translating an mRNA that codes for a secretory or membrane protein will become bound to the ER when an initial signal peptide on the polypeptide it is synthesizing is bound by an SRP (signal-recognition particle), which then attaches to an ER receptor protein.

14.8. **a.** Silent: a nucleotide-pair substitution producing a codon that still codes for the same amino acid.

b. Missense: a nucleotide-pair substitution or frameshift mutation that results in a codon for a different amino acid.

c. Nonsense: a nucleotide-pair substitution or frameshift mutation that creates a stop codon and prematurely terminates translation.

d. Frameshift: an insertion or deletion of one, two, or more than three nucleotides that disrupts the reading frame and creates extensive missense and nonsense mutations.

SUGGESTED ANSWERS TO STRUCTURE YOUR KNOWLEDGE

1. Messenger RNA (mRNA) carries the code that specifies an amino acid sequence from DNA to ribosomes, where the RNA's sequence of nucleotides is translated into a polypeptide.

   Transfer RNA (tRNA) carries a specific amino acid to its position in a polypeptide based on matching its anticodon to an mRNA codon.

Ribosomal RNA (rRNA) makes up about two-thirds of a ribosome and has specific binding and catalytic functions.

   Small RNAs are part of spliceosomes and play a catalytic role in splicing pre-mRNA.

   The versatility of RNA molecules stems from their ability to base-pair with other RNA or DNA molecules, to form specific three-dimensional shapes by base-pairing within itself, and to act as a catalyst.

2.

	Transcription	Translation
Template	DNA	RNA
Location	nucleus (cytoplasm in prokaryotes)	cytoplasm; ribosomes can be free or attached to ER
Molecules involved	RNA nucleotides, DNA template strand, RNA polymerase, transcription factors	amino acids; tRNA; mRNA; ribosomes; ATP; GTP; enzymes; initiation, elongation, and release factors
Enzymes involved	RNA polymerases, spliceosomes	aminoacyl-tRNA synthetase, ribosomal enzymes (ribozymes)
Control—start and stop	transcription factors locate promoter region with TATA box and start point, polyadenylation signal sequence	initiation factors, initiation sequence (AUG), stop codons, release factor
Product	primary transcript (pre-mRNA)	polypeptide
Product processing	RNA processing: 5' cap and poly-A tail, splicing of pre-mRNA—introns removed by spliceosomes	spontaneous folding, disulfide bridges, signal peptide removed, cleaving, quaternary structure, modification with sugars, and so on.

3. The genetic code consists of the RNA triplets that code for amino acids. The order of nucleotides in these codons is specified by the sequence of nucleotides in DNA, which is transcribed into the codons found on mRNA and translated into their corresponding sequence of amino acids. There are 64 possible mRNA codons created from the four nucleotides used in the triplet code (4^3).

   Redundancy of the code refers to the fact that several triplets may code for the same amino acid. Often these triplets differ only in the third nucleotide. The wobble phenomenon explains the fact that there are only about 45 different tRNA molecules that pair with the 61 possible codons (three codons are stop codons). The third nucleotide of many tRNAs can pair with more than one type of nucleotide. Because of the redundancy of the genetic code, these wobble tRNAs still place the correct amino acid in position.

   The genetic code is nearly universal; each codon codes for exactly the same amino acid in almost all organisms. (Some exceptions have been found.) This universality points to an early evolution of the code in the history of life and the evolutionary relationship of all life on Earth.

4.

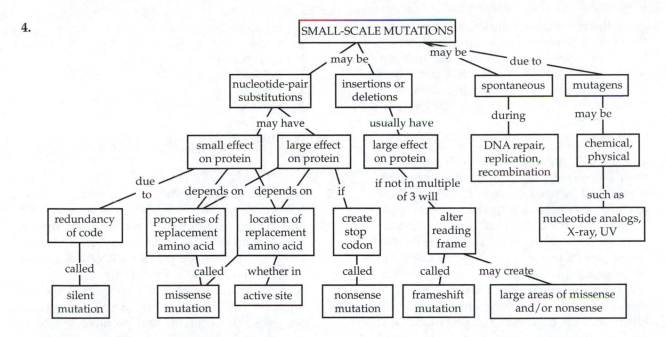

ANSWERS TO TEST YOUR KNOWLEDGE

Multiple Choice:

1. c	5. e	9. c	13. c	17. b	21. d	25. e
2. a	6. c	10. e	14. b	18. b	22. e	26. a
3. b	7. b	11. d	15. e	19. a	23. e	27. d
4. b	8. c	12. c	16. e	20. a	24. c	28. c

CHAPTER 15: REGULATION OF GENE EXPRESSION

FOCUS QUESTIONS

15.1. a. regulatory gene
       b. promoter
       c. operator
       d. genes coding for enzymes
       e. operon
       f. RNA polymerase
       g. active repressor
       h. inducer (allolactose)
       i. mRNA for enzymes for lactose utilization

15.2. a. anabolic; corepressor; on; inactive
       b. catabolic; inducers; off; active

15.3. a. Barr body—one X chromosome is compacted in cells of female
       b. Histone tail deacetylation would decrease transcription because it would make genes in the nucleosome less accessible.

15.4. a. distal control elements in enhancer
       b. activators
       c. DNA-bending protein
       d. promoter

       e. mediator proteins
       f. general transcription factors
       g. TATA box
       h. RNA polymerase II

15.5. a. Many genes code for several different proteins, depending on which exons are combined. Thus, the expression of a particular gene may depend on the regulatory proteins involved in alternate RNA splicing that are present within a cell.
       b. Regulatory proteins may bind to sequences in the 5′ or 3′ UTR and block attachment of ribosomes, thereby decreasing gene expression. Sequences in the 3′ UTR may affect the length of time an mRNA remains intact, thereby either increasing or decreasing gene expression.

15.6. a. A single-stranded miRNA forms a complex with proteins. The complex binds to mRNA with complementary base sequences, and the mRNA is degraded (if the miRNA and mRNA bases are complementary all along their length) or translation is blocked (if the match between bases is less complete).

b. Evidence indicates that they can do both. miRNAs and siRNAs both affect translation by degrading mRNA or blocking translation. Experiments have shown that siRNAs may silence transcription by changing chromatin structure.

15.7. **a.** RT-PCR: The mRNA from different tissue samples could be isolated. Reverse transcriptase would be used to make cDNA of all the mRNA, and PCR using specific primers could amplify only the gene of interest. Running the samples on a gel would show bands only in the tissues that were expressing the gene. Another possibility would be *in situ* hybridization if the tissues you are studying are readily identified in the whole organism or a section taken from the organism. Labeled probes that are complementary to the mRNA transcribed from that gene would identify tissues in which the gene is expressed.

b. DNA microarray assay: As in RT-PCR, mRNA is isolated from different tissues, and cDNA is made and labeled with fluorescent dye. The cDNA is applied to a microarray (single-stranded DNA fragments of the genes of an organism arranged in a grid). The different cDNA will hybridize with the genes that were expressed in the tissue. The intensity with which the hybridized spots fluoresce indicates the relative amount of mRNA that was in the tissue.

f. inactive
g. active
h. corepressor
i. inducer
j. anabolic
k. activate
l. inactivate
m. catabolic
n. cAMP
o. lack of glucose

2. **a.** DNA packing into nucleosomes; histone tail acetylation increases, whereas deacetylation and methylation of tails decreases transcription; methylation of DNA may be involved in long-term inactivation of genes; ncRNAs may promote heterochromatin formation.

b. Specific transcription factors (activators) bind with control elements in enhancers, then interact with mediator proteins and promoter region to form transcription initiation complex; repressors can inhibit transcription.

c. Alternative splicing of primary RNA transcript, 5′ cap and poly-A tail added.

d. Nucleotide sequences in the 3′ UTR affect life span of mRNA, and miRNAs and siRNAs target mRNA for degradation.

e. Repressor proteins and miRNA or siRNA may prevent translation (or short poly-A tail length can allow mRNA stockpiling in ovum); activation of initiation factors begins translation.

f. Protein processing by cleavage or modification; transport to target location; selective degradation of proteins marked with ubiquitin.

SUGGESTED ANSWERS TO STRUCTURE YOUR KNOWLEDGE

1. **a.** operons
b. promoter
c. operator
d. negative control
e. repressor

ANSWERS TO TEST YOUR KNOWLEDGE

Multiple Choice:

1. b	4. a	7. e	10. b	13. b
2. c	5. b	8. c	11. b	14. d
3. e	6. b	9. b	12. e	

CHAPTER 16: DEVELOPMENT, STEM CELLS, AND CANCER

FOCUS QUESTIONS

16.1. A cell is said to be determined when its developmental fate is set. Its series of gene activations and inactivations has set it on the path to express the genes for tissue-specific proteins. When it produces these proteins and develops its characteristic structure, the cell has become differentiated.

16.2. Bicoid mRNA was shown to be localized at one end of the unfertilized egg; later in development, Bicoid protein occurred in a gradient that was most concentrated in the anterior cells of the embryo. Also, injection of bicoid mRNA into various regions of early embryos caused anterior structures to form at those sites.

16.3. DNA methylation and histone acetylation help to regulate gene expression. An adult cell must have these epigenetic changes in its chromatin reprogrammed in order to support normal gene expression during development. The DNA of many cloned embryos has been found to be improperly methylated.

16.4. **a.** Mutations may result in (1) *gene amplification,* in which more copies of the gene are present than normal; (2) *translocation or transposition,* which may bring the gene under the control of a more active promoter or control element; or (3) *point mutation,* resulting in a change in a nucleotide sequence in either a control element that increases gene expression or in the gene that creates a more active or resilient protein.

b. Tumor-suppressor proteins may function in repair of damaged DNA, control of cell adhesion, or inhibition of the cell cycle.

and regulate transcription. Inducers must communicate between cells. They are often proteins that bind to cell surface receptors and initiate a signal transduction pathway involving a cascade of enzyme activations, usually leading to the activation of transcription factors within the target cell.

2. Stem cells are undifferentiated cells that can continually divide to form more stem cells or can differentiate into various types of cells. Embryonic stem cells are cells from the blastula stage or earlier that are able to continually divide and differentiate into many, if not all, different types of cells. Adult stem cells are present in small numbers in various tissues and can divide to form a limited number of cell types. Induced pluripotent stem cells have been transformed from differentiated cells by the introduction of copies of "stem cell" master regulatory genes. These cells may be able to function as ES cells. Yes, we can consider the cells of the apical meristem of shoots and roots to be stem cells in that these undifferentiated, continually dividing cells can develop into all types of plant tissues.

SUGGESTED ANSWERS TO STRUCTURE YOUR KNOWLEDGE

1. Most cytoplasmic determinants are mRNA for transcription factors that are divided by the first few mitotic divisions. They are present in the cells, and their translated product can enter the nucleus

ANSWERS TO TEST YOUR KNOWLEDGE

Multiple Choice:

1. e	**4.** d	**7.** c	**9.** e
2. a	**5.** e	**8.** e	**10.** b
3. c	**6.** b		

CHAPTER 17: VIRUSES

FOCUS QUESTIONS

17.1. **1.** Phage attaches to host cell and injects DNA.
2. Phage DNA forms a circle. Certain factors determine which cycle is entered.
3. New phage DNA and proteins are synthesized and self-assemble into phages.
4. Bacterium lyses, releasing phages.
5. Phage DNA integrates into bacterial chromosome, becoming a prophage.
6. Bacterium reproduces, passing prophage to daughter cells.
7. Large population of infected bacteria forms.
8. Occasionally, prophage exits bacterial chromosome and begins lytic cycle.

a. phage DNA
b. bacterial chromosome
c. new phages
d. prophage
e. replicated bacterial chromosome with prophage

17.2. RNA → DNA → RNA; viral reverse transcriptase, host RNA polymerase

17.3. Viral particles spread easily through plasmodesmata, the cytoplasmic connections between plant cells. As there are no cures for plant viral diseases, reducing the spread of infection and breeding resistant varieties are the best approaches.

SUGGESTED ANSWERS TO STRUCTURE YOUR KNOWLEDGE

1.

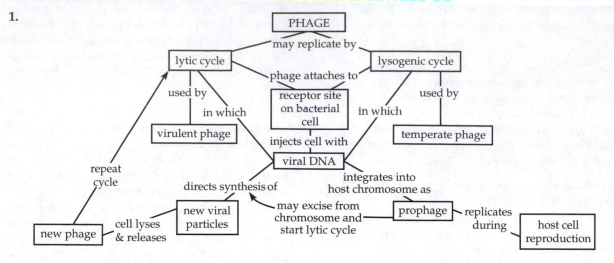

2. **a.** DNA
 b. RNA
 c. protein capsid
 d. host cell
 e. bacterium
 f. lytic or lysogenic cycle
 g. animal
 h. membraneous envelope
 i. reverse transcriptase
 j. plant
 k. plasmids or transposons

ANSWERS TO TEST YOUR KNOWLEDGE

Multiple Choice:

1. b	**4.** a	**7.** b
2. c	**5.** d	**8.** c
3. e	**6.** b	**9.** b

CHAPTER 18: GENOMES AND THEIR EVOLUTION

FOCUS QUESTIONS

18.1. Microbial species—metagenomics enables the sequencing of a mixture of DNA from an environmental sample without the need to grow each species separately in the lab.

18.2. Comparing nucleotide and amino acid sequences may reveal matches with other genes of known function or reveal common amino acid domains for which a function is known and can identify similarities with closely or distantly related species that help trace a species' evolutionary history.

18.3. **a.** the archaean *Archaeoglobus fulgidus* (1,130/Mb); humans (7/Mb)
 b. rice, *Oryza sativa* (40,600); the bacterium *Haemophilus influenzae* (1,700)
 c. a plant, *Fritillaria assyriaca* (124,000 Mb); *H. influenzae* (1.8 Mb)

d. less than 21,000; alternative splicing of exons can allow each gene to code for more than one polypeptide; post-translational processing can also alter polypeptides

18.4. They are first transcribed into RNA; thus, the original retrotransposon remains in place. Reverse transcriptase converts the RNA to DNA, which is inserted into another site in the genome.

18.5. 1. D; 1.5
 2. I; 20
 3. H; 5
 4. B; 44
 5. E; 10
 6. G; 17
 7. C; 15
 8. F; 5–6
 9. A; 3

18.6. The lysozyme gene, which codes for a bacterial infection-fighting enzyme, was present in the last common ancestor of birds and mammals. After their lineages split, the gene underwent a duplication event in the mammalian lineage, and a copy of the lysozyme gene evolved into a gene coding for a protein involved in milk production.

18.7. **a.** Exon shuffling can happen through an error in meiotic recombination, which may occur between two homologous transposable elements, or by the inclusion or tagging along of an exon with a transposable element, which moves a copy of the exon to a new location.
b. Exons often code for domains of a protein. Providing a new domain to a protein may enhance or change its function.

18.8. The homeobox codes for a DNA-binding homodomain, while other domains specific to each regulatory gene interact with transcription factors to recognize particular enhancers or promoters and thus control different batteries of developmental genes.

SUGGESTED ANSWERS TO STRUCTURE YOUR KNOWLEDGE

1. Many non-protein-coding sequences are highly conserved across species, indicating that they perform some important, but as yet unidentified, function. Much of this DNA is transcribed into RNA molecules, some of whose functions are being discovered.
a. Transposable elements include transposons, which are DNA segments, and retrotransposons, which are produced from RNA transcribed into DNA. These transposable elements are very common and move DNA to new locations in the genome.

b. *Alu* elements are about 300 nucleotides long and make up about 10% of the human genome. Some are transcribed into RNA, but their function is unknown. They may provide alternate splice sites for RNA processing.
c. L1 sequences are long retrotransposons that rarely move about. They are found in many introns and may help regulate gene expression.
d. Simple sequence DNAs are highly repetitive tandem sequences found at centromeres and telomeres of a chromosome. They appear to have structural functions in organizing chromatin. Sequences with fewer repetitions are called STRs (short tandem repeats) and are used in genetic profiling.
e. Pseudogenes are remnants of genes that are no longer functional because mutations have altered their regulatory sequences.

2. Chromosome duplication provides additional copies of genes that may undergo mutation and produce new proteins. Rearrangements of chromosome sections, such as duplications, inversions, and translocations, may create reproductive barriers between populations that lead to the formation of new species. Errors during meiosis can lead to duplications of genes or exons, or exchange of exons, providing genetic material that may take on related or novel functions. Transposable elements can facilitate recombination between different chromosomes, can disrupt genes or control elements, and can carry genes or exons to new locations.

ANSWERS TO TEST YOUR KNOWLEDGE

Multiple Choice:

1. c	4. d	7. e	10. d
2. d	5. b	8. e	11. b
3. a	6. a	9. c	12. d

CHAPTER 19: DESCENT WITH MODIFICATION

FOCUS QUESTIONS

19.1. **a.** 1. E f
2. A b
3. B e
4. C d, g
5. F a
6. D c
7. G g
b. a, f, d, b, e, g, c

19.2. Observation 1: Members of a population vary in their inherited traits.
Observation 2: All species can produce more offspring than the environment can support, and many offspring do not survive.

19.3.   **a.** When a new antibiotic is developed, it is effective at killing most pathogens. A few, however, may have an inherent resistance, such as an enzyme like penicillinase that destroys penicillin. As the continued use of the new antibiotic selects against bacteria without an antibiotic-resistance gene, the resistant bacteria can multiply rapidly and the "beneficial" gene becomes more common.

   b. Resistance to multiple antibiotics can evolve if bacteria that are already resistant to one antibiotic are exposed to a different one. The new antibiotic, again, will probably kill most bacteria but "select" any that happen to have a variation that provides resistance to it. The exchange of genes between different strains and even different species of bacteria may also contribute to the evolution of multidrug resistant strains.

19.4.   **a.** biogeography
   b. fossil record
   c. homologies
   d. island species and mainland species or neighboring island species
   e. ancestral and transitional forms
   f. homologous structures
   g. functions and form
   h. molecular comparisons
   i. DNA sequences and proteins
   j. descent from a common ancestor

SUGGESTED ANSWERS TO STRUCTURE YOUR KNOWLEDGE

1. The two main components of Darwin's evolutionary theory are that all of life has descended from a common ancestor and that this evolution has been the result of natural selection. The theory of natural selection is based on several key observations and inferences. Members of a population often vary in their inherited traits. The overproduction of offspring in conjunction with limited resources leads to the unequal reproductive success of those organisms best suited to the local environment. The increased survival and reproduction of the best-adapted individuals in a population leads to the gradual accumulation of favorable characteristics in a population.

ANSWERS TO TEST YOUR KNOWLEDGE

Multiple Choice:

1. c	**5.** d	**9.** e	**13.** c
2. e	**6.** d	**10.** d	**14.** b
3. a	**7.** b	**11.** c	
4. e	**8.** a	**12.** d	

CHAPTER 20: PHYLOGENY

FOCUS QUESTIONS

20.1.   **a.** #1
   b. A
   c. #4
   d. B and C

20.2. Adaptations to different environments may lead to large morphologic differences within related groups, and convergent evolution can produce similar structures in unrelated organisms in response to similar selective pressures. Examples are the varied Hawaiian silversword plants and the similar Australian and North American burrowing moles.

20.3.

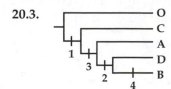

20.4. The four-chambered heart of birds and mammals is analogous, not homologous. It evolved independently in the two groups. An abundance of evidence supports the hypothesis that birds and mammals evolved from different reptilian ancestors.

20.5. Assuming that genes have a fairly constant rate of mutation, in a protein that has a crucial function and whose amino acid sequence is central to that function, mutations that are harmful would be quickly removed from the population. Therefore, fewer mutations would be neutral and remain in the genome. If the exact sequence of amino acids in a protein is less critical to survival, more mutations would be neutral and remain. Thus, in the same amount of time, the sequence of the second gene will change more than the sequence of the more crucial gene.

20.6.
a. Eukarya
b. Archaea
c. Bacteria
d. endosymbiosis of proteobacterial ancestor of mitochondria
e. endosymbiosis of cyanobacterial ancestor of chloroplasts and other plastids

SUGGESTED ANSWERS TO STRUCTURE YOUR KNOWLEDGE

1.
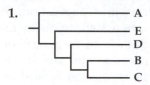

2. In order to reconstruct the evolutionary history of a group of organisms, systematists consider morphological and molecular homologies as well as evidence from the fossil record. Molecular homologies may be identified using computer programs that search for and align common sequences in homologous sections of DNA. Statistical analysis can help sort distant molecular homologies from coincidental homoplasies. Using the tech[...] cladistics, the ingroup of organisms to class[...] compared with an outgroup and with each oth[...] Shared ancestral characters and shared derived characters are used to determine the sequence of evolutionary divergence. Computer programs can determine the most parsimonious tree (requiring the fewest evolutionary changes).

3. A molecular clock is based on the observations that certain regions of DNA evolve (accumulate base changes) at a steady rate. By plotting the number of differences in DNA sequences between groups known to have diverged at specific times, scientists can apply this rate of molecular evolution to infer the timing of other evolutionary divergences.

ANSWERS TO TEST YOUR KNOWLEDGE

Multiple Choice:

1. c	5. d	9. e	13. c
2. d	6. a	10. b	14. b
3. e	7. d	11. b	15. b
4. c	8. e	12. c	16. e

CHAPTER 21: THE EVOLUTION OF POPULATIONS

FOCUS QUESTIONS

21.1.
a. mutation
b. sexual reproduction
c. Prokaryotes and viruses have very short generation times, and a new beneficial mutation can increase in frequency rapidly in an asexually reproducing population. Although mutations are the source of new alleles, they are so infrequent that their contribution to genetic variation in a large, diploid population is minimal. However, the production and union of gametes produces zygotes with fresh combinations of alleles each generation.

21.2.
a. The frequencies of genotypes are 0.49 *BB*, 0.42 *Bb*, and 0.09 *bb* ($^{98}/_{200}$, $^{84}/_{200}$, $^{18}/_{200}$).
b. 0.7 *B*; 0.3 *p*. The 98 *BB* mice contribute 196 *B* alleles, and the 84 *Bb* mice contribute 84 *B* alleles to the gene pool. These 84 *Bb* mice also contribute 84 *b* alleles, and the 18 *bb* mice contribute 36 *b* alleles. Of a total of 400 alleles, 280 are *B* and 120 are *b*. Allele frequencies are 0.7 *B* and 0.3 *b*. Another way to determine allele frequencies is from genotype frequencies: Add the frequency of the homozygous dominant genotype and ½ the frequency of the heterozygote to determine the frequency of *p*. For *q*, add the homozygous recessive frequency and ½ the heterozygote frequency. Or determine *q* from 1 – *p*.

21.3. 0.7; 0.3; 0.49; 0.42; 0.09

21.4.
a. 0.36; 0.48; 0.16. Plug *p* (0.6) and *q* (0.4) into the expanded binomial: $p^2 + 2pq + q^2 = 1$.
b. 0.6; 0.4. Add the frequencies of *AA* + ½ *Aa* to get *p*. Add the frequencies of *aa* + ½ *Aa* to get *q*. Alternatively, to determine *q*, take the square root of the homozygous recessive frequency if you are sure the population is in Hardy-Weinberg equilibrium. The frequency of *p* is then 1 – *q*.
c. These results would suggest that the population is not in Hardy-Weinberg equilibrium and thus is evolving.

21.5.
a. natural selection
b. genetic drift
c. gene flow
d. greater reproductive success
e. small population
f. founder effect
g. bottleneck effect
h. genetic variation between populations

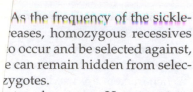

As the frequency of the sickle-
...eases, homozygous recessives
...to occur and be selected against,
...e can remain hidden from selec-
...zygotes.

...gote advantage—Heterozygotes
...from the most severe effects of
malaria and have a selective advantage in areas where malaria is a major cause of death.

SUGGESTED ANSWERS TO STRUCTURE YOUR KNOWLEDGE

1. **a.** The Hardy-Weinberg principle states that al-
    lele and genotype frequencies within a popula-
    tion will remain constant from one generation to
    the next as long as only Mendelian segregation
    and recombination of alleles are involved. This
    equilibrium requires five conditions: The popula-
    tion is large, mutation is negligible, migration is
    negligible, mating is random, and no natural se-
    lection occurs. The Hardy-Weinberg equilibrium
    provides a null hypothesis to enable researchers
    to test for evolution.
    b. $p^2 + 2pq + q^2 = 1$. In the Hardy-Weinberg
    equation, p and q refer to the frequencies of two
    alleles in the gene pool. The frequency of homo-
    zygous offspring is $(p \times p)$ or p^2 and $(q \times q)$ or q^2.

Heterozygous individuals can be formed in two
ways, depending on whether the egg or sperm
carries the p or q allele, and their frequency is
equal to $2pq$.

2. Genetic variation is retained within a population
    by diploidy and balancing selection. Diploidy
    masks recessive alleles from selection when they
    occur in the heterozygote. Thus, less adaptive
    or even harmful alleles are maintained in the
    gene pool and are available should the environ-
    ment change. Balancing selection maintains two
    or more phenotypic forms in a population. In
    heterozygote advantage, two alleles will be re-
    tained in stable frequencies within the gene pool.
    Frequency-dependent selection, in which the
    more common phenotype is selected against by
    predators or other factors, is another type of bal-
    ancing selection.

ANSWERS TO TEST YOUR KNOWLEDGE

Multiple Choice:

1. e	6. a	11. b	16. a	19. e
2. b	7. c	12. d	17. b	20. d
3. d	8. c	13. d	18. c	21. d
4. b	9. d	14. b		
5. b	10. e	15. d		

CHAPTER 22: THE ORIGIN OF SPECIES

FOCUS QUESTIONS

22.1.
  a. post-, reduced hybrid fertility
  b. pre-, gametic isolation
  c. pre-, mechanical isolation
  d. pre-, temporal isolation
  e. post-, hybrid breakdown
  f. pre-, behavioral isolation
  g. post-, reduced hybrid viability
  h. pre-, habitat isolation

22.2.
  a. reproductive isolation
  b. morphological
  c. ecological
  d. smallest group that shares a common ancestor, a single branch on the tree of life

22.3.
  a. 20 chromosomes ($4n = 20$)
  b. 24 chromosomes ($2n = 24$)

22.4.
  a. In allopatric speciation, a new species
  forms while geographically isolated from the
  parent population. In sympatric speciation,
  some reproductive barrier isolates the gene
  pool of a subgroup of a population within the
  same geographic range as that of the parent
  population.
  b. Reproductive barriers may evolve as a
  by-product of the genetic change associated
  with the isolated population's adaptation to a
  new environment, genetic drift, or sexual selection. Sympatric speciation by polyploidy is
  common in plants. A change in resource use
  or sexual selection may reproductively isolate
  a subset of an animal population.

22.5.
  a. Reinforcement: when hybrid offspring are
  less viable or reproduce less successfully, the
  reproductive barriers between the two species may strengthen. Traits that prevent or
  discourage crossbreeding will increase in frequency while traits that allow for mating with
  the other species will decrease in frequency
  as the individuals with those traits produce
  fewer viable offspring.
  b. Fusion: If hybrids are successful in mating
  with each other and with the parent species,

reproductive barriers may weaken. Increased gene flow between the two species may eventually lead to their fusion into a single species.

c. Stability: Hybrids continue to be produced between the two species in the area of their overlap, but the gene pools of both parent species remain distinct. Several cases of such stability among hybridizing species have been documented.

22.6. The total time between speciation events has been estimated to range from 4,000 to 40 million years (with an average of 6.5 million years). Thus, even if the process of speciation is relatively rapid once divergence begins, it may be millions of years before that new species gives rise to another new species.

SUGGESTED ANSWERS TO STRUCTURE YOUR KNOWLEDGE

1. Speciation, which leads to an increase in biological diversity, is the process through which one species splits into two or more species. Microevolution involves changes in allele frequencies in the gene pool of a population as a result of either chance events or natural selection. If the makeup of an isolated gene pool changes enough, microevolution may lead to speciation.

2. Reproductive barriers do not develop in order to create new species; they develop as a side consequence of the genetic changes that occur as a species adapts to a new environment or undergoes its own genetic drift. Gene flow between two species tends to break down reproductive barriers between them. Since gene flow is much less likely between allopatric species, one would predict that reproductive barriers are more likely to evolve. In the case of sympatric speciation by polyploidy, however, reproductive barriers arise in one generation.

3. Punctuated equilibria describes the common pattern seen in the fossil record in which species appear rather suddenly and then seem to change little for the rest of their existence. Thus, evolution seems often to occur in spurts of relatively rapid change interspersed with long periods of stasis.

ANSWERS TO TEST YOUR KNOWLEDGE

Multiple Choice:

1. c	4. c	7. e	10. d
2. d	5. c	8. c	11. a
3. e	6. b	9. b	12. e

CHAPTER 23: BROAD PATTERNS OF EVOLUTION

FOCUS QUESTIONS

23.1. 17,190 years. Carbon-14 has a half-life of 5,730 years. In 5,730 years the ratio of C-14 to C-12 would be reduced by ½; in 11,460 years it would be reduced by ¼; and in 17,190 years it would be reduced by ⅛ (three half-lives; $3 \times 5{,}730$).

23.2. The Himalayas; India was once part of the southern landmass Gondwana, which formed from the split of Pangaea around 135 million years ago. Gondwana and Laurasia separated into present-day continents around 65.5 million years ago, at the end of the Mesozoic.

23.3. a. Early mammals were restricted in size and diversity because they were probably outcompeted or eaten by the larger and more diverse dinosaurs. After dinosaurs (except for birds) became extinct, mammals diversified and filled the vacated niches.

b. The rise in large predators in the Cambrian explosion, the diversification of terrestrial plants and tetrapods, and the adaptive radiation of insects that ate or pollinated plants are all examples.

c. The Hawaiian Archipelago is a series of relatively young, isolated, and physically diverse islands whose thousands of endemic species are examples of adaptive radiation following multiple colonization and speciation events.

23.4. a. This study provided experimental evidence that a specific change in the sequence of a developmental gene could produce a major evolutionary change—the suppression of legs associated with the origin of a six-legged insect from a crustacean-like ancestor with many legs.

b. The coding sequence of the developmental gene was identical in the two populations;

the gene was not expressed in the ventral spine region of developing lake sticklebacks. Thus, the loss of ventral spines results from a change in gene regulation, not from a change in the gene itself.

23.5.   a.   The successful species that last the longest before extinction and generate the most new species will determine the direction of an evolutionary trend, just as the individuals that produce the most offspring determine the direction of adaptation in a population.

b.   Both microevolution and macroevolution are primarily driven by natural selection resulting from interactions between organisms and their environments. Evolutionary trends are ultimately dictated by environmental conditions; if conditions change, an evolutionary trend may end or change direction.

SUGGESTED ANSWERS TO STRUCTURE YOUR KNOWLEDGE

1.   *Plate tectonics:* When continents form one supercontinent, many habitats are destroyed or changed. As continents change latitude, their climates either warm or cool. The separation of continents creates major geographic separation

events, making possible allopatric speciations. Indeed, continental drift helps explain much of biogeography.

Mass extinctions result in the loss of many evolutionary lineages. The resulting empty ecological roles, however, may be exploited by species that survived extinction.

Adaptive radiations are multiple speciation events that fill newly formed or newly emptied niches. Thus, adaptive radiations may follow mass extinctions, the colonization of new regions, or the evolution of novel adaptations. The adaptive radiation of one group may provide new resources (such as food or habitat) that support the radiation of other groups.

2.   New morphological forms may evolve as a result of mutations in developmental genes or changes in the regulation of such genes. Such changes may affect the rate or timing of development or the spatial arrangement of body parts (e.g., *Hox* genes).

ANSWERS TO TEST YOUR KNOWLEDGE

Multiple Choice:

1. e	4. d	7. b	10. b
2. c	5. e	8. b	11. b
3. e	6. a	9. e	12. d

CHAPTER 24: EARLY LIFE AND THE DIVERSIFICATION OF PROKARYOTES

FOCUS QUESTIONS

24.1.   Reproduction that faithfully passes genetic information on to offspring is a characteristic of life. But a cell needs metabolic machinery to provide the energy and building blocks needed to replicate its genetic information and reproduce. Thus, both self-replicating molecules that encode genetic information for metabolism and the resulting metabolic processes are required for life to have evolved.

24.2.   Photoautotroph—uses light energy and CO_2 to synthesize organic compounds; the cell is a heterocyst, specialized for nitrogen fixation.

24.3.   a.   spheres (cocci), rods (bacilli), or spirals
b.   0.5–5 μm in diameter
c.   cell wall made of peptidoglycan; gram-negative bacteria also have outer lipopolysaccharide membrane; archaea lack peptidoglycan; sticky capsule for adherence;

fimbriae and pili for attachment or genetic exchange
d.   flagella; may show taxis to stimuli
e.   infoldings of plasma membrane may function in cellular respiration; thylakoid membranes in cyanobacteria
f.   circular DNA molecule with little associated protein found in nucleoid; may have plasmids with other genes
g.   binary fission; rapid population growth; rapid adaptive evolution

24.4.   a.   circular chromosome
b.   F plasmid, R plasmid
c.   mutation
d.   transformation
e.   naked DNA
f.   transduction
g.   bacteriophage (phage)
h.   conjugation
i. F^+ donor cell and F^- recipient cell
j.   adaptation to environment/evolution

24.5. **a.** Chloroplasts likely evolved from cyanobacteria.

b. Metagenomics is a technique of analyzing genetic material sampled directly from the environment. Before its development, researchers could study only those species that could be cultured in the lab. This technique has led to the discovery of new species and even new branches in prokaryotic phylogeny.

24.6. Chemoautotrophic bacteria form the basis of the food chain in vent communities, which can consist of hundreds of eukaryotic species.

SUGGESTED ANSWERS TO STRUCTURE YOUR KNOWLEDGE

1. The abiotic synthesis of small organic molecules; the joining of small molecules into macromolecules; the packaging of these molecules into membrane-bound protocells; the origin of self-replicating molecules, probably RNA, that made inheritance and natural selection possible

2. Prokaryotes are adapted to live in the diverse habitats on (and in) Earth. They have a wide range of metabolic abilities, including the nutritional categories of chemoautotrophy and photoheterotrophy. Their genetic diversity, along with their short generation time, huge populations, and mechanisms of genetic exchange, enable rapid adaptation to changing environments. The formation of biofilms consisting of multiple species and their many symbiotic associations are further examples of the remarkable abilities of prokaryotes.

3. Decomposers—recycle nutrients; bioremediation: sewage treatment, clean up oil spills.
Nitrogen fixers—provide nitrogen to the soil by bacteria in nodules on legume roots or by cyanobacteria.
Mutualism—intestinal bacteria, some digest food and produce vitamins.
Biotechnology—production of antibiotics, hormones, and many useful products.

ANSWERS TO TEST YOUR KNOWLEDGE

Fill in the Blanks:

1.	cocci	6.	endospore
2.	nucleoid	7.	biofilms
3.	Gram stain	8.	exotoxins
4.	fimbriae	9.	bioremediation
5.	taxis	10.	anaerobic respiration

Multiple Choice:

1. b	4. e	7. e	10. d	13. a
2. c	5. b	8. c	11. b	14. d
3. c	6. c	9. a	12. d	15. e

CHAPTER 25: THE ORIGIN AND DIVERSIFICATION OF EUKARYOTES

FOCUS QUESTIONS

25.1. **a.** The inner membranes of both mitochondria and chloroplasts have proteins that are homologous to those in the plasma membranes of prokaryotes. They replicate by similar splitting processes; contain a single, circular DNA molecule similar to the bacterial chromosome; and their ribosomes are more similar to prokaryotic ribosomes than to eukaryotic ribosomes.

b. Because all eukaryotes have mitochondria (or remnants of them), but not all have plastids, it appears that mitochondria evolved first, with plastids later evolving in one lineage of the ancestral heterotroph.

c. The four membranes indicate secondary endosymbiosis of a green alga (whose chloroplasts originated from the primary endosymbiont, a cyanobacterium). The inner two membranes belonged to the green alga's chloroplast. The third membrane is from the engulfed alga's plasma membrane, and the outer membrane is from the heterotrophic eukaryote's food vacuole. The vestigial nucleus indicates that this process occurred relatively recently, and its genetic sequences point to a green algal origin.

25.2. *Paramecium* is in the clade ciliates in the subgroup Alveolata in the supergroup the "SAR" clade.
a. cilia
b. oral groove, leading to cell mouth
c. food vacuoles, combine with lysosomes
d. contractile vacuole
e. release of wastes at region that functions as an anal pore

25.3. Evidence indicates that chromatophores were derived from a different cyanobacterium than the one from which all other plastids were derived—a remarkable second case of primary endosymbiosis.

25.4. As food is depleted, individual amoeboid cells congregate into a slug-like, motile mass. Some cells form a stalk that supports asexual fruiting bodies, which produce resistant spores.

25.5. Higher sea surface temperatures increase the differences between warm and cold layers of water, reducing the upwelling of cold, nutrient-rich waters from below that contribute to the growth of marine producers. A reduction in this base of marine food webs would affect all consumers in the ecosystem, including important fish species. If diatom populations decrease, less CO_2 would be "pumped" from the atmosphere to the deep ocean floor.

25.6. a. Frequent global changes in the surface proteins of these parasites enable them to evade the host's immune system.
b. The apicoplast appears to have originated by secondary endosymbiosis of a red algal cell, which had originally obtained its plastid from a cyanobacterium. Thus, drugs targeting metabolic pathways of the apicoplast, which is prokaryotic in origin, may not affect metabolic pathways of humans, which would not have those pathways.

SUGGESTED ANSWERS TO STRUCTURE YOUR KNOWLEDGE

1. a. cyanobacterium
   b. primary endosymbiosis
   c. red alga
   d. green alga
   e. secondary endosymbiosis
   f. dinoflagellates
   g. stramenopiles
   h. euglenids

ANSWERS TO TEST YOUR KNOWLEDGE

Matching:

1.	C	3.	B	5.	H	7.	A
2.	F	4.	G	6.	E	8.	D

Multiple Choice:

1.	d	4.	e	7.	e	9.	a
2.	c	5.	b	8.	b	10.	c
3.	d	6.	d				

CHAPTER 26: THE COLONIZATION OF LAND BY PLANTS AND FUNGI

FOCUS QUESTIONS

26.1. Benefits: bright, unfiltered sunlight; plentiful CO_2 and soil minerals. Challenges: lack of structural support and relative scarcity of water.

26.2. a. gametes
   b. mitosis
   c. zygote
   d. mitosis
   e. sporophyte
   f. spores
   g. meiosis
   h. gametophyte

26.3. The fungus provides phosphate ions and other minerals to the plants; the plants supply the fungi with organic nutrients.

26.4. The haploid mycelia of many fungi form structures that produce spores asexually. In sexual reproduction, two mycelia may fuse (plasmogamy) and their nuclei then fuse (karyogamy), producing diploid cells. These zygotes undergo meiosis to produce haploid spores. Both asexual and sexual spores are haploid, because they are either produced from a haploid mycelium or by meiosis from a diploid cell.

26.5. Lignified xylem provides the mechanical support for aerial growth. Vascular tissues convey water and minerals to photosynthesizing leaves and transport sugars throughout the plant. Roots both absorb water and minerals and anchor the plant. Tall plants could outcompete shorter plants for access to sunlight, and their spores could disperse farther.

26.6. Resistant, airborne, or animal-carried pollen grains can transport the male gametophyte, even over long distances, to the female gametophyte. A germinating pollen tube delivers sperm directly to an ovule, eliminating the need for a moist environment for sperm to swim to reach eggs.

26.7. a. integument ($2n$)
   b. megaspore (n)
   c. megasporangium ($2n$)

d. seed coat (from integument) ($2n$)
e. food supply (from female gametophyte tissue) (n)
f. embryo (new sporophyte) ($2n$)

26.8. a. The four floral organs that make up a flower are sepals, petals, stamens, and carpels.
b. The parts of a seed are an embryo surrounded by food enclosed in a seed coat.
c. A fruit is a mature ovary that may be modified to help disperse seeds, enlisting the aid of wind or animals.

26.9. Lichens, associations between green algae or cyanobacteria and fungi, help colonize bare rock. Mycorrhizae increase plant growth. Both lichens and mycorrhizae were probably critical in the movement of plants, and thus animals, onto land. Endophytes may increase a plant's resistance to disease or herbivores or increase its tolerance of heat or drought. As decomposers, fungi recycle nutrients in an ecosystem. However, fungi also spoil food and rot wooden structures.

SUGGESTED ANSWERS TO STRUCTURE YOUR KNOWLEDGE

1. The adaptations of plants to a terrestrial habitat include alternation of generations with a protected and nourished embryo, sporopollenin-walled resistant spores produced in sporangia, apical meristems that allow plants to continue to grow and exploit terrestrial resources, a waterproof cuticle, and stomata for gas exchange. The bryophytes have rhizoids that help anchor them to the ground. Vascular plants have vascular tissues strengthened with lignin, which allow them to grow tall and enable water and nutrient transport between roots and leaves. In seed plants, reduced gametophytes are protected from drying out and UV radiation. Pollen grains allow fertilization to occur without water. In addition, the protected ovules develop into seeds containing an embryo with its food source.

2. Fungi form mutualistic associations with plant roots that increase the efficiency with which plants absorb nutrients. Fossil evidence and molecular analysis of genes required for plants to form mycorrhizae indicate that these associations probably helped plants colonize land. Fungi are also partners in lichens, which are able to colonize bare rock and contribute to soil formation. As decomposers, fungi return nutrients to the environment for future plant growth.

3. a. land plants
   b. nonvascular plants (bryophytes)
   c. vascular plants
   d. seedless vascular plants
   e. seed plants
   f. liverworts
   g. mosses
   h. hornworts
   i. monilophytes (ferns, horsetails, whisk ferns)
   j. gymnosperms
   k. angiosperms
   1. origin of land plants (about 470 mya)
   2. origin of vascular plants (about 425 mya)
   3. origin of extant seed plants (about 305 mya)

ANSWERS TO TEST YOUR KNOWLEDGE

Multiple Choice:

1. c	5. a	9. e	13. d	16. c
2. e	6. e	10. a	14. e	17. b
3. b	7. d	11. c	15. a	
4. d	8. c	12. a		

CHAPTER 27: THE RISE OF ANIMAL DIVERSITY

FOCUS QUESTIONS

27.1. Sponges lack tissues, although they have several different cell types. They are filter feeders with saclike bodies. Flagellated choanocytes draw water in through pores, and mucus on the projections (collar) of these cells traps food particles. Cnidarians have tissues and thus are eumetazoans. Their tentacles capture prey and pass it into their gastrovascular cavity, a digestive compartment with a single opening. They have simple muscles, nerves, and sensory structures.

27.2. (1) The emergence of new predator–prey relationships (increased locomotion and protective shells) may have triggered various evolutionary adaptations. (2) The accumulation of atmospheric oxygen may have supported the active metabolisms of mobile and larger animals. (3) The origin of the *Hox* complex of regulatory genes may have facilitated the evolution of the diverse body plans that appear during the Cambrian explosion. All three of these may have played a role in the Cambrian explosion.

27.3. Sponges have no symmetry to their bodies, no true tissues, and no digestive tract—and thus no body cavity between it and the body wall. Cnidarians have radial symmetry and two germ layers: ectoderm and endoderm. They also have no coelom. Bilateral animals have bilateral symmetry, three germ layers (mesoderm is the third) that form true tissues, and most have a body cavity between their complete digestive tract and body wall.

27.4. The three arthropod characteristics are a segmented body, which allows for regional specialization; a hard exoskeleton; and jointed appendages.

27.5.
  a. Vertebrates
  b. Gnathostomes
  c. Osteichthyans
  d. Lobe-fins
  e. Tetrapods
  f. Amniotes
  g. chondrichthyans (sharks, rays)
  h. ray-finned fishes
  i. amphibians (frogs, salamanders)
  j. reptiles (turtles, snakes, crocodiles, birds)
  k. mammals

27.6. *Tiktaalik* has the fish characters of scales, fins, and gills and lungs, but the tetrapod characters of neck, ribs, flat skull, eyes on top of skull, and bones in its front fins with the basic tetrapod pattern of humerus, radius and ulna, and wrist bones. It did not have limbs with digits and, thus, is considered to be a lobe-fin, not a tetrapod.

27.7.
  a. chorion—functions in gas exchange
  b. amnion—encases embryo in fluid, prevents dehydration, and cushions shocks
  c. allantois—stores metabolic wastes, functions with chorion in gas exchange
  d. yolk sac—contains yolk; blood vessels transport stored nutrients into embryo

27.8.
  a. Some examples include orchids with a very long floral tube and a pollinating moth with an equally long proboscis; flowering plants and their specific pollinators; and thickened shells of periwinkle shells exposed to invasive crabs (documented in the scientific skills exercise) and probably enhanced ability of crabs to crack such shells.
  b. The origin of new species can provide new food sources for predators or parasites.

SUGGESTED ANSWERS TO STRUCTURE YOUR KNOWLEDGE

1.
  a. Porifera (sponges): asymmetrical; lack true tissues
  b. Eumetazoa: "true" animals; have true tissues
  c. Cnidaria: radial symmetry; two germ layers
  d. Bilateria: bilateral symmetry; three germ layers
  e. Deuterostomia: acorn worms and echinoderms are invertebrates; chordates include the vertebrates
  f. Lophotrochozoa: bryozoans, molluscs, and annelids
  g. Ecdysozoa: nematodes and arthropods

ANSWERS TO TEST YOUR KNOWLEDGE

Multiple Choice:

1. c	**4.** d	**7.** a	**9.** b
2. b	**5.** c	**8.** d	**10.** c
3. b	**6.** b		

CHAPTER 28: PLANT STRUCTURE AND GROWTH

FOCUS QUESTIONS

28.1. This plant is a eudicot. Evidence for this conclusion comes from the netlike (nonparallel) veins in the leaves, the taproot, and flower with five parts. Not visible characteristics are the vascular tissue arranged in a ring in the stem, the pollen grains with three openings, and the two cotyledons in a seed.
  a. reproductive shoot (flower)
  b. apical bud
  c. node
  d. internode
  e. vegetative shoot
  f. leaf
  g. blade
  h. petiole
  i. axillary bud
  j. stem
  k. taproot
  l. lateral roots
  m. shoot system
  n. root system

28.2. **a.** sclerenchyma (fibers and sclereids), tracheids, and vessel elements
b. sieve-tube elements (as well as the dead cells listed in **a.**)

28.3. **a.** Plants exhibit **indeterminate growth,** continuing to grow as long as they live. Animals, as well as some plant organs such as leaves and flowers, have **determinate growth** and stop growing after reaching a certain size.
b. Primary growth is the elongation of shoots, branches, and roots and results from cell division within apical meristems. Secondary growth is produced by lateral meristems (vascular cambium and cork cambium) and results in the thickening of roots and shoots of woody plants.

28.4. **a.** epidermis (dermal)—root hairs for absorption
b. cortex (ground)—transport of minerals and water from root hairs to stele; intercellular spaces allow *extracellular* diffusion of water, minerals, and oxygen; food storage
c. vascular cylinder (vascular)—transport
d. endodermis—regulates water and mineral movement into stele
e. pericycle—origin of lateral roots
f. xylem—water and mineral transport
g. phloem—nutrient transport

28.5. **a.** cuticle—waxy covering that reduces evaporative water loss
b. upper epidermis—tightly packed protective outer covering of cells
c. palisade mesophyll—photosynthesis
d. spongy mesophyll—photosynthesis; loosely packed cells surrounding air spaces
e. guard cells—regulate opening and closing of stoma
f. xylem—transport of water and minerals from roots
g. phloem—transport of carbohydrates from leaves to roots and growing parts of plant
h. vein—vascular tissue surrounded by ring of bundle-sheath cells that regulate movement of materials into and out of vein

i. stoma—pore that allows exchange of CO_2 and O_2 between air spaces of leaf and outside

28.6. H, F, A, B, G, D, C, E

SUGGESTED ANSWERS TO STRUCTURE YOUR KNOWLEDGE

1. Except for leaves and flowers, which stop growing when they reach maturity (determinate growth), plant roots and shoots continue to grow throughout the life of a plant. Although certain characteristic forms have been favored by natural selection in species adapted to particular environments, the specific environment in which a plant grows greatly influences its individual body form. For example, a plant may grow taller or produce different numbers or shapes of leaves depending on its light exposure. This developmental flexibility helps compensate for a plant's immobility.

2. This drawing is a eudicot stem, indicated by the ring of vascular bundles and the central pith. It represents primary growth because there is no layer of vascular cambium between the layers of xylem and phloem.
   a. sclerenchyma
   b. phloem
   c. xylem
   d. cortex
   e. pith
   f. vascular bundle
   g. epidermis

ANSWERS TO TEST YOUR KNOWLEDGE

Multiple Choice:

1. c	**4.** b	**7.** e
2. d	**5.** d	**8.** c
3. b	**6.** e	**9.** c

Matching:

1. I	**4.** E	**7.** D	**9.** H
2. F	**5.** C	**8.** G	**10.** B
3. J	**6.** A		

CHAPTER 29: RESOURCE ACQUISITION, NUTRITION, AND TRANSPORT IN VASCULAR PLANTS

FOCUS QUESTIONS

29.1. **a.** At low light intensities, horizontal leaf orientation would maximize exposure to sunlight. In bright sun, a vertical orientation would shield leaves from too high a light intensity and allow light to penetrate to lower leaves.

b. The branching pattern of roots can increase to take advantage of the local availability of nitrate, as can the production of nutrient transporters within the root cells.

29.2. **a.** $\psi_p = 0$, $\psi_s = -0.6$, $\psi = -0.6$ MPa
b. $\psi_p = 0.6$, $\psi_s = -0.6$, $\psi = 0$ MPa. The turgor pressure of the cell that develops with movement of water into the cell finally offsets the solute potential of the cell. No net osmosis occurs; the water potentials of both cell and solution are equal and = 0 MPa.
c. $\psi_p = 0$, $\psi_s = -0.8$, $\psi = -0.8$ MPa. The cell would plasmolyze, losing water until the solute potential of the cell would equal that of the bathing solution (both -0.8 MPa). There would be no turgor pressure.

29.3. **a.** macronutrient; component of chlorophyll, cofactor and activator of enzymes
b. macronutrient; component of nucleic acids, phospholipids, ATP, some coenzymes
c. micronutrient; component of cytochromes, cofactor for chlorophyll synthesis

29.4. **a.** Chlorosis is the yellowing of leaves due to a lack of chlorophyll, often indicative of a deficiency of magnesium.
b. You would expect to find a deficiency in the younger, growing parts of a plant.

29.5. **a.** The N-P-K ratio is the percentage of nitrogen, phosphorus, and potassium that is in a fertilizer.
b. A soil pH of 5 or lower makes Al^{3+} more soluble, increasing absorption of these toxic ions that stunt root growth.

29.6. A loam, which is a mixture of sand, silt, and clay, has enough large particles to provide air spaces and enough small particles to retain water and minerals. Cations bind to soil particles and are made available to roots by cation exchange. Humus in a soil helps to retain water and provide a steady supply of mineral nutrients.

29.7. **a.** nitrogen-fixing bacteria
b. ammonifying bacteria
c. nitrifying bacteria
d. denitrifying bacteria
e. nitrate and nitrogenous organic compounds

29.8. **a.** symplastic
b. apoplastic
c. Casparian strip
d. xylem vessels
e. vascular cylinder (stele)
f. endodermis

g. cortex
h. epidermis
i. root hair

29.9. **a.** The loss of water vapor from the air spaces through the stomata causes evaporation from the water film coating the mesophyll cells, leading to the negative pressure that draws water up from the roots.
b. Evaporation from the water film increases the surface tension of the curving air–water interface. This tension, or negative pressure, produces a gradient of water potentials that extends from the leaf to the root.
c. Water molecules hold together due to hydrogen bonding, and a pull on one water molecule is transmitted throughout the column of water.
d. Water molecules adhere to the hydrophilic walls of narrow xylem cells, helping to support the column of water against gravity.

29.10. In order to provide sufficient CO_2 for photosynthesis, stomata must be open, and large amounts of water will be lost to transpiration. A sunny, windy, dry day will increase the rate of transpiration, perhaps to the point that insufficient water is available in the soil. Due to a lack of water, guard cells may lose turgor (and also be signaled by abscisic acid) and stomata will close, slowing transpiration. In reducing excessive water loss, however, the plant mesophyll cells now receive insufficient CO_2, and the photosynthetic rate will decline.

29.11. **a.** ATP
b. proton pump
c. H^+
d. sucrose
e. cotransporter
The upper portion is the apoplast (cell wall); the lower portion represents the symplast (inside of the cell).

SUGGESTED ANSWERS TO STRUCTURE YOUR KNOWLEDGE

1. Solutes may move across membranes by passive transport when they move down their concentration or electrochemical gradient. Transport proteins, which speed this passive transport, may be specific carrier proteins or selective channels. Active transport usually involves a proton pump that creates a membrane potential as H^+ ions are moved out of the cell. Cations such as K^+ may now move through specific channels down their electrochemical gradient. In cotransport, the inward diffusion of H^+ down its concentration gradient through a cotransporter moves a solute against its concentration gradient.

2. Root nodules are mutualistic associations between legumes and nitrogen-fixing bacteria of the genus *Rhizobium*. These bacteria receive nourishment from the plant and provide the plant with fixed nitrogen. Mycorrhizae are mutualistic associations between the roots of most plants and various species of fungi. The fungi stimulate root growth, increase surface area for absorption, and secrete antibiotics that may help protect the plant. In return, the fungi receive nourishment from the plant.

3. Both these types of symbiotic relationships involve chemical recognition between specific plant species and bacterial or fungal species. Both are mutualistic interactions.

4. The cohesion–tension hypothesis explains the ascent of xylem sap. Tension created by the evaporation of water and the resulting curvature of the air–water interface lowers the water potential in the leaf and sets up the bulk flow of water. The cohesion of water molecules transmits the pull resulting from transpiration throughout the water column, and the adhesion of water to the hydrophilic walls of the xylem vessels aids the flow.

A pressure flow mechanism also explains the movement of phloem sap, but the plant must expend energy to create the differences in water potential that drive translocation. The active accumulation of sucrose in sieve-tube elements lowers water potential at the source end, resulting in an inflow of water. The removal of sugar from the sink end of a phloem tube is followed by the osmotic loss of water. The resulting difference in pressure between the source and sink end of the phloem tube causes the bulk flow of phloem sap.

ANSWERS TO TEST YOUR KNOWLEDGE

Multiple Choice:

1.	c	8.	b	15.	d	22.	d
2.	b	9.	b	16.	c	23.	a
3.	e	10.	e	17.	d	24.	c
4.	a	11.	a	18.	e	25.	b
5.	b	12.	e	19.	d		
6.	e	13.	a	20.	d		
7.	a	14.	e	21.	b		

CHAPTER 30: REPRODUCTION AND DOMESTICATION OF FLOWERING PLANTS

FOCUS QUESTIONS

30.1.
  a. stamen
  b. anther
  c. filament
  d. carpel
  e. stigma
  f. style
  g. ovary
  h. sepal
  i. receptacle
  j. ovule
  k. petal

30.2. a.

Whorl	Genes Active	Organs in Normal Flower	Genes Active Mutant *A*	Organs in Mutant *A* Flower
1	*A*	sepals	*C*	carpels
2	*AB*	petals	*BC*	stamens
3	*BC*	stamens	*BC*	stamens
4	*C*	carpels	*C*	carpels

b. Gene *A* would be expressed alone in each whorl, producing a flower with four whorls of sepals.

30.3.
  a. The male gametophyte consists of a tube cell and a generative cell (which moves into the tube cell and will divide to form two sperm). These cells and the spore wall that encloses them constitute a pollen grain.
  b. The female gametophyte is the embryo sac, often containing seven cells (one of which is an egg) and eight nuclei.
  c. Double fertilization conserves resources by allowing for nutrient development only when fertilization has occurred.

30.4.
  a. seed coat
  b. epicotyl
  c. hypocotyl
  d. cotyledons with endosperm
  e. radicle
  f. cotyledon (scutellum)
  g. endosperm
  h. epicotyl
  i. hypocotyl
  j. radicle

  k. coleorhiza
  l. coleoptile

30.5. The coleoptile pushes through the soil, and the shoot tip is protected as it grows up through the tubular sheath.

30.6. **a.** Buoyant fruits may be dispersed by water; winged fruits may "fly" seeds to new destinations; seeds in edible fruits may be dispersed in feces; prickly seeds may stick to animals, and some animals may "plant" seeds in their underground caches.
  b. Hormonal interactions induce softening of the pulp, a change in color, and an increase in sugar content.

30.7. **a.** Apomixis allows for the dispersal value of seeds, which in this case are clones of the parent plant.

b. Selfing is self-fertilization, in which a plant's sperm fertilizes its own eggs. Although offspring are not as genetically diverse, selfing is an advantage in certain crop plants because it ensures that each ovule will develop into a seed.

30.8. In both vegetative reproduction and self-fertilization, offspring have only one parent. But in vegetative reproduction, all offspring are genetically identical clones. Self-fertilization involves meiosis and random fertilization of gametes. Thus, whatever genetic variation is present in a diploid parent plant is recombined in offspring, producing more genetic variation in offspring.

30.9. Burning of fossil fuels releases CO_2 that was removed from the atmosphere millions of years ago. CO_2 that is released by the burning of biofuels is absorbed for photosynthesis as new biofuel crops grow.

SUGGESTED ANSWERS TO STRUCTURE YOUR KNOWLEDGE

1. One version of the life cycle of an angiosperm is presented in the following figure. See also textbook Figure 30.5.

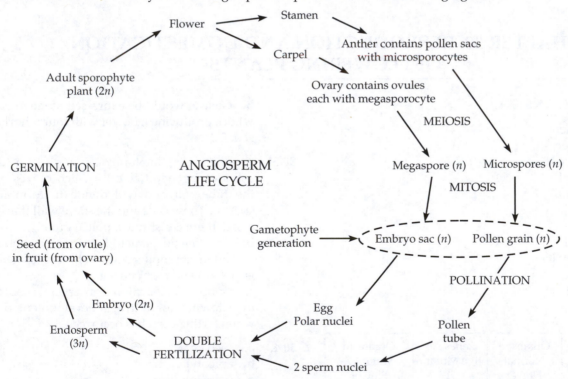

2. *Sexual:* Advantages include increased genetic variability, which provides the potential for adaption to changing conditions, and the dispersal and dormancy capabilities provided by seeds. Disadvantages are that the seedling stage is very vulnerable, sexual reproduction is very energy intensive (since producing flowers, pollen, and fruit takes energy and most seeds and seedlings don't survive), and genetic recombination may separate adaptive traits.
Asexual: Advantages include the hardiness of vegetative reproduction and the maintenance of genetically well-adapted plants in a given environment. Disadvantages relate to the advantages of sexual reproduction: There is no genetic variability from which to "choose" should conditions

change, and there are no seeds for dormancy or dispersal.

3. *Benefits:* Reduce use of chemical pesticides; remove weeds with herbicides instead of erosion-producing tillage; increase nutritional value of crops; produce disease-resistant crops.

   Dangers: Introduce allergens into human foods; harm nontarget species; create hybrid "super-weeds" due to crop-to-weed transgene escape; produce other unanticipated harmful results that cannot be stopped once GMOs are released into the environment.

ANSWERS TO TEST YOUR KNOWLEDGE

Fill in the Blanks:

1. ovary
2. sporophyte
3. biofuel
4. embryo sac
5. radicle
6. epicotyl
7. coleoptile
8. scion
9. microspore
10. callus

Multiple Choice:

1. e
2. a
3. d
4. d
5. e
6. a
7. a
8. e
9. b
10. d
11. b
12. e

CHAPTER 31: PLANT RESPONSES TO INTERNAL AND EXTERNAL SIGNALS

FOCUS QUESTIONS

31.1.
a. proton (H^+) pumps
b. lower
c. expansins
d. osmotic uptake of water
e. increased turgor pressure
f. loosened walls (increased expandability of walls)

31.2.
a. cytokinins
b. ethylene
c. abscisic acid
d. auxin
e. brassinosteroids
f. gibberellins

31.3. The relative amounts of red and far-red light are communicated to a plant by the ratio of the two forms of phytochrome. Canopy trees absorb red light, and a shaded tree will have a higher ratio of P_r to P_{fr}, which induces the tree to grow taller. Direct sunlight shifts the ratio toward the P_{fr} form, stimulating branching and inhibiting vertical growth.

31.4. Free-running periods are circadian rhythms that vary from exactly 24 hours when an organism is kept in a constant environment without environmental cues.

31.5.
a. not flower
b. flower
c. flower
d. not flower
e. not flower
f. flower
g. flower
h. not flower
i. not flower
j. flower

In the next-to-last light regimen, a flash of far-red light cancels the effect of the red flash, and the red flash in the last one cancels the effect of the far-red flash.

31.6.
a. The types of tropisms that a stem exhibits are positive phototropism and negative gravitropism.
b. The growth mechanism that produces the coiling of a tendril is thigmotropism, described as mechanical stimulation causing unequal growth rates of cells on opposite sides of the tendril.

31.7.
a. A plant may reduce transpirational water loss by stomatal closing and wilting, or produce heat-shock proteins that stabilize plant proteins
b. A plant may respond to an unusually cold and wet fall through changing its membrane lipid composition by increasing unsaturated fatty acids; in addition, air tubes may develop in cortex of root

31.8.
a. Insects may harm plants by eating plant tissues and reducing the photosynthetic capacity of plants, as well as by causing wounds through which pathogens can enter a plant.
b. Insects may benefit plants by pollinating them, as well as by eating other insects that are eating plant tissues (such as parasitoid wasps).

SUGGESTED ANSWERS TO STRUCTURE YOUR KNOWLEDGE

1. Some agricultural uses of plant hormones are cytokinin spray used to keep flowers fresh; ethylene used to ripen stored fruit; synthetic auxins used to induce seedless fruit set; gibberellins sprayed on Thompson seedless grapes; and 2,4-D (auxin) used as herbicide.

2.

```
          ┌──────────────────┐
          │  PHOTOPERIODISM  │
          └──────────────────┘
                  is a
          ┌──────────────────┐
          │   physiological  │
          │   response to    │
          │ day/night length │
          └──────────────────┘
               involved in
   ┌──────────┐              ┌──────────────────┐
   │ control of│             │    control of    │
   │ flowering │             │  bud dormancy,   │
   └──────────┘              │ seed germination │
                             └──────────────────┘
   depends on                      such as
    critical
   ┌────────────┐
   │ night length│  measured by
   └────────────┘
     can have                ┌───────────┐
                             │ biological│
                             │   clock   │
                             └───────────┘
  ┌──────────┐ ┌──────────┐  may be set by   ┌────────────┐
  │ short-day│ │ long-day │                  │detects light│
  │  plants  │ │  plants  │  ┌────────────┐  │and triggers │
  └──────────┘ └──────────┘  │ phytochrome│  │   plant     │
                             └────────────┘  │ responses   │
   require      require                      └────────────┘
                             2 forms are        form that
  ┌──────────┐ ┌──────────┐  photoreversible
  │night that│ │night     │  ┌───┐ red light ┌────┐
  │ exceeds  │ │shorter   │  │Pr │ converts  │Pfr │
  │critical  │ │than      │  └───┘   to      └────┘
  │dark      │ │critical  │     far-red converts
  │period    │ │dark      │          to
  └──────────┘ │period    │
               └──────────┘
```

3. Specific resistance begins with the recognition of a pathogen molecule (effector) by a specific plant resistance (R) protein, which triggers a signal transduction pathway that initiates a hypersensitive response and production of an alarm signal. The hypersensitive response includes production of antimicrobial molecules (phytoalexins and PR proteins), sealing off of infected areas, and destruction of infected cells. The distribution of the alarm signal stimulates distant cells to produce salicylic acid, activating a systemic acquired resistance. Molecules are produced that protect the cell against a diversity of pathogens for several days.

ANSWERS TO TEST YOUR KNOWLEDGE

True or False:

1. False—change *abscisic acid* to *gibberellin*, or use *The removal of abscisic acid*.
2. True
3. False—change *thigmomorphogenesis* to *triple response*; or use *Thigmomorphogenesis is a change in growth form in response to mechanical stress*.
4. True
5. False—change *circadian rhythm* to *photoperiodism*; or change *day or night length* to *on a 24-hour cycle*.
6. True
7. False—change *A virulent* to *An avirulent*

Multiple Choice:

1. b	5. a	9. b	12. a
2. c	6. b	10. e	13. d
3. e	7. c	11. c	14. c
4. b	8. b		

CHAPTER 32: HOMEOSTASIS AND ENDOCRINE SIGNALING

FOCUS QUESTIONS

32.1. Types of vertebrate connective tissue include loose connective tissue, which holds organs in place; fibrous connective tissue, which makes up tendons (attach muscles to bones) and ligaments (connect bones at joints); adipose tissue, which stores fat; blood, with red blood cells (oxygen transport) and white blood cells (defense); cartilage, which provides flexible support; and bone, which forms the hard skeleton. Blood has a liquid matrix, whereas bone has a solid matrix.

32.2. a. Activated cooling mechanisms are vasodilation of surface blood vessels and activation of sweat glands.
b. Mechanisms activated when body temperature falls below the set point include vasoconstriction of surface vessels and shivering by contraction of skeletal muscles.

32.3. Hormones are released and travel more slowly throughout the body in the bloodstream. Nerve impulses travel rapidly through neurons directly to target cells. Hormone signals last longer and usually mediate responses of longer duration than the shorter responses to nervous signals. The endocrine system coordinates more gradual functions such as growth, reproduction, and metabolism. The nervous system regulates rapid responses such as locomotion and behavior.

32.4. In negative feedback, the response reduces the initial stimulus, thereby turning off the response. Examples of negative feedback are the rise in pH in the intestine in response to bicarbonate release, which removes the stimulus and stops the release of secretin by the intestine, and the blocking of the release of TRH and TSH by increasing levels of thyroid hormone. In positive feedback, the stimulus is reinforced, leading to an even greater response, which often drives a process to completion. Oxytocin's stimulation of milk secretion in response to a baby's sucking is an example of positive feedback regulation.

32.5. **a.** isoosmotic
b. conformer but may regulate specific solutes
c. ammonia
d. hypoosmotic
e. drink water to compensate for loss to hyperosmotic seawater; gills pump out salt; little urine excreted
f. ammonia diffuses across gills, may also produce urea
g. hyperosmotic
h. copious dilute urine to compensate for osmotic gain; may pump salts in through gills
i. ammonia
j. hyperosmotic
k. body coverings reduce evaporation; drinking and eating moist foods
l. uric acid

32.6. **a.** Substances removed from the blood during filtration are water, salts, nitrogenous wastes, glucose, vitamins, and other small molecules.
b. Substances that remain in the capillaries are blood cells and large molecules such as plasma proteins.

32.7. **a.** Bowman's capsule
b. glomerulus
c. proximal tubule
d. descending limb, loop of Henle
e. ascending limb, loop of Henle
f. distal tubule
g. collecting duct

1. HCO_3^-, NaCl*, H_2O, nutrients*, and K^+ are reabsorbed from the proximal tubule. H^{+*} and NH_3 are secreted into the proximal tubule.
2. Water is reabsorbed from the descending limb of the loop of Henle.
3. NaCl diffuses from the thin segment of the ascending limb and is pumped out* of the thick segment.
4. NaCl*, H_2O, and HCO_3^{-*} are reabsorbed from the distal tubule. K^{+*} and H^{+*} are secreted into the tubule.
5. NaCl* is reabsorbed; urea and H_2O diffuse out of the collecting tubule.

* indicates active transport.

32.8. **a.** Juxtamedullary nephrons; the longer the loop of Henle, the more hyperosmotic the urine can become.
b. Mammals and birds; loops of Henle are especially long in those taxa inhabiting dry habitats.

32.9. **a.** high blood osmolarity
b. ADH, antidiuretic hormone
c. distal tubules and collecting ducts
d. JGA, juxtaglomerular apparatus
e. renin
f. adrenal glands
g. aldosterone
h. Na^+ and water reabsorption

SUGGESTED ANSWERS TO STRUCTURE YOUR KNOWLEDGE

1. All organs are covered with epithelial tissue, and internal lumens, ducts, and blood vessels are lined with epithelia. Most organs contain some types of connective tissue. Muscle tissue would be present if the organ must contract. Nervous tissue innervates most organs.

2. Endotherms, who warm their bodies with metabolic heat, have thermoregulatory mechanisms that enable them to maintain a constant internal temperature across a wider range of environmental temperatures. Even though ectotherms may use behavioral mechanisms to reduce temperature fluctuations, they may still be exposed to greater internal fluctuations as environmental temperatures change.

3. In a simple hormone pathway, a stimulus triggers the release of a hormone, whose effect is to reduce the stimulus, which, by negative feedback, stops triggering the release of the hormone. Secretin, released into the bloodstream in response to an acid

pH, triggers the release of bicarbonate from the pancreas, which raises the pH of the duodenum. The resulting rise in pH removes the stimulus for secretin release.

4. A hormone cascade pathway involves a hormone sequence in which the hypothalamus secretes releasing hormones that stimulate the anterior pituitary to secrete a tropic hormone, which then stimulates target cells to secrete hormones that produce some physiological or developmental effect. For example, in response to nervous stimulation in response to a decrease in thyroid hormone level, the hypothalamus secretes TRH, which stimulates the anterior pituitary to secrete TSH, which acts on the thyroid gland to stimulate release of thyroid hormone. Thyroid hormone then acts by negative feedback to block the further release of TSH and TRH.

5. Two solutes, NaCl and urea, contribute to the high osmolarity of the medulla. As the filtrate moves up the ascending limb, NaCl first diffuses and is then pumped out, contributing to the gradient of NaCl in the medulla. As the filtrate flows through the collecting duct, some urea diffuses out and also contributes to the osmotic gradient. As the filtrate in the collecting duct moves through the high osmolarity of the medulla, water flows out by osmosis, creating a more concentrated urine.

ANSWERS TO TEST YOUR KNOWLEDGE

Multiple Choice:

1. d	6. e	11. c	16. c	20. e
2. b	7. d	12. b	17. d	21. d
3. d	8. d	13. a	18. c	22. d
4. a	9. b	14. a	19. e	
5. d	10. b	15. e		

CHAPTER 33: ANIMAL NUTRITION

FOCUS QUESTIONS

33.1. Vegetarians must eat a combination of plant foods that are complementary in amino acids, such as corn and beans.

33.2. **a.** Calcium and phosphorus are required by vertebrates for bone and tooth construction.
b. Undernutrition results from a deficiency of chemical energy—a diet that lacks sufficient calories. Other types of malnutrition result from the lack of essential nutrients, such as essential amino acids (protein deficiency), essential fatty acids, vitamins, or minerals.
c. The most common type of malnutrition among humans is protein deficiency, as a result of relying on a single food source such as rice or corn, which are incomplete proteins.

33.3. The carbohydrates, proteins, nucleic acids, and fats that make up food are too large to pass through cell membranes, and the molecules in food are not identical to those an animal assembles for its own functions and tissues.

33.4. An alimentary canal can have specialized compartments for the sequential processing of nutrients. Because food moves in one direction, an animal can ingest more food while still digesting a previous meal.

33.5. **a.** Polysaccharides (starch and glycogen) broken down in the mouth by salivary amylase to smaller polysaccharides and maltose; hydrolysis continues until salivary amylase is inactivated by the low pH in the stomach.
b. Proteins are hydrolyzed by pepsin to smaller polypeptides in the stomach.

33.6. **a.** The digestion of starch and glycogen into disaccharides is continued by pancreatic amylase. Disaccharidases, which are enzymes produced by the epithelium, split disaccharides into monosaccharides.
b. Protein digestion is completed in the small intestine by pancreatic trypsin and chymotrypsin, enzymes specific for peptide bonds adjacent to certain amino acids, and by pancreatic carboxypeptidase; as well as enzymes of the brush border: carboxypeptidase and aminopeptidase, which split amino acids off from opposite ends of a polypeptide; and by dipeptidases.
c. Pancreatic nucleases are a group of enzymes that hydrolyze DNA and RNA into their nucleotide monomers. Nucleotidases, nucleosidases, and phosphatases dismantle nucleotides.
d. The digestion of fats is aided by bile salts, which coat or emulsify tiny fat droplets so they do not coalesce, leaving a greater surface area for pancreatic lipase to hydrolyze the

fat molecules into glycerol, fatty acids, and monoglycerides.

33.7. Carnivore dentition is characterized by sharp incisors, fanglike canines, and jagged premolars and molars. Herbivores have broad premolars and molars that grind plant material, and incisors and canines modified for biting off vegetation. Carnivorous vertebrates often have large, expandable stomachs. The longer intestine of an herbivore facilitates digestion of plant material, and its cecum may contain cellulose-digesting microorganisms.

33.8.
   a. insulin
   b. glucagon
   c. uptake of glucose
   d. store glucose as glycogen
   e. pancreatic islets
   f. hydrolyze glycogen and release glucose
   g. blood glucose level

SUGGESTED ANSWERS TO STRUCTURE YOUR KNOWLEDGE

1.
   a. salivary glands—produce saliva
   b. oral cavity—teeth and tongue mix food with saliva

c. pharynx—throat, opening to esophagus and trachea
d. esophagus—peristalsis moves bolus to stomach
e. stomach—churns food with gastric juice
f. small intestine—digestion and absorption
g. large intestine (colon)—absorbs water, compacts feces
h. anus—exit of alimentary canal
i. rectum—stores feces until expelled
j. pancreas—produces enzymes, bicarbonate, and hormones
k. gallbladder—stores bile from liver
l. liver—produces bile; processes nutrients; detoxifies

ANSWERS TO TEST YOUR KNOWLEDGE

Multiple Choice:

1. e	**5.** e	**9.** b	**13.** c	**16.** a
2. a	**6.** d	**10.** c	**14.** b	**17.** b
3. b	**7.** c	**11.** b	**15.** b	**18.** c
4. d	**8.** a	**12.** d		

Matching:

1. J	**3.** C	**5.** H	**7.** I
2. E	**4.** D	**6.** B	**8.** F

CHAPTER 34: CIRCULATION AND GAS EXCHANGE

FOCUS QUESTIONS

34.1.
   a. Hemolymph is basically the interstitial fluid that bathes body tissues and is circulated through open-ended vessels. Blood refers to the circulatory fluid in a closed circulatory system, which remains within vessels and exchanges materials with body cells via the interstitial fluid.
   b. Open circulatory systems take less energy to operate. Higher blood pressures enable more efficient delivery of nutrients and O_2 in closed circulatory systems, which also allow for greater regulation of blood distribution.

34.2.
   a. In the **single circulation** typical of fish, blood is pumped only once. It flows first through the gill capillaries, through a vessel to the systemic capillaries, and then back to the heart. In a **double circulation,** blood is pumped twice as it circulates through the body: once to the gas exchange organs, from which it returns to the heart, and again to the systemic circuit.

   b. The amphibian heart has three chambers: two atria and a ventricle. A ridge in the ventricle helps to direct oxygen-rich blood from the left atrium into the systemic circuit, and oxygen-poor blood from the right atrium into the pulmocutaneous circuit. In the four-chambered heart of mammals, the right side of the heart receives and then pumps oxygen-poor blood to the lungs, and the left side of the heart receives and then pumps oxygen-rich blood to the systemic capillaries.

34.3.
   a. capillaries of head and forelimbs
   b. left pulmonary artery
   c. aorta
   d. capillaries of left lung
   e. left pulmonary vein
   f. left atrium
   g. left ventricle
   h. aorta
   i. capillaries of abdominal organs and hind limbs
   j. inferior vena cava

k. right ventricle
l. right atrium
m. right pulmonary vein
n. capillaries of right lung
o. right pulmonary artery
p. superior vena cava

The pulmonary veins, left side of the heart, and aorta and arteries leading to the systemic capillary beds carry oxygen-rich blood. See text Figure 34.5 for numbers. Note that the aorta splits and oxygen-rich blood travels either to the head region or lower body, returning to the heart through the superior or inferior vena cava, respectively. Thus, numbers 7–10 do not represent the sequential flow of blood.

34.4. **a.** atrioventricular valve
b. semilunar valve
c. SA (sinoatrial) node
d. AV (atrioventricular) node

34.5. **a.** elastic artery walls
b. contraction of ventricle (systole)
c. diastole
d. systole
e. diastolic pressure (lower number)
f. systolic pressure (higher number)
g. resistance
h. vasoconstriction
i. vasodilation
j. nerves, hormones (NO, endothelin)

34.6. A reduction in plasma proteins would reduce the osmotic pressure of the blood, which counters the blood pressure forcing fluid from capillaries. The result would be an increased loss of fluid from capillaries and fluid accumulation in body tissues.

34.7. **a.** plasma
b. blood electrolytes
c. osmotic balance, buffering, muscle and nerve functioning, regulation of membrane permeability
d. plasma proteins
e. buffers, osmotic factors, lipid escorts, antibodies, clotting factors
f. nutrients, wastes, respiratory gases, hormones
g. erythrocytes (red blood cells)
h. transport O_2; help transport CO_2
i. leukocytes (white blood cells)
j. phagocytes
k. immunity
l. platelets
m. blood clotting

34.8. Lifestyle choices correlated with increased risk of cardiovascular disease are smoking, not exercising, and a diet rich in animal fats or *trans fats*.

34.9. RDS in infants is caused by an absence of surfactants, which are not produced by fetal lungs until about 33 weeks of development.

34.10. **a.** water; gill; gill cover (operculum) and mouth pump water into mouth, across gills, and out of body
b. air; tracheoles of tracheal system; body movements compress and expand air tubes
c. air (and water); lungs and moist skin; positive pressure breathing: fill oral cavity, exhale, close nostrils, raise floor of oral cavity, and expand lungs
d. air; alveoli in lungs; negative pressure breathing: lower diaphragm and raise ribs, thereby increasing volume and decreasing pressure of lungs; air flows in

34.11. This graph shows the Bohr shift. The normal pH of blood is 7.4. The dissociation curve for hemoglobin is shifted to the right at a lower pH, meaning that at any given partial pressure of O_2, hemoglobin is less saturated with O_2. A rapidly metabolizing tissue produces more CO_2, which lowers blood pH, and hemoglobin will unload more of its O_2 to that tissue.

SUGGESTED ANSWERS TO STRUCTURE YOUR KNOWLEDGE

1. **a.** aorta
b. left pulmonary artery
c. left pulmonary veins
d. left atrium
e. left ventricle
f. right ventricle
g. inferior vena cava
h. atrioventricular valve
i. semilunar valve
j. right atrium
k. superior vena cava

Blood flow from venae cavae → right atrium → right ventricle → pulmonary artery → lung → pulmonary vein → left atrium → left ventricle → aorta. Check your answer with Figure 34.6 in the text.

2. Nasal cavity → pharynx → through opening to larynx → trachea → bronchus → bronchiole → alveoli → interstitial fluid → alveolar capillary → hemoglobin molecule → venule → pulmonary vein → left atrium → left ventricle → aorta → renal artery → arteriole → capillary in kidney

ANSWERS TO TEST YOUR KNOWLEDGE

Multiple Choice:

1. c	**4.** c	**7.** b	**10.** e	**13.** b	**16.** d	**19.** d	**22.** d	**25.** c
2. c	**5.** e	**8.** d	**11.** d	**14.** e	**17.** c	**20.** a	**23.** e	**26.** b
3. d	**6.** a	**9.** b	**12.** e	**15.** e	**18.** a	**21.** d	**24.** e	

CHAPTER 35: THE IMMUNE SYSTEM

FOCUS QUESTIONS

35.1.
a. phagocytic white blood cells
b. large phagocytes that engulf microbes
c. white blood cells that attack parasitic worms with enzymes
d. white blood cells located in skin that move to lymph nodes and stimulate adaptive immunity
e. white blood cells that attack the body's infected or cancerous cells
f. release histamine to initiate inflammatory response
g. causes vasodilation and increased permeability of blood vessels
h. enzyme that attacks bacterial cell walls; found in tears, saliva, and mucus
i. proteins released by virus-infected cells that stimulate neighboring cells to produce substances that inhibit viral replication
j. set of blood proteins that cause lysis of microbes and are involved with innate and adaptive defenses
k. signaling proteins that enhance an immune response and promote blood flow to injury site

35.2. B cell antigen receptors recognize epitopes of intact antigens that are either molecules on the surfaces of infectious agents or molecules free in the body. T cell antigen receptors recognize pieces of antigens that have complexed with an MHC molecule inside a cell and are then presented on the cell surface.

35.3.
a. During development of B cells and T cells, the *V* and *J* gene segments of one light-chain allele and one heavy-chain allele are randomly recombined. The joining of these diverse light and heavy chains produces a huge number of different antigen-binding specificities among B and T cells.
b. As lymphocytes mature in the thymus or bone marrow, they are tested for self-reactivity, and those that react against self-components are inactivated or destroyed.

c. Once an antigen receptor binds its antigen, the selected cell divides into a clone of effector cells and memory cells specific for that antigen.
d. Immunological memory is an enhanced response to a previously encountered foreign molecule that arises from the presence of memory B and T cells, which rapidly form clones of effector cells upon re-exposure to that specific antigen.

35.4.
a. antigen-presenting cell
b. pathogen
c. antigen fragment
d. class II MHC molecule
e. T cell antigen receptor
f. accessory protein (CD4)
g. helper T cell
h. cytokines stimulate helper T cells, cytotoxic T cells, and B cells
i. cytotoxic T cell
j. B cell
k. cell-mediated immunity (attack on infected cells)
l. humoral immunity (secretion of antibodies by plasma cells)

35.5.
a. Activated helper T cells release cytokines that stimulate cytotoxic T cells, B cells, and themselves.
b. A cytotoxic T cell attached to an infected body cell releases proteins that disrupt the cell membrane and initiate apoptosis of the infected cell.

35.6.
a. HIV destroys helper T cells, thereby crippling the humoral and cell-mediated immune systems and leaving the body unable to fight opportunistic diseases such as Kaposi's sarcoma and *Pneumocystis* pneumonia.
b. HIV is readily passed through unprotected sex and needle sharing. The frequent mutational changes during replication generate drug-resistant strains of HIV. And frequent mutational changes in surface antigens have made development of an effective vaccine difficult.

SUGGESTED ANSWERS TO STRUCTURE YOUR KNOWLEDGE

1.

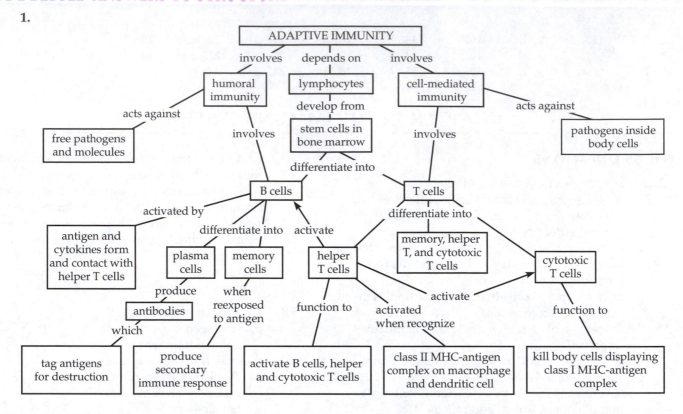

2. An antibody is a Y-shaped protein that consists of two identical light polypeptide chains and two identical heavy chains, held together by disulfide bonds. The amino acid sequences in the variable sections of the light and heavy chains in the arms of the Y account for the specificity in binding between antibodies and epitopes of an antigen. The two arms of an antibody allow it to cross-link antigens, which enhances phagocytosis. The constant region of the antibody tail determines its effector function; there are five types of antibodies.

3. Macrophages and dendritic cells phagocytose microbes in response to broad categories of molecules (innate immunity), but they also present peptide antigens in their class II MHC molecules to helper T cells. This interaction, along with synthesis of cytokines by these phagocytes, helps to activate helper T cells. The complement system functions in both innate and adaptive immunity. Some activated complement proteins promote inflammation or stimulate phagocytosis. The complement system may lyse microbes that are bearing antigen–antibody complexes.

ANSWERS TO TEST YOUR KNOWLEDGE

Multiple Choice:

1.	e	**5.**	d	**9.**	c	**13.**	d	**17.**	b
2.	e	**6.**	a	**10.**	a	**14.**	d	**18.**	e
3.	c	**7.**	e	**11.**	e	**15.**	e	**19.**	a
4.	b	**8.**	b	**12.**	c	**16.**	c	**20.**	d

CHAPTER 36: REPRODUCTION AND DEVELOPMENT

FOCUS QUESTIONS

36.1. **a.** Adaptive advantages that asexual reproduction may provide include perpetuation of successful genotypes in stable habitats; elimination of need to locate mate; and production of a larger number of offspring over time since all offspring can reproduce.

b. Adaptive advantages that sexual reproduction may provide include varying genotypes and phenotypes of offspring, which may enhance reproductive success and adaptation of a population to fluctuating environments.

c. It is sexual reproduction because it involves union of egg and sperm, and offspring will be genetically varied due to genetic recombination during meiosis and fertilization.

36.2. **a.** One mechanism is "courtship" behaviors that trigger the release of gametes, leading to the fertilization of the eggs of one female by one male.

b. Another mechanism is environmental signals (temperature, day length) or chemical signals that trigger *spawning*, the release of gametes from individuals clustered in the same area.

36.3. **a.** vas deferens
b. erectile tissue of penis
c. urethra
d. glans
e. prepuce (foreskin)
f. scrotum
g. testis
h. epididymis
i. bulbourethral gland
j. prostate gland
k. seminal vesicle
l. oviduct
m. ovary
n. uterus
o. clitoris
p. opening of vagina
q. vagina
r. cervix

36.4. **a.** Spermatogenesis occurs continuously as spermatogonia continue to divide from puberty onward. A female's supply of primary oocytes appears to be established before birth.
b. Each meiotic division produces four sperm, but only one egg (along with polar bodies that disintegrate).
c. Spermatogenesis is an uninterrupted process. Oogenesis occurs in stages: prophase I before birth; after puberty, meiosis I and II up to metaphase II in a maturing follicle just before ovulation; and completion of meiosis II (in humans) when a sperm cell penetrates the oocyte.

36.5. **a.** GnRH (gonadotropin-releasing hormone)
b. FSH (follicle-stimulating hormone)
c. LH (luteinizing hormone)
d. androgens (primarily testosterone)
e. negative feedback loops involving testosterone and inhibin

36.6. **a.** LH
b. FSH
c. follicular phase
d. ovulation
e. luteal phase
f. estradiol
g. progesterone
h. menstrual flow phase
i. proliferative phase
j. secretory phase

36.7. **a.** Specific binding between receptors on the plasma membrane and proteins on the acrosomal process assures that only a sperm of the correct species fertilizes an egg.
b. Fast block to polyspermy caused by membrane depolarization during acrosomal reaction and slow block to polyspermy caused by formation of the fertilization envelope during cortical reaction assure that only one sperm will fertilize an egg.

36.8. **a.** future ectoderm
b. mesenchyme cells (future mesoderm)
c. future endoderm
d. archenteron (future digestive tube)
e. anus (from blastopore)

36.9. **a.** HCG is secreted by the embryo. It maintains the corpus luteum (and thus progesterone) in the first trimester.
b. Oxytocin is secreted by the posterior pituitary. It stimulates uterine contractions during birth and triggers release of milk during nursing.

36.10. **a.** prevent release of gametes—sterilization, combination birth control pills (or progestin injection, or minipill)
b. prevent fertilization—abstinence, rhythm method, condom, diaphragm, progestin injection or mini-pill, *coitus interruptus*, spermicides, IUD
c. prevent implantation of embryo—morning after pills, IUD
The most effective methods are abstinence, sterilization, and chemical contraception. Least effective are the rhythm method and *coitus interruptus*.

SUGGESTED ANSWERS TO STRUCTURE YOUR KNOWLEDGE

1. Path of sperm from formation to fertilization

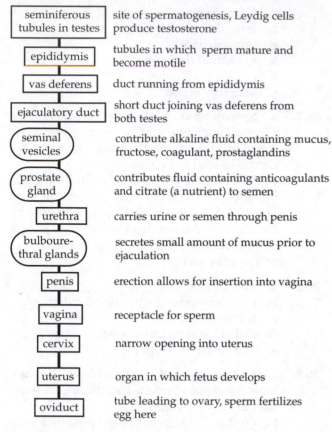

seminiferous tubules in testes	site of spermatogenesis, Leydig cells produce testosterone
epididymis	tubules in which sperm mature and become motile
vas deferens	duct running from epididymis
ejaculatory duct	short duct joining vas deferens from both testes
seminal vesicles	contribute alkaline fluid containing mucus, fructose, coagulant, prostaglandins
prostate gland	contributes fluid containing anticoagulants and citrate (a nutrient) to semen
urethra	carries urine or semen through penis
bulbourethral glands	secretes small amount of mucus prior to ejaculation
penis	erection allows for insertion into vagina
vagina	receptacle for sperm
cervix	narrow opening into uterus
uterus	organ in which fetus develops
oviduct	tube leading to ovary, sperm fertilizes egg here

2. **a.** GnRH is a releasing hormone of the hypothalamus that stimulates the anterior pituitary to secrete FSH and LH.
   b. FSH stimulates growth of follicles.
   c. The LH surge is caused by an increase in GnRH production that resulted from increasing levels of estradiol produced by the developing follicle.

d. The LH surge induces the maturation of the follicle and ovulation, and it transforms the ruptured follicle to the corpus luteum.
e. LH maintains the corpus luteum, which secretes estradiol and progesterone.
f. High levels of estradiol and progesterone act on the hypothalamus and pituitary to inhibit the secretion of LH and FSH.
g. The lack of LH causes the corpus luteum to disintegrate. Thus, the production of estradiol and progesterone ceases, which allows LH and FSH secretion to begin again. The drop in estradiol and progesterone also causes arteries in the endometrium to constrict and the uterine lining to disintegrate, leading to menstruation.

3. Birth control pills, which contain a combination of synthetic estrogens and progestin, act by negative feedback to prevent the release of GnRH by the hypothalamus and FSH and LH by the pituitary, thereby preventing the development of follicles and ovulation.

ANSWERS TO TEST YOUR KNOWLEDGE

Fill in the Blanks:

1. budding
2. pheromones
3. parthenogenesis
4. hermaphrodite
5. cloaca
6. estrous cycle
7. *in vitro* fertilization
8. urethra
9. vasocongestion
10. menopause

Multiple Choice:

1. d	5. b	9. c	13. b	16. b
2. d	6. e	10. a	14. e	17. c
3. b	7. c	11. e	15. a	18. b
4. a	8. b	12. c		

CHAPTER 37: NEURONS, SYNAPSES, AND SIGNALING

FOCUS QUESTIONS

37.1. **a.** dendrites
b. cell body
c. axon hillock
d. axon
e. synaptic terminal. The impulse moves from the axon hillock along the axon to the synaptic terminals. Neurotransmitter is released into the synapse with another cell at the synaptic terminal.

37.2. **a.** K^+ inside the cell; Na^+ outside the cell
b. The inside of the cell is more negative compared to outside the cell.

c. The membrane potential became more negative and closer to E_K; the membrane could be less permeable to the influx of positive sodium ions, thus sodium channels may have closed and/or additional potassium channels opened.

37.3. **a.** sodium channel
b. potassium channel
c. K^+
d. Na^+
e. sodium inactivation loop
f. membrane potential (mV)
g. time

h. resting potential
i. threshold
j. action potential
1. Resting state: Voltage-gated Na$^+$ and K$^+$ channels are closed.
2. Depolarization: Some Na$^+$ channels open; depolarization opens more Na$^+$ channels. If threshold is reached, action potential is triggered.
3. Rising phase: Most Na$^+$ channels open and Na$^+$ influx makes inside of cell positive.
4. Falling phase: Na$^+$ channels become inactivated, K$^+$ channels open, K$^+$ leaves cell, and inside of cell becomes negative.
5. Undershoot: Na$^+$ channels closed (many are still inactivated), K$^+$ channels still open, and membrane becomes hyperpolarized. As these channels close, membrane returns to resting potential.
See textbook Figure 37.11 for location of the numbered phases on the graph.

37.4. Refer to textbook Figure 37.13 to check your drawing.

37.5. **a.** presynaptic membrane
b. Ca^{2+} flowing in through a voltage-gated calcium channel
c. synaptic cleft
d. synaptic vesicle containing neurotransmitter
e. postsynaptic membrane
f. ligand-gated ion channel with bound neurotransmitter
g. closed ligand-gated ion channel

37.6. **a. Temporal**
b. Spatial

37.7. The type of postsynaptic receptor and its mode of action determine neurotransmitter function. Binding of acetylcholine to metabotropic receptors in heart muscle activates a signal transduction pathway that makes it more difficult to generate an action potential, thereby reducing the strength and rate of contraction. The acetylcholine ionotropic receptor in skeletal muscle cells opens ion channels and depolarizes the muscle cell membrane.

SUGGESTED ANSWERS TO STRUCTURE YOUR KNOWLEDGE

1.

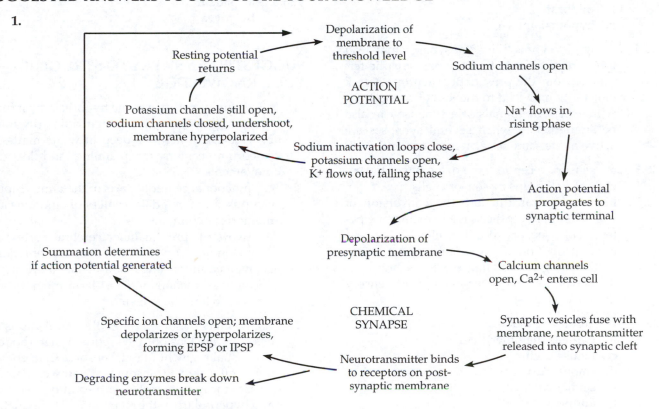

ANSWERS TO TEST YOUR KNOWLEDGE

Multiple Choice:

1. d	**3.** b	**5.** c	**7.** e	**9.** c	**11.** b	**13.** d	**15.** a
2. e	**4.** d	**6.** b	**8.** a	**10.** c	**12.** c	**14.** b	

CHAPTER 38: NERVOUS AND SENSORY SYSTEMS

FOCUS QUESTIONS

38.1. Oligodendrocytes are glia that myelinate axons in the CNS. Schwann cells are glia that myelinate axons in the PNS. Myelination increases the conduction speed of action potentials.

38.2.
a. efferent neurons
b. afferent (sensory) neurons
c. autonomic nervous system
d. motor system
e. sensory receptors
f. enteric
g. parasympathetic
h. sympathetic
i. slows heart, conserves energy
j. skeletal muscles

38.3.
a. cerebrum 5
b. thalamus 3
c. cerebellum 1
d. medulla 4
e. pons 2
f. midbrain 7
g. hypothalamus 6

38.4. A high level of activity at a synapse may lead to the recruitment of additional synaptic terminals from the presynaptic neuron. A lack of activity may lead to the loss of connections with a neuron. Signals at a synapse can also be strengthened when several synapses are active at the same time on a postsynaptic cell.

38.5.
a. sensory reception: the detection of the energy of a stimulus by sensory cells
b. sensory transduction: the conversion of stimulus energy to a change in membrane potential of a sensory receptor cell
c. transmission: the passage of an action potential along sensory neurons to the brain
d. perception: the interpretation of sensory input by the brain

38.6.
a. auditory canal
b. malleus (hammer)
c. incus (anvil)
d. stapes (stirrup)
e. semicircular canals
f. auditory nerve
g. cochlea
h. round window
i. Eustachian tube
j. oval window
k. tympanic membrane (eardrum)

1. The auditory canal and middle ear, which houses malleus, incus, and stapes, are filled with air.
2. The cochlea, semicircular canals, and also the utricle and saccule are filled with fluid.
3. The cochlea houses the organ of Corti.
4. The semicircular canals, extending in three spatial planes, detect angular movements.

38.7.
a. sclera
b. choroid
c. retina
d. fovea (center of visual field)
e. optic nerve
f. optic disk (blind spot)
g. vitreous humor
h. lens
i. aqueous humor
j. pupil
k. iris
l. cornea
m. suspensory ligament

38.8.
a. small
b. fovea

SUGGESTED ANSWERS TO STRUCTURE YOUR KNOWLEDGE

1.
a. group of neurons in the hypothalamus; functions as a biological clock for circadian rhythms
b. clusters of neurons deep in white matter of cerebrum; centers for planning and learning movements
c. functional center in parts of thalamus, hippocampus, and amygdala; centers of emotions and memories in humans
d. two structures in inner cerebral cortex; involved in limbic system and memory; amygdala is involved in forming emotional memories; hippocampus functions in short-term memory and transfer to long-term memory

2. Light enters the eye through the pupil and is focused by the lens onto the retina. When rhodopsin (visual pigment in rods) absorbs light energy, it sets off a signal transduction pathway that decreases the receptor cell's permeability to sodium and hyperpolarizes the membrane. The reduction in the release of the neurotransmitter glutamate by a rod cell may either polarize or hyperpolarize connected bipolar cells, which may then generate action potentials in ganglion cells—the sensory neurons that form the optic nerve. Horizontal

and amacrine cells provide for integration of visual information within the retina.

3. Sound waves are collected by the *pinna* and travel through the *auditory canal* to the *tympanic membrane*, where they are transmitted by the *malleus, incus,* and *stapes*. Vibration of the stapes against the *oval window* sets up pressure waves in the perilymph in the *vestibular canal* within the *cochlea*, from which they are transmitted to the *tympanic canal* and then dissipated when they strike the *round window*. The pressure waves vibrate the *basilar membrane*, on which the *organ of Corti* is located within the *cochlear duct*. Tips of the hair cells arising from the organ of Corti are embedded in the *tectorial membrane*. When the basilar membrane vibrates, the hair cells bend, triggering a depolarization, release of neurotransmitter, and initiation of action potentials in the sensory neurons of the *auditory nerve*. Axons of the auditory nerve carry impulses to auditory areas of the *cerebral cortex*, where information on pitch and volume are integrated and the perception of sound is generated.

ANSWERS TO TEST YOUR KNOWLEDGE

Multiple Choice:

1. a	5. b	9. c	13. c	17. d
2. e	6. d	10. d	14. c	18. c
3. e	7. d	11. b	15. c	19. e
4. b	8. c	12. e	16. b	20. d

CHAPTER 39: MOTOR MECHANISMS AND BEHAVIOR

FOCUS QUESTIONS

39.1. a. sarcomere
  b. Z line
  c. M line
  d. thin filaments (actin)
  e. thick filament (myosin)
  The Z lines will be closer together as the overlap between thick and thin filaments increases. This sliding of filaments occurs by the following repeated sequence: Binding of ATP releases myosin heads from the actin filament; hydrolysis of ATP converts the head to a high-energy form that binds to actin, forming a cross-bridge; myosin head returns to low-energy configuration, pulling the thin filament toward the center of the sarcomere.

39.2. a. Three distinct types of skeletal muscle fibers are slow-twitch oxidative, fast-twitch oxidative, and fast-twitch glycolytic.
  b. Fast-twitch glycolytic has few mitochondria, has little myoglobin, and produces brief, powerful contractions.
  c. You would expect slow-twitch oxidative to be present in the highest proportion in postural muscles

39.3. a. Hydrostatic skeleton: no extra materials needed, good shock absorber; not much protection or support for lifting animal off the ground; cnidarians, flatworms, nematodes, and annelids
  b. Exoskeleton: quite protective; in arthropods, must be molted in order for the animal to grow, restricts size; most molluscs and arthropods
  c. Endoskeleton: various types of joints allow for flexible movement, good structural support; may not be very protective; sponges, echinoderms, chordates

39.4. Mammals are mostly nocturnal; birds are diurnal.

39.5. a. imprinting
  b. associative learning
  c. cognition
  d. spatial learning
  e. social learning

39.6. If a male's parental care resulted in his greater reproductive success, then the genes for his behavior would increase in frequency in a population. If he were parenting offspring that included those of other males, then he would not pass his "parental care" genes on to the next generation in a higher proportion because he was also helping to perpetuate the "non-parental care" genes of other males.

39.7. Raising animals in the lab enables researchers to search for evidence of genetic bases of behavior by controlling for environmental influences.

39.8. a. a parent or a sibling
  b. The individual shares more genes in common with a parent or a sibling. The coefficient of relatedness is 0.5, whereas it is only 0.125 for a cousin. Thus, with Hamilton's rule $rB > C$, the benefit to the recipient is discounted less by a larger value of r.

SUGGESTED ANSWERS TO STRUCTURE YOUR KNOWLEDGE

1. A motor neuron releases acetylcholine into the synapse with a muscle fiber, initiating an action potential that spreads into the center of the muscle cell along the transverse tubules. The action potential changes the permeability of the sarcoplasmic reticulum membrane, which releases Ca^{2+} into the cytosol. The calcium binds with troponin, moving tropomyosin and exposing the myosin-binding sites of the actin molecules. Heads of myosin molecules in their high-energy form bind to these sites, forming cross-bridges. The myosin heads bend, pulling the thin filament toward the center of the sarcomere. When ATP binds to myosin, the cross-bridge breaks. Hydrolysis of the ATP returns the myosin to its high-energy form, and it binds farther along the actin molecule. This sequence continues as long as there is ATP and until calcium is pumped back into the sarcoplasmic reticulum, when the tropomyosin–troponin complex again blocks the actin sites.

2. The scientific study of animal behavior seeks to identify those behaviors that are innate and genetically programmed as well as those that are a product of experience and learning. Fixed-action patterns are clearly developmentally fixed. In other cases, genetics may set the parameters for an organism's behavior; however, experience can modify behavior, and learning is clearly evident.

3. According to the concept of fitness, an animal's behavior should help to increase its chance of survival and production of viable offspring. Survival behaviors that are genetically programmed would be most likely to be passed on, and natural selection would refine innate behaviors for foraging, migrating, mating, and care of offspring. The evolution of cognitive ability would help individuals deal with novel situations. Social behaviors and communication may increase an individual's fitness. Parental investment and certainty of paternity may result in differences in mating systems and parental care that influence an individual's reproductive success. Altruistic behavior may be explained on the basis of kin selection; the inclusive fitness of an animal increases if its altruistic behavior benefits related animals who share many genes.

ANSWERS TO TEST YOUR KNOWLEDGE

Multiple Choice:

1.	a	5.	b	9.	a	13.	c	17.	a
2.	d	6.	a	10.	d	14.	b	18.	c
3.	b	7.	c	11.	b	15.	e	19.	d
4.	d	8.	b	12.	b	16.	e	20.	d

CHAPTER 40: POPULATION ECOLOGY AND THE DISTRIBUTION OF ORGANISMS

FOCUS QUESTIONS

40.1. a. South-facing slopes in the northern hemisphere receive more sunlight and are warmer and drier than north-facing slopes.
   b. Air temperature drops with an increase in elevation, and high-altitude communities may be similar to communities at higher latitudes.
   c. The windward side of a mountain range receives much more rainfall than the leeward side. The warm, moist air rising over the mountain releases moisture, and the drier, cooler air absorbs moisture as it descends the other side.

40.2. a. desert
   b. temperate grassland
   c. tropical forest
   d. temperate broadleaf forest
   e. northern coniferous forest
   f. arctic and alpine tundra

40.3. __4__ benthic   __6__ pelagic   __1__ littoral
   __5__ aphotic   __3__ photic   __2__ limnetic

40.4. a. dispersal: an area may be beyond the dispersal ability of a species
   b. biotic factors: the presence of predators, herbivores, parasites, or competitors may restrict a species' range
   c. abiotic factors: sunlight, water, oxygen, temperature, salinity, and soil characteristics may determine whether a species can inhabit an area

40.5. a. Type I is typical of populations that produce relatively few offspring and provide parental care, such as humans and many large mammals.
   b. Type II has a constant death rate over the life span of the organisms, such as found in some rodents, annual plants, invertebrates, and lizards.

c. Type III is typical of populations that produce many offspring, most of which die off rapidly, such as many fishes and marine invertebrates, and long-lived plants.

40.6. The r_{max} is the maximum per capita rate of increase possible for a species, when a population is in an unlimited environment and members can reproduce at their physiological capacity.

40.7. **a.** exponential growth; $dN/dt = r_{max}N$
b. logistic growth; $dN/dt = r_{max}N(K − N)/K$; K is 1,500

40.8. **a.** An organism has limited resources to divide between growth, survival, and reproduction.
b. r-selected; K-selected

40.9. **a.** Density-dependent factors that may limit population growth include competition for resources, availability of territories, disease, accumulation of toxic wastes, predation, and intrinsic limiting factors.
b. Some abiotic factors that may cause population fluctuations are extremes in weather, natural disasters, and fires.

SUGGESTED ANSWERS TO STRUCTURE YOUR KNOWLEDGE

1. **a.** Ecology is the study of how organisms interact with their abiotic and biotic environments.
b. The interactions of organisms with their environment can result in changes in the gene pool of a population, or evolution. Interactions occurring within an ecological time frame translate into adaptations to the environment that are evident on the scale of evolutionary time.

2. **a.** Biomes are characteristic ecosystem types—usually identified by the predominant vegetation for terrestrial biomes or the physical environment for aquatic biomes—which range over broad geographic areas.
b. Convergent evolution, common adaptations of organisms of different evolutionary lineages to similar environments, accounts for similarities in life-forms within geographically separated biomes.

3.

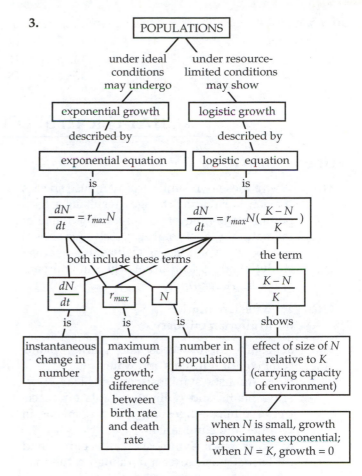

4. Reproductive success is measured in the number of offspring that survive and live to reproduce. Many "choices" are available in life history traits: age at first reproduction, number of reproductive episodes, number of offspring produced as a time, parental investment in size of offspring or care. There are always trade-offs between reproduction and survival due to limited energy budgets. In addition, different environments (with varying degrees of stability and diverse biotic and abiotic factors) and different population densities (both influencing K- or r-selection) create different selection pressures on a population. Thus, there is no one "best" reproductive strategy.

ANSWERS TO TEST YOUR KNOWLEDGE

Matching:

1. C	**3.** D	**5.** A	**7.** B
2. H	**4.** E	**6.** G	**8.** F

1. b	**4.** c	**7.** e	**10.** c	**13.** b	**16.** b	**19.** c	**22.** d	**24.** b
2. d	**5.** a	**8.** e	**11.** d	**14.** a	**17.** e	**20.** d	**23.** a	**25.** c
3. b	**6.** d	**9.** c	**12.** e	**15.** c	**18.** c	**21.** a		

CHAPTER 41: SPECIES INTERACTIONS

FOCUS QUESTIONS

41.1. Where these two similar spiny mouse species coexist, behavioral changes result in a temporal partitioning of resources. The niche of the naturally nocturnal *A. russatus* (as shown by laboratory studies as well as this removal study) changes to diurnal when in competition with *A. cahirinus*.

41.2. **a.** **Batesian mimicry**
b. **Müllerian mimicry**

41.3. **a.** competition: weeds and garden plants compete for nutrients and water
b. mutualistic symbiosis: mycorrhizae, flowering plants and pollinators, ants on acacia trees, cellulose-digesting microorganisms in termites and ruminants
c. commensal symbiosis: cattle egrets and cattle that flush insects (although cattle may benefit when ectoparasites are eaten or when the birds warn of predators)
d. parasitism: a type of symbiosis in which endo- and ectoparasites feed in or on host; tapeworms, ticks
e. predation: animal predators killing prey
f. herbivory: herbivores eating parts of plants
g. facilitation: black rush (*Juncus gerardii*) prevents salt buildup and oxygen depletion in salt marshes by shading soil and transporting oxygen to its roots

41.4. **a.** Pool 1: $H = -(0.4 \ln 0.4 + 0.3 \ln 0.3 + 0.3 \ln 0.3) = -(-0.37 + -0.36 + -0.36) = 1.09$

Pool 2: $H = -(-0.25 + -0.23 + -0.23 + -0.23) = 0.94$

b. Pool 2 has more species and thus a greater species richness. Pool 1 has the higher diversity index.

41.5. The $+/-$ cascade that would be needed to end with a decrease in algae would require an increase in zooplankton, which would require a decrease in primary predators as a result of an increase in top predators. More top predators could be added to the lake, or primary predators could be removed.

41.6. Soil nitrogen levels begin quite low but rise during the alder stage due to the symbiotic nitrogen-fixing bacteria associated with alder.

41.7. More interspecific interactions would have had time to develop, and this longer span of evolutionary time would allow for more speciation events to have occurred.

41.8. **a.** As the number of species increases, the rate of immigration of new species would be predicted to decrease and the rate of extinction (due to competitive exclusion) to increase.
b. The island with the higher number of species would probably be larger and closer to the mainland.

41.9. They are testing for the presence of the H5N1 strain of avian flu virus in migrating waterfowl to monitor the disease's possible entry into North America.

SUGGESTED ANSWERS TO STRUCTURE YOUR KNOWLEDGE

1. **a.** competition $(-/-)$
   b. resource partitioning and character displacement
   c. slightly different niches
   d. keystone predator
   e. herbivory
   f. mutualism $(+/+)$
   g. facilitation
   h. food chain or food web

2. **a.** No two species with the same niche can coexist permanently in a habitat; the more competitive species will cause the local elimination of the other. The evidence of interspecific competition in the past can be seen in resource partitioning and character displacement.
   b. Each trophic level controls the next higher level; adding nutrients will increase the biomass of all other trophic levels.
   c. Community organization is controlled from the top by predation (or herbivory), with a

cascade of $+/-$ effects down the trophic levels with changes in predator numbers.

d. Most communities do not reach a stable climax community but are constantly changing in composition due to the effects of disturbances.

e. The greatest species diversity will be found in communities subject to moderate levels of disturbance. Low levels may allow competitively dominant species to exclude other species, and high levels may exclude slow-growing or slow-colonizing species from the community.

f. The rates of immigration and extinction on "islands" of habitat are affected by the size of the island, the closeness to the "mainland," and the number of species currently on the island. When these rates are equal, an equilibrium number of species is reached.

ANSWERS TO TEST YOUR KNOWLEDGE

Multiple Choice:

1. d	4. b	7. b	10. c	13. e
2. c	5. e	8. e	11. d	14. b
3. d	6. b	9. d	12. d	15. c

CHAPTER 42: ECOSYSTEMS AND ENERGY

FOCUS QUESTIONS

42.1. The mineral nutrient may limit production in that ecosystem.

42.2. Whereas energy makes a one-way trip through ecosystems, chemical elements move through the trophic levels and detritivores and are recycled back to producers.

42.3. **a.** Some ecosystems with high primary production are tropical rain forest, coral reef, swamp and marsh, and estuary.

b. Some ecosystems with low primary production are desert, tundra, and open ocean.

c. The open ocean covers 65% of Earth's surface area.

d. Upwellings in these cold seas bring nitrogen and phosphorus to the surface. The lack of these nutrients limits production in many tropical waters.

42.4. **a.** Much of a bird's or a mammal's assimilated energy is used for respiration to maintain a warm body temperature and is not available for net secondary production (growth and reproduction).

b. $\frac{1}{1000}$ (10% of 10% of 10%) or 0.1%

42.5. Water is essential to all organisms, and its availability in terrestrial ecosystems influences primary production and decomposition. Carbon forms the backbone for all organic molecules, which are essential to all organisms. Nitrogen is a component of amino acids and nucleic acids and is often a limiting plant nutrient. Phosphorus is found in nucleic acids, phospholipids, and ATP, and as a mineral in bones and teeth.

42.6. **a.** Nutrients cycle the fastest in tropical rain forests, because decomposition proceeds rapidly in the warm, wet climate, and nutrients are rapidly assimilated by new growth.

b. Nutrients cycle slowly in lakes and oceans; lack of oxygen slows decomposition and, unless there are upwellings, nutrients in sediments are not available to producers. Decomposition is also very slow in cold, wet peatlands.

c. Nutrients are not recycled and may leave the ecosystem through runoff.

42.7. Prokaryotes that can metabolize uranium are being used to treat contaminated groundwater. Plants that can fix nitrogen are often used in biological augmentation to enrich nutrient-poor soils and facilitate the recolonization of native species.

SUGGESTED ANSWERS TO STRUCTURE YOUR KNOWLEDGE

1. Primary production is the energy assimilated by the producers in an ecosystem in a given period. Primary production is limited in aquatic ecosystems by the penetration of light and the availability of nutrients, especially nitrogen and phosphorus. In terrestrial ecosystems, primary production is influenced by temperature and precipitation (reflected as actual evapotranspiration), as well as by nutrients such as nitrogen and phosphorus.

2. Secondary production is the production of new biomass from the energy contained in a consumer's food. Production efficiency is reduced by the energy

used for cellular respiration, and is lowest for endotherms. Trophic efficiency is the percentage of the energy stored in one tropic level that is transferred to the next trophic level. It is usually around 10%.

3. Photosynthetic and chemosynthetic organisms assimilate inorganic elements and compounds and incorporate them into organic compounds. Animals obtain their organic compounds by eating producers or other consumers. Elements are returned to inorganic form by the processes of respiration, excretion, and decomposition.

ANSWERS TO TEST YOUR KNOWLEDGE

Multiple Choice:

1. e	4. d	7. b	10. a	13. e
2. c	5. e	8. c	11. e	14. c
3. c	6. d	9. c	12. c	

CHAPTER 43: GLOBAL ECOLOGY AND CONSERVATION BIOLOGY

FOCUS QUESTIONS

43.1. Many examples are provided in the text.
a. About 70% of coral reefs have been damaged by human activities; 40–50% of the reefs could be lost in the next few decades. About a third of all marine fish species utilize these reefs.
b. The introduction of the brown tree snake to Guam resulted in the extinction of multiple species of birds and lizards.
c. Commercial harvest and/or illegal hunting have reduced populations of whales, the African elephant, and many fishes.
d. It may take decades for aquatic ecosystems to recover from the damage done by acid precipitation, and damage continues to forests in central and eastern Europe.

43.2. Smaller, because usually not all individuals in a population successfully breed. Conservation programs attempt to sustain the effective population size above MVP in order to retain genetic diversity in a population.

43.3. The loss of genetic variation within a small population due to inbreeding and genetic drift can force it into an extinction vortex in which the population grows smaller and smaller. Promoting migration between small populations or introducing individuals from other populations to increase genetic variation is proposed as an urgent conservation need.

43.4. This proactive approach relies on early detection of population decline, identification of the species' habitat needs, testing to determine which factor is contributing to the decline, recommending corrective measures, and monitoring results.

43.5. Corridors promote dispersal between populations and may be essential to species that migrate between different habitats. However, they may also contribute to the spread of diseases.

43.6. Large reserves are required for large, far-ranging animals that require extensive habitats. They also have proportionally less border area and thus have fewer edge effects. An advantage of smaller reserves that collectively have the same area as a large one is the slower spread of disease within a population.

43.7. Global warming may be slowed by agreements on an international strategy to reduce CO_2 emissions through increased energy efficiency, use of renewable solar and wind power, use of nuclear power, and reductions in deforestation.

43.8. The ultimate carrying capacity of Earth may be determined by food and water supplies, other resources, degradation of the environment, or several interacting factors. When and how we reach a stable population size is an issue of great social and ecological consequence. Damaging Earth's ecosystems through excessive resource use or the buildup of wastes may eventually lower the carrying capacity.

SUGGESTED ANSWERS TO STRUCTURE YOUR KNOWLEDGE

1. The destruction of habitat is probably the greatest threat to biodiversity. As humans alter and disrupt natural ecosystems, many species no longer have the physical or community resources necessary for survival. Introduced species can disrupt their new community by preying on or outcompeting

local organisms. Because these organisms do not have the normal biological controls of predators, pathogens, or parasites, their populations may rapidly increase. The harvesting of wild species beyond their ability to reproduce may drive populations to extinction. Global changes to Earth's ecosystems in the form of acid precipitation, ocean acidification, and climate change are an increasing threat to biodiversity.

2. Biodiversity is a natural resource from which we obtain medicines, crops, fibers, and other products; many potentially valuable species will become extinct before they are known to scientists. A loss of biodiversity may disrupt ecosystem processes in harmful ways. Humans have evolved within the context of living communities and may be affected in unknown ways by changes in our ecosystem.

3. Fragmentation of habitats produces more interfaces or *edges* between different ecosystems (between forests and cleared areas, between deserts and housing developments). The species that inhabit edges are able to use both types of ecosystems. As edges proliferate, edge-adapted species may become more dominant than the species in the adjoining habitats. Strips of quality habitat that connect fragmented habitat patches may serve as *movement corridors* that promote dispersal between isolated populations and help maintain genetic variation.

4. It is a measure of the total land and water area required by an entity (person, nation, etc.) to produce all the resources it consumes and to absorb all the wastes it generates.

ANSWERS TO TEST YOUR KNOWLEDGE

Multiple Choice:

1. e	4. b	7. e	10. d	13. c
2. b	5. d	8. a	11. b	14. c
3. c	6. a	9. c	12. b	15. d